WITHDRAWN

D1393872

Surface Mount Technology

Second Edition

4607807

Surface Mount Technology

Principles and Practice

Second Edition

Ray P. Prasad

Ray Prasad Consultancy Group, Portland, Oregon

UNIVERSITIES AT MEDWAY LIBRARY

621
381
531
PRA

KLUWER ACADEMIC PUBLISHERS
BOSTON/DORDRECHT/LONDON

Distributors for North, Central and South America:
Kluwer Academic Publishers
101 Philip Drive
Assinippi Park
Norwell, Massachusetts 02061 USA
Telephone (781) 871-6600
Fax (781) 871-6528
E-Mail <kluwer@wkap.com>

Distributors for all other countries:
Kluwer Academic Publishers Group
Distribution Centre
Post Office Box 322
3300 AH Dordrecht, THE NETHERLANDS
Telephone 31 78 6576 000
Fax 31 78 6576 254
E-Mail < orderdept@wkap.nl >

 Electronic Services <http://www.wkap.nl>

Library of Congress Cataloging-in-Publication Data
Prasad, Ray P.
    Surface mount technology / Ray Prasad. -- 2nd ed.
      p. cm.
    Includes bibliographical references and index.
    ISBN 0-412-12921-3
    1. Surface mount technology. I. Title.
TK7868.P7P7   1997
621.3815'31--dc21   96.40513
                                  CIP

British Library Cataloging in Publication Data Available

Copyright © 1997 by Chapman & Hall
Sixth printing by Kluwer Academic Publishers 2002.

All rights reserved. No part of this publication may be reproduced, stored in a retrieval system or transmitted in any form or by any means, mechanical, photo-copying, recording, or otherwise, without the prior written permission of the publisher, Kluwer Academic Publishers, 101 Philip Drive, Assinippi Park, Norwell, Massachusetts 02061.

"Surface Mount Technology" is intended to present technically accurate and authoritative information from highly regarded sources. The publisher, editors, authors, advisors, and contributors have made every reasonable effort to ensure the accuracy of the information, but cannot assume responsibility for the accuracy of all information, or for the consequences of its use.

Printed on acid-free paper.

Printed in the United States of America

To my wife Peggy
and my children Geeta, Joe, and Kevin
the ones who really made it possible

FOR USE IN THE
LIBRARY ONLY

Brief Contents

Table of Contents

Foreword

A foreword is usually prepared by someone who knows the author or who knows enough to provide additional insight on the purpose of the work.

When asked to write this foreword, I had no problem with what I wanted to say about the work or the author. I did, however, wonder why people read a foreword. It is probably of value to know the background of the writer of a book; it is probably also of value to know the background of the individual who is commenting on the work. I consider myself a good friend of the author, and when I was asked to write a few words I felt honored to provide my view of Ray Prasad, his expertise, and the contribution that he has made to our industry.

This book is about the industry, its technology, and its struggle to learn and compete in a global market bursting with new ideas to satisfy a voracious appetite for new and innovative electronic products. I had the good fortune to be there at the beginning (or almost) and have witnessed the growth and excitement in the opportunities and challenges afforded the electronic industries' engineering and manufacturing talents. In a few years my involvement will span half a century. Having seen technology advancements from single-sided, paper-based phenolic products evolve into complex multilayered structures, I know that this book will be a major addition to any reader's knowledge or understanding of where the industry is and where it is heading.

I first met Ray when listening to a presentation that he made at an IPC annual meeting. I was impressed with his clarity of thought, and the logic that he used to establish the relationships of products and processes. When, at a later date, he volunteered to participate in the development of an IPC program on surface mount land patterns, I knew that his expertise and openness would help to develop a useful standard for the industry.

Up to that time, industry experts were very "closed mouthed" as to how to effectively mount electronic components on the surface of the printed board or interconnecting product. The hybrid industry had been using this technique for years, and although much was known about attaching parts to ceramic substrates, using organic epoxy-glass printed wiring boards for this purpose was not well understood. In addition, the only experts who would share information were from companies engaged in military systems designs.

These organizations were also struggling to understand the characteristics that permitted electronic assemblies to pass the rigorous military stress testing.

With the development of the leadless ceramic chip carrier (LCCC) many engineers felt that we had the answer for a component package that could satisfy the U.S. military's need for hermetically sealed integrated circuits. The LCCC was the component that almost collapsed the use of SMT as we know it today. The design of this ceramic package with its castellated sides to form the solder joint was doomed to failure when surface mounted on an organic substrate. Little was known at that time about the benefit that solder volume, between the part and the board, provided to reduce the strain caused by differences in the coefficients of thermal expansion between the part and the platform on which it was mounted. Article after article appeared in the trade journals about solder joint cracking. Some companies began to add leads to the LCCC in an effort to pass the "MIL spec" thermal test cycles.

It was in this atmosphere that Prasad volunteered to head up the industry effort to develop a set of criteria to produce hardware that was reliable, and producible in high volume. He had help of course, but he provided the main catalyst to take what was known, compared to what was assumed, and capture the reasonable explanations as to how problems could be avoided. Industry members listened to him, both at meetings and during the presentations that he made at national conferences. His credentials were respected because of the companies for which he worked, and his expertise was acknowledged through the manner in which he presented the ideas and methods needed to produce high quality electronic assemblies using surface mount technology.

The U.S. Secretary of Defense had just been told by his advisors that "surface mounting was a technology at risk." What was not understood was that when designing a product one had to manage all the characteristics of the product in order for it to function in its intended environment. To establish a strong SMT infrastructure, the industry formed the Surface Mount Council, a group of individuals dedicated toward promoting the orderly implementation of surface mount technology, and the standardization needed to develop the infrastructure. Prasad was one of the twenty individuals asked to serve on the council and has served since its inception in 1986.

Over the years Ray Prasad has been a contributor. He authors technical articles, volunteers to help industry representatives, participates in industry standardization, and has a continuous drive toward process improvement.

This book is Prasad's way of reaching a larger audience. One should read the rationale and the explanations that are provided. They are meant to help individuals either starting in SMT or well along the road to gain insight into the process steps for developing high quality, first pass yield electronic assemblies. Will everyone agree with the rationale presented? Probably not. Our industry is made up of many experts. They do not always agree, but the laws of physics and mathematical equations do not lie. One should ask: Why do we do this? or Why is that particular characteristic important? Prasad explains these conditions and provides the explanations in a plausible manner.

Now that Prasad is an independent consultant, he has the freedom to help his clients, without an employer questioning his loyalty. I have no intention of promoting him over other key industry experts; however, having known him for many years I know his motivation, probably better than most. He loves his wife Peggy and their children and enjoys spending some quiet time with his family. He can now schedule this family activity and still serve the industry.

Ray Prasad is dedicated to service. Enjoy his work and count yourself lucky if you have an opportunity to meet him face to face or share in some of his experiences. His growth mirrors the industry growth. He is a major contributor and will continue to enhance the industry's capability to implement sound design and manufacturing processes for surface mount technology and beyond.

Dieter Bergman
Director Technology Transfer
IPC Incorporated

Preface

Surface mount technology (SMT) is not a technology of tomorrow but a technology of today which makes it possible to produce state-of-the-art miniaturized electronic products. Although SMT is a mature technology, it is also constantly evolving. This is evidenced by the increasing popularity of relatively newer packages, such as fine pitch, ball grid array (BGA), and chip scale packaging (CSP).

As widespread as SMT is today, getting decent yield in products using SMT can be a frustrating experience if one goes about SMT implementation by trial and error. In order to take advantage of this technology, a complete SMT infrastructure must be put in place. This requires not only investment in capital equipment but also considerable investment in human resources and company-specific documentation. Most companies make the capital investment, but only a very few invest in the training and documentation that are the backbone of a strong internal SMT infrastructure.

Leading companies such as Intel, my most recent employer, have made the needed investment to create a strong internal infrastructure and keep their customers on the leading edge of technology. Intel is known for the worldwide dominance of its microprocessors, but fewer people know that Intel is a leading turn-key supplier of personal computers (PCs) for many large OEMs (original equipment manufacturers). Intel is a dominant supplier of not only the *computer inside* but the *computer outside* as well.

Who is responsible for putting the SMT infrastructure, external (generally standards) and internal, in place? In addition to the industry organizations such as the Institute of Interconnecting and Packaging Electronic Circuits (IPC), the Electronics Industries Association (EIA), and the Surface Mount Council, it requires the active support of universities (for training future engineers), company management, and practicing engineers.

This book is an attempt to meet the needs of this varied audience. In addition to its use by the universities as a textbook for courses on packaging in general and SMT design and manufacturing in particular, managers can use the information to manage the risk in SMT implementation to achieve their goal of faster time to market, higher yield, and lower cost. Practicing engineers can use this book to solve (and prevent) day-to-day design and manufacturing problems.

To accomplish these challenging objectives, I have not only culled information from published materials, I have also depended on input from my colleagues and such industry organizations as the IPC, the EIA, and the Surface Mount Council. But the underlying basis for this book has been my firsthand experiences in implementing this technology at various companies, including Boeing, Intel, and many client companies (I have been a consultant since 1994).

In a fast-changing technology like SMT, it is very easy for information to become obsolete even before a book is published. For this reason, I have concentrated on the basic principles and practice of SMT. Formulas and theories are discussed where necessary to give a better understanding of the subject. Very little emphasis is given to the details of manufacturing equipment because they are constantly changing. Instead, emphasis is placed on non-equipment-dependent materials and process variables that control manufacturing yield. However, Appendix B does provide equipment selection criteria that I believe will always be valid.

The book is divided into three parts. The first part gives an overview of the technology including emerging technologies such as chip-on-board (COB) and multichip modules (MCM). I also discuss the inherent technical and risk management issues that must be addressed for an effective implementation of SMT, either in-house or at a subcontractor. Since there is a major trend towards outsourcing, Chapter 2 provides a detailed discussion of criteria for selection of an outside SMT assembly subcontractor to meet the specific needs of an OEM.

The second part of the book focuses on the SMT components, including fine pitch and BGA. Here let me say a few words about BGA. Looking at various conference programs and industry literature, one might get a distinct impression that BGA is a technology separate from SMT. I do not consider it as such. In this book BGA is treated as just another surface mount component with some unique features and issues. Hence BGA-specific issues are addressed throughout this book in the applicable chapters. Otherwise all the process parameters equally apply to BGA. In the second part I also cover substrate fabrication, surface finish, solder mask requirements, and design issues such as land pattern design and design for manufacturability critical for producing cost-effective surface mount assemblies. Readers will find practical information on design for manufacturability with reasons behind each of the guidelines. It is my aim that if the reasons change with the passage of time, the reader can make a determination about the validity of those guidelines and can use the basic concepts even if a great change in the technology occurs.

The third part of the book covers SMT materials and manufacturing process issues. It is not enough to design the product for manufacturability. There are two other major areas that control product cost and yield. They are SMT materials such as adhesives, solder paste, and flux, and manufacturing processes such as printing, placement, reflow and wave soldering, cleaning, including no-clean, and statistical process control. These subjects take up the second half of the book and are covered in great detail.

Even if the design is right, and the materials and processes are properly controlled, there will be some variation in the quality of the product. And there will be some defects that require rework. Why? The industry has not been able to achieve zero defect in through-hole technology even after decades of effort. SMT, or any technology, is not perfect. So the third part of this book also covers the non-value-added process steps such as test, workmanship, inspection, and rework.

Writing and revising this book has been more of a challenge than I originally imagined, but it has been a very educational experience for me. It is my hope that this book will make a useful contribution to a better and balanced understanding of this technology and its proliferation and will play some role, minor as it may be, in building a strong external and internal SMT infrastructure.

Acknowledgments

To the extent that I may have been successful in my stated goal of writing a book useful to academia, technical management, and practicing engineers, it would not have been possible without the support of many of my friends and colleagues in the industry. Even though I have made significant changes in this edition, this book still rests on the foundation laid during its first edition. So I would like to thank again the many people who gave me the benefit of their expertise in the first edition. In particular, the contributions of Intel SMT team members Dr. Raiyomand Aspandiar, George Arrigotti, and Alan Donaldson were very significant and are still valid for the second edition. I also want to thank the Intel management at the time for their support, especially my former boss Fred Burris.

For the second edition also, I have benefited greatly from the advice and insights of many industry friends like Bill Kenyon of the Global Center for Process Change, Martin Freedman of AMP, and David Bergman and Dieter Bergman of IPC. Many people reviewed various chapters of the first edition and gave me valuable suggestions for additions and deletions for the second edition. Some of those who gave generously of their time are Foster Gray of Texas Instruments, Dr. Tom Kennedy of Solectron, Jim Bergenthal of Kemet, Greg Munie of Lucent Technology, Don Elliot of Elliot Technology, Phil Zarrow of ITM, and Les Hymes of Les Hymes & Associates. Their suggestions were very helpful to me. However, I am solely responsible for the final decisions (and any shortcomings) on the selection and the content of the subject matter in this book.

I want to thank Margaret Cox of Intel, who saved me considerable time by preparing the majority of the illustrations for this second edition. These illustrations, so important in clarifying the material for the reader, benefited greatly from her professional touch. Thanks, Margaret, for giving up your personal time and many of your weekends.

And finally, I want to sincerely thank another Margaret, my wife Peggy, who not only provided moral support, but is the only other person besides me who has read the entire manuscript (more than once) and made innumerable suggestions for clarity and improvement.

I also want to thank my three children, Geeta, Joe, and Kevin, for their support and patience with me while I was working long hours on this book. I am especially appreciative of Joe and Kevin, who forced me to take breaks from my office for bike rides with them after school and on weekends. Those short breaks were good not only for fresh air but for fresh ideas as well.

December 1996

Ray P. Prasad
Ray Prasad Consultancy Group
Portland, Oregon

About the Author

Mr. Ray P. Prasad is the founder of Ray Prasad Consultancy Group, a technology consulting firm which specializes in helping companies establish a strong internal SMT infrastructure. His firm helps companies set up their in-house SMT lines and develop in-house training and documentation critical to improving their design for manufacturability and process yield.

Before starting his consulting practice in 1994, Mr. Prasad was the SMT Program Manager at Systems Group of Intel Corporation in Hillsboro, Oregon. He and the rest of the SMT team were responsible for the development and implementation of SMT at Intel's Systems Group in Oregon, Puerto Rico, Singapore and Ireland. As part of this effort, he developed and taught in-house and design and manufacturing courses to Intel engineers. His experience at Intel also includes managing the Intel Pentium Pro™ and MCM (10 chip MCM-D with flip chip: silicon on silicon) package programs. He is a recipient of Intel's achievement award and was recognized by Intel CEO Andy Grove for establishing the SMT infrastructure at Intel.

Prior to moving to Intel in 1984, he was a Lead Engineer at Boeing Aerospace Company in Seattle, Washington and was responsible for introducing SMT into Boeing airplanes in 1981—when SMT was in its infancy.

Mr. Prasad is a registered Professional Metallurgical Engineer. He received his MS in Materials Science and Engineering and MBA from the University of California at Berkeley in 1974 and BS in Metallurgical Engineering from the Regional Institute of Technology, Jamshedpur, India in 1970. He has presented papers at various conferences, has published extensively in various technical publications in the U.S. and overseas and is a popular workshop leader at national and international conferences.

A long-time member of SMTA and IPC, he is chairman of the BGA committee J-STD-013 "*Implementation of BGA and other high density technology*". He is the past chairman of the Surface Mount Land Pattern (IPC-SM-782) and Package Cracking (IPC-SM-786, now renamed J-STD-020) committees. He is a recipient of the IPC President's award for his contribution to the advancement of electronics industry.

He has been a member of the Surface Mount Council since its inception in 1986. The council coordinates standards for the electronics industry. Mr. Prasad is a columnist for *SMT* magazine and also serves on its advisory board. Mr. Prasad can be reached by telephone at 503-297-5898 or by email at smt solver@rayprasad.com.

Part One

Introduction to Surface Mounting

Chapter 1

Introduction to Surface Mount Technology

1.0 INTRODUCTION

The present methods of manufacturing conventional electronic assemblies have essentially reached their limits as far as cost, weight, volume, and reliability are concerned. Surface mount technology (SMT) makes it possible to produce more reliable assemblies at reduced weight, volume, and cost. SMT is used to mount electronic components on the surface of printed circuit boards or substrates. Conventional technology, by contrast, inserts components through holes in the board. This deceptively simply difference changes virtually every aspect of electronics: design, materials, processes, and assembly of component packages and substrates.

The surface mount concept isn't new. Surface mounting has its roots in relatively old technologies such as flat packs and hybrids. But the design and manufacturing technologies used previously generally are not applicable to the surface mounting done today. The current version of SMT requires a complete rethinking of design and manufacturing, along with a new SMT infrastructure to develop and sustain it.

Electronics assembly originally used point-to-point wiring and no substrate at all. The first semiconductor packages used radial leads, which were inserted in through holes of the single-sided circuit boards already used for resistors and capacitors. Then Fairchild invented dual in-line packages (DIPs) with two rows of leads. These packages could be wave soldered along with other components such as resistors, capacitors, vacuum tubes, or vacuum tube socket leads.

In the 1950s, surface mount devices called flat packs were used for high reliability military applications. They can be considered the first surface mount packages to be mounted on printed circuit boards. However, the flat-pack devices had to be mounted too close to the board surface, required gold-plated leads, were very costly and required discrete soldering (hot bar). They did not become popular, but they are still used in mostly military and some commercial applications. In the 1960s, flat packs were replaced with DIPs, which were easier to insert and could be wave soldered en masse.

Another significant contribution to today's SMT lies in hybrid technology. Hybrid components, which became very popular in the 1960s, are still widely used. They feature surface mount devices soldered inside a through-hole or surface mount ceramic package body. Hybrid substrates are generally ceramic, but plastic hybrids have been used to reduce cost in some commercial applications. The ceramic and plastic hybrids shown in Figure 1.1 have surface mount devices mounted on the outside. Hybrids may have devices mounted inside the package as well. Also, most of the passive components (resistors and capacitors) used today were originally made for the hybrid industry. Placement and soldering techniques developed for the hybrid industry have also become a part of today's SMT. While the hybrid industry has made a significant contribution to SMT, hybrid substrates are so small compared with the substrates used for surface mounting that the design, manufacturing, and test methodologies are drastically different. But there is no doubt that SMT has profited immensely from the developmental work in hybrids.

Figure 1.1   Some examples of ceramic and plastic hybrids. (Courtesy of Intel Corporation.)

Today's SMT also has its roots in extensive developmental work done in the United States for military applications. For example, to shrink package size for larger pin counts, the military needed hermetically sealed devices with leads on all four sides. This led to the development of leadless ceramic chip carriers (LCCCs) in the late 1960s.

LCCCs proved to have their own problems, however: They require expensive substrates with matching coefficients of thermal expansion (CTEs) to prevent solder joint cracking due to CTE mismatch between the ceramic components and the glass epoxy substrates. The military has spent considerable human and financial resources in developing acceptable substrates for the LCCCs, but the results have been less than satisfactory.

While the electronics industry in the United States was preoccupied with developing substrates for LCCCs for a very limited military market, the Japanese and Europeans were responding to a much wider consumer market. In the 1960s, N.V. Philips, a European company, invented surface-mountable Swiss outline packages for the watch industry. These packages are known today as small outline (SO) or small outline integrated circuit (SOIC) packages. In the early 1970s the Japanese started building calculators using plastic quad packs, which are similar to flat packs but have leads on all four sides. The term "quad" is used for packages that have surface-mountable leads on all four sides, bent down and out like gull wings. This lead configuration is susceptible to bending and damage during handling and shipping. Even though the quad packs provided space savings because of their smaller size compared with their through-hole counterparts, the fragile gull wing leads required special handling to prevent lead bending. Package style has a big impact on manufacturing technologies such as handling, soldering, cleaning, and inspection.

The Japanese also perfected the current version of surface mount technology on high volume consumer products, such as radios, televisions, and video cassette recorders. For the most part, these products require passive surface mount devices for their analog circuits and are wave soldered. Passive components have terminations instead of leads, and they are considerably smaller. Figure 1.2 compares commonly used surface mount passive components with their conventional leaded equivalents.

Whereas the 1970s saw widespread usage of passive devices in consumer products in Japan, the 1980s were characterized by the use of active devices worldwide in mass applications such as telecommunications and computers. In the United States, the electronics industry is driven by digital circuits for telecommunications, computer, and military markets. The electronics assemblies for the military markets primarily use LCCCs (for up to 44 pins) and leaded ceramic chip carriers (LDCCs) for more than 44 pins. The industrial markets use mostly SOICs and plastic leaded chip carriers (PLCCs).

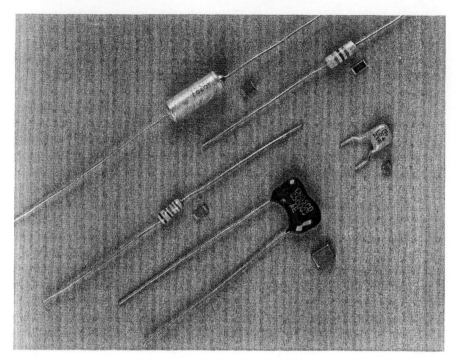

Figure 1.2 Surface mount passive components and their conventional equivalents.

PLCCs have leads on all four sides, like quad packs, but each lead is bent down and under like a "J"; thus PLCCs are also referred to as the J-lead packages. The J leads are less prone to damage during handling and provide better real estate efficiency on a substrate. A PLCC package and its conventional leaded equivalent are compared in Figure 1.3. PLCCs were developed to solve the problem of CTE mismatch between the package and the substrates by providing J leads to take up any strain during the component's use. They are the most commonly used packages in the United States. The Japanese have also adopted the PLCC packages, but generally for meeting the needs of U.S. markets.

These days there are varieties of surface mount components with varying lead pitches (pitch is defined as the distance between lead centers) in use on the same board. The most common 50-mil pitch packages are PLCCs and SOICs. For higher pin counts, finer pitch (25-mil pitch or less) packages are used. We discuss package types further in Section 1.7. Various types of surface mount components are discussed in greater detail in Chapter 3.

Having briefly reviewed the history of SMT, let us discuss the advan-

Figure 1.3   Surface mount active component [a plastic leader chip carrier (PLCC)] and its conventional equivalent.

tages and disadvantages, the technical concerns, and the future of this technology.

1.1   TYPES OF SURFACE MOUNTING

Since many components aren't yet available for surface mounting, SMT must accommodate some through-hole insertion. For this reason, "surface mount assembly" is an incomplete description. The surface mount components, actives and passives, when attached to the substrate, form three major types of SMT assembly—commonly referred to as Type I, Type II, and Type III, as shown in Figure 1.4. The process sequences are different for each type, and all the types require different equipment.

The Type III SMT assembly contains only discrete surface mount components (resistors, capacitors, and transistors) glued to the bottom side. The Type I assembly contains only surface mount components. The assembly can be either single-sided or double-sided. The Type II assembly is a combination of Type III and Type I. It generally does not contain any active surface mount devices on the bottom side but may contain discrete surface mount components glued to the bottom side.

The complexity of SMT assembly is increased if large fine pitch (over 208 pins with 0.5-mm pitch), ultra fine pitch (pitch under 0.5 mm), QFP (quad flat pack), TCP (tape carrier package), or BGAs (ball grid arrays) and very small chip components (0603 or 0402 or smaller) are used on these assemblies along with conventional (50-mil pitch) surface

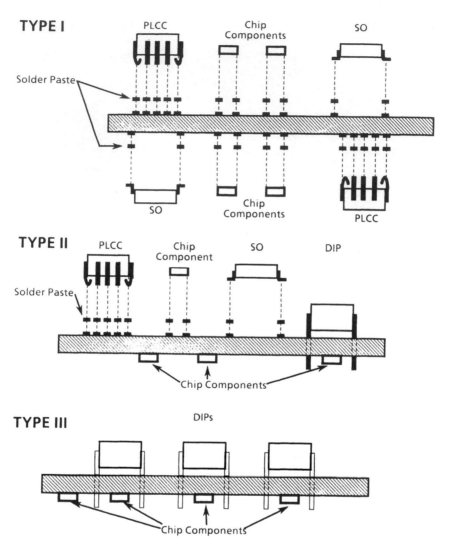

Figure 1.4 The three principal surface mount assemblies in surface mount technology.

mount packages. Assemblies containing these types of packages will be referred to as Type IC, Type IIC, and Type IIIC to indicate their complexities.

We should note that if the pitch is greater than 20 mils, it is referred to in mils, but if the pitch is 20 mils or less, it is referred to in mm (millimeters) in conformance with agreements of standard setting bodies such as the EIA (Electronics Industries Association) and EIAJ (EIA Japan).

Type I and Type II become Type IC and Type IIC if large fine pitch (over 208 pins), ultra fine pitch or UFP (lead pitch of 0.4 mm or below), TCP, BGAs, and 0603 (60 mils long and 30 mils wide), 0402 (40 mils long and 20 mils wide), or smaller chip components (resistors or capacitors) are used. Type IIIC is applied to assemblies if SOICs and small chip components (0603 or 0402 or smaller) are used on the secondary side, with only through-hole components on the primary side. Figure 1.5 shows Type IC, Type IIC, and Type IIIC assemblies.

This description of SMT Types I, II, and III is by no means universal, but it is the most commonly used in the industry. For simplicity and consistency these definitions have been used throughout this book.

The process sequence for Type III SMT is shown in Figure 1.6. First the through-hole components are autoinserted and clinched using existing through-hole insertion equipment. Next the assembly is turned over and adhesive is applied. Since the leads of the autoinserted components are clinched, they do not fall off the board. Then the surface mount components are placed by "pick-and-place" machine(s), the adhesive is cured in a convection or infrared oven, the assembly is turned over, and both leaded and surface mount components are wave soldered in a single operation. The discrete components on the bottom side of the board are held in place by the adhesive during wave soldering. If autoinsertion equipment is not used, the leads are not clinched and the process sequence must be reversed: the adhesive is dispensed first; the discrete surface mount components are set in place; the adhesive is cured; the assembly is turned over and all the through-hole components manually inserted; and the assembly is wave soldered, cleaned, and tested.

Many companies wave solder SOICs, but this is not a recommended process because flux may seep through the lead frame material during wave soldering. Such seepage, in turn, will cause die corrosion, thus compromising the reliability of the device. If an SOIC is glued to the bottom of the assembly and wave soldered, we will classify the assembly as Type IIIC. Such assemblies must be evaluated for reliability before use.

The process sequence for Type I SMT is shown in Figure 1.7. A Type I assembly uses no through-hole components. First solder paste is screened, then components are placed, and the assembly is baked in a convection or infrared oven to drive off volatiles from the solder paste. (Pastes are also available that do not require baking.) Finally the assembly is reflow soldered and cleaned (assuming no-clean paste is not being used). For double-sided Type I SMT assemblies, the board is turned over and the process sequence just described is repeated. The solder joints on

(a)

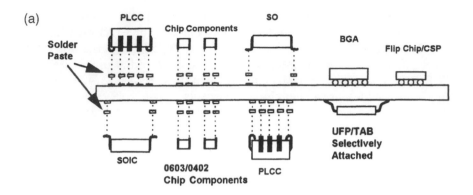

(b)

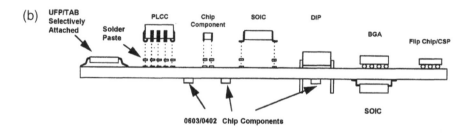

(c)

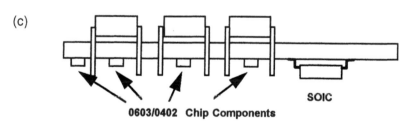

Figure 1.5   (a) Type IC SMT, (b) Type IIC SMT, (c) Type IIIC SMT.

the top side of the board are reflowed again. During the second reflow, the components are held in place by the surface tension of previously reflowed solder. Note that the assembly must first be reflowed on the top side of the board for the surface tension of solder to prevent components from falling off when the assembly is turned over during the second reflow.

The process sequence for a Type IC assembly does not generally change; it is the same as for Type I. However, if UFP packages (with lead pitches of 0.4 mm and below) are used, they may be soldered by the hot bar process at the end of the reflow process step shown in Figure

ERSITIES AT MEDWAY LIBRARY

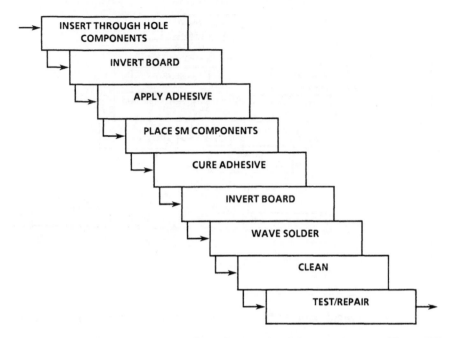

Figure 1.6 Typical process flow for underside attachment (Type III SMT).

1.7. Many companies have problems printing solder paste if the assembly contains UFPs. The use of BGAs in Type IC assemblies does not change the process sequence shown in Figure 1.7. Because BGAs have wider pitches (generally 50 mils), paste printing is relatively easy.

The process sequence for Type II SMT is shown in Figure 1.8. Since a Type II assembly is a combination of Type I and Type III, it uses all the processes needed for both. It is the most difficult assembly to manufacture because it has the most process steps. The Type II assembly goes through the process sequence of Type I followed by the process sequence for Type III. Note that a Type II assembly may not have discrete surface mount components on the bottom side. In that case, the Type III process sequence of dispensing and curing adhesive and placing surface mount components on the bottom side will be omitted, but the assembly will still use two soldering operations: reflow soldering for the active surface mount components on the top and wave soldering for the through-hole components also on the top side.

The process sequence for Type IIC assembly is affected in the way just discussed for Type IC because only the process related to the Type I portion of the Type II process is affected. In other words, as in Type

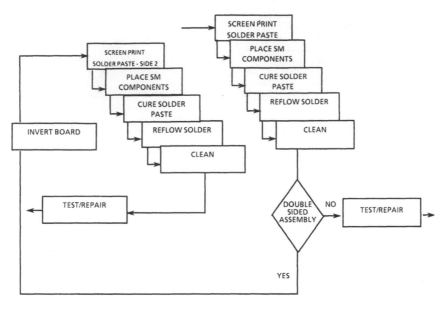

Figure 1.7   Typical process flow for total surface mount (Type I SMT).

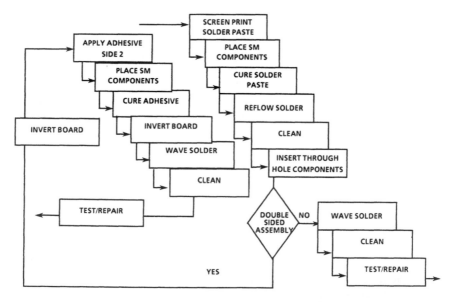

Figure 1.8   Typical process flow for mixed technology (Type II SMT).

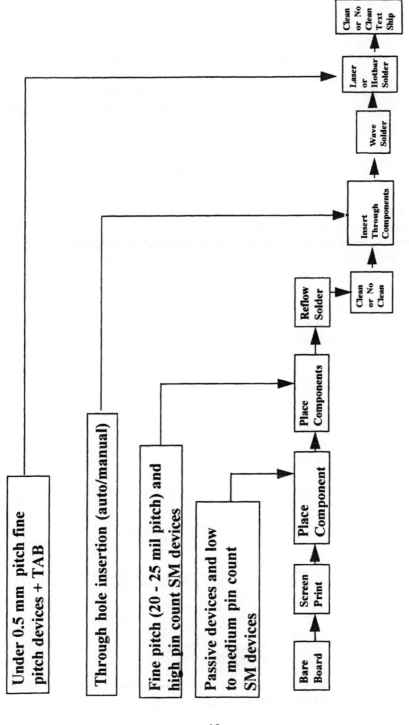

Figure 1.9 Process sequence for Type IIC assembly with fine pitch and TAB.

13

IC, if the Type IIC assembly has UFPs, the hot bar process step may be used after the wave soldering step as shown in Figure 1.9.

The process details for all three types of SMT assembly—adhesive, solder paste, placement, soldering, and cleaning—are covered in Chapters 8, 9, 11, 12, and 13, respectively. And finally, every type of assembly is inspected, repaired if necessary, and tested. The details of inspection, test, and repair are discussed in Chapter 14.

1.2 BENEFITS OF SURFACE MOUNTING

Because surface mount components are small and can be mounted on either side of the board, they have achieved widespread usage. The benefits of SMT are available in both design and manufacturing.

Among the most important design-related benefits are significant savings in weight and real estate and electrical noise reduction. As shown in Figure 1.10, surface mount components can weigh as little as one-tenth of their conventional counterparts [1]. This causes a significant reduction in the weight of surface mount assemblies. Weight savings is especially important in aircraft and aerospace applications. Because of their smaller size, surface mount components occupy only about one-half to one-third

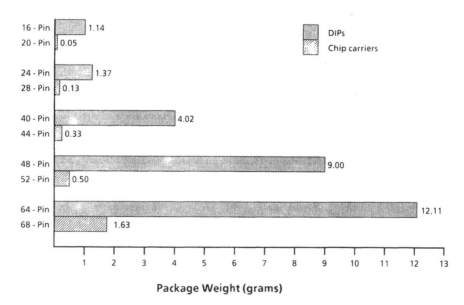

Figure 1.10 **The weight of various lead count chip carriers for SMT compared to the weight of an equivalent DIP [1].**

of the space on the circuit board. Figure 1.11 compares the areas occupied by DIP, PLCC, and SOIC packages of different pin counts.

The size reduction in active components is a direct function of lead pitch, a term that describes the distance between the centers of adjacent leads. The active surface mount components have leads spaced at 0.050 inch (50 mil) centers or less. In other words, their lead pitch is at 50 mil centers or less. The packages that have lead pitches of less than 50 mils are generally referred to as fine pitch packages. The lead pitches of commonly used fine pitch packages are 33 mils (0.8 mm), 25 mils (0.63 mm), and 20 mils (0.5 mm). As mentioned earlier, packages with lead pitches of 20 miles and below are referred to in millimeters (mm) only. Packages with pitches of less than 0.5 mm are referred to as ultra fine pitch (UFP) packages. For comparison, the lead pitch for DIPs is 100 mils. As the lead pitch gets smaller, the reduction in real estate requirement becomes more significant. Figure 1.12 shows a family of surface mount and through-hole packages with different lead pitches. See Chapter 3, Figures 3.13 and 3.14 for package areas of higher pin count surface mount packages with varying lead pitches.

Size reduction is also a function of the mix of surface mount and

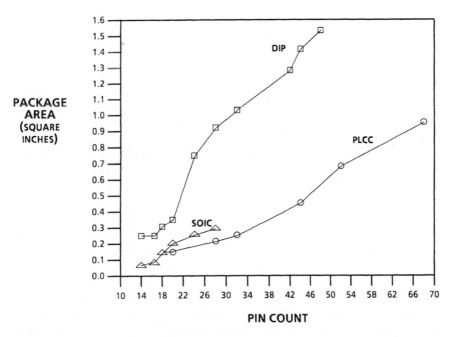

Figure 1.11 **A comparison of package area with conventional (DIP) and surface mount (PLCC and SOIC) packages.**

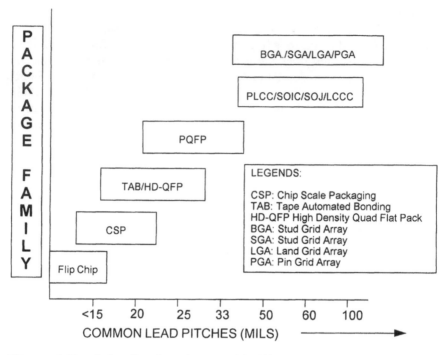

Figure 1.12   A family of packages with different lead pitches.

through-hole components on the board. Since not all components are available in surface mount, the actual area savings on a board will depend on the percentage of through-hole mount components replaced by surface mount components. Depending on the component mix, the three types of surface mount assemblies discussed earlier provide different levels of benefits. For example, the real estate savings are much smaller in a Type III SMT board, since only discrete components are replaced with surface mount counterparts. The real estate savings in a Type III board typically varies from 5 to 10%.

At the other extreme, the greatest real estate savings is accomplished in a Type I SMT board, which contains all components in surface mount. Since surface mounting also allows double-sided mounting, the real estate savings will be about twice as much in a double-sided Type I assembly. For example, 8 megabytes (8 MB) of 4 MB dynamic random access memory (DRAM) with conventional DIP components will require about 4 × 4 inches of printed circuit board area. The same board with 100% surface mount components can accommodate 16 MB in a single-sided board or 32 MB in a double-sided memory board.

Type II SMT is a mixture of through-hole and surface mount active (integrated circuits) components and passive components (resistors and capacitors). Since the availability of surface mount components has been constantly improving, so has the real estate savings for Type II assemblies. For example, Figure 1.13 shows that in a single-sided, mix-and-match board with approximately 50% surface mount components, the real estate savings is about 20%. When the component availability in surface mount improved to 80%, the real estate savings jumps to 35%, as shown in Figure 1.14.

SMT also provides improved shock and vibration resistance as a result of the lower mass of components. Also, because of their shorter lead lengths, the surface mount packages have lower parasitics (undesirable inductive and capacitive elements) than their through-hole counterparts, as shown in Table 1.1. This results in reduced propagation delays as shown in Figure 1.15 [1]. A reduction in parasitics also reduces package noise. For example, in a memory board with DIP memory devices on the top side and surface mount capacitors on the bottom side, about 25% noise reduction can be expected when compared with its conventional

Figure 1.13   An example of 20% area savings in an experimental Type II SMT board where only 50% of the components are surface mount. (Courtesy of Intel Corporation.)

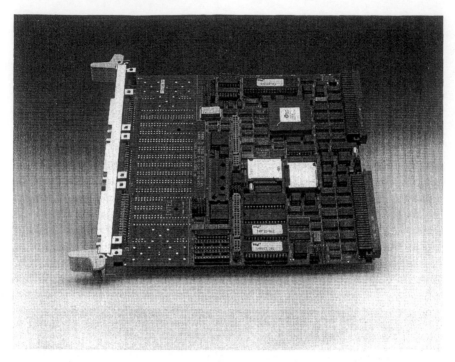

Figure 1.14   An example of 35% area savings in an experimental Type II SMT board where 80% of the components are surface mount. (Courtesy of Intel Corporation.)

version with all through-hole mount components. As a result, only about half as many decoupling capacitors may be necessary to provide effective decoupling. Improved electrical performance becomes even more significant and critical as clock speeds increase. (Refer also to Chapter 5, Section 5.9.)

In addition to the foregoing examples of design benefits of a surface mount assembly, SMT provides many manufacturing benefits. These include reduced board cost, reduced material handling cost, and a controlled manufacturing process. Routing of traces is also improved because there are fewer drilled holes, and these holes are smaller. The size of a board is reduced and so is the number of drilled holes. A smaller board with fewer drilled holes will naturally cost less. If the functions on the surface mount board are not increased, the increased interpackage spacings made possible by smaller surface mount components and a reduction in the number of drilled holes may also reduce the number of layer counts in the printed circuit board, thus further lowering board cost.

Surface mounting is more amenable to automation than is through-

Table 1.1   Reduced inductance in surface mount devices

TYPE OF PACKAGE	INDUCTANCE (nH)
DIP, 14 pin	3.2 − 10.2
SOIC, 14 pin	2.6 − 3.8
PLCC, 20 pin	4.2 − 5.0

hole mounting. It even eliminates some process steps. For example, different types of autoinsertion equipment (such as DIP inserter, radial inserter, axial inserter, and sequencer) are used for different through-hole mount components. This requires setup time for each piece of equipment. When SMT is used, the same pick-and-place equipment can place ALL types of surface mount components, thus reducing setup time. In addition, stock can be kept on the placement machine, further decreasing setup time.

Higher product quality results from the consistency of automated placement, improved process control, and the use of smaller components. The tape and reel, tube, or waffle tray component packaging used with SMT permits large quantities of components to be issued to the production area without the need to sequence or repackage components, thus providing an additional benefit.

What do all these manufacturing benefits mean in terms of cost? Depending on the application, the cost saving is 30% or better in placement time alone. Additional savings can be realized through the reductions in material and labor costs that are associated with automated assembly. Moreover, because repairs are fewer, less scrap is generated. However, reduced product costs may elude the novice SMT user. Skill and experience are necessary to secure packaged components at reasonable prices in a configuration that matches the assembly capability, to implement and control the process, and to design the printed board for optimum manufacturability. Once these skills have been learned, costs should decrease.

In through-hole assemblies, damage during repair is a major concern if the process variables are not properly controlled [2]. As discussed in Chapter 14, the repair of surface mount assemblies, contrary to popular belief, is easier and less damaging to the assembly than the repair of through-hole assemblies.

In SMT there are some less tangible benefits as well. For example, the pick-and-place machines tend to be extremely quiet compared with autoinsertion equipment, thus providing a less noisy work environment. Also, because surface mount components are smaller, the need for warehouse space can be significantly reduced. Smaller size and weight of components and finished assemblies also mean lower shipping and handling costs.

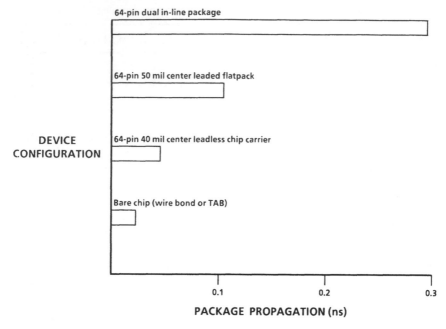

Figure 1.15 Comparison of the propagation delays of surface mount and DIP packages [1].

Does this mean that SMT assemblies always cost less? No. Actually, quite the reverse may sometimes be true—for a very good reason, as explained in Section 1.4 (When to Use SMT).

1.3 SMT EQUIPMENT REQUIRING MAJOR CAPITAL INVESTMENT

Through-hole and surface mount assemblies (Types I, II, and III) have three major process steps in common: component mounting, soldering, and cleaning. To understand the major differences in the requirements for equipment for through-hole and surface mount assemblies, let us refer to Figure 1.16, which lists the major pieces of equipment required for conventional through-hole mount (THM) as well as all types of surface mount assemblies. It is not the intent of Figure 1.16 to show the specific process sequence in different types of SMT, but only to highlight the major equipment requirements for different assemblies.

We supplement the legend to Figure 1.16 by noting that for through-hole assemblies, insertion, wave soldering, and cleaning equipment are

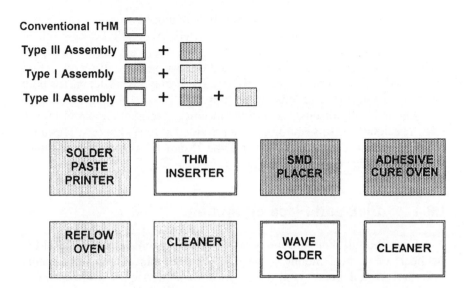

Figure 1.16   Major equipment for Type II SMT assembly.

shown. The cleaning equipment may be either aqueous or solvent type. (If no-clean solder paste or flux is used, a cleaner will not be necessary.) For Type III, only two additional types of surface mount equipment are required: the component placement machine and an adhesive curing oven. For the assembly using Type I SMT, none of the through-hole equipment shown in the double box is required.

A completely through-hole process and a surface mount process have nothing in common. For a Type I assembly, all the equipment shown in the single boxes of Figure 1.16 is required. If, however, the placement equipment for Type III SMT already exists, the additional equipment needed for Type I will be a solder paste screen printer, a reflow soldering machine (vapor phase or infrared), and a cleaner. If an infrared oven is used for curing adhesive for Type III SMT, it can also be used for the reflow soldering. (Infrared is defined broadly here to mean mostly infrared, mostly convection, or a combination of the two. See Chapter 12 for a detailed discussion of reflow machines.) At this time the convection reflow system is most widely used and the vapor phase reflow system the least.

For Type II SMT, it is a different matter, however. All the equipment (single and double box) shown in Figure 1.16 is required. This means that if one has the capability for Type II SMT, the capabilities for through-hole, Type III, and Type I automatically exist.

In Figure 1.16, we have shown two cleaners (one after reflow for Type I and one after wave soldering for Type III). Only one cleaner may be sufficient. Or, as noted earlier, no cleaner would be necessary if no-clean solder paste or wave flux is being used. A detailed discussion of cleaning is given in Chapter 13. Similarly, two ovens are shown in Figure 1.16 (one for Type III adhesive cure and one for reflow). If a convection or infrared reflow oven is used, only one oven may be sufficient for both steps when volume requirements are not high.

Let us briefly discuss the surface mount equipment shown in Figure 1.16. The details on this equipment are provided in later chapters. Through-hole equipment is also required for Type II and III assemblies, but their description is beyond the scope of this book.

1.3.1   Pick-and-Place Equipment

Even though there are many types of equipment required for SMT, the heart of surface mount is the machine that places the components onto the printed board land areas prior to soldering. These machines generally constitute about 50% of the total cost of the complete line.

Some surface mount placement machines are very versatile, capable of placing many different components, while others are dedicated to a few component types. Placement machines use vacuum pickup tools to hold the components, and many also provide vision-assisted alignment. In general, placement machines offer better speed, accuracy, and flexibility than through-hole insertion machines or other earlier generation place-ment systems.

Fine pitch (0.5 mm) and ultra fine pitch (0.4 mm–0.3 mm) placement machines require greater dexterity and precision placement capability. Some of these machines cut and form the leads of the IC package at the time of placement in order to avoid component lead damage due to mishandling. This feature increases process accuracy and precision capa-bilities. The new array type packages are somewhat more forgiving be-cause the pitch of the array is large and can allow more liberal tolerance conditions. However, chip scale package mounting in the array format requires good placement accuracy.

The number of machines required to adequately assemble all surface mount components varies greatly depending on the type of components being assembled and the through-put desired by the manufacturer. Dedi-cated surface mount component placement equipment can achieve the greatest through-put. Versatility and flexibility in the ability of the equip-ment to handle a variety of different components reduces the speed and effectiveness of a single machine for meeting all the needs of the assem-bly function.

Some companies also use dedicated robots for placement. The equipment is generally very inexpensive, but the software and hardware development costs can be considerable. Robotic placement should be considered for components for which pick-and-place equipment is not widely available. Also, robotic placement implies that a company wants the ultimate in flexibility and has sufficient engineering resources to develop the necessary hardware and software. Examples of the placement equipment discussed in this section are shown in Chapter 11.

1.3.2   Solder Paste Screen Printer

A screen printer is needed to screen solder paste onto the printed circuit board (PCB) before placement of surface mount components. Screen printers have been widely used by the PCB industry for screening solder mask. This equipment has also been extensively used by the hybrid industry for screening solder paste. Different equipment, however, is used for the screening of solder mask and solder paste. The cost of screen printers can vary widely, depending on their degree of automation and the sizes of boards they can handle.

Printing systems are available in three configurations: manual, semi-automatic, and fully automatic. The printing equipment may be table mounted, stand-alone, or in-line. Many semi-automatic printers offer manual vision alignment capability, while fully automatic printers offer automatic vision alignment capability.

Stencils are preferred over screens because of better image accuracy, volume control, and longer service life. Screens are not in common use today. Stainless steel is the most common stencil material, but other choices include brass, alloy 42, and molybdenum. Chemical etching is commonly used to fabricate stencils. Several other methods have been developed to aid fine pitch applications, including plating (electroless nickel), electrodeposition, laser cutting, and electropolishing. Stepped thickness stencils have been developed to apply various thicknesses of solder paste to different component types on a substrate. Fine pitch printing requires very accurate alignment of the stencil apertures to the land patterns. Process and equipment for screen printing are discussed further in Chapter 9.

1.3.3   Curing/Baking Oven

For Type III surface mount, components glued to the bottom of the board require curing before wave soldering. In addition, components placed on solder paste in Type I and Type II assemblies achieve reduced solder defects (such as solder balls) by being baked before reflow soldering. An oven is required for both adhesive curing and solder paste baking operations. The same or different ovens may be used for adhesive curing and baking of solder paste. If an infrared (IR) oven or a convection oven is used for reflow soldering, it is not necessary to have separate adhesive and solder paste baking ovens. The same IR or convection oven can be used for adhesive curing, solder paste baking, and reflow soldering. The decision to combine solder paste baking, solder reflow, and adhesive curing in the same oven depends on volume requirements. It also means that a decision has been made to use IR or convection instead of other reflow soldering processes such as vapor phase, laser soldering, or convection belt soldering. (We should note here, as mentioned earlier, that today the convection oven is more widely used than any other type of oven.)

1.3.4   Reflow Soldering Equipment

The selection of reflow equipment is the second major decision, after pick-and-place equipment has been chosen. As we discuss in Chapter 12, there are many reflow soldering methods. Each has advantages and disadvantages, and none of them is perfect for every application. Equipment cost, maintenance cost, and yield are some of the leading deciding factors. The most widely used reflow processes are vapor phase and infrared. The vapor phase process is very versatile in that it can be applied to assemblies of any shape, and various models are available in both batch and in-line types. The capital costs of the vapor phase and infrared processes are comparable, but the operation cost may be higher for vapor phase. In addition to cost, many other technical issues that have an impact on frequency of solder defects should be considered in selecting a particular reflow soldering process.

Infrared equipment has been widely used in the hybrid industry, but at much higher temperatures, for firing thick film deposits on ceramic substrates. The same equipment was initially tried for reflow soldering on glass epoxy substrates with terrible results: burning, charring, and warpage of boards were fairly common. Now the industry has made tremendous progress in redesigning heating elements to eliminate earlier problems. Effective IR equipment generally applies nonfocused heat (more convection energy than radiation energy). In addition to eliminating burning and warpage, this mode of heat application reduces the sensitivity of heat energy to the color of components and boards. The near-IR process, which relies more on infrared radiation than on convection energy, also has been used very successfully for reflow soldering. The development of a reflow profile for each product is key to obtaining optimum yields no matter which reflow system is used. With an appropriate profile, any of the systems will work. The pros and cons of convection IR versus near-IR versus vapor phase reflow soldering are discussed in detail in Chapter 12.

1.3.5   Solvent Cleaning

After soldering, the electronic assemblies must be cleaned to remove the flux. The subjects of flux and cleaning are closely intertwined. The selection of solvent and cleaning equipment depends on the flux used and the cleanliness requirements.

The cleaning process has received considerable attention because of efforts to eliminate chlorofluorocarbons (CFCs) and other ozone depleting chemicals. CFCs were banned at the end of 1995. Electronic assemblies are presently soldered using various fluxes whose residues may require cleaning after the soldering process. The use of rosin-based fluxes has mandated a cleaning step for any manufacturer who performs bed-of-nails testing, because the rosin acts as an insulator and, additionally, it can transfer sticky residues to the problems, making them ineffective for electrical testing.

Military specifications have always required cleaning of Mil-approved RMA or RA fluxes and pastes, with ionic contamination testing to assure that the remaining residues do not deteriorate the assembly. Others require cleaning of rosin flux for cosmetic reasons. Cleaning is also required prior to conformal coating. So those who must use rosin pastes or flux use alternate solvents (non-ozone depleting solvents), semi-aqueous solutions, and saponifying detergents.

In most commercial applications, water soluble solder pastes and fluxes are used. They are generally cleaned with de-ionized water. And many companies have chosen to use low residue, no-clean pastes and fluxes that require no cleaning at all.

No matter which alternative is chosen, there is a cost associated with new equipment and material purchases for the chosen alternative. Numerous cleaning technologies are available. One must choose the cleaning process based on the product being built and the cost, performance, and environmental impact of the options. The use of alternative cleaning methods must be coupled with an understanding of the flux type being used. See Chapter 13 for a detailed discussion of flux and cleaning.

1.3.6   Wave Soldering Equipment

Wave soldering is the most widely used method for soldering through-hole components. Components glued to the bottom of a mixed-technology board also require wave soldering, although the required wave geometry is different from that for through-hole components. In the most common wave geometry, the dual-wave type, one wave is turbulent and the other wave is smooth, as in conventional technology soldering.

The need for wave soldering equipment can also be met by simply retrofitting the conventional solder pot with the dual pot—at a small fraction of the cost of a new machine. A completely new machine is not necessary.

Another wave geometry that is becoming popular for surface mounting is a vibrating single wave known as the omega wave. Like the dual-wave geometry, it helps in reducing solder defects. The agitation provided by the dual or vibrating wave dislodges trapped flux gas that forms during wave contact and forces solder into areas with poor wetting angles. Proper component orientation is also very important.

Some companies have successfully used drag soldering equipment and techniques. Ideally, the wave or drag soldering process should be limited to soldering ceramic surface mount resistors and capacitors. Small outline transistor (SOT) components are difficult to wave or drag solder.

Wave and foam fluxers are generally used to apply flux to the board. Spray fluxers are being used to apply very controlled amounts of flux, especially when no-clean fluxes are used. The development of an appropriate solder profile is important for best results. See Chapter 12 for a detailed discussion of this subject.

1.3.7   Repair and Inspection Equipment

Equipment unique to surface mounting includes inspection and repair/rework equipment. Inspection equipment for surface mounting has not matured yet. There are two main types of automated inspection equipment on the market: x-ray and laser. However, most companies depend on visual inspection at 2 to 10X, using either a microscope or magnification lamp.

Repair/work equipment uses conductive tips or hot air for removal and replacement of components. Conductive tooling, such as soldering iron tip attachments, is very inexpensive. The tips come in different configurations for different components but have limited applications. At a slightly higher cost, there are other conductive tips with built-in heating elements. At a much higher cost, hot air repair/rework tools are available that use nozzles of different sizes for components of different types and sizes. The hot air repair tooling is most commonly used for surface mounting because it is ideal for damage-free repair.

Even in-line (and expensive) repair systems are available on the market. However, when selecting repair/rework equipment, it must be kept in mind that repair/rework tooling does not need to be an automated, expensive in-line system designed for high volume repair. If a large number of boards is to be repaired, automated repair equipment is not the answer. Instead, the process should be controlled to keep the defect rate low. See Chapter 14 for details on repair, inspection, and process control.

1.4   WHEN TO USE SURFACE MOUNTING

With all the benefits SMT has to offer, it might seem foolish not to use it in every product. For many technical reasons, however, it is not appropriate for every application. Even when it is, SMT requires a complete new infrastructure to implement. The issue of infrastructure is discussed in Chapter 2. The major technical issues that must be overcome are discussed briefly in Section 1.5 of this chapter. The rest of the book is devoted to resolving those technical issues. Since surface mount assembly processes cannot be performed manually, and since the processes are relatively high speed, thousands of assemblies can be built before the problem is detected. It has the potential to produce more scrap than conventional through-hole technology for the following reasons:

- The processes are incredibly high speed.
- They must be performed by machines.
- The equipment must be thoroughly characterized. Characterization can be defined as understanding all parameters that affect the equipment's performance.
- Vendors may say it is easy—it is not.
- Most large companies have assigned engineers to optimize; small companies "learn as they go."
- "Learn as you go" is not a real option, since revenue or product schedule or both may be adversely impacted.
- Expect to invest considerable engineering resources in process characterization, training, and documentation.
- With the advent of fine pitch, UFP, BGA, 0604, and 0402 resistors and capacitors and the widespread use of no-clean flux, the inherent problems are simply compounded.

We stated at the outset that SMT should not be considered for an application if the performance, weight, functionality, and real estate requirements can be met by the conventional technology. One must not use SMT just for the sake of using a new technology. The SMT infrastructure requires a considerable investment in human resources and capital equipment. One might ask, however, whether companies that have already invested in an SMT infrastructure, and happen to have conventional design and manufacture capability, should switch to SMT for all new designs. The answer is no; conventional boards of lower functionality will be cheaper than SMT boards of higher functionality. Thus, if the needs of the marketplace can be met without jeopardizing the future, one should stick to conventional technology for such products.

Since the customer is buying the functionality, not the technology, one must carefully analyze the marketplace before deciding to use SMT. Some of the important questions that should be asked are: How much real estate is needed, and which type of SMT best meets those needs? Should Type I, II, or III SMT be used or will Type IC, IIC, or IIIC be needed? Are all the components available?

Many components are not available in surface mount format. For example, many very large scale integrated (VLSI) and electromechanical devices, such as sockets and connectors, are not widely available. Some connectors and sockets are available in surface mount but tend to take up more real estate than conventional devices, thus defeating one of the purposes of surface mounting. However, surface mount connectors do provide better routing efficiency, since they help reduce the number of via holes. In addition, some components may have a cost premium, although this situation is improving rapidly. Because of cost premiums and the newness of the technology, SMT boards tend to be more expensive, especially in the beginning stages, when yields are generally much lower.

A surface mount board costs more than a through-hole board in most applications because through-hole boards are rarely redesigned just to achieve cost reduction. In most new designs, instead of reducing board size, designers put added functions on the same board. This increases cost per board, since the same board will contain more components, which constitute a significant portion of its cost. The board cost will also be higher, since a denser board with more functions than a through-hole board will require finer lines and spaces while maintaining or even increasing the layer count. If the technology requires fine pitch, UFP, or BGA, the cost may be further increased.

However, the cost per function will be significantly less, since instead of designing and building two through-hole assemblies, only one surface mount board will be needed. See Chapter 5 (Section 5.5: Cost Considerations) for more details on cost.

1.5 TECHNICAL ISSUES IN SURFACE MOUNTING

Many technical issues in SMT must be resolved before this technology can be implemented in a product. There are only two very simple differences between surface mount technology and through-hole technology. The surface mount components are smaller and they mount on the surface of the board. These simple differences, however, create technical concerns that affect every aspect of the design and process of building an electronic assembly. These technical issues in SMT are summarized in Figure 1.17. Let us briefly discuss them, not to provide an answer but simply to gauge the extent of the problem.

Let us start with a discussion of components. As far as components are concerned, the main issues are price and availability. These matters are interrelated and are addressed in Chapter 3. Another major issue associated with components is package reliability.

Like conventional through-hole dual in-line packages (DIPs), surface mount plastic components absorb moisture during storage. However, when surface mount packages are reflow soldered at 200° to 240°C, the moisture expands and may crack the package. This phenomenon is commonly referred to as the popcorn effect. The same plastic molding compound is used for DIPs, but DIP packages do not crack during soldering simply because during wave soldering, the plastic bodies of DIP packages do not experience the high temperatures that surface mount packages undergo during reflow soldering. In wave soldering only DIP leads are heated to high temperatures.

BGA packages are even more susceptible to moisture-related cracking because the BT (bismalide triazine) resin used in BGA absorbs even more moisture than the conventional molding compounds used in other packages.

Not all surface mount plastic packages are susceptible to package cracking. Susceptibility here depends on package design, die size, strength

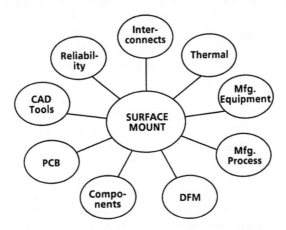

Figure 1.17   Key technical issues in surface mount technology.

and thickness of the molding compound, and moisture content of the package. Baking components before reflow solves this problem. See Chapter 5 for details.

The ceramic packages used in military applications are not plagued by such cracking because ceramic does not absorb moisture. However, ceramic packages have their own problems. Military applications require hermetically sealed ceramic devices (leadless ceramic chip carriers or LCCCs), but these cannot be used on glass epoxy substrates. To prevent cracking of solder joints, the coefficients of thermal expansion of surface mount packages and the substrates should match. Even when the X and Y CTEs of the substrates match, the CTE in the Z direction may be different. This may cause cracking in the via or plated through hole in the substrates. The danger is eliminated by selecting an appropriate substrate material. However, cracking of a plated through hole in the substrate is a matter for concern. Commercial applications eliminate solder joint cracking problems by using surface mount packages with compliant leads. See Chapters 4 and 5.

Let us refer again to the technical issues presented in Figure 1.17. Solder joint reliability, computer-assisted design (CAD) tools, thermal management, and interconnect design rules are also affected when SMT is used. For example, in conventional assemblies, the plated through hole provides a strong mechanical connection with a considerable safety margin. In surface mounting, the safety margin for the mechanical strength of a solder joint is not as great because there is no plated through hole.

CAD tooling for SMT is also one of the critical issues in SMT. The CAD system should have the capability for autorouting and on-line design rule checking. Most CAD systems are adequate for the Type III layout and for single- or double-sided memory boards, but not for the design of double-sided logic boards with or without blind/buried vias. By laying out pilot boards, one can determine whether the existing CAD system will be adequate or whether it must be replaced. The CAD system must also be capable of producing placement data output in a format acceptable to pick-and-place machines.

Since SMT makes it possible to increase the function per square unit of area, the power density per square unit also increases. If a thermal problem already exists on a board, SMT will compound it. For the increased functional density, SMT will also require finer line width and spaces for routing of boards. This increases the potential of signal cross coupling. These issues are addressed in Chapter 5.

The land pattern design, which is unique to SMT, is covered in Chapter 6. Here we note simply that in addition to proper selection of components and substrates, proper land pattern design is essential if reliable solder joints are to be provided.

SMT also requires good design for manufacturability (DFM in Figure 1.17). This technology is not really suited for manual placement and soldering. As explained in detail in Chapter 7, designing for manufacturability is very critical for good yield.

SMT also has an impact on the manufacturing processes and equipment. A major capital investment is required to implement this technology.

Another major issue in manufacturing is testing. The straightforward solution to testing problems—namely, to provide access to all critical test points—will offset some of the increased density benefits of SMT. However, this also requires major capital investment since automated test equipment (ATE) may cost more than the total cost of an entire SMT line. If this investment is not made, the yield needs to be very high so that excessive resources are not wasted in debugging a failed product.

In SMT manufacturing, issues of materials (adhesive, solder paste) and processes (solderability, placement, soldering, cleaning, and inspection/repair) are unique to the technology. Among manufacturing issues, adhesive, solderability, and cleaning are more critical than technical problems in placement, solder paste, or soldering. For example, if adhesive is cured too rapidly, voids generated during cure may absorb flux when the assembly is wave soldered. This can cause serious cleaning problems, especially if aggressive fluxes are used. If low residue or no-clean flux and solder paste are used, the defects may increase significantly since "no-clean" is not a drop-in process. The solutions to materials and process problems are discussed in Chapters 8 through 14.

Reading this section, one may get the feeling that we are almost in a development phase in SMT. However, this is not true. None of the technical issues discussed above poses a serious impediment to using SMT as a production-worthy manufacturing process. Considerable progress in each of the issues discussed above has been made. We are essentially talking about improving and fine-tuning the technology rather than any fundamental concern. To keep things in proper perspective, it should be noted that even in conventional through-hole technology, not all the technical problems have been totally resolved. For example, the industry is still struggling with solderability, repair/rework, inspection, and solder defect problems even after decades of widespread use of through-hole technology.

1.6 TRENDS IN SURFACE MOUNTING

There are many benefits and concerns in surface mount technology, but the benefits far outweigh the concerns. Even though many issues have not been totally resolved, the electronics industry, having recognized the importance of SMT, is moving toward solutions. For example, various industry organizations are addressing the surface mount standards issue. Nor should the role played by through-hole technology in the future of surface mount be neglected. For example, the through-hole technology has matured to the point at which no further significant improvements in cost effectiveness are possible. Moreover, its component tooling costs have been amortized by the component suppliers, and new investments generally are made in the SMT areas. The suppliers of surface mount components have started to recover their investment in tooling for SMT.

The cost structure of SMT has nearly caught up with conventional technology. For example, most passive and memory devices are available at no extra cost. This is not surprising, since these devices are typically used in large numbers on electronics assemblies of all types. But logic devices also will catch up with conventional products very soon. Indeed, many vendors do not charge a premium for surface mount components over through-hole components. Component availability in surface mount is improving.

Other developments in surface mount technology are under way. For example, certain high pin count devices are now available only in surface mount. When the count exceeds 84 pins, they must be in fine pitch (lead pitch at 25 mils and under) as well as in surface mount. This trend is going to continue, resulting in finer and finer lead pitches, until eventually the industry will get rid of the package entirely and mount the die directly on the board, as discussed in Section 1.7 (The Future). However, that switch to direct chip will only be in niche markets. In the interim, fine pitch and ultra fine pitch packages are being replaced by BGAs because they simplify the manufacturing processes [3].

This is not to suggest that through-hole technology will disappear altogether. It won't, just as hand soldering did not disappear, even after wave soldering became commonplace. Many high pincount, high performance microprocessors will continue to be in through-hole PGA (pin grid array) for various reasons (see Chapter 3). Surface mount technology will complement, not replace, the conventional technology. Still, the decision to convert to SMT is a gigantic step. Many companies wonder when and how they should convert. This problem resembles the question faced by the industry when wave soldering began to replace hand soldering. Because of the technical problems with icicles and bridging in wave soldering in the 1950s, the industry was really debating whether to replace hand soldering with wave soldering. Today it would be unthinkable to go back to hand soldering for volume production. Similarly, today the use of SMT is widespread, and the process is universally accepted for almost all applications.

Companies using SMT will improve their productivity and responsiveness to change in the business cycle and will be involved in the interconnection of high pin count devices that are more difficult to design, assemble, and test. State-of-the-art, cost-effective miniature products are easier to market. It is safe to say that SMT is not a fad; it is here to stay. SMT is not a technology of tomorrow; it is the technology of today. It keeps our customers on the leading edge of their business.

1.7   THE FUTURE

Surface mount technology is the current frontier of packaging high pin count components. Even though it provides substantial design and manufacturing benefits, its potential is far from being fully realized. Surface mount packages, as small as they may be compared with their conventional through-hole counterparts, can be shrunk in size considerably by decreasing their lead pitches (lead center to lead center spacing). The lead pitch of standard surface mount packages is at 50 mil (0.05 inch) centers. Fine pitch packages (32, 25, and 20 mil pitches), UFP (0.4 mm and lower lead pitch), BGA, and chip scale packaging (CSP) are also available for better real estate efficiency. This trend has lead us to 10 mil center packages; for example, those used in microprocessors in TCPs for notebook, lap top, and palm top computers. Real estate requirements on printed circuit boards can be further reduced only by mounting the die (singular) or dice (plural) directly on the substrate: the so-called chip-on-board (COB) technology.

As shown in Figure 1.18 [4], the COB technology encompasses wire bond, TAB (tape-automated bonding) and flip chip technologies. Since

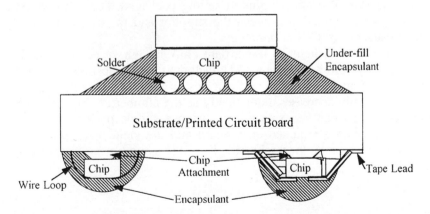

Figure 1.18   Chip-on-board technologies: wire bond, TAB, and flip chip [4].

Table 1.2 Comparison of areas occupied by various packages [3]

PACKAGE TYPE	DIP 24	SOP 24	CC 24	TAB 24	BARE CHIP
Area (mm²)	472.44	157.7	124.99	50.12	37.12
Relative area	12.73	4.25	3.37	1.35	1

the die is considerably smaller than the packages of any lead pitch, the real estate savings provided by COB can be substantial. Table 1.2 summarizes the real estate occupied by packages of four types.

The COB technology has been widely used in consumer products, especially in Japan. The Japanese have essentially dominated this type of consumer market, but COB is also being used in this country (for example, in calculators and high end notebook computers). The driving force for COB is the need for miniaturization (size and weight) in products such as smart cards, wristwatches, cameras, calculators, and thermometers, just to name a few. One of the requirements for this technology is high volume, because very few devices are used per board. Since specialized tooling and equipment are required, volume must be high to justify the capital expenditures.

There are two main concerns about COB technology: the availability and reliability of dice. Unless a company is vertically integrated, it must rely on the very limited supply of bare chips that is available from outside semiconductor suppliers. Semiconductor suppliers are often reluctant to ship bare dice, although this situation has been changing over time. One reason is that they do not like to disclose the yield data on their wafers. Also, the loss of profit margins in selling packaged dice is a deterrent. Thus, close cooperation between the supplier and the user is necessary for the success of COB technology.

Since it is difficult to test a bare chip at high speed for complete functionality before placement and attachment to the substrate, enhancing the yield of an assembled board can pose a serious problem. The yield of the assembly decreases dramatically as the number of chips per board increases. This can be explained by using a simple probability theory. When n chips are mounted on a board and the yield of assembling an individual chip is Y_p, the total yield (Y_t) of the assembly can be expressed as follows:

$$Y_t = (Y_p)^n$$

Figure 1.19 plots total yield (Y_t) against total number of chips (n) on the substrate. As is obvious from the figure, for a small decrease in

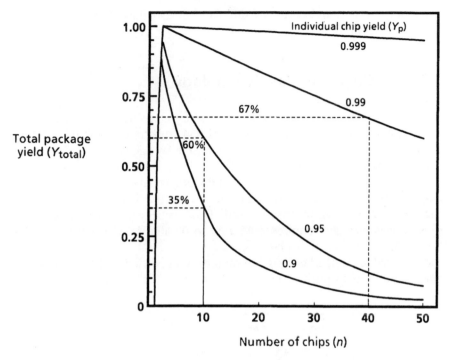

Figure 1.19   Impact of number of chips and their individual yield on the total yield of the assembly in chip-on-board (COB) technology.

the assembly yield of an individual chip (from 95% to 90%), the final yield of the assembly drops rapidly (from 60% to 35%) for 10 chips per board. If the number of chips per board is higher, the decrease in total yield is drastic. Improved passivation technology has increased chip reliability, but they can be damaged easily during shipping and handling, which can affect the final yield. The above formula can also be used to predict the yield during placement or soldering where n is the total number of components to be placed or soldered.

There is also some concern about the long-term reliability of the assemblies because they no longer enjoy the protection provided by packages. Silicon or epoxy resin encapsulation is applied and then cured over the chips after assembly. But these glob-type encapsulants are not adequate for high performance and high wattage devices. For this reason, their application should be considered only where field reliability requirements are not stringent. Repair of bare chips is also a matter of concern.

As mentioned earlier, the COB technology is divided into three main categories: chip-and-wire, tape-automated bonding (TAB), and flip chip [4]. TAB is also referred to as tape carrier package (TCP). Some people refer only to chip-and-wire as COB. We discuss these categories in the following sections.

1.7.1 Chip-and-Wire Technology

In chip-and-wire technology, a bare integrated circuit chip is attached to any substrate, including glass epoxy (FR-4) substrate. The wire bonding is done by ball-wedge (Figure 1.20) or wedge-wedge (Figure 1.21) bonding processes.

In ball-wedge bonding, the ball and wedge shape is dependent on the capillary (tool used for bonding) and on whether the thermocompression or thermosonic bonding process is used. This process is generally used for plastic packages and requires gold wire to be able to withstand potential bond wire corrosion in a plastic package.

In wedge-wedge bonding, aluminum wire is used. A hermetic package is required to prevent bond wire corrosion. The loop height (H in

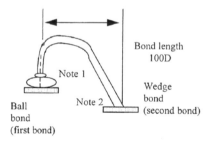

Note 1   Ball shape dependent on capillary and whether thermocompression or thermosonic bonding process was used.

Note 2   Wedge shape somewhat dependent on capillary type and shape and whether thermocompression or thermosonic bonding process was used.

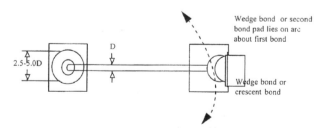

Figure 1.20   Top and side views of ball-wedge bonds.

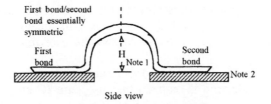

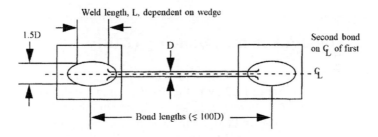

Note 1   Loop height, H, is generally lower than in the ball-wedge processes.

Note 2   Flatness of weld dependent on wedge shape and ultrasonic bonding parameters.

Figure 1.21   Top and side views of wedge-wedge bonds.

Figure 1.21) is generally lower than in the ball-wedge process. The flatness of the weld depends on the wedge shape and ultrasonic bonding parameters. Aluminum wire bonding allows finer pitch bonding than is possible with gold wire.

The stress generated by the coefficient of thermal expansion mismatch between the bare silicon chip (CTE of nearly 2.6 ppm/°C) and the FR-4 (CTE of about 14 ppm/°C) is not of concern. The ductile wire leads are compliant and can take up the stress without breaking.

Before IC placement, a conductive adhesive (generally silver epoxy paste) is screened on the substrate; it is cured after IC placement. After adhesive cure at about 150°C, die bond pads are wire bonded to the pads on the substrate.

The die is placed using a pick-and-place machine, as in hybrids. It is presented to the machine either in waffle packs (plastic trays segregated into squares or rectangular cavities) or in wafer form. In waffle packs, each cavity holds one die. Only good dice are packed in the waffle packs. Sometimes the entire silicon wafer is supplied, and the user cuts it into individual dice and presents them to the pick-and-place machine. The advantage of this method is minimization of scratches and other forms of damage to the silicon chips during shipping. When using wafers, the user or the supplier must mark the bad dice. This means that the supplier must agree to divulge his wafer yield and the user must have dicing or sawing capability.

Chip-and-wire technology affords reductions in cost and in real estate. The basic cost reduction is realized because bare IC chips on substrates are used. There is no cost for packaging of the dice. Since the silver paste used in chip and wire is very expensive, solder paste is substituted for silver paste to reduce cost in some applications. For using solder paste, however, a film of gold is required on the backside of the die, which tends to offset any cost saving. This also puts an extra burden on the board manufacturer, who is not accustomed to plating dense, wire-bondable gold on a printed circuit board.

1.7.2   Tape-Automated Bonding (TAB)

In the schematic view of chip-and-wire and tape-automated bonding (TAB) presented in Figure 1.22, note the difference in height requirements for wire bond and TAB. Figure 1.23 shows a schematic of TAB [5]. The main advantage of TAB, or TCP, over bare die for wire bond or flip chip (discussed next) is that it provides known good die (KGD) because TAB devices can be tested for complete functionality at speeds similar to those for conventionally packaged devices. Other advantages of TAB or TCP over COB technology is that TAB provides a lower profile on the PC board as shown in Figure 1.22. Moreover, since wire lengths are longer in chip-and-wire, the lead inductance tends to be about 20% higher. This makes TAB better than chip-and-wire for good electrical properties, especially at high frequencies. TAB is also a faster process, since inner (or outer) leads can be gang bonded.

Gang bonding is typical of the processes used to bond inner leads. However, outer leads are generally bonded either individually, one side at a time, or all four sides simultaneously. Gang bonding decreases bonding time and cost. TAB also makes denser interconnection possible since it allows bonding to finer pitch pads than is possible in wire bonding. Because of this feature, if TAB is used instead of a standard wire bonding process, even the die size can be shrunk for many devices. Most important of all, TAB, when packaged in individual carriers, allows pretesting and burn-in of devices before assembly. This increases product reliability and reduces cost, since there is less rework.

TAB can be packaged in either individual carriers like camera slides (to allow pretesting) or on a reel (no pretesting possible). In either case a tape is used, which acts as the transport medium for the die. The typical

TAB

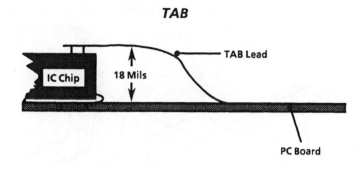

WIRE BONDED

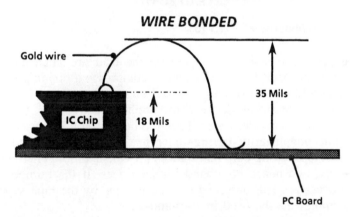

Figure 1.22   Comparison of the COB and tape-automated bonding (TAB) technologies.

TAB tape in 35 mm format shown in Figure 1.24 resembles a movie tape. TAB tapes are standardized by JEDEC (Joint Electron Device Engineering Council) in 35, 48, and 70 mm formats. Tapes come in single, double, and triple layers. The conductors are etched on the tape to match the inner and outer lead pads. The tape is unsupported in its center to accommodate the die. For the inner lead bonding (attaching the chip to the tape), two methods are used: bumped tape-automated bonding (BTAB) or bumped die. The bumps elevate the leads from the die and prevent shorting to the edge of the die.

   Copper, solder, and tin are used for bumping. In bumping a die, die pads are plated with gold. Bumps on tape are formed by etching a thicker

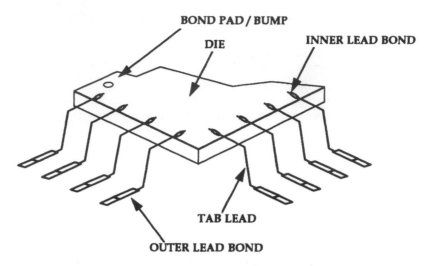

BOND PAD / BUMP

DIE

INNER LEAD BOND

TAB LEAD

OUTER LEAD BOND

Figure 1.23 **Schematic of TAB [5].**

layer of copper so that a bump is left on the lead tip. Then the tip is coated with other metal layers. If a thermocompression bonding process is scheduled, gold is used on the tape and on the die pads.

If bumped dice are used, the chip supplier electroplates gold pads on the die before sawing the wafer. The gold bumps on IC chips compensate for some nonplanarity during inner lead bonding. Bumped chips require extra steps during chip fabrication, provide additional hermeticity and reliability, and hence command higher prices. If the bumped tape (BTAB) process is used, bumping of tape is done by the tape supplier; the die resembles the die used in chip-and-wire.

The inner lead bonding is either high speed solder or a thermocompression bonding process. In the solder process, an intermetallic tin-gold bond is achieved between the gold bumps on the chip and tin-plated tape leads at relatively low temperature (284°C) and pressure (40–75 g per lead). A typical thermocompression cycle includes preheating of the chip at 200°C before inner lead bonding at about 500°C for 1 to 3 seconds at about 100 g per lead force, for a lead 3 mil wide and a 4 by 4 mil bump. The pressure is very critical at the interface area.

The outer lead bonding process in TAB is similar to the reflow soldering operation in SMT. However the lead pitch is much smaller. Some of the high end microprocessors are now available in UFP with lead pitches of 0.25, 0.2, and 0.10 mm. For outer lead bonding, the thermocompression bonding or conductive adhesive bonding process can also be used. In thermocompression bonding, the functions of the placement and soldering machines are combined in one piece of equipment.

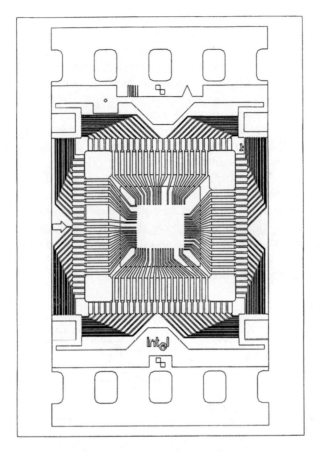

Figure 1.24   Typical 35 mm format tape design for TAB. (Courtesy of Intel Corporation.)

Such equipment excises the outer leads from the tape, aligns the tape visually on the PC board, and then solders it to the board by thermocompression bonding. Generally the hot bar process for soldering is used. See Chapter 12 for the recommended thermal and mechanical profile for TAB and Chapter 6 for the recommended land pattern design for TAB.

1.7.3   Flip Chip or Controlled Collapse Bonding

Flip chips enable designers to reach the ultimate circuit density since they eliminate both the bond wires and packages. In the flip chip process, the chip is turned or flipped over (hence the name) and mounted directly on the substrate. It is interconnected by solder pads on the die and the substrate. Figure 1.25 is a schematic representation of a flip chip on a substrate. Figure 1.26 provides a magnified view of the flip chip connection on the substrate. The most widely known application of the flip chip process is IBM's thermal conduction module containing up to 133 chips [6].

The flip chip process requires very tight control, but can be significantly faster than any other chip-on-board (COB) technology. Of course, as with SMT, the placement speed depends on the type of placement equipment. Figure 1.27 shows an example of a dedicated pick-and-place machine for flip chips. A closer view of the placement head is shown in Figure 1.28. Such machines are much slower than the pick-and-place machines used in SMT. However, the pick-and-place machines used for flip and die attach must have vision capability with up- and down-looking cameras. They can pick up dice from either waffle/gel packs or tape and reel feeders.

There are other side benefits to the flip chip process. Since the solder

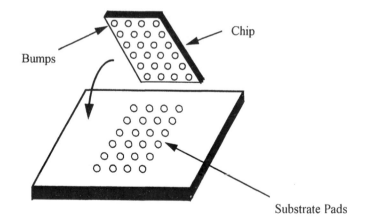

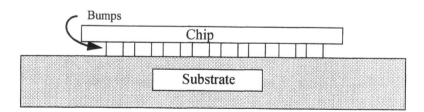

Figure 1.25   Flip chip on a substrate.

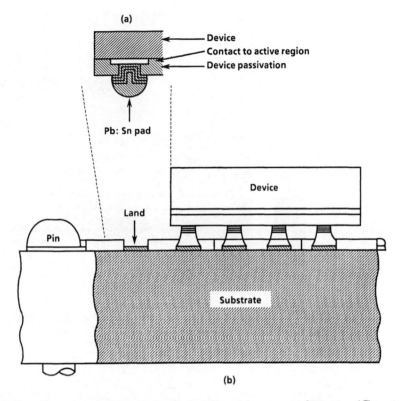

Figure 1.26   A magnified view of a flip chip on a substrate. (Courtesy of Intel Corporation.)

pads can be provided at any location on the die, a die design that is very efficient with respect to die size can be used. The flip chip process is also a more reworkable process than either TAB or COB. But flip chip is not a truly chip-on-board technology, since the "board" must have a coefficient of thermal expansion value compatible with that of the silicon chip to prevent solder joint cracking. This problem is similar to those faced by the military when using leadless ceramic chip carriers on incompatible substrates. The CTE mismatch problem in flip chip is less critical in comparison, however, since the silicon die is much smaller than an LCCC. The smaller the die size, the less severe is the stress on the solder joint due to CTE mismatch between the die and the substrate. Solder joint stress due to CTE mismatch can be further reduced by using a silicon substrate.

The flip chip process has some similarities to the SMT processes and also some differences. The placement, soldering, and cleaning pro-

Figure 1.27 Pick-and-place equipment for flip chip. (Courtesy of Kulicke & Soffa Industries.)

cesses are identical; thus one is concerned about process issues similar to those found in SMT—but at a miniature level. The differences lie in the fact that the manufacturing environment resembles a semiconductor operation rather than a surface mount line.

We now discuss the major process steps in a flip chip assembly. First is the preparation of solder bumps on dice and substrate. This is a critical operation, and the solder bump size and diameter must be tightly controlled. It is because of this preparation step that the process is also

Figure 1.28 Magnified view of placement area of the pick-and-place machine shown in Figure 1.27. (Courtesy of Kulicke & Soffa Industries.)

referred to as the controlled collapse process. That is, the name reflects the need to maintain an appropriate gap between the chip and the substrate. This is done by depositing in vacuum a controlled amount and geometry of solder on dice before the wafer is sawed. The composition of the solder is also tightly controlled.

The next step is to do a wafer sort (reject bad dice), saw the dice, and package them in a transport medium (generally waffle pack) that can be presented to the pick-and-place machine. The placement machine is very similar to the ones used in SMT, but it has a vision system and tighter placement accuracy. Before the placement of chips, a very low activity flux is applied on the substrate to provide tackiness and to prevent chip movement. The assembly is reflow soldered. It is cleaned and inspected (usually by process control and destructive pull testing on a sampling basis since solder joints cannot be seen), and epoxy is injected between the die and the substrate to improve solder joint reliability.

Thus the manufacturing issues in flip chip (tackiness of flux, accuracy of placement, part movement and reflow soldering profiles and cleanliness) are very familiar to people working in the SMT field. The types of defects in flip chip are similar to defects encountered in SMT—wrong part (die), defective chip, open, partial wetting, dewetting, or non-wetting.

In flip chip, one of the key processes is bumping of the die. There are three options: bumping the die, bumping the substrate, or bumping both. Bumping the die provides a higher yield than bumping the larger substrate since defect density is a function of size. One would choose to bump the substrate when bumped dice are not available, or when the chemical process is too aggressive for the dice. Both die and substrates are bumped when higher stand-off is required for better reliability. This is important when the die is large and there is a reliability concern due to CTE mismatch between the die and the substrate.

Solder bumping for flip chip is done by evaporation. Using photolithographically etched molybdenum metal masks, lead and tin are sequentially evaporated from solder pellets. Lead evaporates more than tin due to differences in vapor pressure, so this is a good process for high lead (90 to 97% lead) solder. The bump size and diameter must be tightly controlled to maintain the desired gap. The solder composition is high temperature solder (90/10, 95/5, or 97/3 Pb/Sn).

Before the deposition of solder, a base limiting layer (BLM) is electroplated on the pad. The BLM consists of Cr-Cu over Al pads. Final plating is done with gold using liquid or dry film resist as in PCB fabrication. This process is capable of producing very fine pitch bumps.

The main BLM requirement is adhesion to base metal (aluminum). Ti or Cr serves as the glue layer and both stick to Al and polyimide. (Sticking to polyimide is also important for sealing an unpassivated die area.) After Cr/Cu, copper, then gold, is electroplated. The function of Cr/Cu is to prevent exhaustion of Cu consumed during the formation of the intermetallic layer of solder joints. The thickness of copper needs to be tightly controlled. Too thin a copper layer can be consumed as Cu-Sn intermetallic which has very poor solderability (hence the solder bump will not solder and fall off). Too thick a copper layer is not desirable either. Excess copper will require higher reflow temperature and can produce a very thick Cu-Sn intermetallic layer. Cu-Sn intermetallic is required for adhesion, but too much thickness produces a brittle and less reliable solder joint.

Solder bumping by evaporation is a relatively slow and expensive process. It is not suitable for evaporating low-cost eutectic (63 Sn/37 Pb) solder bump, which is required for low cost laminate and flex circuits (e.g., MCM-L discussed in the next section). Solder plating is used by some companies as an alternative to the evaporation process. However, it is difficult to control alloy composition and bump height uniformity.

Another method of bumping solder known as flex on cap (FOC) was developed by Delco Electronics for automotive applications and is a relatively cheaper process than the evaporation method. In this process solder paste is printed using a metal stencil. There is also a difference in the BLM layer in the FOC process. Instead of using Cr-Cu or Ti-W-Cu as in the evaporation method, Ni-Cu-Au is used as the BLM layer over the aluminum pad on the silicon. Users of this system claim it to be a low stress BLM layer.

The main disadvantage of this process is the pitch. The density of bumps in the evaporation process can be down to 6 mil pitch. But in a paste printing process, the minimum pitch with good yield is generally around 10 mils. This may not be a disadvantage because a much higher density will be difficult to route in a laminate material. The tighter pitches used in the evaporation methods are generally intended for mounting on silicon substrate where fine pitch routing is not a problem. Also, as mentioned earlier in this section, there is some reliability concern due to mismatch between the silicon die and the laminate material. However, Delco has demonstrated good reliability in flip chips mounted on laminate substrates.

Solder bumping is not the only option for flip chip interconnection. Instead, conductive adhesive is used by some companies such as Fujitsu for flip chip interconnection. In this process, instead of bumping the pad, conductive adhesive is dispensed followed by die placement and adhesive curing.

There are various guidelines that should be followed when designing flip chips. For example, flip chips should not be used if the operating temperature exceeds 100°C (or 125°C when underfill is used) due to lower creep strength at elevated temperatures. Applications above 100°–125°C require evaluation. Also, all flip chip diameters should be uniform; any variation in diameter impairs the chip attach. Larger balls suspend the chip and prevent smaller ones from reflowing.

The main disadvantage of flip chip, in addition to the KGD (known good die) concern, is a lack of die availability. Also the CTE mismatch on nonsilicon substrates may pose reliability concerns. But generally, an underfill epoxy is used to improve CTE related reliability concerns.

1.7.4 Multichip Module

So far we have provided some background on various types of packages. These packages are not simple containers for silicon. They perform very important functions such as supplying power to the chip, dissipating heat, providing interconnection to other devices, and protecting the silicon from the environment. Yet, while performing these functions, they end up as bottlenecks in the performance of the silicon. For example, while the silicon designer talks about performance in picoseconds, the computer performance is still measured in nanoseconds. Thus, the packages "cost" the silicon about 1000X degradation in performance.

How does this happen? The unwanted capacitance, resistance, and inductance, generally referred to as the "parasitics" of the bond wire (inside the package) and lead length (outside the package) are the source of the problem. The longer this distance, the higher the parasitics and the greater the degradation in performance. This is why the industry has been moving from through-hole to SMT to fine pitch to UFP to BGA to chip scale packaging, and finally to direct chip attach (flip chip) as just discussed. In each of these stages of development, the package parasitics drop because the wire bond and lead lengths continue to decrease—until one moves to flip chip. Since there are no wire bonds or package leads, the flip chip provides the ultimate in performance and densification as package parasitics disappear almost entirely.

But a single chip does not make a computer system. It needs to be interconnected to other chips in the system through a mother board. And since there are still delays from the interconnection path between different chips, the full potential of the silicon's performance remains unrealized. This is where the multichip module (MCM) comes in. MCM is a substrate that can interconnect multiple bare chips with the shortest signal paths possible and thus increase system performance significantly.

With MCMs there are three assembly choices, each for various types of substrates. Regardless of assembly, however, the MCM's classification is based on the substrate selected. They are generally referred to as MCM-L, MCM-C, MCM-D, MCM-L/D, and MCM-C/D.

MCM-Ls generally consist of only laminate materials with finer PCB traces and small vias for use in cost-sensitive applications. The MCM-C features a cofired ceramic substrate, a technique long used in the hybrid circuit and preferred in applications requiring high thermal dissipation and hermiticity. MCM-D is the ultimate in providing dense interconnections because trace widths in microns, not mils (25 microns equal one mil), are easily achieved using aluminum conductors on polyimide dielectric. While MCM-D can give the best performance, it is also the most expensive. MCM-L/D and MCM-C/D are variations of MCM-L and MCM-C with an additive process for depositing finer circuit features.

Once it has been decided what type of substrate to use, the next logical question is which assembly technology will be used to bond the bare silicon to the selected substrate. The commonly used three assembly choices are TAB, wire bond, and flip chip. The key issues with each

Table 1.3   A summary of assembly choices for MCM

	PROs	CONs
Wire Bond	• Versatile • Proven technology	• Difficult at <4 mil pitch • High inductance limits performance
TAB	• Testable before assembly • Improved inductance for multilayer TAB	• Need bumped die • OLB difficult at <15 mil pitch • Bigger die-to-die spacing • Rework difficult
Flip Chip	• High IO count • Superior electrical characteristics	• Need CTE matching • Infrastructure lacking

assembly process have been discussed earlier and are summarized in Table 1.3.

The packages provide one useful function that bare die is unable to at this time—pretesting of silicon at speed. Bare die used for wire bond and flip chip cannot be tested at speeds even though much progress is being made in this area. So if a company does not have the internal infrastructure for bare die testing, flip chip and wire bond assembly methods may not be the right choice because the final yield will drop significantly even if the yield of an individual die is fairly high. Refer to Figure 1.18.

TAB has the unique advantage of allowing complete testing of silicon before it is committed to the substrate and therefore provides KGD. At the same time it allows a reduction in die size because of narrower bond pitches than are possible in wire bonding. TAB should be used in applications where KGD is very important because the silicon is too expensive to throw away. But one must keep in mind that because of the fanout required for TAB, the module size will be fairly large. Thus, when using TAB, real estate conservation must not be a major concern [5].

In general, there is no real *technology* hurdle in the widespread use of MCM. However, it is generally more expensive. So MCMs are primarily used in applications where performance is critical and the die is readily available.

1.8   SUMMARY

Surface mount is a packaging technology that mounts components on the surface of a printed wiring board instead of inserting them through holes in the board. It provides state-of-the-art, miniature electronics products at reduced weight, volume, and cost.

SMT is rooted in the technology of flat packs and hybrids of the 1950s and 1960s. But for all practical purposes, today's SMT can be considered to be a continually evolving technology. Currently, the use of fine pitch, ultra fine pitch (UFP), and ball grid arrays (BGAs) are becoming ever more common. Even the next level of packaging technologies such as chip-on-board (COB), tape-automated bonding (TAB), and flip chip technologies are gaining widespread acceptance. Multichip modules (MCMs) using wire bond, TAB, or flip chip are used to achieve the highest performance possible but with a cost premium.

For electronics companies an entirely new internal and external infrastructure is necessary to develop and sustain the SMT technology. It is prudent for almost every electronics company to begin developing its own internal SMT infrastructure. Those who do not will risk their very survival.

Certainly many technical issues must be taken into account and thoroughly understood before converting to SMT. Later chapters will discuss the details of design, materials, processes, and equipment issues for both military and commercial applications to take advantage of this emerging technology.

REFERENCES

1. Technical Staff, ICE Corporation, Surface Mount Packaging Report, *Semiconductor International,* June 1986, pp. 72–77.

2. Prasad, P. "Contributing factors for thermal damage in PWB assemblies during hand soldering." *IPC Technical Review,* January 1982, pp. 9–17.

3. Joint Industry Standard ANSI-J-Std-013. Implementation of Ball Grid Array and Other High Density Technologies. IPC, Northbrook, IL.

4. Lau, John, ed. *Chip on Board Technologies for Multichip Modules.* Chapman and Hall, NY.

5. Doane, Daryl Ann, Franzon, Paul D., eds. *Multichip Module Technologies and Alternatives, The Basics.* Chapman and Hall, NY.

6. Joint Industry Standard ANSI-J-Std-012. Implementation of Flip Chip Technology. IPC, Northbrook, IL.

Chapter 2

Implementing SMT In-house and at Subcontractors

2.0 INTRODUCTION

To implement any new technology, one needs strong infrastructures, both internal and external. Examples of external infrastructure are industry standards and a stable supplier base for equipment and materials. The development of an external infrastructure is dependent upon the market and it takes time to fully develop it. SMT is now in a stable state with the external infrastructure firmly in place.

Examples of internal infrastructure are the manufacturing line, training, and documentation. Clearly a good internal infrastructure (process and design rule development, validation, documentation, and the development and training of manufacturing capability) is necessary to develop and sustain the technology. The internal infrastructure is especially important, and it is management's task to understand the technical issues and put the resources in place to develop it. We should note that an understanding of infrastructure issues is important whether the assembly is being manufactured inhouse or at a subcontractor.

There is an increasing trend toward outsourcing of electronics manufacturing. Management must decide whether to implement this technology in-house or move the manufacturing operations to a subcontractor. And, should implementation be in one phase or several phases? These decisions will be made based on the type of SMT (I, II, or III), the stability of the technology, the available capital, and the maturity of the internal infrastructure.

In this chapter we will discuss the effective implementation of SMT. We will also discuss the reasons for outsourcing or not outsourcing and various aspects of subcontractor selection and qualification in the areas of business, technology, quality, and manufacturing. The focus of this chapter is on the management aspects of SMT rather than the technical details of SMT issues; those are addressed in various chapters in the rest of the book.

2.1   SETTING THE IMPLEMENTATION STRATEGY

Before proceeding further, let us briefly review the definition of the various types of SMT discussed in Chapter 1. Type I is all surface mount, either single or double sided. No through-hole parts are used. It is common mostly for memory modules. Type II is mixed assembly with both SMT and through hole. Type III is minimal SMT with chip components glued to the board on the bottom side only. Type II assembly is the most common today and will remain so for the foreseeable future because not all components are available in surface mount.

As the push for further miniaturization continues, these assemblies become more complex. In this book we will use the terminology Type IC, Type IIC, and Type IIIC to indicate complexity. Type I and II become Type IC and IIC if large fine pitch (over 208 pins with 0.5 mm pitch), ultra fine pitch (pitch under 0.5 mm), QFP (quad flat pack), TCP (tape carrier package), or BGAs (ball grid arrays) and very small chip components (0603 to 0402 or smaller) are used on these assemblies along with conventional (50 mil pitch) surface mount packages. Type IIIC applies to assemblies if SOIC (small outline integrated circuits) and SOTs (small outline transistors) and very small chip components (0603 or 0402 or smaller) are used on the secondary side (and only through-hole components on the primary side).

Note that if the pitch is above 20 mils, it is referred to in mils, but if the pitch is 20 mils or below it is referred to in mm (millimeters) to conform to agreements of standard setting bodies such as the EIA (Electronics Industries Association) and EIAJ (EIA Japan). It is important to keep these definitions in mind because we will refer to them repeatedly throughout this book.

SMT is commonly referred to as a leading edge technology, and almost every aspect of design and process in building an electronic module with SMT is pushed to its very limits, especially if complex SMT (Types I/II/IIIC) is being implemented.

SMT requires its own design rules, materials, processes, and equipment. Thus it is highly critical that the surface mount board be packaged so that it is not only electrically, thermally, mechanically, and environmentally reliable, but producible with zero defects. This is definitely a difficult goal, considering the reality that many companies are still struggling with excessive numbers of defects in conventional through-hole mount assemblies that have had the benefits of design and process refinement for decades. Therefore, achievement of denser, more reliable SMT assembly with zero defects is a difficult goal, but certainly worth aiming for.

Many engineering issues (see Figure 1.17 in Chapter 1) in both design and manufacturing SMT assemblies need to be resolved. The technical issues fall into two major categories: design and manufacturing, as briefly discussed in Chapter 1 (Section 1.5). Here we briefly review some of the design and manufacturing problems that management must understand and overcome by putting the appropriate resources in place.

Because of the differences in types of packages, processes, and solder joints for SMT, the reliability window through which conventional boards pass is inadequate for SMT. Stress in solder joints is introduced by mechanical handling, vibration, soldering processes, and temperature fluctuations in service. Thermal and mechanical cycling tests for solder joint reliability should be defined. This will help determine optimum pad sizes of active and passive components that allow a maximum number of traces between pads without compromising solder joint reliability.

Increased interconnect density in surface mount boards, resulting from using finer lines and spaces, may potentially increase signal cross-talk levels. Plans should be made to evaluate the placement of critical traces in the ground and power planes. In addition to providing the design rules for reducing cross-talk, one should also qualify vendors with the capability for producing fine line boards. Design rules should be established and implemented in work station tools for circuit simulation capability. Refer to Chapters 3 through 7 for details on the resolution of design issues.

Manufacturing and quality assurance organizations should be concerned with the issues of design and process compatibility and control of machine and nonmachine variables to achieve zero defects in SMT assemblies. There are standards for some processes but not for others.

Also, new SMT equipment and processes are continually being introduced in the marketplace. Other conditions to be considered include the following: (a) the design of manufacturing equipment is constantly changing, (b) the "best" cleaning methods, including no cleaning, will always be debated, (c) the equipment vendor base is slowly maturing, and (d) it is hard to predict which vendor will still be around a few years from now to service the equipment. Again, these issues are better resolved for standard SMT than for assemblies with UFP and BGAs. It is not easy to weave through these issues and prevent equipment obsolescence.

Most of the materials and processes for SMT differ also from those of conventional technology, even as they play a critical role in SMT. The first goal should be to establish the generic materials and process requirements that should not depend on any specific equipment purchased and installed by manufacturing. The requirements should be based on determining the critical variables in adhesives, fluxes, pastes, solderability, soldering, cleaning/no cleaning, and repair/rework, and their impact on reliability and manufacturability. Any materials and process requirements should be equally valid for in-house manufacturing and an outside assembly house. The technical details of materials, process, and equipments are the focus of Chapters 8 through 14.

Given the technical issues involved in the design and manufacturing of SMT assemblies, where should one begin? The most complex type of board has the biggest payoff in terms of real estate and cost per function savings, but it is not the place to start. The easiest types of surface mount boards should be introduced first, before moving onto more difficult types. It is very hard to come up with answers if one tries to resolve all the issues for the most difficult type of SMT, namely Type II. It will become even more difficult if UFP packages (0.5 mm center lead pitch) are added to the board. The strategy for implementation of SMT should satisfy four major requirements:

1.   The product requires the inherent advantages of SMT.
2.   The packages are available.
3.   The assembly and testing capabilities are ready.
4.   SMT is introduced only to the degree necessary.

That is, SMT should be implemented in a logical manner in various stages. It should be an evolutionary and not a revolutionary process. A gradual introduction of technology allows development of a coordinated approach for design guidelines and CAD tools, packages (availability, style, reliability), assembly (process, equipment), and testing.

It is fortunate that SMT is modular, and hence can be implemented in various phases. There are various flavors of SMT. We have standard SMT (50 mil pitch), fine pitch SMT (25 and 20 mil pitch), and ultra fine pitch SMT (below 0.5 mm pitch). High density quad flat pack and tape carrier package (TCP) or TAB (tape-automated bonding) are examples of ultra fine pitch packages. And there are array SMT packages such as ball grid array (BGA), and chip scale packaging and finally direct chip attach (flip chip). Each type of SMT and related technology has its own set of problems, design rules, and process parameters.

The finer the pitch, the more difficult the implementation. With reduced pitch, precision becomes more intense and process windows tighten. As a result, lower yield and higher costs are encountered if design is not precise and processes are not very tightly controlled. Those who started successfully in standard SMT have found it easier to move on to the next level of SMT.

The suggested order of implementation is Type III SMT, Type I memory, Type II, and then Type IIC. Once a decision has been made to install SMT capability, Type III SMT is the logical first step for a company before making a full commitment to SMT. Type III entails the lowest possible risks in both technology and capital investment. Capital investment is minimum because the pick-and-place machine and the adhesive dispensing and curing apparatus are the major pieces of new equipment required for a Type III SMT line. It impacts only components and their placement in a significant way.

If the need for high density memory exists, it is also easier to implement this technology if the capability for Type III exists. It is a different matter for Type II SMT, however. The design and assembly of a type II logic board is more complex than that for either Type III or Type I memory. The Type II logic boards also tend to have many programmable logic devices (PLDs). They usually require manual handling for programming, which tends to cause lead bending (coplanarity). (However, there are PLDs available that can be programmed during the final test and do not need manual programming.) Type II assemblies also require two soldering processes (reflow and wave soldering) and two cleanings, unless no clean flux or paste is being used.

Chip-on-board, tape-automated bonding, and flip chip technologies should be attempted last. These technologies have many similarities to surface mount but require entirely different design and manufacturing strategies, as indicated in Chapter 1.

One certainly has the option of introducing all types of SMT products simultaneously, provided the resources exist to resolve all the SMT issues and build a stable infrastructure. However, this not only requires a substantial investment in capital and personnel but also poses a higher risk.

A gradual progression into different levels of surface mount technology allows the advantage of closely managing its introduction, while spreading the costs of such a large program over a longer period of time. This approach also helps build a strong infrastructure to support the technology.

2.2   BUILDING THE SMT INFRASTRUCTURE

Both external and internal infrastructures are necessary for successful implementation of SMT. There is not too much an organization can do about parts of the external infrastructure, such as components, design, and manufacturing equipment availability. These factors are mostly driven by the market conditions and are well established now. Other external infrastructure issues, such as component, process, and design rule standardization, are set by industry organizations. Active participation in standards organizations by both user and supplier member companies helps develop and maintain meaningful standards that benefit all concerned.

2.2.1 Developing the Internal SMT Infrastructure

In-house materials, process and design specifications, and ready access to trained personnel are some of the elements of a strong internal SMT infrastructure. The most effective way to develop an internal SMT infrastructure is to form an SMT team. The team members should be drawn from functional areas such as packaging, R&D, CAD, manufacturing, testing, quality, components, reliability, purchasing, and design engineering. The team should be headed by a program manager with both technical and program management experience. It is critical that the team have the full support of top management. The first task of the program manager should be to select team members with specific technical skills to address the issues summarized previously in Figure 1.17.

2.2.1.1   Writing the Plan

The first task of the team should be to write an SMT program plan that addresses the various technical issues and establishes major milestones. The plan should be approved and supported by management. The program manager also needs full support from the managers of the technical personnel on the team.

It is essential that there be one person on the team, preferably the team leader, who is very knowledgeable about the technology. If this is not possible, which is generally the case when a company is trying to get into SMT, someone should be brought in either on a part- or full-time basis to help train the team members. This will ensure that the team is headed in the right direction and help avoid costly mistakes.

The team should hold regular meetings (preferably weekly at a set day and time) to review implementation of the issues as outlined in the plan. The meetings will also serve as a forum for training team members on specific issues. Having the core team involved in every aspect of SMT implementation ensures that all the issues related to design, equipment, and process will be properly covered.

2.2.1.2   Management Review

A monthly management review should be held with all managers of the team members participating. This review process is necessary for good communication and course correction by management as deemed necessary. The review also acts as a motivating factor for team members. It is not uncommon for most of the work to be completed just before the management review. The review establishes frequent deadlines for the team and prevents excessive schedule slippage because the team members do not necessarily report to the team leader.

Before writing the plan, the team members should visit other companies and attend conferences to familiarize themselves with SMT issues. There are some professional conferences and even societies that are targeted at SMT. In addition to tutorials at national and regional conferences, in-house SMT courses are available in the industry. Such courses may be a good place to start to learn about the technology.

2.2.1.3 Hands-on Experience

Reading published material and attending outside training classes and conferences are no substitute for hands-on experience. Learning about SMT is a lot like learning to ride a bike. You can read about it and talk about it, but unless you ride on it and fall a few times, you will never learn. An SMT plan is not realistic if it does not take into account a few mistakes. The goal should be to prevent fatal mistakes, not all mistakes.

A successful SMT implementation strategy should include plans for an in-house SMT laboratory. This can be accomplished at a relatively low cost. Whenever possible, the laboratory equipment and processes should be compatible with the manufacturing equipment and processes used for production. In the beginning stages, the lab serves many functions: it helps in minimizing the risk in SMT implementation, provides prototyping capability for engineering and test boards, and allows hands-on training for in-house personnel.

An SMT lab helps prevent expensive mistakes. For example, it is very useful in developing and characterizing processes to help in the final selection of appropriate manufacturing equipment. The lab can also serve as a backup facility, especially for prototyping actual products in the future. Even when a full-fledged manufacturing line has been set up, the lab can meet short-term prototyping needs if production equipment is down.

The next step in the implementation plan should be to develop and characterize some SMT processes. Test boards with varying design rules (different sizes and patterns, varying interpackage spacing) should be designed and built. They can be used for developing appropriate design rules and characterizing processes such as screen printing, placement, soldering, cleaning, and repair. The test boards can also be used to characterize materials, such as solder paste, adhesive, and solder mask.

2.2.1.4 Process Selection

There are two main criteria to consider when selecting the final process and design variables: solder joint reliability and solder defects. The materials, process, and design variables that give the best results in terms of yield and reliability for the desired application should form the basis for process and design specification. The workmanship standards should define the requirements, and the design and process specifications should establish rules to ensure that those requirements are met. Four different categories of specifications are required to control the quality and reliability of the final products.

1. *Material specification.* Material specifications for such surface mount materials as adhesive, solder paste, components, and PCBs are generally intended to control the quality of incoming materials. See Chapters 3 (components), 4 (substrates), 5 (baking of surface mount plastic packages before reflow to prevent package cracking), 8 (adhesive), 9 (solder paste), and 14 (workmanship standards) for guidelines in developing applicable specifications. These specifications are intended for both users and suppliers.

2. *Non-equipment-dependent process specifications.* Non-equipment-dependent process specifications are for internal use. They establish procedures to meet the requirements for producing assemblies with ever-improving quality. Examples include guidelines and rules for the application of adhesive and solder paste, and process parameters for placement, soldering, cleaning, and repair.

    The non-equipment-dependent process specifications establish the process "windows." As long as the process is maintained within the specified "window," the solder joints will be metallurgically sound, providing reliable solder joints, no matter what equipment is used.

3. *Equipment- and process-dependent specifications.* The specifications in this group, though also for internal use, are specific to the equipment used in manufacturing. They depend very heavily on equipment and will play a key role in achieving zero defects in the final product. Each piece of equipment should be extensively characterized to develop this specification. Some examples are: pressure, nozzle size, and temperature to control adhesive dot size in an adhesive dispensing machine and individual panel settings, and conveyor speed to achieve a desired adhesive cure or solder reflow profile in a reflow oven. The equipment-dependent specifi-

cations must meet the windows defined under the materials and process specifications.

4. ***Design rules and guidelines.*** The design rules and guidelines are for internal use and also for vendors. No matter how good the materials and process parameters are, if the board is not designed for manufacturability, zero defects can never be achieved. In this book, design rules and guidelines are discussed in connection with system design considerations (Chapter 5), land pattern design (Chapter 6), and design for manufacturability, testing, and repair (Chapter 7).

The four groups of specifications listed above form the key to internal infrastructure. The next step is to ensure that they are followed both in letter and spirit; otherwise all the time devoted to writing those specifications will have been wasted. Such compliance can be achieved only with the support of top management.

2.2.1.5   Training

Development of specifications and top management support are not enough, however. Effective SMT implementation calls for the development of in-house training courses in SMT manufacturing and design. There are many reasons for having in-house training classes. Generally the specifications do not include the details of why they were established. In the training class, the backup material and rationalization for the rules and guidelines can be explained. If people understand the reasons behind the rules, implementation is much easier. In addition, in-house classes are a quick and cost-effective way to train newly hired personnel and provide updates on changes in the technology. In-house training classes are directly applicable to the specific needs of the organization.

For a company to realize its full potential in a new technology, the whole organization must be brought up to speed. To be successful, one must invest in people (training) and in documentation (design for manufacturability, process guidelines and workmanship standards, etc.). In-house training, taught by either internal or external instructors, also helps improve internal communication by having people from different organizations in the same room discussing the issues most relevant to the company.

For successful implementation of SMT, the existence of a core group of in-house "experts" in SMT is also necessary but it is only a beginning. If in-house expertise does not exist, it needs to be developed gradually. This is where many companies make mistakes. They will invest in state-of-the-art equipment but not in people and then wonder why they are having all kinds of process problems. Employees' skills do matter. Unless process and design rules exist and personnel across all functional organizations are properly trained, there will not be a strong infrastructure to support and sustain SMT in production volumes.

2.2.2   Influencing External SMT Infrastructure

The external SMT infrastructure refers to component availability and standardization; maturing of design, process, and equipment technology; availability of experienced technical personnel; existence of a mature vendor base for fabrication and assembly of boards; and existence of industry standards on boards, materials, and process and equipment specifications. Standardization is almost mandatory for surface mount technology because of the need for automation. Indeed, adaptation of SMT has been slower than expected partly because of the lack of standards.

Standards make the market grow at a faster rate than it would without them. The proliferation of various packages offers an example. Users generally think that only the component suppliers are responsible for the proliferation of standards, but users have also contributed to the proliferation by accepting packages any way they can get them, with or without standards.

A good standard benefits both users and suppliers. For example, if the package size tolerances are tightly controlled (within the requirements of the standard), the user can properly design the land pattern and program his/her pick-and-place machine. The user can use the same design for all suppliers of that package. Suppliers also benefit because, as long as packages meet the standard, they can meet the needs of all customers. A good standard is a win-win situation for everyone.

There is much to be done in the international arena of standards, but much progress has been made in the United States. As far as surface mounting is concerned, two main organizations set standards in this country: the Electronics Industries Association (EIA) and the Institute of Interconnecting and Packaging Electronic Circuits (IPC), a technical society.

The IPC, with membership mostly from users, sets standards for printed circuit boards, assemblies, and interconnect design and manufacturing processes as well as acceptability and process control. Various IPC committees undertake many activities on surface mount land patterns, design guidelines, component mounting, soldering and cleaning methods, and requirements to promote surface mounting. Everyone is welcome by the IPC to participate in their activities. Developing their standards does take time, but the standards are very meaningful and relevant because all serious inputs are considered before they are released.

EIA is a trade organization of electronics component manufacturers. Various committees of EIA establish outlines for passive and electrome-chanical components, and the JC-11 committee of the Joint Electron Device Engineering Council (JEDEC), part of EIA, is chartered to establish a mechanical outline for packages for active devices. Similarly, various "P" or parts committees of EIA are responsible for setting standards for the mechanical outlines of passive devices.

How are standards set by IPC and EIA? Refer to the IPC and EIA standardization flow charts in Appendix A. Let us further discuss the standardization process of EIA to clarify their process better. EIA maintains a very definite distinction between standards and registration. Both EIA and JEDEC committees essentially approve standards in the same manner. We will use the JEDEC procedure to illustrate the standardization process.

For JEDEC there are two levels of standards: JEDEC registered and true standard. Any manufacturer can submit an outline for registration. If two-thirds of the committee members vote in favor, the package is registered but is not considered a standard. A committee member need not be in attendance to vote, but all "no" votes must be accompanied by an explanation so the committee may make an informed decision. The committees welcome constructive input, and an attempt is made to resolve all "no" votes.

To become a true standard, a package must be registered and all conflicts resolved. After the package has been registered, the proposal is submitted to a second group called the JEDEC council. If two-thirds of the council members approve, the package becomes a standard. Very few packages are true standards: LCCC, PLCC, and SOIC are some examples.

One note of caution about JEDEC/EIA standards: these documents should not be treated as procurement specifications. The standards tend to stress loose tolerances and many may not meet the requirements of manufacturing or process specifications. Nevertheless, in practice, many people use JEDEC/EIA/EIAJ standards as procurement specifications. For this reason, the responsible committees are attempting to tighten toler-ances.

Some very large companies, by their sheer size, tend to dictate standards. But the task is too big even for them, and it is very important for manufacturers at all levels to get involved in the committees of IPC and EIA to set the direction for the electronics industry. By actively participating in EIA and IPC committees and subcommittees, member companies can influence standards. The committees generally welcome input from various sectors of the industry to develop useful and balanced standards. Various surface mount standards applicable to SMT are shown in Appendix A.

It has been a difficult process to coordinate the standards set by EIA and IPC. The activity of one group directly affects the other as far as surface mounting is concerned. For example, the setting of land patterns for surface mount (IPC-SM-782) by IPC was like laying out railroad tracks without consulting the manufacturer of railway cars. In 1986 IPC and EIA joined forces and formed the Surface Mount Council, to coordinate various surface mount standards. The Surface Mount Council does not replace either EIA or IPC and does not set any standards. It brings the users and developers of standards together for better coordination and points out new SMT issues that need industry focus.

The membership of the council consists of chairmen of EIA and IPC committees, and representatives from users, suppliers, equipment manufacturers, the Department of Defense, and the army, navy, and air force. Any company can become an associate member of the Surface Mount Council, which has as its objective the promotion of surface mount technology by promoting development of coordinated standards. An example of an industry standard developed, at the direction of the Surface Mount Council, in the mid '90s is ANSI-J-STD-013, Implementation of Ball Grid Array and Other High Density Technologies, chaired by the author.

Other organizations, such as the International Society for Hybrid and Microelectronics (ISHM) and the International Electronics Packaging Society (IEPS) are working to resolve technical issues. In addition, organizations solely dedicated to SMT have been formed. The Surface Mount Technology Association (SMTA) and the Surface Mount Equipment Manufacturers Association (SMEMA) are working with the Surface Mount Council to promote surface mount and related technologies.

Currently the trend is towards joint industry standards with the cooperation of IPC, EIA, and the Department of Defense. Examples of some joint standards are J-STD-001 (Workmanship), J-STD-002/3 (Component and board solderability), J-STD-004 (Solder paste), J-STD-006 (Flux), and J-STD-013 (Ball Grid Arrays).

Active participation in standards organizations by both user and supplier member companies helps develop and maintain meaningful standards that benefit all concerned. Those who are involved in the electronics industry should consider not only joining the relevant standards organizations but should also consider active participation in order to help develop a strong external SMT infrastructure.

2.3   SETTING IN-HOUSE
MANUFACTURING STRATEGY

Setting manufacturing strategy is one of the key items to be decided in the early stages of conversion to SMT. There are various options depending on budget and manpower resource constraints. One may decide to have a full surface mount line in-house or depend on an outside assembly house at least during the initial stages of SMT implementation. Both approaches have their strengths and weaknesses. Going to an outside assembly house will delay capital investment but will make close coordination and feedback between design and manufacturing difficult. Close coordination, though important in complex logic board designs, may be less significant in a memory product.

In-house SMT manufacturing capability is good for the long term but it requires considerable investment in capital equipment and personnel. The cost can vary from a few hundred thousand dollars to millions. Generally, the cost of the pick-and-place machine constitutes about half of the total equipment cost of an SMT line. Refer to Chapter 1 (Section 1.3) for a brief description of various SMT equipment.

Once the decision has been made to go ahead with the installation of an in-house manufacturing line, there are many variables to consider. The most important involve the types of products to be built, the maximum board size, fixed or variable form factors, product mix and volumes, and types of surface mount assembly. These variables will determine the types of equipment to be purchased and their total cost, as well as the lead time and the manpower resources necessary to select, purchase, install, and debug the line.

Another important item to consider is whether the line will be brought up for all types of surface mount at one time or in stages. The lead time on some of the equipment can be from 3 to 6 months after placement of the purchase order, excluding the time for capital budget authorization and capital equipment survey and selection. Also, time must be allowed for installing and debugging the line and training personnel. Thus at least a year should be allowed to set up and debug a manufacturing line and about 6 months to set up a surface mount lab.

An alternative to installing an in-house manufacturing line may be to go to an established subcontract assembly house to gain quicker access to a manufacturing line. However, this route is not as quick as is often believed. Proper selection of a subcontract assembly house takes about 3 months for a conditional qualification and an additional few months, depending on volume, for dock to stock that requires no incoming inspection or approved qualification. The issues in selection and qualification of a subcontract assembly house are discussed next.

2.4   SELECTION OF AN OUTSIDE ASSEMBLY HOUSE

Outside assembly houses are also referred to as vendors by many companies. The preferred terminology today is subcontractor or supplier. I personally prefer supplier because that is what they are—suppliers of key products and technologies. The term vendor or subcontractor does not convey the same meaning or purpose.

In any event, before delving into the details of supplier selection and qualification, we should note that very few companies have formal supplier qualification procedures as is suggested in this book. Most companies generally select a supplier based on price. There is nothing wrong with this approach, at least in the beginning, but focusing on price alone may prove to be a costly and very time-consuming option in the long run.

In this section we will discuss a formal approach to supplier qualification. This is a time-consuming process in the short run but it saves time in the long run. It is intended to select a long-term supplier who will become your partner. The supplier can be thought of as an extension of or a replacement for your in-house manufacturing capability to provide a lower cost and/or higher quality option for your product.

Before getting into the details of supplier selection and qualification, let us briefly discuss reasons for using and not using suppliers. Then we will discuss the details of formal supplier qualification.

2.4.1   Reasons for Not Using Suppliers

There are some disadvantages in going to an outside assembly house. For example, one can lose control of manufacturing technology that may be critical in the future. Also, when the product is new and going through many changes, as is common in the early stages of a new design, it may be difficult to coordinate changes because of the physical distance and possibly poorer communication between the customer and the supplier. This can delay debugging of a new product design and may potentially increase time to market.

Even if you plan to outsource your manufacturing, you still need in-house expertise to select, evaluate, and manage the supplier. But you will never gets hands-on experience in the manufacturing technology that may prove to be critical in the future.

If one plans to use a new leading edge manufacturing technology that still requires some development, it may be better to develop that technology in-house than at a supplier. Otherwise, it will be difficult to bring the technology in-house in the future unless your engineers are at the supplier learning the intricacies of the technology first hand as it is being developed.

Subcontracting is a very competitive business with fairly low margins. So very few suppliers can afford to have highly qualified process engineers on their staff with the necessary understanding of the fundamentals of materials and process issues to get to the root cause of problems when they arise (and they do).

There are many subcontract houses whose sales and marketing personnel may give the impression that their assembly capability is excellent. However, some of these suppliers may not have the maturity necessary to build all types of products, especially complex Type II (Type IIC) in large volumes and with consistent quality over time. And they may not have skilled and experienced technical personnel. And even the very best suppliers will be challenged if the product requires assembly of packages such as BGA, TAB, 0603, and 0402 discrete devices.

To build good quality products requires not only the right type of equipment but also a solid infrastructure of skilled personnel, as well as extensive documentation and process control and equipment characterization. Characterization of process and equipment, which is necessary for higher yield, requires an investment in engineering resources that has not been made by all suppliers.

There are a surprising number of suppliers who are financially weak. And quite a few of them may not be operating under the law as far as dealing with hazardous waste, air quality, and employer-employee relationships. They may not carry sufficient insurance and this poses a risk to your product.

On the other hand, there are many excellent suppliers who will not only meet but exceed your requirements and ship quality products. You may have to pay more. However, price is relevant only when quality and support are equivalent. Let us discuss reasons for going to a subcontractor and then we will discuss how to select the right supplier for your product.

2.4.2   Reasons for Using Suppliers

In-house manufacturing capability requires a considerable investment in capital equipment and personnel. The cost can vary from a few hundred thousand dollars to millions. Developing an in-house SMT infrastructure (design, manufacturing, training, and documentation) also requires considerable investment. A good subcontractor has already made this investment and has the line ready.

Going to an outside assembly house has the advantage of delaying the purchase of capital equipment. For many companies, the equipment selected for the in-house manufacturing line will not have a track record of performance because only a very few have been installed in the field. This may be a significant consideration, especially when the technology is in a transition phase and the chances of purchasing the "wrong" equipment may be high. Thus, the best way to get into surface mounting may be to start with a subcontract assembly house.

There are many suppliers who have the capability to do Type I, II, and III SMT with advanced technologies such as ultra fine pitch and BGA. Their quality is good and some have even won prestigious quality awards. It could take years to achieve the same level of maturity in-house.

Since subcontract assembly houses have a strong core competency in manufacturing, the customers can focus on their core competency— product design, development, and marketing. And some contract manufacturers have added design support and technology development resources to their staff so they can offer full-service, one-stop concurrent engineering capability.

A large investment in SMT lines will necessitate high utilization. If demand for the company's products falls off, layoffs and plant closures may result. If a supplier is used instead, one can simply delay or cancel future orders for assemblies.

Contract manufacturers have their assembly capability in place and thus the product designer can take full advantage of this capability. Time to market is minimized since the line is ready to go. We should note that time to market is the key to success in a really competitive marketplace. Being a few months ahead of the competition can improve margins significantly, and in some cases the very survival of the company may be at stake.

I believe we have covered both sides of the issue in some detail. Each company has to make its own decision as to what is the best strategy for its products. But the trend is very clear. There is phenomenal growth (over 20% a year) in the subcontract assembly business. Indeed, there are now over 1400 subcontract assembly houses in North America alone.

Obviously, though, not all of these 1400 companies have made the investment necessary to develop a good internal infrastructure. This makes it even more important to take the time to select carefully the supplier who can become your long-term partner and an extension of your company. In the following sections we will discuss the details of selecting and qualifying a supplier.

2.4.3   Evaluation and Qualification of Suppliers

One must devote considerable effort to properly evaluate a supplier. Sometimes even more human resources may be necessary than would be the case if the product were built in-house. Some suppliers can be qualified very rapidly, but for others the process can take up 2 years. And under the best of circumstances, it should take 6 to 8 months to get a supplier on approved status (dock to stock with no incoming inspection).

A thorough evaluation of the supplier will prevent product crashes at the outside facility during the production phase when one can least afford them. At the same time, an extensive evaluation puts the supplier on notice that he can enjoy continued business only if he adheres to the user's requirements.

Selection and qualification of subcontracted assembly is a major task, and sufficient time should be allowed for it. First a strategy to assess and qualify the SMT assembly supplier must be put in place. The evaluation strategy, stressing the importance of the "partnership" between the supplier and the user, should hinge on a methodology that rates the ability and maturity of the supplier in critical areas after specific periods of time and quantities of SMT boards manufactured. Based on the results of this rating, an approved qualification status can be granted to the successful candidate. Such a strategy will weed out immature suppliers at the prototype stage rather than at the production stage, when manufacturing capability is critical.

In the following sections we will focus on issues involved in evaluating, selecting, and qualifying SMT assembly suppliers.

2.4.4   Stages of Supplier Qualification

It is advisable for the user to form a team before beginning to qualify suppliers. The supplier qualification team should be staffed by personnel who have expertise in critical areas, such as technology, manufacturing, test, procurement, and quality assurance. The first job of the team should be to develop a supplier qualification procedure and questionnaires for the survey. The procedure for supplier qualification should be divided into various stages, such as Preliminary Survey, Evaluation status, Conditional status, and Approved status. These stages are defined in the following subsections.

2.4.4.1 *Preliminary Survey*

All suppliers under consideration should be surveyed over the phone to assess their capability. If a supplier is located in close proximity, the Preliminary Survey may involve a visit to the premises by one or two members of the team. If an overseas supplier is being considered, a site visit would be very expensive, so as much data as possible should be gathered before putting the supplier on the evaluation list.

For the Preliminary Survey, the team should use a Preliminary Survey questionnaire, which should be a short version of the extensive question- naire document. Section 2.4.6 suggests some questions. The Preliminary Survey questionnaire is only intended for use in a phone survey or on an initial visit and should not be sent to the supplier.

The purpose of the Preliminary Survey is to select suppliers who appear to meet some fundamental requirements such as pricing, technol- ogy, and market focus. This is very important because if your focus is primarily on price, there is no need to waste your and the supplier's time on other aspects of qualification. Also the supplier should be large enough in terms of total revenue that your company's products do not constitute more than 10–20% of their total revenue.

On the basis of data obtained during the preliminary survey, the qualification team should narrow the list of candidates. This list should be no longer than four or five since the team must visit these sites to audit their facilities. A supplier not exhibiting sufficient capability and maturity to warrant potential Evaluation status should be dropped from further consideration.

2.4.4.2 *Evaluation Status*

The suppliers selected for Evaluation status must have been surveyed at least by phone. It is absolutely essential that the supplier qualification team visit potential suppliers with a prepared questionnaire. This detailed questionnaire should be sent to the supplier before the visit. During the visit, the team should fill out the questionnaire.

If the supplier appears to meet all the requirements, the final step to complete the evaluation will be to build either some prototype boards or a very mature board to assess the quality. This is very important because the design and process may not be compatible. The results will indicate where corrective action is necessary, including redesign of the product.

2.4.4.3 *Conditional Qualification Status*

Based on the results of supplier site visits and building of a few prototype boards during the evaluation stage, the list of potential suppliers can be narrowed to one or two final candidates, depending on the supplier's capacity and the company's volume requirements.

Two final candidates should be selected for Conditional qualification who will build production boards. Purchase orders for several different production lots (six lots minimum if possible) should be placed with both of the conditionally qualified suppliers. While the suppliers are on Conditional qualified status, long-term agreements can be negotiated but should not be signed until after the supplier has achieved Approved qualification status.

2.4.4.4   *Approved Qualification Status*

The qualification requirements get stiffer as the supplier moves from Evaluation to Approved status. For example, to get an Approved status, the supplier should not only demonstrate continuous improvement during the first few lots but should also maintain the higher rating for several lots thereafter. A sudden decrease in quality in any given area will flag an impending problem and may require careful reevaluation by the user.

To achieve Approved status, suppliers who are on Conditional qualified status must build at least six different production lots. Suppliers who demonstrate continuous improvement over the first three lots (minimum assembly quantity of 50) and consistent quality over the final three lots should be put on Approved status.

Once a supplier achieves Approved status, incoming quality should be checked on a sample basis only. An "Approved status" implies no or minimum incoming inspection of the assembly. However, source inspection should be performed as necessary. Once the supplier achieves Approved status, final contracts can be negotiated and signed.

2.4.5   Supplier Rating

During the various qualification stages, the supplier should be rated in four critical categories: technology, manufacturing, quality assurance, and business issues. The qualification team prepares a checklist of issues in each category for supplier rating. In each area the types of checklist items should be such that the answers will permit a clear understanding of the supplier's capability, maturity, and viability.

Minimum points required and maximum points possible should be assigned for each category during each stage of qualification. The assigned points should vary for each category in different stages of qualification depending upon the importance to the evaluation team. As a guideline, technology should be of paramount importance during the evaluation stage. Even the best of manufacturing equipment or a strong financial position will not compensate for a supplier's failure to demonstrate sound technological know-how. However, during Conditional and Approved stages of qualification, quality, business issues, and manufacturing become more important.

One needs to develop scoring guidelines and minimum and maximum points suggested for each stage of qualification. Each company must come up with its own scoring guidelines to meet its own specific requirements. However, keep in mind the following points when rating a supplier in different categories.

When evaluating a supplier in the business area, look for compatibility in management philosophy and company culture. The senior executive's leadership and personal involvement in developing and sustaining an environment for excellence are also important.

When evaluating a supplier's technological capability, look for in-house expertise in the technical areas. Are there qualified engineers? Do they understand the fundamentals of materials and process issues? Are they committed to developing new technologies as required for your current and future products?

When rating their manufacturing capability, look at the way the line is set up, the material and product flow, and the cleanliness of the facility. Is there sufficient capability to meet your needs? Are the machines dedicated for high volume or flexible to meet different requirements?

When evaluating their quality program try to determine if management actively and visibly participates in customer focused quality efforts. Do their quality plans stress prevention rather than detection?

The questions discussed in the next section are intended to solicit answers vital to the rating of suppliers.

2.4.6   Questionnaires for Rating of Suppliers

The key to selecting a supplier is asking the right questions. This requires expertise in the technology that is being considered. What this implies is that even if you plan to outsource, you must have in-house expertise to ask the right questions. And herein lies a problem. Most companies like to outsource because they do not have or do not want to develop in-house technical capability. However, unless you understand the technology you cannot judge the competency of the supplier you are assessing. And, of course, if the company has in-house expertise, the right people must be on the supplier evaluation team.

The team should develop a questionnaire including several pages in each of the areas under evaluation: technology, business, manufacturing, and quality. If your product requires BGA and fine pitch capability, you must ask detailed questions to evaluate whether a supplier understands the intricacies of the processes involved. This is the only way to judge the capability of the supplier. Let us discuss some of the questions that should be asked of these categories. We should note that the answers to the questions suggested in the sections to follow (Sections 2.4.6.1 through 2.4.6.4) are intended to help in establishing the scoring guidelines to be developed by the OEM (original equipment manufacturers) on various issues suggested in Section 2.4.5.

2.4.6.1 Business Questions

Questions about business are aimed at determining the supplier's financial stability, long-term viability as a business, pricing policy, and so on. One typical point of debate between the supplier and the user centers around the question of which workmanship standards should be followed to ascertain product quality and acceptability. It should be decided at the Evaluation stage if the workmanship standard used will be that of the supplier, the user, or an industry standard, such as the IPC-A-610/J-STD-001 workmanship standards and the applicable revision of this standard. Other pertinent questions include the following:

- What is the financial condition of the supplier?
- Does the supplier guarantee quality?
- Does the supplier check parts and take responsibility for incorrect parts in consignment jobs?
- Does the supplier meet all the applicable environmental and labor laws?
- What is his business focus—quick-turn, consignment, or turnkey? Or all three?
- Has the supplier achieved any quality awards?
- Is the supplier qualified to ISO 9000 or any other quality standard?
- Does the supplier have in-house test capability? If not, does he outsource testing? If testing is outsourced, does the supplier take full responsibility for the final product?
- Will the supplier be responsible for damaged boards? What is the supplier's policy for compensating the user for damaged boards and their potential impact on the user's product schedule, and what is the procedure for implementing this policy?
- Will the supplier guarantee a certain percentage of capacity in the future, or is the user likely to be dropped if more lucrative business opportunities materialize?

- Does he carry sufficient insurance?
- How is the commitment to delivery date made? And how is the progress tracked?
- What is his preference—a high pin count plastic BGA or a very high pin count PQFP. Why?
- Will the board assembly cost be different whether a PBGA or a PQFP is used on the board?

After the supplier has reached the Approved status, an agreement should be drawn to freeze the "recipe" used by the supplier to manufacture the user's product, so that consistent quality is maintained over time. Some suppliers will object to providing and freezing of the "recipe" which they claim as proprietary know-how. This is really not a valid objection, however, because a finely tuned process such as a solder profile may be product specific, hence not proprietary at all. The idea here is not to induce the supplier to give away proprietary information but to discuss the technical issues in the beginning and to have a clear understanding even before Conditional qualification status is granted.

2.4.6.2 *Technology Questions*

Let us discuss some specific questions that could be asked in assessing a supplier's technological capability. For example, to gauge the engineering experience of a potential supplier's personnel, the following questions may be asked:

- How many engineers are working on manufacturing process development and how many engineers are working in production?
- What is the experience of manufacturing personnel?
- What is the average monthly volume of product being built?

The following questions could be used to gauge the supplier's understanding of materials and process development:

- What are the properties (compositions, metal content, particle size range, etc.) of the solder paste used?
- What is the solder application method (stencil or screen)? Why was this method chosen?
- What is the paste deposit thickness? Does this thickness requirement change if fine pitch is used on the same board? What is the approach (differential stencil thickness versus micro modification of stencil aperture) for applying paste on a board with standard surface mount and fine pitch? Has the supplier looked into the

implications of each approach? Does he get heel fillet in fine pitch without getting insufficient fillets in standard components?

- What is the reflow soldering method used (vapor phase, infrared, convection, or combinations thereof)?
- What are typical time-temperature profiles on the board surface and at a solder joint?
- Does he develop a unique profile for each board? How?
- Have the reflow processes been compared for manufacturing yields?

To determine the forward-thinking nature of a contractor one can ask:

- What cleaning method and solvent are used?
- How is the cleanliness of product boards monitored?
- What sort of repair/rework equipment is used, and what are the time-temperature profiles for repairing each surface mount device type?

The idea here is to establish whether the supplier has done extensive process evaluation and whether he or she understands the importance of critical materials and process variables on product quality and reliability. It will be obvious within the first few minutes of the discussion if there is someone in the company who understands these issues or if they are using some of the materials such as adhesive and paste based solely on the recommendation of the material supplier.

I should emphasize very strongly here that there are many suppliers who do understand these technical issues far better than many customers or OEMs (original equipment manufacturers). We should also note that the intent of these questions is not to dictate the process to the supplier but to assess his understanding and capability. The customer should focus mainly on the end requirements. Let the supplier worry about how best to meet those requirements.

To determine if the supplier has any plan to develop capabilities for newer technologies such as BGA and ultrafine fine pitch packages (if your product requires them now or in the future) you should ask detailed questions related to that technology. Some examples of questions related to BGAs are discussed below:

- How many members are in your BGA process development team? What are their areas of technical expertise?
- What do you see as the major concerns in PBGA and how do they compare to 304 pin PQFP?

- What is your experience (poor, fair, good, excellent, etc.) relating to BGA and PQFP on issues such as package planarity (potato chip effect), popcorning, bake and bag, handling/test, lead coplanarity, and rework?
- What are the process parameters (paste printing, placement, reflow, etc., as applicable) that are critical? How do you control those process parameters?
- What is your rework process for BGA (please describe as applicable such as PBGA, CBGA, TBGA, etc.) and high pin count PQFP?
- Do you have a rework process to salvage a PBGA package (process for putting the eutectic balls back on)?
- Do you prefer copper defined pad or solder mask defined pad? Why? Do you prefer via in pad or tear drop design? Why? Did you conduct any evaluation on PCB routing for different pitch, pad sizes, line width/space, etc., and their impact on layer count? Did you experiment with different sizes of pad and their impact on process yield?
- Do you have any reliability data on BGA? If yes, please describe the test condition, failure criteria, type of BGA used, and reliability results.

Questions along these lines should be asked if any other new technology is of interest such as tape–automated bonding (TAB), flip chip, or any other new technology that you may need in the future. The idea here is to go to a supplier who will grow with you. After you have put in all the resources in developing this supplier, you won't want to go to an entirely new one if your needs change.

2.4.6.3 Manufacturing Questions

The questions that are to determine the manufacturing capability of a supplier should relate to the capacity and utilization of the manufacturing equipment, material control and handling methods, equipment operating procedures, equipment maintenance and calibration schedules, and operator training. Questions on these and similar matters should cover each type of equipment in use at the supplier's facility.

The type of equipment used by a supplier can give a good idea about the firm's capability. The manner in which the equipment is organized is a good indicator of the contractor's awareness of manufacturing processes. Good quality, in-line equipment is preferable to batch equipment for automation. If quick turn is important to you, is the supplier set up for quick turn? Does he use in-line production equipment for prototype jobs? If yes, does it create delivery slippage on quick turn? As a general guideline, if in-line production machines are used for "quick-turn" prototype work as well, the supplier is not set up for quick turn. You can expect to have delivery problems either on the quick turn or production lot or both. Also, ask what other equipment was considered and why the firm bought a particular brand and model.

Observe the performance of the equipment in operation. How often is it down for repair? How accurate is the pick-and-place equipment? Does the X-Y table move during placement? Does that cause misplacement? Is the pick-and-place data transferred electronically or is the "teach in" method used for programming. "Teach in" is a sure sign of future problems. How closely can the solder profile be controlled? Is there a unique profile for each product? If not, why not?

Does the screen printer provide good and consistent print? How effective is the soldering and cleaning equipment in providing good and clean boards? Batch versus in-line cleaning for surface mounting should be closely investigated. With tighter spacings among components and lower standoff heights above board, in-line cleaning systems with sufficient nozzle pressure are very effective in providing boards free of solder balls and flux residues. If no-clean solder paste and wave flux are used, how are solder balls avoided or removed? Has the touch-up rate increased with no-clean flux and paste? How was the transition made to no-clean? No clean is not a "drop in" process. It requires considerable groundwork to switch over to no-clean technology.

And finally, one must note that having the best equipment is no guarantee of excellent yields. Does the supplier know what the yields are? Is process control used to monitor and control defects? Have all the equipment and processes been characterized? Does documentation exist for operation and maintenance of all equipment? And are the operators knowledgeable about the documentation?

2.4.6.4   Quality Assurance Questions

Generally most supplier problems are encountered in the quality assurance area. Some sample questions for assessing whether the supplier can deliver consistent quality over time and whether the firm has good process control as well as documentation in place to meet commitments are given below:

- Are areas sensitive to electrostatic discharge (ESD) clearly identified?

- Do all personnel and visitors wear approved ESD protective clothing and wrist straps while in the ESD-sensitive area?
- Are all work surfaces, dissipative or conductive, grounded through a resistor per specification?
- Are conductive or antistatic containers issued to transport ESD-sensitive materials from one location to another?
- Are ESD-sensitive areas and containers periodically audited for conformity to documented ESD guidelines?
- Are documented workmanship standards maintained? (This is not an issue if the user's workmanship standards are to be used.)
- Is a controlled document control procedure in place?
- Is a minimum distribution list for new specifications maintained?
- Is there a controlled method to assure removal of previous revisions of documents?
- How is defect logging/archiving and failure analysis accomplished?
- Are work instructions, production equipment, and appropriate work environments in place to accomplish the major SMT processes?
- Are the workmanship standards available to assemblers and inspectors?
- Is a corrective action system maintained, and does it provide preventive measures also?
- Is there a feedback loop so that the problem does not recur?
- Is statistical quality control (SQC) used in the production area? This may be one of the most important questions of all. Refer to Chapter 14 for a discussion of SQC.

2.4.7   Supplier Management or Partnership

The subcontracting environment is such that there is more demand than supply for contract manufacturing. And there are a limited number of suppliers with manufacturing expertise, quality control, and financial strength. The need for more advanced technologies is increasing. However, only very few suppliers have the capability.

In such an environment, what should be the focus? Partnership with a supplier or management of a supplier? Partnership implies a mutual stake in the outcome, whereas management implies a master-slave relationship. Given the supply and demand environment and the need for more advanced technologies, partnership is the key to successful supplier relationships.

A good fit in technology, management, and business focus is also critical for a long-lasting partnership. Each party needs what the other offers and offers what the other needs. In other words, you have complementary strengths. There are various other implications of partnership. It implies the following actions:

- Choose the supplier for the long haul.
- Choose as few suppliers (components, boards, and assembly) as possible.
- Put long-term agreements in place.
- Share your vision, business strategy, and technology roadmaps with your partner.
- Treat supplier personnel as part of your design and manufacturing team.
- Seek vendor input at the schematic stage.
- Ensure DFM (design for manufacturability) compatibility with the supplier's processes.
- Get supplier buy-off before releasing new designs.
- Build the relationship slowly but steadily.

In many ways this implies a break from the common business philosophies of most corporations. For example, do not develop relationships quickly and break them easily. Cost will continue to be important for everyone. But in a partnership you work together to reduce cost and share in the outcome. If new technology is needed, do not jump to another supplier. If you have shared your vision and product and technology roadmaps, this situation should not arise. This means you give enough lead time to allow the supplier to develop the technology that is needed in the future. Everyone talks about partnership but few really practice it. Successful resolution of cost issues and the need for new technologies to the satisfaction of both parties are the true tests of partnerships.

In summary, one should decide carefully the reasons for outsourcing. They could be lower cost, supplier's expertise, or postponement of capital investment. And definitely watch out for problems with cost, quality, and delivery. In some cases, issues related to technology transfer from the supplier may have to be addressed in the beginning.

Once the decision is made to outsource, set up a team to select and quality suppliers. Conduct detailed evaluations before giving Conditional and Approved status. And once the supplier is on Approved status, view his failure as your failure. Share your vision in technology and business strategies with your partner.

And finally keep this in mind: Every supplier is not for every customer, but there is always a best supplier for every customer.

2.5 MANAGING THE RISK: PILOT TO PRODUCTION

Taking risk is an integral part of developing competitive products using new technology. Even when an old technology is used, not all new products make it to the marketplace. The chances of survival with new technology are even lower because of the added risk. The point here is not to avoid a new technology like surface mount but to carefully assess the inherent risk before committing a product to it. One should not use new technology for the sake of technology; rather the goal should be to provide functionality and quality to the customer at a low cost within the available market window. In other words, the new technology must fulfill the customer's need. *What this means is that designers should use SMT only if the need is demonstrated.*

Once it has been decided that surface mount is the most desirable way to meet the customer's needs and gain a competitive advantage, steps should be taken to carefully assess and manage the risks. Yet before the use of the new technology can be considered, the infrastructure for the new technology must be in place. What does this mean?

Certain basic questions must be asked. For example, are the components available and qualified? Has the design tool been validated? Are the manufacturing locations ready? (They could be either in-house or at a subcontract assembly house.) Has the documentation for process and design been released? Are trained personnel available? And finally, has any product been designed and assembled using this technology?

The risk will be the greatest if the product is committed before the technology has been validated on a nonproduct board, because the technology, manufacturing, and reliability problems will be unknown. For example, the following circumstances must be anticipated: (a) the first product may have to go through multiple redesigns, (b) the materials and process problems in manufacturing may cause very poor yield, and (c) the product may not be testable. Under any of these conditions, there may be an adverse impact on schedule and revenue.

To minimize the risk of introducing a new technology in the manufacture of a new product, development and validation of the technology should be conducted in phases. First, the material, process, and design issues should be resolved on nonproduct boards or technology pilot boards, preferably on a production line. The use of the nonproduct board serves to flush out technical problems in design, manufacturing, and testing and to assess the risk associated with the technology itself. The process and design variables that are found to affect the quality, reliability, and manufacturability of the board should be documented and incorporated in the process and design specifications.

Next the technology should be validated in a manufacturing environment on an actual product. Only a minimum number of products, preferably one or two, should be committed at this time because not all problems in a manufacturing environment can be foreseen. One limits the number of products at this time simply to validate the process and design rules in a production environment without incurring significant risks. Thus the company's revenue will not suffer significantly if serious problems are discovered along the way.

Although some technical problems are bound to emerge no matter how carefully the technology is implemented, the key concern should be to avoid major problems. For example, even when the technology is sound, the vendor base may be unstable, and there may be problems in the areas of receiving components and boards on time. This is very common. Thus, time should be allowed in the project schedule for startup problems in manufacturing and vendor deliveries. After successful completion of the first round of real products, the technology can be released for additional new products.

A successful release of technology is defined as an actual product being built in a manufacturing environment at a very low rate of defects per unit and a high first-pass acceptance. There are no absolute numbers on defect rate and first-pass acceptance. One must be able to demonstrate process capability. A decision here depends on the product complexity and the level of sophistication of the company. The defects per unit or the first-pass acceptance numbers should be comparable to those for products using conventional technologies with which the company has considerable experience.

If no problems are encountered, additional products may be introduced. The risk on the third and fourth products will be almost negligible. After a certain volume of products has been built at a predefined quality level consistently, widespread usage of the new technology can begin.

The approach just discussed should be repeated for each category of new technology boards. For surface mount, the technology should be divided into four categories for the purpose of minimizing risk. In ascending order of risk incurred, these are Type III logic, Type I memory, Type II logic (or Type I single-sided logic). And as discussed earlier the fourth step should be to attempt complex SMT—Types I/II/IIIC. (See Chapter 1 for definitions.) We should also note that the complexity further increases if the number of types of packages on the same board increases. For design, testing, thermal management, component availability, and the related process issues for various types of packages, the distinction of SMT in various categories is very important.

After some initial experience with successful releases, the level of difficulty in implementing the additional phases should be lower even for the more complex technologies. One to two years should be allowed to develop and validate the full surface mount process and design capability. Additional time should be allowed if complex types of SMT (Types I/II/IIIC) are being implemented.

Most important of all, good working relations with suppliers will play a critical role in the management of risk and the successful implementation of new technologies.

There is one more item that should be kept in mind when putting a plan in place to manage the risk of SMT implementation. Despite the incentive of simplifying the process by using all surface mount components on a board, the industry cannot get away from through hole completely for a long time to come.

Major semiconductor companies will continue to supply high performance microprocessors in the through-hole PGA (pin grid array) package for a long time to come. There are various reasons for this. Users like the option of socketing. Larger devices such as connectors and sockets require stronger joints and through-hole joints are stronger than surface mount joints. Also, socketing reduces the import duty on mother boards assembled overseas since duty is based on the value of the boards. (See Chapter 5, Section 5.2.1 for details). So, even though surface mount is the wave of the future, you must still plan to deal with all the issues related to complex Type II mixed assembly technology to reduce your risk.

2.6   SUMMARY

For the successful implementation of surface mount technology, the first order of business for management consideration is to obtain a clear outline of the outstanding technical issues. Then an SMT team consisting of a selected group of individuals from all functional organizations should be assembled to address those issues. The team should be headed by a program manager who not only is qualified technically but also has the full support of top management.

Because of such constraints imposed by the external infrastructure as component availability and the evolutionary nature of the technology, SMT should be developed and introduced in stages. This allows for gradual building of the internal infrastructure and the spreading out of capital investment over a longer period of time.

In some cases subcontract assembly houses may be a better option than in-house manufacturing to meet either the short- or long-term needs and to help postpone capital investment. But extreme care should be taken in the evaluation and selection of a subcontractor. The successful candidate must meet stringent technology, manufacturing, business, and quality requirements to prevent "vendor crashes" at a critical moment. Appropriate selection of a subcontractor requires a considerable investment in human resources.

Even after the technology has been evaluated and debugged, it should be implemented on a few selected products first. Only when the technology has been validated on some "shared" technology and product pilots should it be used across product lines. Such caution is important in reducing risk and preventing multiple failures across product lines, which could adversely affect the company's revenue.

Part Two

Designing with Surface Mounting

Chapter 3

Surface Mount Components

3.0 INTRODUCTION

Surface mount devices, active or passive, are functionally no different from their conventional through-hole counterparts. Thus the design and electrical function of an internal device is not unique to surface mounting, hence beyond the scope of this book. What is different in surface mounting is the packaging of those devices. Surface-mounted devices (SMDs) or components (SMCs) provide greater packaging density because of their smaller size.

Among many other benefits of surface mounting, the real estate savings is of paramount importance. Since reduced size is the key benefit of SMT, there is a continuous demand for smaller sizes. This has lead to the widespread use of fine pitch (20 and 25 mil pitch) and ultrafine fine pitch (0.5 mm or 20 mil and lower pitch) and ball grid arrays (BGAs). Even chip scale packaging (package size not more than 1.2 times die size) and direct chip attach are becoming more and more common to achieve further densification. All these surface mount packages affect not only the real estate on the board but also the electrical performance of the device and the assembly. Moreover, due to component packaging differences, the parasitic losses such as capacitance and inductance in surface-mounted devices are considerably less than those for the through-hole technology.

The component packages, in addition to saving real estate and providing better electrical performance, serve many other functions. They protect the devices within them from the environment, provide communication links, remove heat, and offer a means for handling and testing.

It is safe to say that when it comes to component packaging, the world of surface mounting is much more complex than that of conventional through-hole mount technology. In this chapter we discuss commonly used passive and active components. We discuss packaging for shipping and handling and procurement specifications for surface mount packages.

We should note that "surface mount package" is defined broadly here to include all types of surface mount packages. There are certainly some differences among standard surface mount technology (50 mil pitch) and fine pitch, ultra fine pitch, and BGA technologies. In this chapter we discuss the details of all these packages.

3.1 SURFACE MOUNT COMPONENT CHARACTERISTICS

Certain characteristics are common to all the many surface mount package types. For example, they all mount on the surface of the substrate instead of protruding through the plated through hole, as is the case in through-hole mount technology (THMT) boards. This means that the solder joint, which provides both mechanical and electrical connections, is very important for reliability of the assembly.

The surface mount packages are designed to meet the requirements of two major types of applications: commercial and military. The commercial applications with benign environments can use nonhermetic packages. The operating temperature requirements are generally from 0 to +170°C.

Military or high reliability applications designed for severe environments require the use of hermetic packages in the −55°C to +125°C range. Of course there are other environments between these extremes for which the designer must choose appropriate packages to meet the reliability requirements.

The hermetic chip carrier is expensive and is intended for high reliability products. It requires substrates of matching coefficient of thermal expansion and is still prone to solder joint failure during thermal cycling.

The surface mount package bodies (and leads) also see much higher temperatures during soldering than the through-hole package bodies. This makes these packages more susceptible to moisture related cracking. They must be designed with this requirement in mind. Because of their smaller size, it is very difficult to provide part markings on them, especially on passive components. If the devices do get mixed, they must be positively identified or thrown away. This almost mandates packaging of these devices to facilitate automated placement.

The packages also mount very close to the surface of the substrate, hence have relatively less clearance off the substrate. To achieve the required cleanliness, therefore, very good process control is necessary.

Since surface mount leads (not package bodies) experience lower temperatures in soldering than through-hole leads, the solderability test methods and requirements need to be different. The surface mount package bodies undergo higher temperatures, as mentioned earlier, but the surface mount leads or terminations see lower soldering temperatures as compared with DIP leads. This means that the surface mount leads must pass even more stringent solderability requirements. This problem is also compounded by the fact that today much less active fluxes are used for surface mount assemblies.

3.2   PASSIVE SURFACE MOUNT COMPONENTS

The world of passive surface mounting is somewhat simpler. Monolithic ceramic capacitors, tantalum capacitors, and thick film resistors form the core group of passive devices. The shapes are generally rectangular and cylindrical. The surface mount versions of these devices have gained wide acceptance because they occupy half the space when mounted on the top of the substrate. When mounted on the bottom of the substrate, as is the case in Types II and III SMT boards, they utilize the space otherwise not occupied at all. (See Chapter 1 for definitions of Type I, II, and III SMT assemblies.) The mass of these devices is about 10 times lower than that of leaded devices, and the existence of terminations instead of leads provides design benefits of better shock and vibration resistance and reduced inductance and capacitance losses.

The use of surface mount resistors and capacitors has been very extensive in Japan (consumer electronics) and in the automotive electronics industry in the United States. Because of this extensive usage, surface mount resistors and capacitors are less expensive than through-hole axial components.

The surface mount resistors and capacitors come in various case sizes to meet the needs of various applications. As the size shrinks from 1206 (120 mils long and 60 mils wide) to 0805 (80 mils long and 50 mils wide) to 0603 (60 mils long and 30 mils wide) to 0402 (40 mils long and 20 mils wide), the requirements for increased placement accuracy put more demands on the placement equipment industry. We should note with some relief that as small as these devices are, the competition has driven the passive component industry to supply part markings on these parts. Three character EIAJ (Electronic Industries Association, Japan) marking codes are common down to 0603 case sizes.

While there is a trend towards shrinking case sizes, larger case sizes are also used if capacitance requirements are large. However, the use of larger ceramic capacitors such as 1825 (180 mils by 250 mils) should be minimized to avoid CTE (coefficient of thermal expansion) mismatch problems on commonly used glass epoxy FR-4 boards.

These devices come in both rectangular and tubular (MELF: metal electrode leadless face) shapes, with the latter at lower cost but also some what lower reliability. Because MELFs are more difficult to place, the total cost of usage may be lower for the rectangular components. Thus automated placement using tape and reel packaging is highly recommended.

Wraparound terminations are preferred over flat terminations (no solderable terminations on the side) for good solderability, and plated terminations are preferred over solder-dipped terminations for accurate placement. Nickel barrier underplating, especially for wave soldering applications, should be required to prevent dissolution (leaching) of solderable coating in the solder bath.

Problems such as misalignment and tombstoning of parts during reflow soldering and leaching during wave soldering are of concern when passive surface mount devices are used. Solutions to these problems are addressed in Chapters 6 and 7 (design), 10 (metallurgy of soldering), and 12 (soldering). In the sections that follow, we discuss passive components of various types in further detail.

3.2.1   Surface Mount Discrete Resistors

There are two main types of surface mount resistors: thick film and thin film. Thick film surface mount resistors are constructed by screening resistive film (ruthenium dioxide based paste or similar material) on a flat, high purity alumina substrate surface, as opposed to depositing resistive film on a round core as in axial resistors. The resistance value is obtained by varying the composition of resistive paste before screening and laser trimming the film after screening.

In thin film resistors the resistive element is nichrome film that is sputtered on the substrate instead of being screened on. The construction details of a surface mount resistor are shown in Figure 3.1.

Figure 3.1 shows a resistive element on a ceramic substrate with protective coating (glass passivation) on the top and solderable terminations (tin-lead) on the sides. The terminations have an adhesion layer (silver deposited as thick film paste) on the ceramic substrate, and nickel barrier underplating followed by either dipped or plated solder coating. The nickel barrier is very important in preserving the solderability of terminations because it prevents leaching (dissolution) of the silver or gold electrode during soldering.

The resistive layer on the top surface dissipates the heat and should always face away from the substrate surface. The passivation layer is

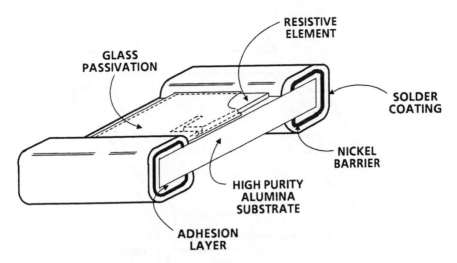

Figure 3.1 The construction details of a surface mount resistor.

very brittle, however, and should not be probed with hard points such as test probe points during testing. Damaging the passivation layer could expose the resistive layer to the environment with the chance of further damage to the resistive element, thus degrading the resistor.

Thick film surface mount resistors are available in various tolerances (1, 5, 10, and 20%). The thin film resistors are made for very high precision circuits that require very close tolerances (<1%). There can be a substantial difference in cost depending on the tolerance requirement. For example, resistors with 1% tolerance generally cost twice as much as 5% tolerance resistors. Tighter tolerance components, which are expensive, should be avoided unless circuits really require them.

Resistors come in 1/16, 1/10, 1/8, and 1/4 watt ratings in 1 ohm to 100 megohm resistance in various sizes. The commonly used designations for sizes (0402, 0603, 0805, 1206, 1210, etc.) are established by EIA specification IS-30. The EIA size refers to dimensions in hundredths of an inch. See Figure 3.2 for typical dimensions of commonly used resistors. It should be noted that the dimensions vary from manufacturer to manufacturer, but generally speaking, 1/16, 1/10, 1/8, and 1/4 watt resistors come in EIA sizes 0402/0603, 0805, 1206, and 1210, respectively. The 1206 size with an 1/8 watt rating is the most commonly used size. Zero ohm resistors (jumpers) are also available in the 1206 size. See Table 3.1 for wattages for various sizes of resistors.

As shown in Table 3.1, 0603 and 0402 case sizes are used for 1/16 watt resistors. They are used where real estate constraints are severe.

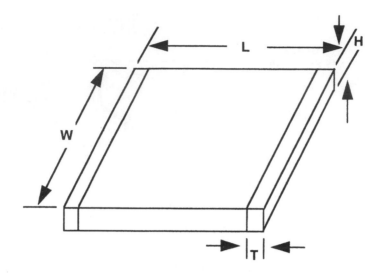

DIMENSIONS IN INCHES				
DESIGNATION	LENGTH NOM (L)	WIDTH NOM (W)	HEIGHT MAX (H)	TERMINATION MIN (T)
R0402	0.040	0.020	0.016	0.004
R0603	0.060	0.030	0.024	0.006
R0805	0.080	0.050	0.028	0.006
R1206	0.120	0.060	0.028	0.010
R1210	0.120	0.100	0.028	0.010
R2010	0.200	0.100	0.028	0.014
R2512	0.250	0.120	0.028	0.014

Figure 3.2 Typical dimensions of surface mount resistors.

Smaller sizes are more expensive and more difficult to place and solder than larger 0805 and 1206 sizes. Their use should be avoided unless real estate and circuit requirements justify their use.

A surface mount resistor has some form of colored resistive layer with protective coating on one side and generally a white base material on the other side. Thus the outside appearance offers a simple way to distinguish between resistors and capacitors. Capacitors have one color (generally gray, blue, or brown) on both sides. It must be emphasized that there is no standard on color. However, there is a height difference

Table 3.1 Case size versus wattage for surface mount resistors.

EIA CASE SIZE	WATTAGE (WATTS)
R0402	1/16
R0603	1/16
R0805	1/10
R1206	1/8, 1/4
R1210	1/4
R2010	1/2
R2512	1.0

between these components, ceramic resistors being about half as thick as ceramic capacitors.

It is not practical for capacitors and resistors to be the same size even though there have been moves in the EIA to make them the same. There are two reasons for this. The capacitance value is determined by the thickness. So less thickness means less capacitance for a given dielectric constant of the material. And making resistors thicker will add substantially to the cost.

As discussed in Chapter 6, resistors and capacitors require different land pattern designs because of height differences even when they have the same widths and lengths. The other critical dimension for both resistors and capacitors is the width of terminations. The EIA dimensions for resistors are shown in Figure 3.2.

As noted earlier, three digit codes are used for part markings. There is generally not an absolute need for part markings, however, because these devices are most widely supplied on a tape and reel. Also, some pick-and-place equipment can test capacitors and resistors before placement without compromising placement speed, especially if only the first few parts on a reel are tested for correct values. This is very helpful to assure that the correct value is placed in the specified placement machine feeder slot.

3.2.2 Surface Mount Resistor Networks

The surface mount resistor networks or R-packs are commonly used as replacements for series of discrete resistors. This saves real estate and placement time. The currently available styles are based on the popular SOIC (small outline integrated circuits) package, but the body dimensions vary. They generally come in 16 to 20 pins with 1/2 to 2 watts power per package. SOICs are discussed later (Section 3.4.2).

The most commonly available body dimensions are as follows: 0.150 inch body, known as the SO package, with 8, 14, and 16 pins; a 0.220 inch body width known as SOMC, with 14 and 16 pins; and a 0.295 inch version, the SOL, with 16 and 20 ins. Some vendors are even supplying body widths of 0.410 inch. The dimensions of SO and SOL are based on SOIC packages and have the same dimensions. The body dimensions based on SOIC package dimensions are likely to gain more user acceptance because of the popularity of SO packages and commonality in land pattern design and feeders for the placement machines.

Because R-packs are used relatively less often, they tend to be much more expensive. Resistor networks are unlikely to become as inexpensive as the equivalent amount of discrete chip resistors. Network resistors should be used only when matching of tolerance is critical in the circuit. In addition to price, the surface mount R-packs can have lead coplanarity problems (Section 3.7.1). Less than 0.04 inch coplanarity is now generally required by most users. The problem of lead coplanarity can be avoid in R-packs by using flat chip arrays—relative newcomers to resistor network packaging. The flat chip arrays come in arrays of various case sizes, such as the popular 1206, connected together.

3.2.3   Ceramic Capacitors

In a high frequency circuit application, it is important to physically locate the capacitor as close to the high speed device as possible and to keep the lead length to a minimum, to minimize circuit inductance. The surface mount capacitor is ideal for such a purpose because it does not have any leads and can be placed underneath the package on the opposite side of the board. In very critical and high speed applications such as DRAMs (dynamic random access memory), some capacitors can be placed directly underneath the surface mount packages on the same side. There are capacitors available only 18 mil thick that can fit underneath a leaded surface mount package. Since the solder joints cannot be inspected, the process yield must be 100% for such applications.

Surface mount capacitors are used for both decoupling applications and for frequency control. As the frequency of microprocessors has increased, the use of capacitors for frequency control has become more critical in high volume SMT applications such as computers and telecommunications. On a typical board, the total number of surface mount capacitors is generally more than that of any other type of component.

Multilayer monolithic ceramic capacitors are used because of improved volumetric efficiency. In the multilayer ceramic capacitor, the electrodes are internal and are interleaved with ceramic dielectric. The alternate electrodes are exposed at the ends and connected to the end termination.

The construction of a multilayer ceramic capacitor is shown in Figure 3.3. The end terminations of the capacitors are similar in construction to

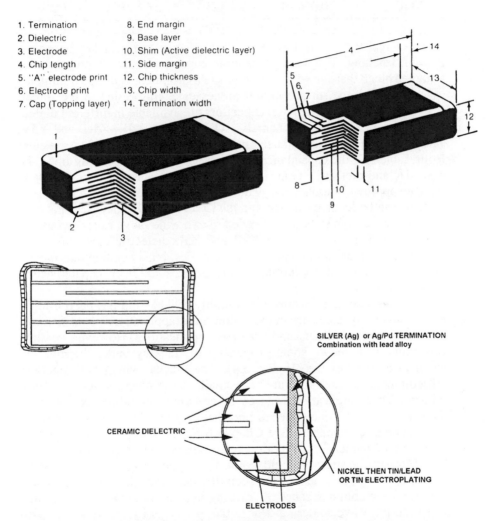

1. Termination
2. Dielectric
3. Electrode
4. Chip length
5. "A" electrode print
6. Electrode print
7. Cap (Topping layer)
8. End margin
9. Base layer
10. Shim (Active dielectric layer)
11. Side margin
12. Chip thickness
13. Chip width
14. Termination width

SILVER (Ag) or Ag/Pd TERMINATION
Combination with lead alloy

CERAMIC DIELECTRIC

NICKEL THEN TIN/LEAD
OR TIN ELECTROPLATING

ELECTRODES

Figure 3.3 Construction detail of ceramic capacitor with exposed alternate electrodes at the ends and connected to the end termination (top figures). The cross-sectional views (bottom figures) show the details of end termination materials.

Table 3.2 Capacitance range for different dielectric materials and case sizes

	NPO	X7R	Y5V	Z5U
CASE SIZES	0402–2225	0402–2225	0402–1210	0603–2225
CAPACITANCE RANGE	0.5 pF–0.056 µF	100 pF–2.2 µF	1000 pF–0.47 µF	0.001 µF–4.7 µF

those of resistors, with a silver adhesion layer and a nickel barrier to prevent leaching. This kind of ceramic capacitor construction results in a rugged block that can withstand a harsh environment and such treatments associated with the surface mount processes as immersion in the solder.

Monolithic surface mount capacitors are available in different dielectric types per EIA RS-198, namely COG or NPO, X7R, Z5U, and Y5V. They have different capacitance ranges as shown in Table 3.2. As is clear from Table 3.2, the capacitance range varies widely depending upon the case size and dielectric. One should refer to the latest product catalogue from an approved supplier for available capacitor values, dielectric types, voltages, and tolerances. In general, the COG or NPO dielectric capacitors are used when high stability is needed over a wide range of temperatures, voltages, and frequencies. The X7R and Z5U dielectric capacitors have poorer temperature and voltage characteristics, but since they are mostly used for bypass and decoupling applications, stability of capacitance is not very important.

The surface mount capacitor is highly reliable and has been used in high volumes in under-the-hood automotive applications. It also has a proven history in military and aerospace applications. However, ceramic capacitors have shown some propensity to cracking under thermal and mechanical stresses (see Figure 3.4). The cracks, which are generally difficult to see because they are very small, can grow in service and cause failures. There are various causes of cracking, including excessive or uneven solder fillet due either to poor land pattern design or to improper orientation of components. See Chapter 6 for a discussion of land pattern design and Chapter 7 for recommended component orientation.

Mechanical stresses could come from poor handling or excessive pressure during placement or too high thermal shock during wave soldering. The most common cause of cracking in ceramic capacitors is thermal shock during wave soldering. Poor quality control by the vendor is also a major factor, although the capacitor industry has improved the thermal robustness of capacitors in recent years. As shown in Figure 3.3, a ceramic capacitor contains a metallic electrode, termination material, and ceramic dielectric. Each has a different coefficient of thermal expansion. For

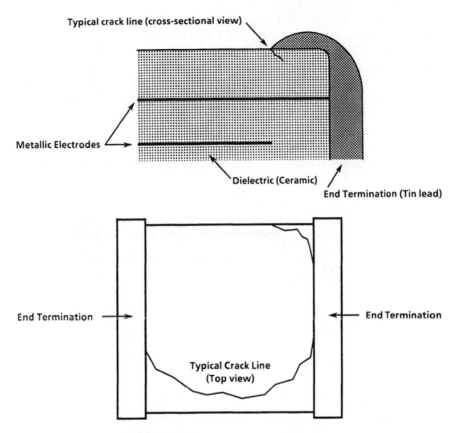

Figure 3.4   Cracks in multilayer ceramic chip capacitors: cross-sectional view (top) and top view (bottom).

example, the CTEs of the internal electrode, the tin-lead termination and nickel over silver, and the ceramic, respectively, are 16, 18, and 9.5 to 11.5 ppm/°C [where ppm = microinch (mil) per inch]. During preheat and soldering, the electrodes and terminations heat up more rapidly than the ceramic body. If the expansion differential is too rapid due to a steep thermal profile, the stresses generated by CTE mismatch can crack the component, especially if the component thickness is greater than 0.040 inch.

Thermal shock can be minimized by gradually preheating the board before the assembly goes over the solder wave and using capacitors that meet the thermal profile being used. Refer to Chapter 14 (Section 14.3.2: Component-Related Defects).

If the design is correct, if component quality is properly controlled by the vendor, and if the user maintains the top side preheat and solder pot temperatures at 220 to 240°F and 475°F (for eutectic solder), respectively, cracking should not be a problem. Note that we have talked about maintaining top side preheat at 220°F to 240°F. This means that the bottom side temperature will be 240°F to 260°F. Also note that the Z5U ceramic capacitors are more susceptible to cracking than the X7R. Since their costs are comparable, X7R capacitors should be used if possible.

✳ As discussed earlier and shown in Table 3.2, the capacitance values of the capacitors are related to their case sizes and dielectric constants. The most commonly used case size is EIA 1206 (0.120 inch by 0.060

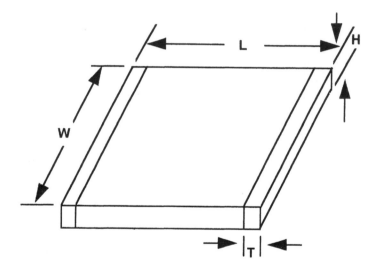

DESIGNATION	LENGTH NOM (L)	WIDTH NOM (W)	HEIGHT MAX (H)	TERMINATION MIN (T)
C0402	0.040	0.020	0.024	0.004
C0603	0.060	0.030	0.034	0.008
C0805	0.080	0.050	0.050	0.010
C1206	0.120	0.060	0.054	0.010
C1210	0.120	0.100	0.054	0.010
C1812	0.180	0.120	0.054	0.010
C1825	0.180	0.250	0.054	0.010

Figure 3.5 The dimensions (in inches) of different case sizes of ceramic capacitors.

inch nominal). The newer case sizes are 0402 and 0603. The selection of either lower or higher case size should be considered when the required capacitance value is not available in the desired dielectric type.

Figure 3.5 shows the case sizes for ceramic capacitors. The component termination width is the most critical dimension for soldering and land pattern design. The components from vendors that provide closer component tolerances and meet other procurement requirements (discussed in Section 3.8) will provide assemblies with higher yield.

The most widely used packaging for ceramic capacitors is 8 mm tape and reel. When establishing procurement specifications, the objective should be to minimize the case sizes and part numbers (capacitance values). This will provide increased purchasing leverage and will allow for reduced inventory. In addition, the number of feeders for the pick-and-place machine in manufacturing can be reduced.

3.2.4 Tantalum Capacitors

For capacitors, the dielectric can be either ceramic or tantalum. Surface mount tantalum capacitors offer very high volumetric efficiency or a high capacitance-voltage product per unit volume and high reliability.

One of the most important considerations in selecting tantalum capacitors should be the construction of the end terminations, which come in two major configurations—welded sub contacts (unmolded) and with wrap-under lead contacts (plastic molded). The welded stub causes problems during placement, and these capacitors come with gold terminations, which can embrittle solder joints. Now they are rarely used so we will not discuss them in this book.

The wrap-under lead capacitors, commonly called plastic molded tantalum capacitors, have leads instead of terminations and a beveled top as a polarity indicator. (See Figure 3.6) These capacitors are constructed using a sintered tantalum anode and a solid electrolyte. The construction detail of molded plastic tantalum capacitors is shown in Figure 3.7 (top figure). The capacitance range available in molded capacitors is shown in Table 3.3.

There are no soldering or placement concerns when using the molded plastic tantalum capacitors. Their case sizes fall into two main categories called standard and extended range. Each category has four different case sizes as shown in Figure 3.7 (see bottom figure and table).

Standardization to EIA sizes has become more apparent in recent years, although some variation in dimensions may exist from supplier to supplier. This can cause confusion in land pattern design. It is important to check the dimensions of the packages before designing the land pattern.

Polarized

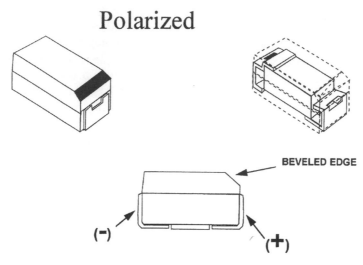

BEVELED EDGE

(−)   (**+**)

Figure 3.6   Molded plastic tantalum capacitor.

The concept discussed in Chapter 6 for land pattern design is recommended.

The capacitance values for tantalum capacitors vary from 0.1 to 100 μF and from 4 to 50 V dc in different case sizes. As the demand for higher capacitance values in smaller case sizes increases, the manufacturers continue to find ways to increase the ranges available. So, as mentioned earlier, one should refer to the manufacturer's literature for the capacitance range for a given voltage and case size. Tantalum capacitors should never be used above the rated voltage and should never be used in reversed polarity. If these precautions are ignored, the devices may blow up when powered. They can, however, withstand immersion in the wave soldering process.

Some suppliers have found tantalum capacitors to be susceptible to moisture induced cracking during soldering (wave, vapor phase, or infrared). The susceptibility to cracking depends on the level of moisture absorbed, which in turn depends on the relative humidity, time, and temperature of component storage.

If we refer back to Figure 3.7 (top figure), moisture easily permeates through the encapsulant and the conductive coatings during storage. In addition, the core electrolyte (fine particles of tantalum surrounded by tantalum pentoxide and manganese oxide) is very porous and easily absorbs moisture. When the moisture absorption reaches a threshold limit, the moisture will convert to steam during soldering and cause the component to

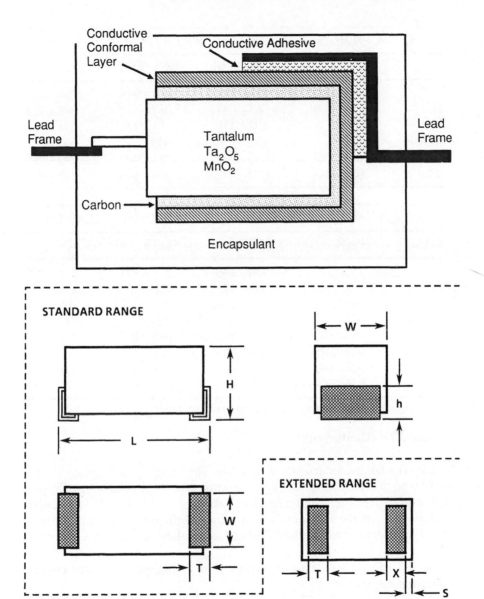

Figure 3.7 The construction details of a molded tantalum capacitor (top) and the case sizes (inches) and designations for plastic molded tantalum capacitors with standard and extended capacitance range (bottom).

SIZE CODE/STANDARD CAPACITANCE RANGE

	[A] 3216	[B] 3528	[C] 6032	[D] 7343
L	.118–.134	.130–.146	.224–.248	.268–.299
H	.055–.071	.067–.083	.087–.110	.098–.122
W	.05–.071	.102–.118	.114–.138	.157–.181
M Min	.028	.028	.040	.040
*Y	.043–.051	.083–.09	.083–.09	.090–.098
T*	.020–.043	.020–.043	.020–.043	.020–.043

SIZE CODE/EXTENDED CAPACITANCE RANGE

DIM	3518	3527	7227	7257
L	.130–.146	.130–.146	.272–.295	.272–.295
H*	.067–.083	.067–.083	.098–.122	.118–.146
W	.063–.079	.095–.118	.095–.118	.205–.244
M Min	.028	.028	.040	.047
Y*	.063–.071	.095–.102	.095–.102	.213–.228
X*	.024–.040	.024–.040	.031–.047	.031–.047
T*	.028–.043	.028–.043	.051–.067	.051–.067
S*	.016–.024	.016–.024	.024–.032	.024–.032

*Dimensions are tailored to facilitate standard land pattern design.

Figure 3.7   (Continued)

crack. The basic mechanism of moisture absorption and cracking is very similar to that of plastic package cracking discussed in detail in Chapter 5, Section 5.7.1. The solution requires baking of components before soldering to drive off the moisture. For specific details on the susceptibility of the component to cracking and the proposed solution, users should contact

Table 3.3   Case size versus capacitance for surface mount plastic molded tantalum capacitors

EIA CASE SIZE	CAPACITANCE (μF)
A (3216)	0.1–10 μF
B (3528)	0.22–22 μF
C (6032)	0.68–68 μF
D (7343)	3.3–100 μF

their supplier. Also, like ceramic capacitors, tantalum capacitors should be preheated gradually before wave or reflow soldering.

Tantalum capacitors are available with or without marked capacitance values in bulk, in waffle packs, and on tape and reel. Correct polarity orientation is critical and for this reason tape and reel packaging is preferred for handling. They are packaged in 8 mm and 12 mm tape widths.

3.2.5   Tubular Passive Components

The cylindrical devices known as metal electrode leadless faces (MELFs) are used for resistors, jumpers (zero ohm resistors), ceramic and tantalum capacitors, and diodes. They are cylindrical and have metal end caps for soldering.

MELF resistors are thick or thin film deposits on a tubular ceramic that is then spiraled just like a leaded ceramic resistor. The body construction for a tubular capacitor is a hollow tube with electrodes on the inside and outside of the tube. The capacitance ranges are limited. MELF diodes have a bumped die inside. The construction detail of a MELF diode device is shown in Figure 3.8.

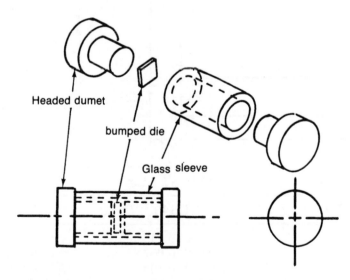

Figure 3.8   Construction of metal electrode leadless face (MELF) diode [1].

Since MELFs are cylindrical, the resistors do not have to be placed with resistive elements away from the board surface as is the case with the rectangular resistors discussed earlier. In addition, MELFs are less expensive than the rectangular devices. Like the conventional axial devices, MELFs are color coded for values.

MELFs do have their problems—they tend roll away during handling after placement in uncured adhesive or paste or even during reflow. To prevent movement during reflow, some people use a notch in the land pattern (see Chapter 6), although notches have their own problems (lack of end fillets if MELF is misplaced along the component length). The difficulty in placement may also require a special pipette in the placement machine. When considering the added care required in using MELFs, the small savings in the initial component cost may not be considered worthwhile. The devices are generally packaged on tape and reel. MELF diodes are identified as MLL 41 and MLL 34. MELF resistors are identified as 0805, 1206, 1406, and 2309. Their dimensions are shown in Figure 3.9.

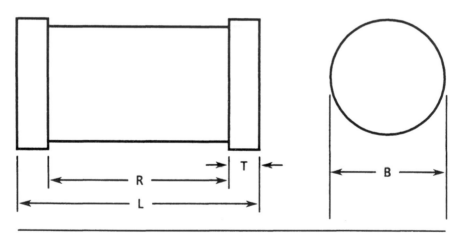

	CASE STYLE	
DIM (IN)	MLL34 (50D-80)	MLL 41
L	0.130–0.146	0.189–0.205
B	0.059–0.067	0.096–0.100
R	0.098–0.102	0.146–0.181
T	0.011–0.022	0.014–0.020

Figure 3.9   Body dimensions (inches) of MELF components [1].

3.3 ACTIVE COMPONENTS: CERAMIC PACKAGES

Surface mounting offers considerably more types of active and passive packages than are available in through-hole mount technology. For example, in the DIP, there are only three major body widths: 300, 400, and 600 mils, in 100 mil center pitch. The size of the package and the lead configuration are the same for both ceramic and plastic bodies. The world of surface mounting is considerably more complex in comparison.

Figure 3.10 sets forth the hierarchy of chip carriers available for surface mounting. There are two main categories of chip carriers; ceramic and plastic. The plastic chip carriers are primarily used in commercial applications. Just as in dual-in-line packages, the surface mount chip carriers are available in many constructions and materials. The ceramic packages provide hermeticity and are used primarily in military applications.

The most commonly used type of chip carrier in military applications is of leadless construction. Its main disadvantage, and it is a very significant one, is that any mismatch between the coefficients of thermal expansion of the package and the substrate will cause solder joint cracking because

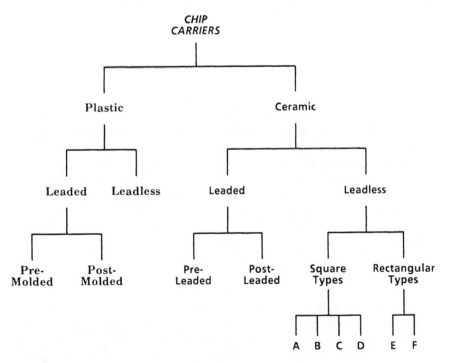

Figure 3.10 Hierarchy of chip carriers.

there are no leads to take up the stresses. This requires the use of expensive substrates with matching CTEs for leadless chip carriers. This issue is discussed in detail in Chapter 4.

Then the industry came up with plastic leaded chip carriers (PLCCs) with compliant J leads to solve the problem of solder joint cracking for nonhermetic commercial applications. The small outline integrated circuits (SOIC) packages also are used. They are more space efficient for pin counts of 20 and under. The Japanese, however, have not embraced the J-lead configuration for either lower or higher pin count packages.

Now the U.S. industry has turned around and is offering the high pin count, fine pitch packages in gull wing lead configurations, as the Japanese have done. However, as we will see later, the U.S. fine pitch packages have bumpers at the corners to protect the leads from damage. This design is superior to that of the Japanese fine pitch packages because, for example, there are fewer problems in manufacturing. See Section 3.4.5.

There are also discrete active devices known as small outline transistors (SOTs), which have three gull wing leads and come in various configurations. There are also other discrete diode devices, generally with four leads, both gull wing and the J type. In the sections that follow we discuss the various categories of active surface mount component packages in some detail.

3.3.1   Leadless Ceramic Chip Carriers

Ceramic chip carriers usually are constructed from a 90% to 96% alumina or beryllia base. Like DIPs, these carriers are constructed in single or multiple layers and can be used in the operating temperature range of −55° to 125°C. They are hermetically sealed and provide good environmental protection. The ceramic chip carriers come in leadless and leaded configuration. In this section we discuss leadless chip carriers. The leaded ceramic chip carriers are covered in the next section.

As the name indicates, leadless chip carriers have no leads. Instead they have gold-plated, groove-shaped terminations known as castellations. The castellations provide shorter signal paths, which allow higher operating frequencies because inductive and capacitive losses are lower. Reduced signal lead resistance and improved power dissipation also are provided by leadless chip carriers.

The list of hermetic leadless ceramic chip carrier types alone is very long. For example, in the leadless ceramic package, there are two predominant pitches: 50 mil centers and 40 mil centers. This does not include the 20, 25, and 33 mil center packages for fine pitch. The variety does not end here. The 50 and 40 mil center packages break down further

Table 3.2   The designation of ceramic packages and their pin counts
(*Source:* **JEDEC Standard, Publication 95). Also refer to Appendix A.**

DESIGNATION	CHIP CARRIER	PIN COUNT
MS-002	Leadless, Type A	28, 44, 52, 68, 84, 100, 124, 156
MS-003	Leadless, Type B	28, 44, 52, 68, 84, 100, 124, 156
MS-004	Leadless, Type C	16, 20, 24, 28, 44, 52, 68, 84, 100, 124, 156
MS-005	Leadless, Type D	28, 44, 52, 68, 84, 100, 124, 156
MS-006	Leaded, Type A	24 only
MS-007	Leaded, Type A	28, 44, 52, 68, 84, 100, 124, 156
None	Leadless, Type E	28, 32
None	Leadless, Type F	18, 20
MS-009	Leaded, Type B	16, 20, 24, 32, 40, 48, 64, 84, 96
MS-014	Leadless, 40 mil pitch	16, 20, 24, 32, 40, 48, 64, 84, 96

into Types A, B, C, D, E, and F. The ceramic packages of various types have unique designations (MS002, MS003, etc.) given by JEDEC. Table 3.4 shows the designation of packages and their pin counts for each package type.

The leadless ceramic chip carriers can be divided into different families depending on the pitch of the package. The most common is the 50 mil (1.27 mm) family. The castellations of these packages are at 50 mil centers. There are also 40, 25, and 20 mil families of leadless chip carriers. The 20 and 25 mil center families are discussed under Fine Pitch Packages in this chapter (Section 3.4.5). The 50 mil center family is most widely used, but we should keep in mind that the 40 mil family is also a part of the JEDEC standards. These packages are used by the military for high density applications. The 40 mil center family came into usage before the 50 mil center family.

The leadless ceramic chip carriers in the 50 mil center family come in square and rectangular configurations. Their dimensions are shown in Figure 3.11, which also shows the dimensions of plastic leaded chip carriers. It is worth noting that even though there are many types of plastic

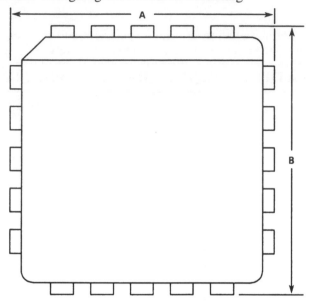

PACKAGE DIMENSION (INCH)				
	LCCC PACKAGE		PLCC PACKAGE	
PIN COUNTS LCCC/PLCC	WIDTH (A) MAX	LENGTH (B) MAX	WIDTH (A) MAX	LENGTH (B) MAX
Rectangular Package				
18-Short	.300	.440	.327	.467
18-Long			.327	.527
22			.327	.527
28	.360	.560	.395	.595
32	.460	.560	.495	.595
Square Package:				
16	.310	.310		
20	.360	.360	.395	.395
24	.410	.410		
28	.460	.460	.495	.495
44	.660	.660	.695	.695
52	.760	.760	.795	.795
68	.960	.960	.995	.995
84	1.160	1.160	1.195	1.195
100	1.360	1.360	1.335	1.335
124	1.660	1.660	1.695	1.695
156	2.060	2.060		

Figure 3.11 Maximum package sizes for leadless ceramic and plastic leaded chip carriers: LCCCs and PLCCs. (Courtesy of Intel Corporation.)

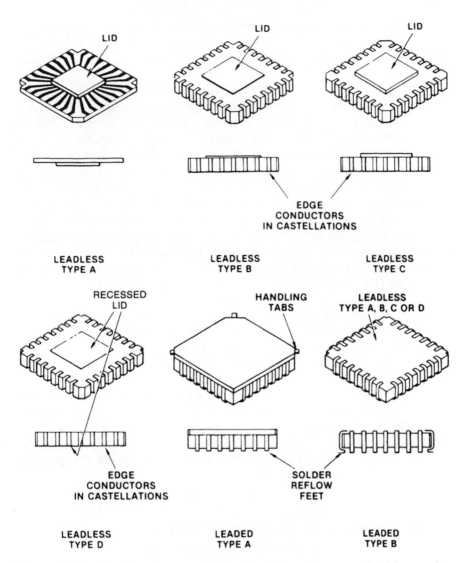

Figure 3.12 **Seating planes and lid orientations for ceramic chip carriers [1].**

and ceramic chip carriers, they have common land patterns, as discussed in Chapter 6.

Why are there so many ceramic package types? Each type is designed for a different application. Some are designed for heat dissipation in air and others for heat dissipation through the substrate. Some can be directly mounted on the board and others are designed for socketing. The packages that can be mounted directly on the board can have their lid either up or down depending on the heat dissipation requirements (dissipation through the air or the substrate).

"Lid up" implies a cavity-up position and "lid down" implies a cavity-down orientation. These terms refer to the backside of the die, which serves as the package's primary heat transfer or heat removal surface. When the lid is up, the backside of the die faces the substrate and the heat is dissipated through the substrate. This kind of package is not suitable for air-cooled systems or for the attachment of heat sinks. It does allow for a larger cavity size, and thus a larger die. Type B and C packages fall into this category. Figure 3.12 shows the lids of Type B and C packages on the opposite side of the seating or mounting plane.

The orientation in which the cavity or lid is down is suitable for air-cooled systems, since the backside of the die faces away from the substrate. Also, heat sinks can be attached to the backside of the die in cavity-down packages. However, the cavity size is reduced, hence the die size. Types A and D fall into this category (Figure 3.12 shows the lids of Type A and D packages on the same side as the seating plane). For socketing, Type A (lid-down position) and Type B (lid-up position) are used; for direct attachment, Type C (lid-up position) and Type D (lid-down position) are used. Thus these four packages meet the requirements of both socketing and direct attachment.

The leadless Type E and F packages are rectangular and are intended for memory devices. They are not high power devices, hence thermal properties are not important. They are generally used for direct attachment in the lid-up position, and the heat is dissipated through the substrate. Refer again to Figure 3.10 for the hierarchy of chip carrier packages by type, A through F.

3.3.2 Ceramic Leaded Chip Carriers (Preleaded and Postleaded)

As discussed earlier, the leadless ceramic chip carriers may cause solder joint cracking if there is mismatch between the coefficients of thermal expansion of the package and the substrate. This problem may be prevented by attaching compliant leads to the leadless packages. (In Chapter 4, Section 4.3 discusses the impact of package type on substrate selection.)

The leaded ceramic carriers are available in both preleaded and postleaded formats. The preleaded ceramic chip carriers have copper alloy or Kovar leads that are attached by the manufacturer. The leads, which are either top brazed or attached to the castellations, can be formed to different configurations, such as J, L, S, or gull wing shape. The preleaded chip carriers are supplied with the leads straight and attached to a common strip. The user cuts the common strip and forms the lead in the desired configuration. This minimizes the bending of leads during shipping and handling.

In postleaded chip carriers, the user attaches the leads to the castellations of the leadless ceramic chip carriers. The leads are generally edge clips supplied on common carrier strips. The leads may be supplied unbent, or preformed. If unbent, the user can bend the leads in the desired configuration, as in the case of preleaded chip carriers. The edge clips are attached to the castellations and then reflowed to accomplish the bond.

When the leadless types discussed earlier (A, B, C, and D) are mounted with leads for direct soldering to the substrate, they are called leaded Type B. The leaded Type B packages cannot be socketed because of the nature of their lead configuration. They are intended for soldering directly to the board. The leaded Type A breaks down further into three different categories: leaded ceramic, premolded plastic, and postmolded plastic. All Type A leaded packages can either be socketed or soldered directly to the substrate.

When leaded ceramic packages are used, their dimensions are generally the same as in plastic leaded chip carriers. Examples of this include JEDEC MS-044 leaded ceramic packages with CERDIP/CERQUAD construction. It should be kept in mind that leaded ceramic chip carriers are generally not used. The most commonly used packages in the industry are leadless ceramic and leaded plastic packages. The reason is very simple. Very few users are set up for the tedious process of attaching leads, which is not suitable for volume production in any event.

The industry is getting ready to supply J or gull wing packages in the same way that plastic packages (PLCC and SOIC) are supplied. However, there is no clear direction in lead configurations for ceramic packages. This may change, however, because the main reason for using leaded ceramic packages is to avoid the expensive and exotic substrates that are necessary for the leadless ceramic chip carriers.

3.4   ACTIVE COMPONENTS: PLASTIC PACKAGES

As discussed earlier, ceramic packages are expensive and are used primarily for military applications. Plastic packages, on the other hand, are the most widely used packages for nonmilitary applications, where hermeticity is not required. The ceramic packages have solder joint cracking problems due to CTE mismatch between the package and the substrate, but the plastic packages are also by no means trouble free.

Some surface mount plastic packages may be moisture sensitive. For example, if they contain moisture above a threshold limit and are subjected to reflow soldering temperatures, the package may crack when the moisture expands. The susceptibility to cracking is dependent on the thickness of the plastic, the moisture content, and the die size. See Chapter 5 (Section 5.7.1) for a discussion of the mechanism of cracking and recommended solutions.

Other constraints for plastic packages include limited temperature range (0–70°C) and nominal environmental protection. The commercial sector of the electronics industry needs the plastic packages for cost effectiveness and reliable interconnection on glass epoxy substrates. The most common plastic packages used in commercial electronics are the discrete transistors known as small outline transistors (SOTs), small outline integrated circuits (SOICs) with gull wing leads, small outline devices with J leads (SOJ), plastic leaded chip carriers (PLCCs) with J leads, and fine pitch devices with gull wing leads known by various names such as mini-packs or plastic quad flat packs (PQFPs).

As pin counts have increased, the pitches of fine pitch devices have been decreasing to accommodate the higher and higher numbers of pins. However, packages with reduced pitches have very fragile leads that introduce all kinds of manufacturing problems. So a relatively new arrival in the packaging field is the ball grid array (BGA). BGAs do not have leads but robust balls. They solve many of the problems created by fine pitch but they introduce some of their own. See Chapter 5 (Section 5.2, Package Drivers) for trade-offs among various types of packages. We will discuss various types of BGA packages in this chapter (Section 3.5). See Chapters 6 and 7 for land pattern and design for manufacturability issues. The manufacturing issues related to BGAs are discussed in the appropriate chapters of Part 3 of this book.

SOTs have unique lead spacing and construction, which we will describe later. The SOICs, SOJs, and PLCCs are 50 mil center pitch packages. In terms of real estate considerations, the most efficient packages are SOICs for pin counts under 20 and PLCCs for pin counts between 20 and 84. For pin counts above 84, the fine pitch packages with lead pitches of 20, 25, and 33 mils are most common. For even higher pin counts, tape-automated bonding (TAB) devices are necessary. Figure 3.13 plots real estate efficiencies of packages of various types for surface mounting. Figure 3.14 shows the real estate requirements for various types of packages including BGA.

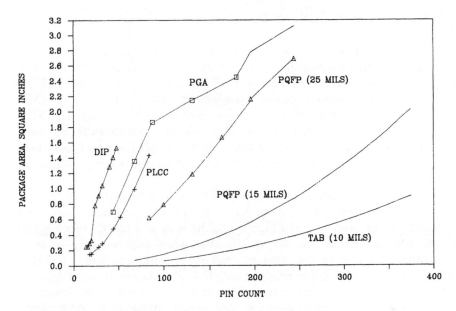

Figure 3.13 Real estate efficiencies of packages of various types for surface mounting.

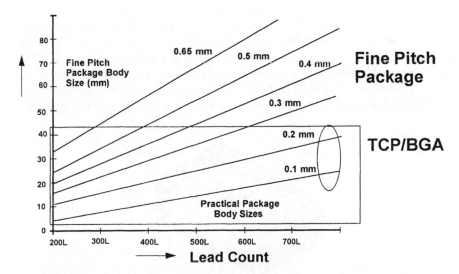

Figure 3.14 Real estate requirements for various types of high pin count packages.

3.4.1   Small Outline Transistors

The small outline transistors are one of the forerunners of active devices in surface mounting. They are three- and four-lead devices. The three-lead SOTs are identified as SOT 23 (EIA TO 236) and SOT 89 (EIA TO 243). The four-lead device is known as SOT 143 (EIA TO 253). These packages are generally used for diodes and transistors. The SOT 23 and SOT 89 packages have become almost universal for surface mounting small transistors. Even as the use of high pin count complex integrated circuits is becoming widespread, the demand for various types of SOTs and SODs continues to grow.

SOT 23 is the most commonly used three-lead package, with internal construction as shown in Figure 3.15. This package can accommodate a maximum die size of 0.030 inch by 0.030 inch and can dissipate up to 200 mW in free air and up to 350 mW when attached to a ceramic substrate. It comes in low, medium, and high profiles as shown in Figure 3.16, to meet the differing needs of both hybrid and printed board applications. The higher profile package is more suitable for printed board application because it provides better cleaning. The dimensions of the package are given in Figure 3.17. SOT 23 (TO 236) devices are packaged in 8 mm tape width with 4 mm pitch.

The SOT 89 (TO 243) is used for higher power devices. To be able to transfer heat to the substrate more efficiently, this package sits flush on the substrate surface. The outline dimensions of SOT 89 are shown in Figure 3.18. The SOT 89 package can accommodate a maximum die size of 0.060 inch by 0.060 inch. At 25°C it can dissipate 500 mW in free air and 1 W when mounted on an alumina substrate. SOT 89 (TO 243) devices are packaged in 12 mm tape width with 8 mm pitch.

SOD 123 (small outline diode), also referred to as DO 214, is shown in Figure 3.19. It is a two-leaded diode device with a body size approxi-

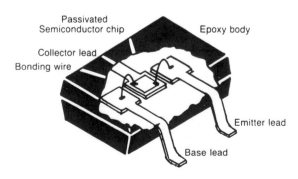

Figure 3.15   Internal construction of SOT 23 [1].

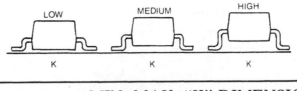

PROFILE	MIN.-MAX. "K" DIMENSIONS (INCHES)
High Profile	.004–.010
Medium Profile	.003–.005
Low Profile	.0004–.004

Figure 3.16   Clearance (in inches) from top of board to bottom of SOT 23 in low, medium, and high profiles [1].

mately the same as that of SOT 23. Power ratings are the same as the 0.5 W MELFs. In many applications, MELFs are being replaced by SOD 123 because it is easier to assemble than the MELFs. It comes in two different configurations—gull wing and lead bent under the package as in the molded tantalum capacitor, as shown in Figure 3.20. The second version is referred to as SMB to distinguish it from the gull wing version. SMBs are higher power plastic diode packages used to house the rectifier and zener die. SOD 123 and SMBs are packaged in 12 mm tape width with 8 mm pitch.

The SOT 143 (EIA TO 253) packages come in the four-lead configuration (Figure 3.21) and are commonly used for RF transistor applications. They are used for dual diodes and Darlington transistors. They are like SOIC packages but with lower (0.002 to 0.005 inch) clearance off the board. The SOT 143 package can accommodate about the same die size (0.025 inch by 0.025 inch) as the SOT 23 package. SOT 143 devices are packaged in 8 mm tape width with 4 mm pitch.

SOT 223 (EIA TO 261), like SOT 143, is used for dual diodes and Darlington transistors. The dimension of the package is shown in Figure 3.22. The devices are packaged in 12 mm tape width with 8 mm pitch.

TO 252 (DPAK) are also used for dual diodes and Darlington transistors. DPAK's power rating is considerably higher than that of most SOTs. It can be as high as 1.75 watts. The dimensions of the package are shown in Figure 3.23. They are packaged in 12 mm tape with 8 mm pitch.

The SOT 23, SOT 89, SOD 123, SOT 143, SOT 223, and TO 252 (DPAK) packages are most commonly supplied in tape and reel per EIA standard RS-481. The most popular are conductive tapes with embossed cavities for the packages. These packages are the only active devices that

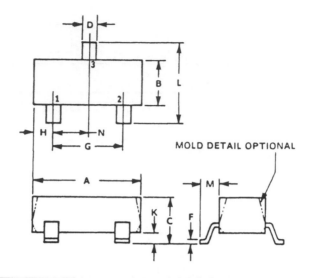

DIMENSION	MIN.-MAX. DIMENSIONS (INCHES)
A	[0.110–0.120]
B	[0.048–0.055]
C	[0.033–0.048]
D	[0.015–0.018]
F	[0.003–0.006]
G	[0.070–0.081]
H	[0.020–0.024]
K	(See Figure 3.15)
L	[0.083–0.098]
M	[0.018–0.024]
N	[0.035–0.040]

Figure 3.17   The dimensions of SOT 23 [1].

are soldered using both wave and reflow (vapor phase or infrared) methods. The other active devices, such as SOICs and PLCCs, are mostly soldered using only reflow soldering methods.

3.4.2   Small Outline Integrated Circuits (SOICs and SOPs)

The small outline integrated circuit (SOIC, or SO for brevity) is basically a shrink DIP package with leads on 0.050 inch centers. It is

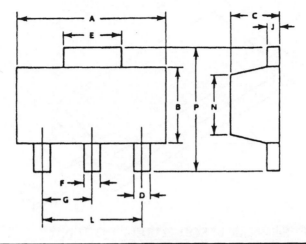

DIMENSION	MIN.-MAX. DIMENSIONS (INCHES)
A	[0.174–0.182]
B	[0.090–0.102]
C	[0.056–0.062]
D	[0.015–0.018]
E	[0.064–0.070]
F	[0.018–0.020]
G	[0.059]
J	[0.014–0.017]
L	[0.118]
N	[0.081–0.089]
P	[0.156–0.167]

Figure 3.18   The dimensions of SOT 89 [1].

used to house larger integrated circuits than is possible in SOT packages. In some cases, SOICs are used to house multiple SOTs.

SOIC contains leads on two sides that are formed outward in what is generally called a gull wing lead. The Swiss watch industry is generally given the credit for originating this package style in the 1960s. However, for the modern electronics industry, N. V. Philips originated this package style in 1971.

In the United States, Signetics, a division of N. V. Philips, is considered to be the pioneer of this package style for digital devices. National Semiconductor committed to this package style after the automotive and telecommunication industries provided support for it. The Japanese refer

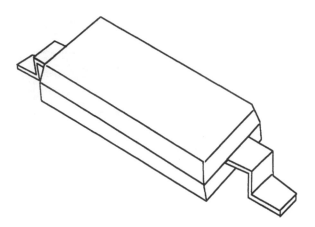

Figure 3.19 Schematic of SOD 123 (EIA DO 214).

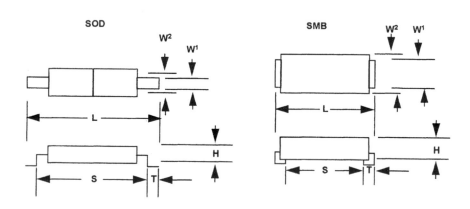

COMPONENT IDENTIFIER	L	S	W1	W2	T	H (max)
SOD 123	0.140–0.155	0.095–120	0.020–0.25	0.55–0.07	0.010–0.025	0.055
SMB	0.210–0.225	0.085–0.130	0.080–0.090	0.130–0.160	0.030–0.60	0.095

Figure 3.20 Dimensions of two different versions of SOD 123 [1].

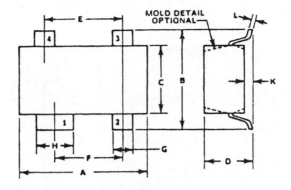

DIMENSION	MIN.-MAX. DIMENSIONS (INCHES)
A	0.110–0.122
B	0.083–0.102
C	0.048–0.067
D	0.033–0.048
E	0.071–0.079
F	0.063–0.071
G	0.015–0.019
H	0.031–0.035
K	0.002–0.010
L	0.004–0.006

Figure 3.21   The dimensions of SOT 143 [1].

to this package as a mini-flat. The mini-flats have dimensions slightly different from the JEDEC dimensions.

Compared with J-lead packages, the SOICs need to be handled more carefully to prevent lead damage. In the beginning, there was no lead coplanarity requirement for SOICs. Now the JEDEC standard on SOICs requires 0.004 inch lead coplanarity, as is the case in PLCCs. The SOICs provide better real estate savings than PLCCs for pin counts under 20, but the solder joints (heel fillet) are difficult to inspect. SOICs are available in 8, 14, 16, 20, 24, and 28 lead counts.

The SOIC packages come in mainly two different body widths: 150 and 300 mils. The body width of packages having fewer than 16 leads is 150 mils; for more than 16 leads, 300 mil widths are used. The 16 lead package comes in both body widths. Figure 3.24 gives SOIC package dimensions.

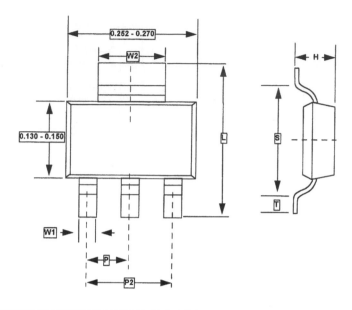

COMPONENT IDENTIFIER	L	S	W1	W2	T	H (max)	P1	P2
SOT 223	0.270–0.290	0.165–0.195	0.025–0.035	0.115–0.130	0.035–0.050	0.070	0.090	0.180

Figure 3.22 The dimensions of SOT 223 [1].

The 150 mil wide body packages are referred to as narrow body, and the 300 mil body packages are called wide body. The 300 mil body width packages are referred to with suffix "L", such as 16 SOL or 18 SOL. The narrow body packages are also referred to as JEDEC MS-012 AA-AC, and the wide body packages are referred to as JEDEC MS-013. There are also SOICs with a body width of 330 mils. They are generally used for static RAMs, which are mostly Japanese packages.

EIAJ has SOPs (small outline packages) at 50 mil pitch that are similar to SOIC with some minor variations. For example, they are slightly wider in body width than the narrow body SOICs but much narrower than the wide body SOICs. SOPs are classified based on their package height as shown in Table 3.5.

At one time there was great concern about the reliability of SOIC packages. The concern had more to do with theoretical than with empirical data. Theoretically speaking, the amount of plastic around the die in an

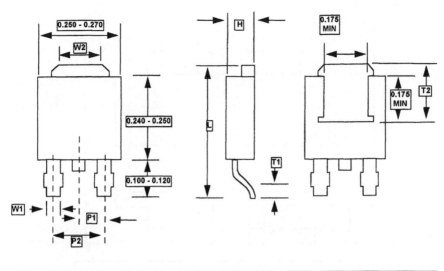

COMPONENT IDENTIFIER	L	W1	W2	T1	T2	P1	P2	H (max)
TO 252 (DPAK)	0.375–0.415	0.030–0.045	0.210–0.220	0.020–0.030	0.190–0.210	0.090	0.180	0.095

Figure 3.23 The dimensions of TO 252 (DPAK) [1].

SOIC package is a fraction of that in a standard DIP. Because of this, thermal problems were expected to be compounded in an SOIC package. In fact, reliability problems have not been encountered because the manufacturers have done such things as using copper lead frame material instead of Kovar or alloy 42 and using improved plastic materials in the package.

The wave soldering process is not very effective on PLCCs, but many companies do use wave soldering for SOICs and SOPs. One should be careful in wave soldering any plastic package (SOIC, SOP, SOT, or PLCC) because package reliability may be compromised at the relatively higher temperatures in the wave. Also, the flux used in wave soldering may seep inside the package. This is possibly due to the different coefficients of thermal expansion between the plastic body and the lead frame. Any such seepage may have an adverse impact on the reliability of the device. If the SOIC or SOP happens to be moisture sensitive, the reliability problem may be even more serious. In Chapter 5, Section 5.7.1 gives details on the moisture sensitivity of plastic packages and the mechanism of cracking during reflow soldering.

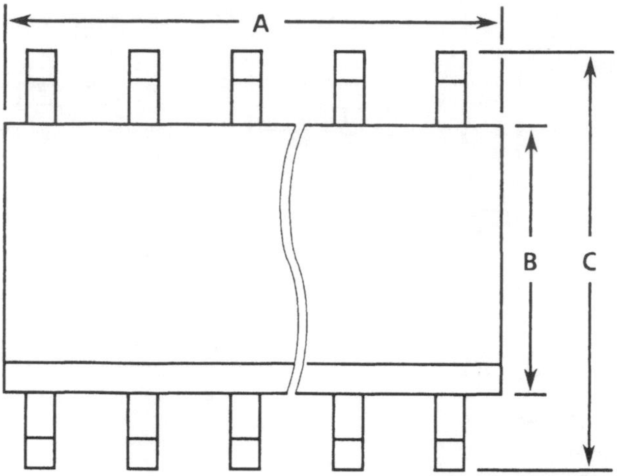

| PIN COUNTS | DIMENSIONS (INCHES) | | |
	A	B	C
	MIN	MAX	MAX	MAX
14*	.336	.344	.150	.250
16*	.385	.394	.150	.250
16L	.397	.413	.300	.420
20L	.496	.512	.300	.420
24L	.598	.614	.300	.420
28L	.697	.713	.300	.420

*Narrow body (150 mils) packages. Others are called wide body (300 mils) packages.

Figure 3.24   The dimensions of SOIC packages [1].

Table 3.5   Classification SOP packages

TYPE	SOP LEAD RANGE	PACKAGE HEIGHT
I	6–14	1.5 mm
II	16–20	2.0 mm
III	22–24	2.5 mm
IV	28–30	3.0 mm
V	32–36	3.5 mm
VI	40–42	4.0 mm

3.4.3   Plastic Leaded Chip Carriers

The plastic leaded chip carrier (PLCC) is a cheaper version of the ceramic chip carrier. The PLCCs are almost a mandatory replacement for plastic DIPs, which are not practical above 40 pins because of excessive real estate requirements. The PLCCs come in a lead pitch of 0.050 inch with J leads that are bent under the packages.

The packages with higher pin count and generally with the gull wing leads of the SOIC are defined as fine pitch quad packages and are discussed later. The PLCC was introduced in 1976 as a premolded plastic package, but it did not become very popular. The postmolded PLCC, pioneered by Texas Instruments, was adopted by the major semiconductor companies such as Intel, Motorola, and National Semiconductor, and the major Japanese manufacturers (for products to be sold in the United States).

This package was originally referred to as postmolded Type A leaded, but JEDEC abandoned this nomenclature in favor of PLCC. A JEDEC task force was organized in 1981 for the registration of PLCCs. The JEDEC outline known as MO-047 for square packages having 20, 28, 44, 52, 68, 84, 100, and 124 terminals was completed in 1984. These packages have an equal number of J leads on all four sides. See Figure 3.25 for dimensions. PLCCs have 84 pins maximum. When the device requires more than 84 pins, fine pitch packages have to be used.

A second registration, known as JEDEC outline MO-052, was completed in 1985 for rectangular packages with pin counts of 18, 22, 28, and 32. The 18 pin PLCCs have two sizes: a shorter one for 64K dynamic random access memory (DRAM) and the longer one for 256K DRAMs. The 28 and 32 pin packages are used for electrically programmable read only memories (EPROMs). (See Figure 3.26 for dimensions.) Since 64K and 256K DRAMs are obsolete, 18 pin PLCCs are rarely used.

The pin numbering system for PLCC is different and should be noted to prevent problems of misorientation in manufacturing. The convention for the location of pin 1 in SOIC is the same as in DIP (starting at the top left-hand corner and progressing counterclockwise). For PLCCs, however, pin 1 is the top center pin when there is an odd number of pins on each side and one to the left of center when there is an even number of pins on each side. In larger PLCCs, pin 1 can also be at the top left-hand corner of the package; this is the corner with the beveled edge. The location of pin 1 is confirmed by a dot in the package, and the progression of the pin count continues counterclockwise.

The leads in PLCCs provide the compliance needed to take up the solder joint stresses and thus prevent solder joint cracking. Since the design of the J lead is intended to provide lead compliance, it is important to ensure that the ends and sides of the leads are not touching the plastic

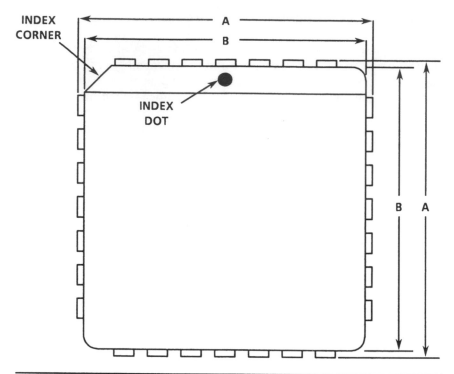

| NUMBER OF | "A" | | "B" | |
TERMINALS	MIN	MAX	MIN	MAX
20	.385	.395	.350	.356
28	.485	.495	.450	.456
44	.685	.695	.650	.656
52	.785	.795	.750	.756
68	.985	.995	.950	.958
84	1.185	1.195	1.150	1.158

Figure 3.25   The dimensions (inches) of square PLCC packages. (Courtesy Intel Corporation.)

body. In PLCCs there is some concern that bent leads touching the plastic body will be constricted in their movement, and hence may be noncompliant. This may happen if the leads are bent during either shipping and handling or planarization.

Planarization, the process of forcing the finished package with formed

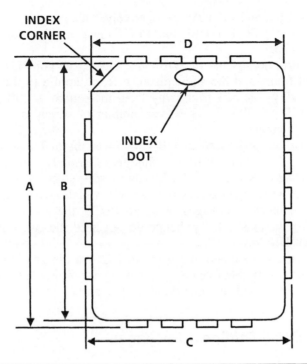

NUMBER OF	"A"		"B"		"C"		"D"	
TERMINALS	MIN	MAX	MIN	MAX	MIN	MAX	MIN	MAX
18[1]	.317	.327	.457	.467	.282	.293	.422	.428
18[2]	.317	.327	.517	.527	.282	.293	.487	.493
28	.385	.395	.585	.595	.347	.353	.547	.553
32	.485	.495	.585	.595	.447	.453	.547	.553

1. Predominately used for 64K DRAM
2. Predominately used for 256K DRAM

Figure 3.26   The dimensions (inches) of rectangular PLCC packages. (Intel Corporation.)

leads through a die, is conducted by some suppliers to control lead coplanarity. Planarization is not a recommended method for controlling lead coplanarity, however. This subject is discussed later.

There is also some concern about the thermal resistance of PLCCs because the surface area for heat dissipation is smaller. This is a more valid concern in SOIC than in PLCC. For example, the junction-to-ambient thermal resistance (Θja) for a 20 pin DIP, a 20 pin SOIC, and a 20 pin

PLCC are 68, 97, and 72° C/W, respectively. This means that SOIC will run hotter than DIP, but DIP and PLCC have a minor difference in thermal resistance.

It is worth noting that in SOIC and PLCC, the combination of the copper lead frame and the short distance from the die to the PC board interface readily conducts heat away from the source. In DIPs, the lead frame is made of alloy 42 or Kovar, neither of which is as thermally conductive as copper.

Why isn't copper used in DIPs to lower thermal resistance? The reason is simple. Although more conductive thermally, copper is not as stiff as alloy 42 or Kovar. The DIP leads must be stiffer to withstand the insertion and withdrawal forces of a socket, which are not encountered by surface mountable packages. Also, in PLCC, the leads are bent and tucked under the plastic body, hence do not need the stiffness required by a straight DIP lead.

It should also be noted that some PLCCs with large die-to-package ratios may be susceptible to package cracking due to moisture absorption. This issue is discussed in some detail in Chapter 5, Section 5.7.1.

3.4.4   Small Outline J Packages

The small outline J (SOJ) package was originally designed to house memory devices efficiently in order to connect data and address lines of multiple devices placed closed to each other. With high capacity memory systems, the real estate savings can be significant since memory is generally used in multiples.

The SOJ packages have J-bend leads like PLCCs, but they have pins on only two sides, as shown in Figure 3.27. This package is a hybrid of SOIC and PLCC and combines the handling benefits of PLCC packages with the routing space efficiency of SOIC packages. The leads of SOJ packages, like those of PLCC, SOIC, and fine pitch, must be coplanar within 0.004 inch. SOJs, like SOICs and PLCCs, come in tube or tape and reel. Tape is preferred for higher volume applications. The lead pitch generally is 50 mils, but some SOJs have 33 mil (0.8 mm) pitch. The SOJ packages are available in 300, 350, 400, and 500 mil body widths. 300 and 400 mil body widths are the most common. The pin counts range from 14 to 42 pins. The body width, pin count, and pitch for SOJs vary, so it is important to check the supplier's literature.

SOJs are commonly used for high density (1, 4, and 16 MB) DRAMs. The low density DRAMs (64 and 256K) come in PLCCs, as discussed earlier, but have been replaced by higher density DRAM packages. When SOJs are used for memory, the leads on each side are split between two

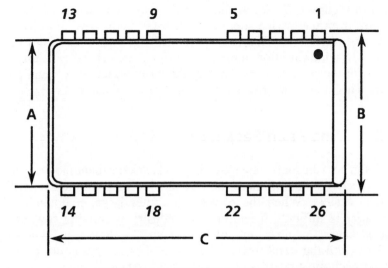

	IMB	4MB
A	0.300 (Nominal)	0.350 (Nominal)
B	0.335–0.345	0.385–0.395
C	0.675–0.680	0.675–0.680

Figure 3.27   The SOJ package and its dimensions (inches).

groups separated by a large center gap. The gap provides a space for traces to pass under the package. However, there are some devices other than DRAMs, such as static DRAMs, that are also housed in SOJs and do not have center gaps.

The 1 MB SOJ (DRAM) package is also referred to as a 26 position, 20 pin package. This means that it could have a total of 26 pins but has only 20 pins. The body width is 300 mils. As shown in Figure 3.27, the pin count goes from 1 to 5; pins 6, 7, and 8 are skipped; then the pin count resumes at 9. On the right-hand side, pins 19, 20, and 21 are skipped, and the pin count resumes at 22 and ends at 26. The body width of the 1 MB DRAM is 300 mils, and it is referred to as EIA package MO-060. The SOJ package for 4 MB DRAMs is available in a 26 position, 20 pin package. However, the body width is 350 mils. The dimensions of 300 and 350 mil body SOJs are shown in Figure 3.27.

It should be noted that 4MB DRAMs also come in 400 mil body widths in 28 position, 24 pin packages. 16 MB DRAMs come in 400 and 500 mil body widths. One of the 500 mil body size packages also comes in 34 pin, 33 mil pitch. All this can be very confusing and hence is not shown in Figure 3.27. So, as mentioned earlier, one needs to check the literature from the device supplier for correct body size, pin count, and layout and pitch data. This is very critical for appropriate land pattern design. The wide variation in package design is one of the most important reasons for thoroughly understanding the basic principles of land pattern design. This subject is covered in great detail in Chapter 6.

3.4.5 Fine Pitch Packages

From both the real estate efficiency and package manufacturing standpoints, packages having more than 84 pins become impractical with 50 mil lead centers. When the package becomes large, maintaining lead coplanarity is difficult. Therefore the industry is moving toward finer pitch packages for pin counts of 84 and up. As shown in Figures 3.13 and 3.14, with the trend toward the usage of larger pin count packages, the lead pitch must decrease to ensure real estate efficiency. Also, when the lead counts increase as they have with the increased usage of application specific integrated circuit (ASIC) devices, finer pitch devices have become very common.

The current standard adopted by all the VLSI manufacturers in the United States is the 25 mil center, gull wing package known as the minipack (Figure 3.28). Even though the J-lead PLCC has been accepted as the industry standard, fine pitch packages have to be of the gull wing type. As lead thickness and width decrease, J-lead packages are harder to manufacture.

The 25 mil pitch (0.635 mm) packages are registered under EIA MO 986 designation. They cover pin counts of 84, 100, 132, 164, 196, and 244. The body sizes of these packages range from 20 mm to 42 mm square.

The most exotic feature of the U.S. version of the fine pitch package is its corner bumpers, which extend about 2 mils beyond the leads to protect them during handling, testing, and shipping. For this reason this package type is usually called a "bumper" pack. Board real estate is not wasted by the bumper because the lands extend at least 10 mils beyond the leads to form solder fillets. These fine pitch packages are available in tube, tray, or tape and reel.

The Electronics Industries Association of Japan (EIAJ) has fine pitch packages known as quad flat packs (QFPs) and SQFP (shrink quad flat pack) with various body sizes, thicknesses, lead pitches, and pin counts. The SQFP is used for applications requiring low height and high density. It is registered under EIAJ-ED-7401 and also under EIA-MO-108. It is agreed by EIA that any package with lead pitch at 0.5 mm and below will be standardized in hard metric.

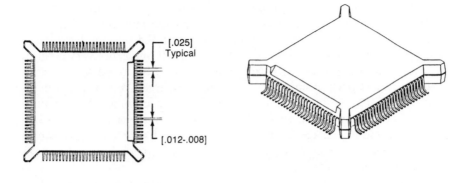

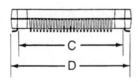

TERMINAL COUNT	C	D
84	.710–.750	.775–.785
100	.810–.850	.875.–885
132	1.010–1.050	1.075–1.085
164	1.210–1.250	1.275–1.285
196	1.410–1.450	1.475–1.485
244	1.580–1.620	1.645–1.655

Figure 3.28   The U.S. version of a fine pitch package with corner bumpers (dimensions in inches). (Intel Corporation.) NOTE: D is maximum terminal dimension and not maximum body dimension. Only dimensions C and D are needed for land pattern design. See Chapter 6.

The EIAJ package styles are rectangular and square. The square SQFP family comes in 13 standard sizes, all of which can come in either a 0.5, 0.4, or 0.3 mm pitch. Therefore, there are 39 configurations of square SQFPs alone. The body size of square packages varies from 5 mm on a side with only 24 pins to 46 mm on a side with 576 pins.

The body sizes of the rectangular packages vary from 5 × 7 mm with 32 pins to 28 × 40 mm with 440 pins. And, like the square SQFP family, the rectangular family also comes in 13 standard sizes, each of which can come in either a 0.5, 0.4, or 0.3 mm pitch. Therefore, there are 39 configurations of rectangular SQFPs also.

Another noteworthy feature of the EIAJ package standard is that two different pin counts are allowed for each package and the package will still meet the standard (e.g., a 5×5 package with 0.3 mm pitch can have either 56 or 48 pins and still meet EIAJ-7404-1). If one is not very careful, incorrect land pattern designs can result. In addition, there is one other feature of EIAJ packages that should be remembered. The overall dimension of a given package can vary enough that a common land pattern may not be possible for a given pin count and pitch. For example, a 160 pin, 25 mil pitch package from two different EIAJ suppliers may require two different land pattern designs because the pin lengths and overall dimensions may be different.

The variations in package dimensions noted above make it very clear that land pattern designs must be based on the dimensions given by the package supplier and not on the dimensions from the EIAJ publication. This is one reason why EIAJ package dimensions are not provided in this book.

The EIAJ packages do not come with corner bumpers and hence require handling and shipping in waffle packs to prevent lead damage. Waffle packs can be a problem in manufacturing. For example, some placement machines require special waffle-pack handlers (to transport the part from the waffle pack to pick-up position). This adds to machine cost. Also, most placement machines have a limited waffle-pack handling capacity. By taking up too much feeder slot, they reduce the feeder slot capacity for tape and reel and tube feeders. The bumpered fine pitch packages do not pose such manufacturing constraints since they can be packaged in tubes and tape and reel, as well as waffle packs (if necessary).

Lead damage has been a concern in all gull wing type leads, but the problem is compounded in fine pitch packages because the leads are thin and quite fragile. They are very susceptible to lead damage and distortion of lead planarity during shipping, handling, and placement. The corner leads are especially prone to damage. Loss of lead planarity in a fine pitch package may not be of concern if the hot bar soldering process is used instead of reflow (vapor phase, IR, or convection) soldering.

There is an added problem in the EIAJ packages since they are very thin and have practically no stand-off. They are desirable for keeping a lower profile but they compound thermal and cleaning problems. The bumpered package, being much thicker (140 mils), enhances package thermal conductivity through the use of heat spreaders. As shown in Table 3.6, the bumpered package with a heat spreader coupled to its copper lead frame drops the junction-to-ambient thermal resistance value of a 100 lead package by approximately 60% relative to the QFP [2]. It is almost impossible to clean under the EIAJ fine pitch packages because they sit almost flush to the board. The stand-off clearance in these packages

Table 3.3 Comparison of thermal resistance of thinner (Japanese) and thicker (U.S.) fine pitch packages [2].

	Θ_{ja} (°C/W)		
LEAD COUNT	QFP (JAPAN)	FINE PITCH (U.S.)	FINE PITCH WITH HEAT SPREADER (U.S.)
64/68	100	60	—
80/84	85	47	35
100	80	44	33
132	—	40	30

is 0–10 mils. Such concerns are not present in JEDEC packages since they have 15–25 mils stand-off.

It is a challenge to accurately place fine pitch packages on the substrate. Use of an accurate placement machine fitted with vision capability is mandatory for fine pitch. The fine pitch packages have thinner leads and require a thinner land pattern design. Other challenges for users of fine pitch packages include getting consistent and good solder paste printing without causing bridging in fine pitch and insufficient solder on standard SMT components. Fine pitch packages are also more difficult to inspect and repair. These issues are addressed in Chapter 9 (printing), Chapter 11 (placement), Chapter 12 (soldering), Chapter 13 (cleaning), and Chapter 14 (inspection and repair).

3.5 BALL GRID ARRAYS (BGAs)

The driving forces for various types of packages, including ball grid arrays (BGAs), are discussed in Chapter 5, and BGA land pattern design is discussed in Chapter 6. The manufacturing issues related to BGAs are discussed in Chapter 9 (printing), Chapter 11 (placement), Chapter 12 (soldering), Chapter 13 (cleaning), and Chapter 14 (inspection and repair). In this chapter we will discuss the construction details of various types of BGAs and their advantages and disadvantages.

When using fine pitch, precision becomes more intense due to reduced pitch and the process windows tighten. The result is lower yield and higher cost if design is not precise and processes are not very tightly controlled. Fine pitch packages with their fragile leads pose serious handling problems. Also it is difficult to print, place solder, and rework fine pitch packages.

BGAs have become popular because of the many problems associated with manufacturing with fine pitch packages. They do not have the same limitations because, instead of fragile leads, they have robust balls at much higher pitches. Refer to Section 3.7.2.2 in this chapter for comparing the features of various types of lead configurations.

Historically, newer technologies such as fine pitch and TAB provide denser packaging at the expense of yield and quality. BGA is an exception. It provides higher density while improving yield. Component electrical performance improves due to reduced lead inductance because propagation delay is a function of lead length. And most important of all, BGAs are compatible with standard SMT processes (paste printing, placement, and reflow). This allows a quicker time for process stability to achieve high assembly yield than is possible when using fine pitch or ultra fine pitch packages.

BGA is an array package like PGA (pin grid array) but without the leads. There are various types of BGAs, but the main categories are ceramic and plastic. The ceramic BGAs are called CBGA and CCGA (ceramic column grid arrays), and the plastic BGAs are referred to as PBGA. These is another category of BGA known as tape BGA (TBGA). The ball pitches have been standardized at 1.0, 1.27, and 1.5 mm pitch. (40, 50, and 60 mil pitch). Table 3.7 provides a summary of ball sizes and compositions for different types of BGAs.

The body sizes of BGAs vary from 7 to 50 mm and their pin counts

Table 3.7   Features of different types of BGAs

FEATURE	CBGA	CCGA	PBGA	TBGA	μBGA*
Ball alloy (Pb/Sn)	90/10	90/10	63/37	90/10	Ni with Au coat
Solder on PCB (Sn/Pb)	63/37	63/37	63/37	63/37	63/37
Ball dia. (in.)	0.035	0.020 D 0.050– 0.087 H	0.030	0.025	0.0035
Pitch (in.)	0.050	0.050	0.040– 0.060	0.040– 0.060	0.012– 0.060

*Note: μBGA is one of the CSPs (chip scale packages) and is the trade mark of Tessera Corporation. See Section 3.6 and Figure 3.33.

vary from 16 to 2400. Most common BGA pin counts range between 200 and 500 pins. BGAs are registered by JEDEC under the following designations:

- TBGA - MO 149
- PBGA - MO 151
- CBGA - MO 156
- CCGA - MO 158

BGAs are very good for self-alignment during reflow even if they are misplaced by 50% (CCGA and TBGA do not self-align as well as PBGAs and CBGAs do). This is one reason for the higher yield with BGAs. Of course the fact that they make it possible to increase the lead pitch to the familiar 40 to 50 mils from the 25 mils and below common in gull wing packages certainly helps in improving yield. Also, the package thickness can be even thinner than for gull wing packages. This is one of the reasons why BGAs are popular for hand-held consumer applications such as pagers.

BGAs appear to be the answer to the high pin count packaging trend since they can accommodate high I/O in a very small form. But the savings in real estate comes at a price. As discussed in Chapter 5, Section 5.5.1, the layer count requirement in a board using such packages is increased, and an increase in layer count means an increase in board cost.

3.5.1 Ceramic Ball Grid Array (CBGA)

The ceramic BGA (CBGA) is also called SBC (solder ball connection) by IBM. Figure 3.29 shows an illustration of a CBGA. The internal connection in the package can be made either with conventional wire bonding or by flip chip. Figure 3.29 shows flip chip bonding inside the package. The package can be either cavity up or cavity down as in the LCCC discussed earlier. The solder balls are high temperature solder (90% lead and 10% tin) with a melting point of 302°C. The balls are attached to the package with eutectic solder (63% Sn, 37% Pb). The ball diameter is 35 mils. The body size of CBGA is from 18 to 32 mm.

The CBGAs are hermetic (do not absorb moisture) and hence are not subject to the popcorn effect as are PBGAs. Also, since the solder balls have a high melting point, they do not melt during rework and, unlike PBGAs, can be reballed for reuse. The balls are 0.035 inch in diameter and provide sufficient stand-off for reliability and cleaning.

The disadvantage of CBGA is that its high thermal mass makes reflow profile development difficult. Also, like the LCCCs discussed earlier, the

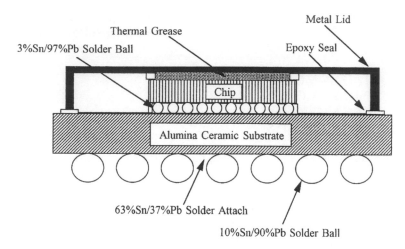

Figure 3.29 **Cross-section of a ceramic ball grid array (CBGA) package.**

CTE mismatch between the package and the board limits useful life. For example, Banks et al. [3] found that the corner joint (farthest distance from neutral point or DNP) fails first. The failure occurs between the ball and the board pad. The CCGAs, on the other hand, generally fail in the columns. We discuss CCGAs next.

3.5.2 Ceramic Column Grid Array (CCGA)

The ceramic column grid array, shown in Figure 3.30, is used for larger packages (32 to 45 mm). Like CBGAs and PBGAs, CCGAs also use multilayer ceramic packages. They are very much like PGAs but with lower pitch and more fragile leads (columns).

The CCGA column diameter is about 20 mils and the column height varies from 0.050 to 0.087 inch. These columns are attached to the package either by eutectic (63% Sn/37% Pb) solder or they are cast in place using 90% Pb and 10% Sn.

Taller columns increase solder joint reliability by taking up stresses created by the CTE mismatch between the package and the board. Banks et al. [3] found CCGAs to be 3 to 5 times more reliable than CBGAs. They found the first failure in CCGA after 2485 thermal cycles (0–100°C). Fifty percent of the samples failed in 2630 cycles. Compare this to 479 cycles for failures in CBGA. An increase of 100 milliohms in solder joint resistance was defined as a failure.

Longer columns reduce electrical performance and increase the over-

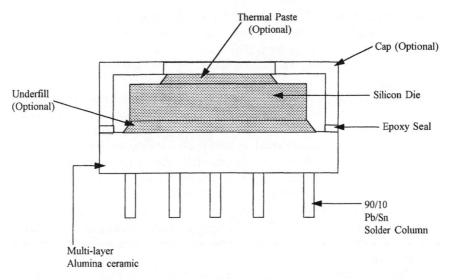

Figure 3.30 Cross-section of a ceramic column grid array (CCGA) package.

all package profile. Also the columns are not as rugged as solder balls and are susceptible to handling damage like fine pitch.

3.5.3 Plastic Ball Grid Array (PBGA)

The plastic ball grid array (PBGA), shown in Figure 3.31, is made of high temperature PCB laminate. It is also known as OMPAC (over molded plastic pad array) by Motorola. This package was originally developed by Motorola and Citizen in 1989. As in CBGA, the internal connec-

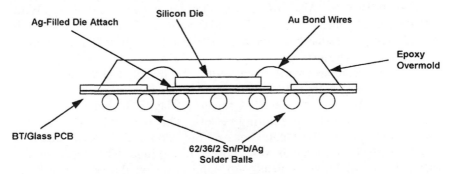

Figure 3.31 Cross-section of a plastic ball grid array (PBGA) package.

tion is made either with wire bond or by flip chip interconnection. For higher thermal performance, a heat slug can be incorporated inside the package. The solder balls are 0.030 inch diameter eutectic 63% Sn, 37% Pb balls (melting point 183°C) or 62% Sn, 36% Pb, and 2% Ag balls (melting point 179°C).

The resins used in PBGAs have a T_g (glass transition temperature) necessary for high temperature stability. BT (bismaleimide triazene) resin is the most commonly used resin for PBGA. Driclad, another resin material developed and patented by IBM, is also used for PBGAs. These materials have a CTE very similar to that of commonly used FR-4 laminate (16–20 ppm) and hence do not pose any solder joint reliability concerns. Driclad is more resistant to moisture, but all PBGA packages are considered to be extremely moisture sensitive. Hence most PBGA packages are susceptible to the popcorn effect (like any plastic package). The failure is generally seen as a crack in the package or delamination in the die attach region.

The PCB layer in the package can be two-layer to multilayer depending upon package complexity. The ball attachment process is as follows. Liquid flux is dispensed on a strip containing multiple packages after they have gone through wire bonding and plastic molding process steps. Then balls of the desired size (generally 0.030 inch) are placed either by gang placement machines or dispensed en masse with a stencil-like fixture. The flux holds the balls in place during reflow in an inert (nitrogen) environment. N_2 helps provide consistent ball quality and keeps them from oxidizing during reflow. However, N_2 is not necessary for final reflow of the package to the PCB. The eutectic solder balls provide "controlled collapse" and self-alignment during reflow, compensating for some misplacement.

PBGA is a generally low cost and low profile package. However, in some cases PBGA packages may cost more than fine pitch packages. The reason is that expensive fine line boards and high temperature resins are used in PBGAs. Also, BGA is a relatively newer technology. With the passage of time, the cost differential between PBGA and fine pitch should disappear as has been the case between through-hole and standard surface mount packages.

There are some major issues with the plastic BGA package. In addition to being extremely moisture sensitive, it is difficult to rework. (See Chapter 14 for details on BGA rework.) The PBGA balls are difficult to reball after they collapse during rework. This is not the case with the high melting point CBGA balls because they do not melt during rework.

The PBGA package is susceptible to warpage. The edges of the package tend to lift up, causing no connection on the outer rows. As expected, the larger PBGA packages are more susceptible to warpage than the smaller packages. This is one of the reasons why the lead coplanarity allowed by JEDEC standard (MO 151) is 0.006 inch for smaller BGAs (1 mm pitch) and 0.008 inch for larger BGAs (1.27 mm and 1.5 mm pitches). The coplanarity requirement for larger BGAs was increased because most package suppliers have difficulty meeting 0.006 inch requirement. (in peripheral components it is 0.004 inch). Warpage is less of an issue if solder paste (as opposed to flux only) is applied for reflow. The package warpage is caused by CTE mismatch between the PBGA substrate and the silicon inside. This problem becomes serious if the die is large.

Unlike CBGA, CTE mismatch does not exist between the PBGA package and FR-4 PCB. However, CTE mismatch between a large die and package body is of concern and can cause solder joints near the corner of the die to crack. Solder balls beneath the corner of the chip generally fail first. However, if balls are depopulated under the die, the corner joints fail first as in CBGA.

Failures in PBGA are seen either at the package-ball interface or PCB-ball interface. Solder joint volume is important to reliability. Hence monitoring of solder paste height as a process control is recommended. Johnson et al. [4] found that failure is a function of die size and package size. The results of their thermal cycle tests (−25° to 100°C) are shown in Table 3.8.

Table 3.8   PBGA reliability data at test condition −25° to 100°C thermal shock test [4]

NUMBER OF PINS	DIE SIZE (mils)	SAMPLE SIZE	FIRST FAILURE	CYCLES TO 50% FAIL
72	270×270×15.5	336	1768	3403
165	437×437×21	336	966	2124
225	389×389×18.7	32	2350	2804

3.5.4   Tape Ball Grid Array (TBGA)

The tape ball grid array (TBGA), shown in Figure 3.32, is another low cost, low profile package. It uses a low dielectric substrate (polyimide) and a two metal layer TAB-type substrate (1 signal and 1 ground). In TBGA, the CTE mismatch issues are nonexistent since adhesive and flexible substrates take up strains. IBM reported no failures in 0–100°C thermal cycling or 22,000 power cycles (25–75°C, 3W). TBGAs can use flip chip or TAB interconnection to make possible a lower die pitch than is possible with wire bond. For example, in wire bond the pitch is generally

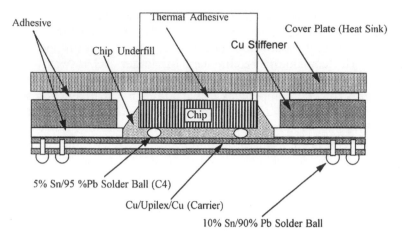

Figure 3.32 **Cross-section of a tape ball grid array (TBGA) package.**

4 mils but with TAB used for TBGA, the pitch can be 3.2 mils. The use of smaller pitch helps in shrinking the die. See Chapter 1 for a discussion on wire bonding versus TAB bonding.

Unlike CBGA, there is no eutectic solder joining the package to the balls. Balls of 0.025 inch diameter, 90% Pb, 10% Sn, are attached to the via pads by partially melting them. Commonly the TBGA pitch is kept at 50 mil center and their body sizes vary from 21 to 40 mm.

One concern with TBGA is that it allows only a single wiring plane in the substrate. Thus, TBGA is essentially restricted to being a single chip package. The other types of BGAs discussed above can accommodate multiple dice because they use multilayer substrates (either ceramic or BT resin substrates). TBGA also requires a gold bumped die, which may not be available in the open market. It is also a moisture sensitive package since polyimide absorbs moisture.

3.6   CHIP SCALE PACKAGING (CSP)

A package that takes no more than 20% additional area than the bare silicon is defined as a chip scale package. What is the driving force for CSP? It provides higher packaging density than BGA but lesser density than is possible with flip chip. However, the CSP assembly process is not as complex as the flip chip process either. For example, it does not pose the handling problems of the bare die used in flip chip. The assembly processes are not only much easier than those for flip chip, they are even compatible with SMT processes. CSPs can be pretested like SMT and reworked like SMT. Their pin count, size, and thermal and electrical performance are close to those of flip chip without the known good die concern of flip chip. Their disadvantages in comparison to flip chip is that they are slightly larger in size and their electrical performance is not as good.

Tessera Micro BGA is one example of CSP. As shown in Figure 3.33, Tessera Micro BGA has the traces on a flex circuit fanning inward to the array of bumps underneath the die. This is the reverse of a TAB device, where the traces fan outward. The package is assembled using modified wire bonding equipment where gold wires are bonded to the chip's pads.

The package has an array of 85 to 90 micron electroplated nickel bumps plated with 0.3 micron gold flash on 25 micron thick polyimide film on a single- or two-layer tape. The second layer, when used, acts as the ground plane. A 150 micron thick high temperature silicone elastomer filled with 50% pure silica between die and polyimide film is used for X-Y-Z compliancy.

Tessera Micro BGA has excellent electrical performance (0.5–0.7 nH) and a good thermal path through the back side of the die to a heat sink. The main issue with this package is that the paste printing and placement problems are somewhat like those for fine pitch packages since its pitch is very small.

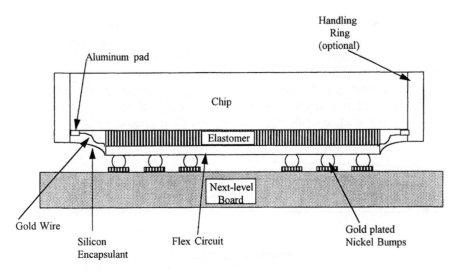

Figure 3.33   Cross-section of a Tessera Micro BGA package.

3.7 MAJOR ISSUES IN COMPONENTS

Many issues in component packaging need to be resolved. Some are technical, some are related to supply and demand, and others are due to lack of standards. In this section we discuss some of the major component issues such as lead coplanarity, lead configuration, lead or termination finish, packing components for shipping and handling, and standardization. Finally, we will discuss component procurement strategy in light of the issues dealt with in this section.

3.7.1 Lead Coplanarity

One of the critical issues in surface mount packages with leads is the coplanarity of leads. "Coplanarity" is defined as lying or acting in the same plane. Mathematically speaking, therefore, coplanarity is the distance of component leads above and below a common plane defined as a plane that passes through the average length of all the leads.

This definition is difficult to implement on the manufacturing floor. Thus noncoplanarity, a simplified term, is the maximum distance between the lowest and the highest pin when the package rests on a perfectly flat surface (see Figure 3.34). This definition represents a package sitting on a PC board on at least three leads.

What this tells us is that even if one lead out of 68 or 84 leads is out of an acceptable range, many leads, not just that particular lead, may not solder properly. In the worst case, coplanarity will cause open solder joints. Let us discuss the solder-open situation first, and then the solder joint weakness caused by lead coplanarity.

The common problem that users encounter because of coplanarity is the phenomenon known as solder wicking; that is, the solder paste wicks up the lead, causing open solder joints. It should be kept in mind

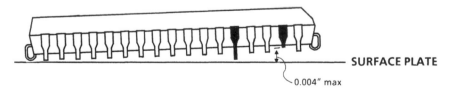

SURFACE PLATE

0.004″ max

(DIMENSIONS NOT TO SCALE)

Figure 3.34 Lead coplanarity in PLCC.

that poor board or lead solderability or uneven and fast heating during the reflow process also can cause open solder joints. See Chapter 12 for a discussion on wicking.

Even if we can reduce the incidence of wicking by depositing thicker solder paste and exercising good process control, coplanarity may cause poor solder reliability. The farther away from the seating plane a lead rests, the more solder volume resides between the lead and the pad. Since the solder itself is weaker than the intermetallic bond at the base metal interface, failure occurs within the solder volume. Hence the concern that PLCC lead coplanarity has a direct effect on the strength of its solder joints [5]. The more solder present between a lead and pad, the weaker the joint will be.

For these two reasons (solder open and solder joint weakness), a user naturally wants all the lead ends to lie perfectly in the same plane, to avoid manufacturing problems.

Component suppliers, however, have a difficult time supplying a perfectly planar package. The JEDEC committee established a maximum coplanarity of 0.004 inch. Is this adequate, however, for preventing open solder joints? With solder paste 8 mils thick, leads with 6 mil coplanarity can be soldered with proper process control. But the author of Reference 5 found that solder joint strength shows a sharp decrease beyond 0.004 inch coplanarity. It should also be kept in mind that even if the packages are supplied in perfect planarity, 1 to 2 mils can be added during shipping and handling, especially if methods other than tape and reel are used.

A 0.004 inch maximum coplanarity (0.006-0.008 inch for BGA) is generally accepted by both users and suppliers, and it is also the JEDEC standard. Since the solder paste thickness varies from 0.006 to 0.010 inch, users can live with 0.004 inch coplanarity. In any event, suppliers have a very difficult time meeting tighter requirements, especially for the large packages. Shadow graphs, vision systems, or gauges are generally used for determining coplanarity. However, measuring coplanarity in a production environment is not an easy task for the user. Honaryar, of the University of Lowell (Massachusetts) Center for Productivity Enhancement, has undertaken the development of a pneumatic sensor device for measuring lead coplanarity just before component placement [6].

All these different methods for measuring coplanarity are either slow, sophisticated and expensive, or still under development. Thus users should insist that their supplier ensure that the 0.004 inch coplanarity requirement is met. It should be noted that the number for acceptable lead coplanarity is not ±0.004 or ±0.002 inch, but 0.004 inch maximum, and it should be specified as such in the procurement specification.

3.7.2   Lead Configuration

The configuration or the shape of leads is important for reasons having to do with manufacturability and reliability. A silicon device may be packaged in various types of packages such as SOIC, fine pitch, and PLCC. These packages are primarily available in four different lead configurations: gull wing, BGA, J lead, and butt or I lead. The most common lead configurations used today are J lead and gull wings, and the newest package to get tremendous industry attention is the ball grid array (BGA). The butt lead is just an interim solution for a package that is not available in surface mount.

Everyone has their favorite lead configuration, but for all practical purposes users have to deal with all of them because they are all used in different devices. Each lead configuration has its advantages and disadvantages, as summarized in Figure 3.35. The determining factors in selecting a particular package over another for a given application should be electrical performance, availability of the package from the vendor of choice, real estate considerations, assembly capability, and most important of all, its impact on product cost and time to market. Let us review each lead style in some detail.

3.7.2.1   Gull Wing

Gull wing leads, which bend down and out, are used in both lower and higher pin count packages. In lower pin count (under 20 pins) devices,

SHAPE OF TERMINAL / ITEM	Gull Wing	J Lead	Butt Lead	BGA
Compatibility with High Pin Package Trend	◇	◯	△	◉
Package Thickness	◇	◯	△	◉
Lead Rigidity	△	◇	◯	◉
Repair/Rework	◯	◇	◯	△
Ability to Self-Align During Reflow	◇	◇	△	◉
Solder Joint Inspection	◯	◇	◯	△
Ease of Cleaning	△	◉	◉	◉
Real Estate Efficiency	△	◇	△	◉

◉ = EXCELLENT   ◇ = VERY GOOD   ◯ = GOOD   △ = POOR

Figure 3.35   Key attributes of different lead configurations.

the gull wing shape is common in SOIC because it is more space efficient than a low pin count J leaded PLCC package. Gull wing is also the most commonly used for high pin count (over 84 pins) packages. This is evident in fine pitch (25 mils pitch and under) PQFP, TAB, or ultrafine pitch (below 20 mils pitch) packages. One of the desirable attributes of gull wing packages is their small body thickness, so critical to applications such as notebook computers.

Gull wing leads are compatible with almost any reflow soldering process such as IR, convection, vapor phase, hot gas, and hot bar. Their ability to self-align during reflow when they are slightly misplaced is fairly good. Boards using these packages require larger space on the board but lower board layer count for signal routing as discussed in Chapter 5, Section 5.5.1. A lower layer count lowers the board cost. The major advantage of gull wing leads is their compatibility with trends toward the finer pitch packages of the future.

Their disadvantage, contrary to popular belief, is inspection because of hidden heel fillet, which is most important for reliability. In gull wing packages, heels of only eight corner solder joints can be inspected. Based on these corner joints, one hopes that the other solder joints are acceptable. The toe fillet, which is easily visible, is not required or even achievable all the time. Achieving toe fillet on a consistent basis is difficult because the leads are cut after plating, thus exposing copper, which is more difficult to solder.

One of the biggest problems with gull wing leads is their fragility and susceptibility to lead damage such as lead coplanarity, lead bending, and sweep especially in fine pitch packages without corner bumpers (Figure 3.28). The fragile leads are the leading cause of most defects in gull wing packages. Such packages generally have very low stand-offs and cause flux entrapment and hence cleaning problems. This is especially true in EIAJ (Japanese) fine pitch packages, which have almost zero stand-off.

A variation of the gull wing was suggested by authors from IBM [7], who provide a via hole in gull wing or J leads. The via is intended to provide an escape path for volatiles and will prevent the internal voids common in reflow solder joints. The via hole is also intended to allow the solder to penetrate through and encapsulate the lead frame, both top and bottom (in addition to side fillets as in gull and J leads); hence the solder joint formed will be stronger. At this point this suggestion remains only a concept, however.

3.7.2.2   *Balls (in BGA)*

In the past, primarily gull wing leads have been used for high pin count packages. However, because of the inherent problems in gull wing devices mentioned earlier, ball grid array packages (BGAs) have become popular. BGAs are array packages like PGAs (pin grid array). However instead of leads in a PGA they use balls underneath the package. Since the fragile leads are replaced with rugged balls, the problems related to bent leads simply disappear. BGA also provides a much shorter signal path compared to fine pitch. This can be very critical in high speed applications.

As summarized in Figure 3.3.5, BGAs have the highest number of positives but they also have some serious problems. For example, the hidden solder joints are difficult to inspect and rework. BGAs are not compatible with the hot bar and laser reflow processes because the balls are hidden from the heat source. Finally, they are extremely susceptible to moisture induced cracking.

3.7.2.3   J Lead

J-lead packages have been the real work horse of surface mount technology. This package is basically an American idea and comes only in 50 mil pitch. It is more popular in the United States than in Japan. The packages start at about 20 pins and are commonly used up to 84 pins. For pin counts over 84, one must move on to different lead configurations such as gull wing or BGA, which allow lower pitches and fine pin counts. J leads cannot be used for pin counts above 84 pins because of the difficulty of forming the leads.

Like BGA, J-lead devices are not compatible with all types of reflow processes such as hot bar. And, like gull wing, they also self-align very well during reflow if slightly misplaced. However, we should note that neither J lead nor gull wing have the self-alignment capability of a BGA. In any event, one should never count on self-alignment as a substitute for accurate placement. J-leaded packages are excellent for cleaning because of higher stand-offs. It is also worth noting that the outer fillets are the major fillets in J lead and hence solder joint inspection is easy in J-lead packages.

J leads are sturdier than gull wing leads, hence stand up better in shipping and handling. However, the J-lead packages have a higher profile than the gull wing packages. This is a disadvantage in applications where all packages must have lower profiles. Overall, as summarized in Figure 3.35, J-leaded packages may not have many positive attributes but they do not have any poor attributes when compared to the other types of lead configurations.

3.7.2.4   Butt or I Lead

Butt leads, also known as I leads, are not commonly used. The advantage claimed for butt leads is the possibility of converting a through-hole component into surface mount by clipping the leads and accomplishing the soldering of all components in one reflow operation. Since not all components are available in surface-mounted form, this is a convenient way to reach the goal of a full Type I surface mount assembly. However, through-hole packages generally are not meant for reflow soldering. If they are to be reflow soldered, they must be able to withstand the reflow soldering temperatures.

Butt mounting of DIPs does not provide any real estate saving but sometimes allows a one-step soldering process and provides some cost saving. This saving may not always be realized, however. In Chapter 7 (Section 7.4) we discuss situations in which it may or may not be appropriate to use butt leads for surface mounting. But are the butt joints reliable? Derfiny and Dody [8] concluded a reliability study on the solder joints formed by butt, J, and gull wing leads, using lead forms made of the same lead frame materials mounted on glass epoxy boards. Pull, shear, and thermal cycling tests were conducted for reliability determination.

While the relevance of the pull and shear tests for solder joint reliability may be debated since the controlling failure mechanism is fatigue, the data give relative reliabilities of different lead configurations. The butt lead joints were found to have 65% less pull and shear strengths than J or gull wings. The test results were corroborated by −40 to +125°C thermal cycling tests. It was also found that the butt lead configuration was more sensitive to process-related handling, placement, and soldering when compared to the J or gull wing lead configuration.

Butt leads, especially those created by cutting DIP leads, should not be selected until the reliability issues have been resolved. It should be noted that DIPs may be damaged during bending operations for J or gull wing shapes or during cutting operations for butt lead shapes.

There is an interesting trend in Japan on butt leads. Some leading Japanese companies are using the butt lead configuration and finding considerably fewer defects (bridging and open joints) when compared to gull wing devices. Such a trend is supported by other researchers who not only preferred butt leads over other types but even suggest that gull wing leads be twisted to form butt [9]. The twisted lead configuration, a variation of the butt lead, will allow more routing space between leads and, by piercing through the solder paste, will prevent paste displacement.

At this time, considerable work is going on in the industry on the advantages and disadvantages of various lead configurations. It should be noted that the concepts suggested in References 7 and 9 are just that—concepts. They have not been adopted by the industry, and it is not clear whether they will ever be used. And the likelihood of their use is considerable diminished with the emergence of BGA. However, they do deserve consideration for the obvious benefits pointed out by their proponents.

3.7.3 Standardization

It cannot be overemphasized that standardization, one of the key issues in promoting any new technology, is almost mandatory for surface mount technology because of the need for automation. In the past, the lack of surface mount standards was a major concern for SMT users. However, considerable progress has been made in this area by the industry, although much remains to be done.

The problem of standardization is exacerbated because Japanese, European, and U.S. manufacturers tend to go their own ways. In many cases duplication of effort by two different standards setting organizations tends to confuse the users. In any event, tolerances in existing standards generally tend to be poor so they should not be used as procurement specifications. The proliferation of various packages was cited in Chapter 2, Section 2.2.2: users have been accepting packages any way they can get them, with or without standards.

The use of tape and reel is another example. Tape and reel packaging is best for accuracy and throughput in component placement. EIA-RS-481 establishes a standard for tape and reel, but it is not without problems. The maximum size of the reel is set at 13 inches. There is no minimum size. Thus a reel can be any size up to 13 inches. The user can generally specify reel size, which should be a function of throughput, placement capability, and the number of components desired on a reel. Since the size and thickness of components vary, different reel diameters are needed to meet different needs.

Since active components have generally been available only in the 13 inch reel size, inventory cost may be a real problem for small users and even large users during developmental stages of a product. Now active components can be purchased on smaller reels but there is a cost premium. Refer to Chapter 11 (Section 11.4) for details on the advantages and disadvantages of feeders of different types and the standardization issues in tape and reel. Also refer to Chapter 2 for a better understanding of how standards are set and how one can influence development of standards through the EIA and the Institute of Interconnecting and Packaging Electronic Circuits.

3.8 COMPONENT PROCUREMENT GUIDELINES

Developing in-house component procurement specifications is important for any surface mount component. There are certain general guidelines, as discussed below, for developing in-house specifications.

- One should deal with a limited number of vendors, establishing specifications in cooperation with the preferred vendors. The current trend is to rely on one vendor.
- Only a limited number of packages that enjoy widespread support should be selected for use. Limiting the number of part types provides such benefits as increased purchasing leverage, reduced inventory cost, and reduced number of feeders for the pick-and-place equipment.
- The components selected should be qualified for performance and reliability, solderability, component tolerance and compatibility with land pattern design, and compatibility with processes and equipment used in manufacturing.

Within the framework of the general guidelines above, some specific requirements should be incorporated into the procurement specifications:

1. Solder-dipped terminations or leads provide good solderability but many compound the coplanarity problem. Hence the tin-lead plated terminations or leads are becoming more widely used. Most users want to minimize the intermetallic thickness by asking that terminations or leads not be fused before shipment.
2. The coating thickness of solder should be 300 microinches (0.3 mil) and solder should be either eutectic or contain 60% to 63% tin. Both types of solder coating are acceptable as long as they meet coplanarity and solderability requirements.
3. The preferred lead frame material for better thermal conduction is copper, and this metal should be required.
4. Component tolerances should be closely specified. Industry standards and tolerances should be specified where acceptable. The component tolerances should be measured after solder dipping or fusing.
5. A nickel barrier underplating of 50 micro inches (0.000050 inch) between the solder termination and the gold or silver adhesion layer should be required to prevent leaching or dissolution of gold or silver during the soldering process. This requirement is necessary primarily for passive components but should be specified for the leads of active or passive devices if precious metal underplating is used.
6. The parts must be able to withstand at least two cycles in the soldering environments used in manufacturing. Specifying the soldering profile is more relevant, but 60 seconds at 215°C in vapor phase or 2 minutes at 230°C in an infrared or convection oven or at least 10 seconds in molten solder at 260°C should be

used. If all these soldering methods are used in manufacturing, the parts should be able to withstand all these temperatures for the specified time without any measurable impact on performance or reliability.

7. The parts must be resistant to solvents at the temperatures used for cleaning (generally 40°C), including exposure to ultrasonic cleaning if necessary. The requirement of 1 minute at a frequency of 40 kHz and a power of 100 W/ft² is sufficient [1]. Ultrasonic cleaning for active devices or any small leaded surface mount part is generally not recommended for fear of breaking the internal wire bonds.

8. Component marking should be required for active devices. For passive devices marking is not necessary, especially if there is a cost premium and if tape and reel packaging is acceptable. Tape and reel should be specified as the packaging format whenever feasible. The embossed plastic reels are preferable to paper or punched reels because the former are less prone to delamination and placement problems. Parts that are sensitive to electrostatic discharge must be supplied on conductive or static dissipative reels.

9. Surface mount standards, even when they exist, are not supposed to be used as procurement specifications. The general intent of most standards is to accommodate the needs of all member companies. This tends to make a given standard too loose. Therefore the industry standards should be used as a guideline, not as a substitute for in-house procurement specifications.

In developing in-house specifications, it is important to keep in mind that they should not require special handling by the suppliers. To do so may increase cost or delivery time or both.

3.9   SUMMARY

Surface mount components are no different from through-hole components as far as the electrical function is concerned. Because they are smaller, however, the SMCs provide better electrical performance. Not all components are available in surface mount at this time; hence the full benefits of surface mounting are not available, and we are essentially limited to mix-and-match surface mount assemblies. The use of through-hole components such as pin grid array for high end processors and large connectors will keep the industry in a mixed assembly mode for the foreseeable future.

While only a few types of conventional DIP packages meet all the packaging requirements, the world of surface mount packages is vastly more complex. The package types and the package and lead configurations available are numerous. In addition, the requirements of surface mount components are far more demanding. SMCs must withstand the higher soldering temperatures and must be selected, placed, and soldered more carefully to achieve acceptable manufacturing yield.

There are scores of component types available for some electrical requirements, causing a serious problem of component proliferation. There are good standards for some components, whereas for others standards are inadequate or nonexistent. Some components are available at a discount, and others carry a premium.

While surface mount technology has matured, it is constantly evolving as well with the introduction of new packages. The electronics industry is making progress every day in resolving the economic, technical, and standardization issues associated with surface mount components.

REFERENCES

1. ANSI/IPC-SM-782. Surface Mount Land Patterns (Configuration and Design Rules). IPC, Northbrook, IL.

2. Knuduson, E. "Surface mount packages for VLSI device." *Proceedings of the SMART III Conference*, January 1987. Technical paper SMT III-33. IPC, Northbrook, IL.

3. Banks, Donald R.; Heim, Craig G.; Lewis, Russ H.; Caron, Alain; and Cole, Marie S. "Second level assembly of column grid array packages." *Proceedings of SMI*, 1993, pp. 92–98.

4. Johnson, Randy; Mawer, Andrew; McGuiggan, Tim; Nelson, Brent; Petrucci, Mike, and Rosckes, Dan. "A feasibility study of ball grid array packaging." *Proceedings of NEPCON E 93*, pp. 413–422.

5. Smith, W. D. "The effect of lead coplanarity of PLCC solder joint strength." *Surface Mount Technology*, June 1986, pp. 13–17.

6. Honaryar, B. "Progress report on a pneumatic inspection device for coplanarity sensing of surface mounting PLCC." *Proceedings of the SMART III Conference*, January 1987. Technical paper III-19, IPC, Northbrook, IL.

7. Kang, S. K., and Moskowitz, P. A. *IBM Technical Disclosure Bulletin*, Vol. 29, No. 4, September 1987, p. 1612.

8. Derfiny, D., and Dody, G. "On optimizing lead form." *NEPCON Proceedings*, 1987, pp. 251–256.

9. Buckley, D. "SMC packaging causes problems." *Electronic Production*, April 1986, pp. 32–33.

Chapter 4

Substrates for Surface Mounting

4.0 INTRODUCTION

The substrate, also referred to as the packaging and interconnecting structure, plays a crucial role in ensuring the electrical, thermal, and mechanical reliability of electronic assemblies. The term substrate can be defined in two categories: (1) laminate substrate, which is used to make the printed wiring board (PWB) or printed circuit board (PCB); (2) constraining substrate, which is used for decreasing the CTE (coefficient of thermal expansion) or increasing the thermal conduction. Unless we are specific, the term substrate could refer to either of these two categories. Before choosing from among the many types of substrate (of either category) that are available for military and commercial applications, however, it is necessary to determine the properties that will be required. Then a substrate material that meets all the requirements in a cost-effective manner can be selected.

The substrate properties that will be required depend on the type of application (commercial or military) and the type of packages (ceramic or plastic). Ceramic packages (generally leadless ceramic chip carriers: LCCCs up to 44 I/O with 50 mil pitch) are used in military applications to provide the needed hermeticity. In addition, as mentioned previously, if there is any difference between the coefficients of thermal expansion of component packages and substrates, the stress induced by CTE mismatch can strain the solder joints and cause fatigue failure after repeated thermomechanical cycles.

There are essentially three approaches to overcoming the stress on solder joints when LCCCs are used on the substrate: (a) developing substrates with matching CTE to prevent the generation of stress on solder joints—this is done by minimizing the CTE differences in the substrate and the package being mounted, (b) developing substrates with compliant top layers that can absorb the stress (not used to any extent), and (c) adding leads to the surface mount packages to take up the stress (when the LCCC I/O exceeds 44 with 50 mil pitch).

Each method has its advantages and disadvantages. Approach (a), the use of a substrate with a compatible CTE, is the most common, but is expensive. Moreover, even when a substrate has compatible CTE values, failures in the via holes are encountered. To prevent via hole cracking, the board thickness is reduced, or the via hole diameter is increased, or a high T_g material with hole size up to 13 mils is used. A modification of approach (a) is to use a substrate material with a lower modulus of elasticity in the vertical direction to prevent via hole cracking. Teflon-based material (e.g., Roger RO 2800) is one of the materials that does this. However, it requires fusion bonding at greater than 800°F and is very expensive.

Approach (b), the use of a substrate with a compliant top layer, has been tried but has not really caught on. Solder joint cracking may be prevented, but the reliability problems are shifted to surface traces and vias. The compliant layer approach has not proven very reliable and hence is not generally used.

Approach (c), the use of leaded packages, allows the selection of relatively inexpensive FR-4 glass epoxy substrates if they meet the thermal and electrical requirements. The paper-based substrates used in many consumer applications are even less expensive than glass epoxy materials. However, even the decision to use leaded ceramic packages does not completely eliminate solder joint cracking problems if the lead material and lead configurations are not very compliant. The use of leaded plastic packages on a glass epoxy substrate is the sure way to prevent reliability problems, but this approach is limited mostly to the commercial sector and is not adequate for military applications. Although there are no easy answers yet in substrate selection, some defense contractors have made considerable investment in finding cost-effective solutions and are making good progress.

In this chapter we discuss some commonly used substrates for the approaches outlined above. However, we will concentrate on the glass epoxy substrate not only because it is most widely used but also because the final process steps in the fabrication of glass epoxy boards are applicable to most other substrates. But first let us discuss some basic concepts. The notions of glass transition temperature and coefficient of thermal expansion are useful no matter what substrate is used.

4.1   GLASS TRANSITION TEMPERATURE (T_g)

Except for the ceramic substrate, almost all substrate laminates contain polymers. Polymers, unlike metals, undergo major structural changes at certain temperatures. The temperature at which the physical structure of the laminate changes from hard and brittle, or glasslike, to soft and rubbery (rubber-like) is called the glass transition temperature T_g. The laminate materials go through a structural change above T_g and lose a considerable amount of their mechanical strength.

The glass transition temperature is a characteristic property of most polymers, hence is unique to a particular polymer material. Next to CTE, T_g is the most important property for the laminate and is critical in the selection of a laminate material.

Stress-deformation curves for an epoxy-glass laminate illustrate the criticality of T_g. As shown in Figure 4.1, at a temperature below T_g, to attain a specific level of deformation in the laminate d_0, a stress F_0 must be applied. If the stress is applied above T_g, a much larger deformation, d_1, will result. This larger deformation will generate considerable warpage and will degrade the reliability of the printed board. Figure 4.1 shows the effect of force on deformation at temperatures above and below T_g. However, like most materials, the laminate also expands with an increase in temperature even if no stresses are being applied. Such expansion increases exponentially above T_g and linearly below T_g.

The value of T_g plays an important part in temperature-related expansion. Let us compare the expansions in two commonly used laminate

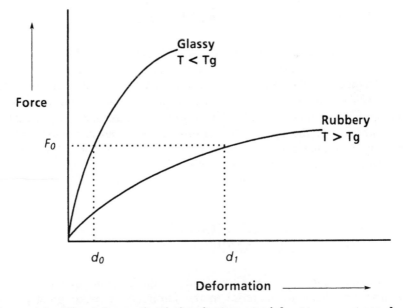

Figure 4.1 **The deformation in laminate materials at temperature above and below their glass transition temperature T_g.**

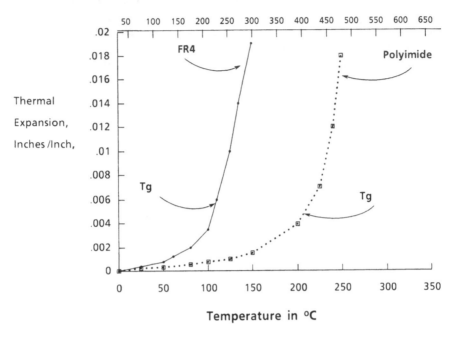

Figure 4.2   The effect of T_g on laminate expansion [1].

materials: FR-4, which has a T_g of about 120°C, and polyimide, which has a T_g of about 230°C. For example, as shown in Figure 4.2, the unconstrained thermal expansion of epoxy glass is about 0.019 inch per inch at 150°C. However, for polyimide it is minuscule in comparison—about 0.001 inch per inch [1]. The reason for this disparity is that at 150°C, the T_g for the epoxy glass laminate has been exceeded and the exponential expansion rise has already set in, whereas for polyimide, the temperature is still much below the material's T_g, hence laminate expansion is still in the relatively low linear range.

4.2   X, Y, AND Z COEFFICIENTS OF THERMAL EXPANSION

Above the glass transition temperature, the laminate materials—that is, resins, not the glass material in the laminate—expand greatly. The thermal expansion values shown in Figure 4.2 are for the free and uncons-trained expansion of the resin ingredient of the laminate. When the laminate is used in a multilayer printed circuit board, however, the other elements of the multilayer board exert constraining effects on the laminate.

The constraining effects are not identical in all dimensions. In the X and Y directions, laminate thermal expansion is constrained by the glass fibers in the laminate, the copper traces and pads on all the layers, and the barrel plating around the plated through hole (Figure 4.3). However, the constraining effect is very small in the Z direction, with only the plated through-hole (PTH) barrel having some effect near the regions around the barrel.

A look at thermal expansion for an epoxy glass laminate based multilayer board will illustrate this point better. Below T_g the coefficient of thermal expansion in the X and Y directions is between 12 and 16 microinches (0.000,016 inch) per inch per degree Celsius (ppm/°C). For the Z direction, the expansion coefficient is 100 to 200 ppm/°C. This is an order of magnitude greater than the CTE for the X and Y directions, since there is no fiber weave to provide restraint.

The Z axis thermal expansion in and around a multilayer PTH board was empirically measured by researchers at IBM [2]. They found that the laminate (50 mils away from the barrel of a plated through hole) and the barrel increased in length at quite similar rates. However, they found a

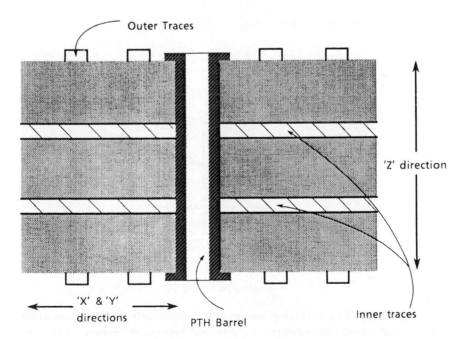

Figure 4.3 **The constraining effects on laminate in a multilayer board.**

sudden, sharp increase in the expansion of both laminate and barrel in the Z direction. The rate of expansion was still linear, but the laminate expansion rate was higher than that of the barrel. Hence, as temperature increases beyond 110°C, the disparity in the Z axis expansion between the laminate and barrel becomes larger. The effect of this phenomenon on a plated through hole is illustrated in Figure 4.4.

As shown in Figure 4.4, the free expansion rate of the plated through-hole barrel (copper and solder) is lower than that of the surrounding laminate. This implies that the barrel is being pulled along by the laminate because the laminate expands faster than the barrel. This strain condition creates tensile stresses in the barrel, and as the temperature increases, the tensile stresses will increase. If and when these stresses exceed the fracture strength of the barrel plating, the plating will fracture, thereby causing

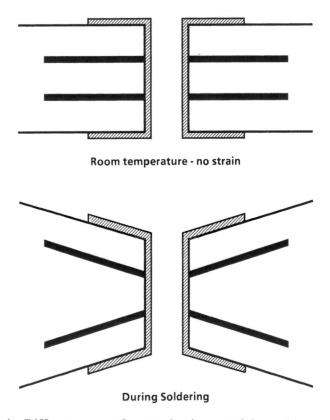

Room temperature - no strain

During Soldering

Figure 4.4   Different rates of expansion in materials used in a multilayer board can cause deformation in internal layers, as shown in the bottom sketch.

board failure. The higher rate of laminate expansion causes a dramatic increase in the stresses within the plated through-hole barrel, inner traces, and pads. These stresses can result in cracks in the barrel, traces, or pads or even in the laminate, especially if the substrate is operated above its T_g. In addition to thermal stress, inadequate ductility of copper also plays a role in the cracking at the knee of the PTH. Due to uneven heating, this problem is made worse during hand soldering than in wave or reflow soldering. See Chapter 14 for a further discussion on the effect of hand soldering on barrel cracking.

4.3 SELECTION OF SUBSTRATE MATERIAL

As indicated in Table 4.1, all the numerous substrate materials available on the market today for military and commercial applications have their advantages and disadvantages [3]. During fabrication, assembly, and service, these materials may be subjected to high temperatures, high humidities, corrosive chemicals, and mechanical stress. They are expected not only to survive these severe conditions but also to perform electrically. Therefore, it is imperative that designers and manufacturing personnel have a good understanding of substrate materials and their properties. This knowledge will enable designers to predict the behavior of materials under fabrication, assembly, and service conditions. Also, in case of a failure, the follow-up analysis requires a knowledge of material properties.

Each of these materials has its own particular characteristics and properties. Consequently, the materials behave differently under varying conditions of temperature, humidity, stress, and so on. For the assembly to function properly, the electrical and mechanical requirements of each of the materials used in substrates must be met under their characteristic operating conditions and environments. The materials used in substrate are as follows:

1. The laminate, which forms the basic structural backbone and determines the mechanical and electrical characteristics of the substrate.
2. The solder mask or conformal coating, which is applied on the outer layers of the board.
3. The metal platings in the plated through holes and via holes, as well as the PTH pads and surface mount lands, which provide electrical interconnections.
4. The inner layer and outer layer copper traces, which form the pathways for the signals to flow from one device to another.

Table 4.1 Advantages and disadvantages of various types of substrates [3]

TYPE	MAJOR ADVANTAGES	MAJOR DISADVANTAGES	COMMENTS
Organic based substrates (epoxy fiberglass)	Substrate size; weight; reworkable; dielectric properties; conventional board processing	Thermal conductivity; X, Y, and Z axis CTE	Because of its high X-Y plane CTE, it should be limited to environments and applications with small changes in temperature and/or small packages
Polyimide fiberglass	Same as epoxy fiberglass plus high temperature Z axis CTE; substrate size; weight; reworkable; dielectric properties	Thermal conductivity; X and Y axis CTE; moisture absorption; brittle resin (prone to cracking); very poor adhesive—low bond strength to copper; Cost 2X–5X of FR-4	Same as epoxy fiberglass
Epoxy aramid fiber woven 108 and 120 cloth	Same as epoxy fiberglass; X-Y axis CTE; substrate size; lightweight; reworkable; dielectric properties	Thermal conductivity; X and Y axis CTE; resin microcracking; Z axis CTE; water absorption; cost 3X to 5X FR-4	Volume fraction of fiber can be controlled to tailor X-Y CTE. Resin selection critical to reducing resin microcracks.
Epoxy aramid paper	Same as epoxy aramid fiber but eliminates resin cracking; thin multilayer board; decreased Z axis CTE	Low strength; Cost 2X–3X of FR-4	Used in conjunction with lightweight, low CTE metal substrate

Polyimide aramid fiber	Same as epoxy aramid fiber; Z axis CTE; substrate size; weight; reworkable; dielectric properties	Thermal conductivity; X and Y axis CTE; resin microcracking; water absorption	Same as epoxy aramid fiber.
Polyimide quartz (fused silica)	Same as polyimide aramid fiber; Z axis CTE; substrate size; weight; reworkable; dielectric properties	Thermal conductivity; X and Y axis CTE; Z axis CTE; drilling; availability; cost; low resin content required	Volume fraction of fiber can be controlled to tailor X-Y CTE. Drill wearout higher than with fiberglass.
Fiberglass/aramid composite fiber	Same as polyimide aramid fiber; no surface microcracks; Z axis CTE; substrate size; weight; reworkable; dielectric properties	Thermal conductivity; X and Y axis CTE; water absorption; process solution entrapment	Resin microcracks are confined to internal layers and cannot damage external circuitry.
Fiberglass/Teflon laminates	Dielectric constant; high temperature	Same as epoxy fiberglass; low temperature stability; thermal conductivity; X and Y axis CTE	Suitable for high speed logic applications. Same as epoxy fiberglass.
Flexible dielectric	Lightweight; minimal concern to CTE; configuration flexibility	Size	Rigid-flexible boards offer tradeoff compromises.

157

Table 4-1 (*Continued*)

TYPE	MAJOR ADVANTAGES	MAJOR DISADVANTAGES	COMMENTS
Thermoplastic	3-D configurations; low high-volume cost	High injection molding setup costs	Relatively new for these applications.
NONORGANIC BASE Alumina (ceramic)	CTE; thermal conductivity; conventional thick film or thin film processing: integrated resistors	Substrate size; rework limitations; weight; cost; brittle; dielectric constant	Most widely used for hybrid circuit technology.
SUPPORTING PLANE Printed board bonded to plane support (metal or nonmetal)	Substrate size; reworkability; dielectric properties; conventional board processing; X-Y axis CTE; stiffness; shielding; cooling	Weight	The thickness/CTE of the metal core can be varied along with the board thickness, to tailor the overall CTE of the composite.
Sequential processed board with supporting plane core	Same as board bonded to supporting plane	Weight	Same as board bonded to supporting plane.

158

Discrete wire	High speed interconnections; good thermal and electrical features	Licensed process; requires special equipment	Same as board bonded to low expansion metla support plane.
CONSTRAINING CORE Porcelainized copper clad invar	Same as Alumina	Reworkability; compatible thick film materials	Thick film materials are still under development.
Printed board bonded with constraining metal core	Same as board bonded to supporting plane	Weight; internal layer registration; delamination; via hole cracking, Z axis CTE	Same as board bonded to supporting plane.
Printed board bonded to low expansion graphite fiber core	Same as board bonded to low expansion metal cores; stiffness, thermal conductivity; low weight	Cost; microcracking; Z axis CTE	The thickness of the graphite and board can be varied to tailor the overall CTE of the composite.
Compliant layer structures	Substrate size; dielectric properties; X-Y axis, CTE	Z axis CTE; thermal conductivity	Compliant layer absorbs difference in CTE between ceramic package and substrate.

Table 4-2 Substrate selection criteria [3].

DESIGN PARAMETERS	TRANSITION TEMPERATURE	COEFFICIENT OF THERMAL EXPANSION	THERMAL CONDUCTIVITY	TENSILE MODULUS	FLEXURAL MODULUS	DIELECTRIC CONSTANT	VOLUME RESISTIVITY	SURFACE RESISTIVITY	MOISTURE ABSORPTION
				MATERIAL PROPERTIES					
Temperature & Power Cycling	X	X	X	X					
Vibration				X	X				
Mechanical Shock				X	X				
Temperature & Humidity	X	X				X	X	X	X
Power Density	X		X						
Chip Carrier Size		X		X					
Circuit Density						X	X	X	
Circuit Speed						X	X	X	

160

Table 4.3   Substrate properties [3]

MATERIAL PROPERTIES

MATERIAL	GLASS TRANSITION TEMPERATURE	XY COEFFICIENT OF THERMAL EXPANSION	THERMAL CONDUCTIVITY	XY TENSILE MODULUS	DIELECTRIC CONSTANT	VOLUME RESISTIVITY	SURFACE RESISTIVITY	MOISTURE ABSORPTION
UNIT OF MEASURE	*(°C)*	*(PPM/°C)* *(note 2)*	*(W/M°C)*	*(PSI × 10^{-6})*	*(At 1 MHz)*	*(Ohms/cm)*	*(Ohms)*	*(Percent)*
Epoxy Fiberglass	125	13–18	0.16	2.5	4.8	10^{12}	10^{13}	0.10
Polyimide Fiberglass	250	12–16	0.16	2.8	4.8	10^{14}	10^{13}	0.35
Epoxy Aramid Fiber	125	6–8	0.12	4.4	3.9	10^{16}	10^{16}	0.85
Polyimide Aramid Fiber	250	3–7	0.15	4.0	3.6	10^{12}	10^{12}	1.50
Polyimide Quartz	250	6–8	0.30		4.0	10^{9}	10^{8}	0.50
Fiberglass/Teflon	75	20	0.26	0.2	2.3	10^{10}	10^{11}	1.10
Thermoplastic Resin	190	25–30		3–4	10^{17}	10^{13}	NA	
Alumina-Beryllia	NA	5–7 21.0	44.0	8.0	10^{14}			
Aluminum (6061 T-6)	NA	23.6	200	10	NA	10^{6}		NA

161

Table 4-3 (*Continued*)

MATERIAL	GLASS TRANSITION TEMPERATURE	XY COEFFICIENT OF THERMAL EXPANSION	THERMAL CONDUCTIVITY	XY TENSILE MODULUS	DIELECTRIC CONSTANT	VOLUME RESISTIVITY	SURFACE RESISTIVITY	MOISTURE ABSORPTION
				MATERIAL PROPERTIES				
Copper (CDA101)	NA	17.3	400	17	NA	10^6		
Copper-Clad Invar	NA Note 1	3–6	150XY/ 20Z	17–22	NA	10^6		NA

Notes: 1. The X and Y expansion is controlled by the core material and only the Z axis is free to expand unrestrained. Where the Tg will be the same as the resin system used.

2. Figures are below glass transition temperature, are dependent on method of measurement and percentage of resin content. NA—Not applicable.

162

5.   The legend ink on top of the solder mask, which serves as component identification.

Among the important properties that must be considered in selecting
substrate materials are cost, performance, and reliability. The principal
substrate properties that should be taken into account in the selection
process are summarized in Table 4.2 [3]. The specific properties of various
substrates commonly used in the industry are shown in Table 4.3 [3].

4.3.1   CTE Compatibility Considerations in Substrate Selection

Although many properties must be considered in the selection of a
substrate material (see Table 4.2), surface mounting has made the coefficient of thermal expansion requirement almost primary as a selection
criterion, especially for military applications that use leadless ceramic
packages. The reason for this is very simple. As illustrated in Figure 4.5,
if the CTEs of the substrate and the packages are different, they will
expand and contract at different rates during power and thermal cycling.
This difference in rates of expansion caused by CTE will stress the joint,
and in extreme cases the solder joint will be fractured.

As stated earlier, the three main approaches used to prevent solder
joint cracking when using leadless ceramic packages are as follows: using
substrates with matching CTEs, using substrates with a compliant top
layer, and using leaded ceramic chip carriers.

Most of the effort in the United States has focused on the first
approach—that is, the use of substrates with compatible CTEs to relieve

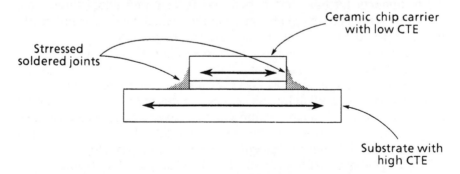

**Figure 4.5   CTE mismatch between substrate and component package
can stress the solder joint.**

stress on the solder joint. Examples of substrates selected for this approach are ceramic (Section 4.4), porcelainized steel (Section 4.4.1), and constraining core substrates such as copper-invar-copper and graphite epoxy (Section 4.5). This approach has not been entirely satisfactory because it is very expensive and solder joint reliability results have been less than desirable.

Even when the solder joint cracking problem is solved by using substrates with compatible X-Y CTE values, failures have been shifted to the plated through holes and via holes. The via failures were initially attributed to an increased CTE in the Z axis direction caused by the constraints of the X-Y axis. However, test data reported by Gray at the Printed Circuits World Conference 1987 [4] show that via hole reliability is a fundamental problem involving the plated through-hole aspect ratio (board thickness divided by drilled hole diameter). The large diameter plated through holes in conventional through-hole technology did not fail because the aspect ratio was less than 2.5.

For the small via holes used in SMT, thin boards are needed to maintain high plated through hole (PTH) reliability. However, there are some practical limits for thin boards, because they also need to be about eight layers because of increased interconnection density. The via hole size may be increased to maintain a low aspect ratio (<3), but larger via holes, which take up routing area, still may not match the Z direction CTE of a copper-plated hole. Stronger copper (tensile strength >42,000 psi) with ductility >20% is recommended for thin (35 mils) eight layer boards for reliable via holes.

Another potential solution may be to use blind and buried vias, illustrated in Chapter 5, Figure 5.4. Since the depth of blind or buried vias is small, smaller via hole sizes can be tolerated without increasing the aspect ratio, hence without compromising via hole reliability. However, the technology for blind and buried vias is very immature at this time. Very few vendors have production capability, which tends to significantly increase the cost of substrates with such a configuration.

A possible modification of the compatible CTE approach is to use a substrate material having a lower modulus of elasticity. This does not require an increase in via hole size to maintain a low aspect ratio, hence gives good via hole reliability. Since the modulus of the substrate material is low, it is soft and cannot rupture via holes even if there is a great deal of Z direction expansion of the substrate material. An example of a substrate material with low modulus of elasticity is the Rogers product RO2800. This material has a vertical CTE of only 24, but its modulus is only one-tenth that of epoxy. This property would make cracking unlikely even with very small via holes. The main problem with RO2800 is that the material is thermoplastic and is not suitable for prepreg. So fusion bonding at 800°F is used. Also, the industry does not have much experience with this approach, which has rarely been tried. At this time there are not enough data on the material.

The second approach, which has been tried, is to use a compliant substrate top layer, thus transferring solder joint stress to the "butterfly" top layers. However cracking of the traces and via holes remains a chief reliability concern. The compliant layer substrate is discussed in Section 4.6.

The third approach is to use leaded instead of leadless ceramic packages on substrates with unmatched CTEs, such that the stress on the solder joint is taken up by the compliant leads. There are many problems with this approach also. For example, the commonly used Kovar or alloy 42 leads are not very compliant and do not completely eliminate solder joint cracking problems unless a substrate with a compatible CTE is used. In addition, the leaded ceramic packages do not address the need for intimate contact between the substrate and the package, to ensure effective cooling. Even when thermal pads are used (always recommended if feasible), effective cooling is difficult to achieve since the package is off the board. For thermal pads to work efficiently, the package should touch the board surface.

For military applications, other approaches that should be considered are design related. For example, solder joint stress may be reduced by reducing the thermal stresses between the substrate and packages. If the package/substrate combination does not change temperature much, failures related to CTE mismatch will be reduced. Thermal cooling aids such as thermal vias, heat sinks, and immersion cooling have been used in this connection. These issues are further discussed in Chapter 5, Section 5.6 (Thermal Considerations) and Section 5.8 (Solder Joint Reliability Considerations).

The safest way to prevent reliability problems without increasing cost is to use plastic leaded packages on glass epoxy substrates. This approach is almost universally used. Any stress generated by minor differences in CTE values is easily accommodated by compliant copper leads in the PLCCs. The main problem associated with this approach is that it is limited mainly to commercial applications where the environment is very benign, e.g. −40° to +55°C. But the commercial use of glass epoxy boards accounts for the majority of the PCB firms' business, and these boards are the most widely used in the industry. Fabrication of glass epoxy substrates is discussed in detail in Section 4.7.

4.3.2 Process Considerations in Substrate Selection

No substrate can withstand all manufacturing environments equally well. Thus, when selecting a substrate, the process to which the material will be subjected must be considered. Surface mount assembly processes can have an adverse impact on the substrate material if extra care is not exercised with respect to the application of thermal and mechanical stresses, the removal of corrosive contaminants from the surface of the board, and so on.

Wave soldering, reflow soldering, repair/rework, and burn-in are assembly processes during which heat, or mechanical stresses, or both are applied to the entire board or to a localized region. Flux and other contaminants are removed during cleaning processes from the board. Most assembly processes are carried out at high temperatures.

Printed circuit boards are subjected to high temperature processes that include baking to drive off absorbed moisture; preheating before soldering, reflow soldering, or wave soldering; and repair/rework of boards and components. During wave soldering, the board reaches an even higher temperature than during baking or reflow soldering, but for a shorter time. The major side effect of these processes is to generate thermomechanical strains in the board as a result of the different materials expanding at different rates as the temperature increases above the ambient. These strain differences cause major stresses at crucial points in the board. The higher the temperature at which these processes take place, the higher the strains and stresses generated in the boards during processing.

Besides the material elongation with temperature, other minor effects are observed at high temperatures. The laminate will creep to a considerable degree if left at high temperatures for an extended time. ("Creep" is defined as the elongation of a material with time at elevated temperatures. Unlike thermal expansion, creep is an irreversible diffusion-controlled phenomenon.)

Shrinkage of the laminate will also occur as volatiles (i.e., moisture) are evolved. If the rate of moisture evolution is high, the resin in the laminate will shrink quickly and cause recession from the plated through-hole walls. Hence, the rate of temperature increases should be restricted. A low rate of temperature rise also diminishes thermal shock in the metal platings. As previously discussed, the laminate around a plated through hole was observed to expand more than the hole barrel and to increase in temperature. This disparity in elongation increased above the glass transition temperature.

Repair and rework generally require the application of heat (with a soldering iron or hot air) to a localized region on the surface of the board. The localized region is usually a pad and/or a component lead. Localized heating is even worse than heating the entire board to a particular temperature because it generates strains and stresses in one particular region. Such stresses are typically higher than those incurred when the entire assembly is heated to the same temperature.

Consider the case of a rework operation that may require the application of an 800°F iron to a hole pad to remove a component lead. The metal platings in the hole barrel and elsewhere are much better conductors of heat than the laminate. When the top of the soldering iron is applied to the pad, the hole barrel will become heated to a temperature higher than that of the surrounding laminate. Close to the barrel, the laminate will expand more than the barrel. But, when compared to the laminate expansion about 50 mils away from the barrel, the expansion of the barrel will be more than the expansion of the laminate.

Thus, the repair/rework operation generates more stress on the board than any other assembly process; therefore it must be undertaken with the utmost care. Otherwise, deformation of internal traces, as shown previously in Figure 4.4, will result. The time of repair/rework or the time of applying a soldering iron or hot air to the localized region should be kept to a minimum. Operator skill and minimum pressure on the pad are also very important. See Chapter 14, Section 14.6.1 for necessary precautions during repair and rework.

4.4   CERAMIC SUBSTRATES

The general appeal of ceramic substrates is that they have the same CTE as the ceramic package and do not pose any problems resulting from CTE mismatch. Ceramic substrates are widely used in hybrid applications, where small size is not a major concern. There are some other differences between the substrates used in hybrids and in surface mounting, however. For example, hybrids use bare or unpackaged ICs.

The commonly used compositions are pure (99%) alumina, 96% alumina, and beryllia. The high purity alumina, which is more expensive, is used only where higher strength is necessary. When high thermal dissipation is required, beryllia is used. The higher thermal conductivity of this metallic oxide, which is electrically insulative, is very useful for high power devices; however, beryllia dust does pose a health hazard.

Both thin film and thick film processes are used for building ceramic substrates. The thin film process is a subtractive method of defining conductors and resistors on a ceramic substrate. In thin film technology a resistive material such as tantalum nitride or nichrome is first deposited in vacuum; then a metallic surface (generally gold) is plated over it. Then, as in the subtractive process used in printed circuit boards, the gold is removed from the areas in which resistors are desired.

Gold wire bonding between IC pads and the gold pattern on a hybrid substrate poses no problem. In surface mounting applications, packaged ICs are used and are soldered to the surface. The gold conductors on the substrate are leached by the material used for soldering, and this may embrittle the solder joints. The industry has achieved some success in using copper instead of gold to prevent the embrittlement problem and to also reduce cost.

The thin film process is expensive, but very fine line widths (0.002 inch) are easily achieved. This somewhat compensates for the failure of thin film technology to accommodate more than a single layer of conductors. The thin film process does allow precise resistance on the substrates, but this is not a real advantage in surface mounting for two reasons. First, very precise resistance values are not necessary in the digital circuits commonly used today. And secondly, if really needed, discrete surface mount resistors with tight tolerances of 1% are available for surface mounting.

In the thick film process, line widths of 0.004 inch are possible, but generally 0.01 inch widths are produced. The thick film process has two general variations. In the conventional process the various conductive, resistive, and insulative layers are screen printed, dried to remove volatiles, and then fired. The firing of the paste is done at a much lower temperature (900°C) than the firing of the ceramic substrate, which is done at 1500°C. As many layers of paste as desired (up to 20 layers) can be sequentially deposited and fired to achieve the specified interconnect.

A variation of the conventional thick film process is the cofired process in which "green" layers of ceramic are printed with conductive, resistive, and insulative inks. Each individual ceramic sheet is dried, whereupon all the layers are stacked and fired simultaneously. This is somewhat similar to the lamination process used in printed circuit boards except that printed board lamination is accomplished at a much lower temperature than is used for cofiring.

The sequential printing and firing of various layers of paste produces undulations or peaks and valleys of resistors and conductors in the conventional thick film process. The cofired process, on the other hand, gives a very flat surface, which is much more desirable for surface mounting. It is not possible to deposit, fire, and trim resistors in the cofired process, however, as can be done in conventional thick film work.

Generally circuits are formed in "multipacks" (several identical circuit boards in panel form), and the individual boards are cut off by scribing or laser. Ceramic substrates are limited by smaller size, higher dielectric constant, and high cost. The most commonly used size is 2 in.^{2}, but sizes up to 7 in.^{2} are possible at an increase in cost.

4.4.1   Porcelainized Steel Substrates

Certain disadvantages of ceramic substrates, such as their smaller size and higher dielectric constant, can be overcome by using porcelainized steel instead of plain ceramic. For example, a 14 inch by 24 inch substrate size is possible with porcelainized steel. A well-known application of porcelainized steel substrates in mass productions is in the Flashbar for Polaroid SX-70 cameras [5].

Like any other substrate, the porcelainized substrate has its advantages and disadvantages. For example, the CTE of a porcelainized substrate is very high (13.3 ppm/°C), rendering the material unsuitable for leadless chip carriers [5]. Where a lower CTE is desired, a substrate of porcelainized copper-clad invar should be used instead of porcelainized steel. Both these substrates have lower dielectric constants than ceramic and can be used in relatively faster circuitry.

The procedure for screening the conductive, insulative, and resistive layers on porcelainized steel is the same as that used in conventional thick film technology, as discussed earlier. The firing of the paste is done at a lower temperature (about 650°C), however, because of the lower melting point of steel.

4.5   CONSTRAINING CORE SUBSTRATES

The constraining core approach for matching the CTEs of a component package and a substrate has served successfully in producing reliable surface mount assemblies using leadless chip carriers. The basic idea is to produce a "sandwich" substrate in which the core has a very low CTE.

The constraining core can be either functional or nonfunctional electrically, and it can be a metal or a nonmetal. Even when electrically nonfunctional, the core acts as a heat sink and a supporting plane. When the core is electrically functional, it also serves as the power and ground planes. The core is predrilled and filled with compatible nonconductive resin before lamination. The outer layers are processed as in conventional printed circuit glass epoxy board fabrication. Figure 4.6 presents construction details of a substrate with a constraining core.

Metal cores of various types are used to restrain the substrate from its normal expansion. Examples are alloy 42, copper-clad molybdenum, Kovar, porcelainized invar, metal matrix composite such as Lanxide (SiC-Al), and copper-invar-copper. The copper-invar-copper type of substrate is more commonly used.

4.5.1   Low CTE Metal Core Substrates

There are various metal substrates with low CTE that are used as constraining cores. Examples are CIC (copper-invar-copper), moly graph-

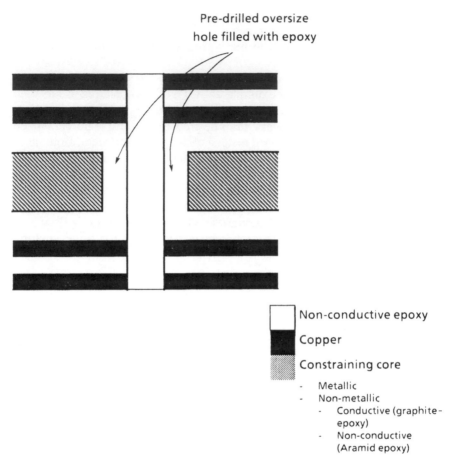

Figure 4.6   Construction detail of a multilayer board with constraining core.

ite moly, Beo, and SiC-Al (metal matrix composite). All these substrates are used to reduce the overall CTE of the board to be compatible with the CTE of the ceramic package. However, CTE is not the only consideration in selecting a particular constraining core material. Any trade-offs, in addition to CTE, should also include thermal conductivity and weight, especially if the application is for avionics.

Among various low CTE constraining core materials, the CIC metal core material pioneered by Texas Instruments is widely accepted in constraining core applications. It is a three-layer conductive metal composite of copper, invar, and copper generally referred to as Cu-In-Cu or CIC substrate [6]. Invar (an abbreviated form of "invariant") is an alloy composed of 36% nickel and 64% iron. It has practically no CTE. When invar is clad with copper, the CTE of the composite can be tailored by varying the thickness of the invar and copper laminates.

Copper-invar-copper consists of pure (99.95%) copper bonded to both sides of an invar layer. Invar alloy contains trace amounts of minor impurities. Reference 7 has established the requirements of a CIC metal foil. To form a substrate like the one shown in Figure 4.6, the copper-invar-copper sandwich is predrilled and then laminated with outer layers as in conventional multilayer epoxy glass board construction. The outer layers are regular copper-clad epoxy laminates and are processed in the same fashion as the usual glass epoxy substrates discussed in later sections.

By varying the thickness and the composition of the Cu-In-Cu plane, a substrate having the desired CTE value can be tailor made. Since copper has a higher CTE than invar, to lower the CTE of the composite laminate core, the thickness of each copper layer is reduced. This also increases the tensile strength of the core. However, as in any other metal, the greater tensile strength is acquired at the expense of ductility. Since the component packages run hotter than the substrate during power cycling, the CTE of the substrate should be slightly higher than the CTE of the package. Hence, depending on the wattage of the package, the CTE of the substrate should be 1 to 2 ppm higher than that of the package. In the beginning, the Cu-In-Cu sandwich contained 20% copper, 60% invar, and 20% copper, where the numbers refer to the percentage thickness of each metal in the core. The 20-60-20 metal core gives a CTE of about 5.2 ppm/°C (9.4 ppm/°F) [7].

A Cu-In-Cu core of 12.5-75-12.5 will give a CTE of 4.6 ppm/°C (8.2 ppm/°F) [7], slightly higher than that of the ceramic package. To achieve this in the total thickness of a 6 mil core, the top and bottom copper layers must be less than 1 mil thick and the invar layer about 5 mils thick.

Excellent solder joint reliability results were found with an 8-84-8 Cu-In-Cu core [8]. Even localized etching can be performed to tailor the CTE of the substrate to both connectors and ceramic chip carrier. The CTE value of the commonly used 8-84-8 core is about 3.9 ppm/°C (7 ppm/°F).

Both problems and successes have been reported in connection with Cu-In-Cu substrates. For example, leadless ceramic chip carriers ranging in size from 20 to 84 input/output connections have been tested to 1500 temperature cycles from −55 to +125°C without failure [8]. The problems that generally have been seen are delamination of the core from the glass epoxy outer layers and cracking of the via holes. A new red oxide copper treatment process has been found to improve adhesion of the Cu-In-Cu layers, hence decreasing the potential for delamination.

The cracking of the plated through holes is generally encountered when epoxy (a nonconductive material) must be used to fill in around the holes to prevent the shorting of circuitry from the conductive cores. In thermal cycling, the larger Z axis expansion of the adhesive material causes failure in the plated through holes. Epoxy filled with glass beads, clay, or mica helps to mitigate the via hole cracking problem.

The problem of via hole cracking can be prevented by substituting the clearance hole with a clearance "ring" in the voltage and ground planes. In other words, the size of the predrilled hole is reduced to provide a minimum amount of adhesive fill. Figure 4.6 shows an oversize predrilled hole filled with epoxy. The resulting "nonfunctional" pad reduces the fill volume and reinforces the plated through hole. Plated through holes using this approach have passed 3800 temperature cycles from −55 to +125°C without failure [8]. One must very carefully determine the minimum size of the via holes even in glass epoxy boards. Cracking of via holes is discussed in detail in Section 4.11.

4.5.2   Graphite Epoxy Constraining Core Substrates

As mentioned earlier, the constraining core can be either a metal or a nonmetal. One core material that has been used at Boeing with the company's patented process is graphite [9]. Graphite has a very low CTE and can be used to produce a composite substrate with glass epoxy that can match the CTE of ceramic. For example, the graphite filament core has a CTE of 0 to 1.6 ppm/°F [9] and the glass epoxy substrate has a CTE of 7.5 ppm/°F. The composite laminate can be tailored, as in CIC substrates, to produce the desired CTE.

A graphite epoxy substrate is constructed essentially in the same fashion as the copper-invar-copper substrate shown previously (Figure 4.6). First the constraining core, consisting of graphite filament reinforced with thermosetting resin, is prepared. Depending on thermal requirements, the resin can be either epoxy or polyimide. Since graphite, like metal, is electrically conductive, larger via holes must be predrilled and filled with nonconductive adhesive during lamination.

The graphite filament can be replaced with nonconductive fibers such as quartz or aramid. Otherwise, nonconductive adhesive must be used to prevent electrical shorting. The adhesive does pose the problem of large Z axis expansion. Thus, as in CIC substrates, a clearance ring instead of a clearance hole should be used; this will reduce Z axis expansion by minimizing the amount of adhesive.

The outer layers of a graphite epoxy substrate are processed like conventional glass epoxy or glass polyimide substrates. Then the constraining graphite core and the outer layers are laminated, drilled, and processed as in conventional substrates. The internal core and the outer layers are bonded with adhesive.

The graphite epoxy substrate generally suffers from short cracks that separate the epoxy from the graphite laminate. This problem, called microcracking, tends to cause a shift in the CTE values. For this reason, the electronics companies that used graphite epoxy originally have switched to Cu-In-Cu. Nevertheless, the graphite epoxy substrate does have some inherent advantages over the CIC substrate. For example, the graphite core, unlike a metal core, is a lightweight material and is used in many structural aerospace applications where weight is a major consideration.

4.6   COMPLIANT LAYER SUBSTRATES

The most common approach to eliminating the CTE differential problems encountered with ceramic packages and glass epoxy substrates is to use ceramic packages and the copper-invar-copper substrates just discussed. This is an expensive option, however, and substrates with a compliant layer offer an intermediate solution at a minor increase in cost. Because the stresses created by the CTE differential are taken up by the compliant or "buttery" top surface, it is possible to greatly reduce the solder joint reliability problem.

Compliant layer substrates are fabricated by the same processes used for the conventional glass epoxy substrates but with a minor difference, illustrated in Figure 4.7. The "compliance" comes from nonreinforced epoxy layers (i.e., layers having no glass fiber) added to one or both sides of the copper layers that surround the epoxy. Epoxy layers on both sides are preferred, to provide balance and to prevent board warpage.

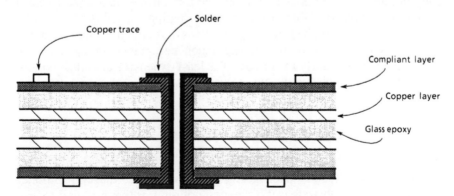

Figure 4.7   Compliant layer substrate construction.

The layers are generally 0.01 inch thick on each side. This dimension is achieved by laminating various sheets of nonreinforced laminates. Generally each sheet is about 0.002 inch thick. The nonreinforced laminate can either be epoxy or polyimide, depending on the base substrate.

The final thickness of the top epoxy layer determines the amount of stress on the solder joint. The higher the thickness, the lower the stress on the solder joint. For example, one researcher has found stresses of 100,000 and 5160 psi/joint in compliant layer thicknesses of 0.001 and 0.007 inch, respectively [10]. Naturally, the lower the stress on the solder joint, the more thermal cycling it can withstand.

Satisfactory reliability data on compliant layer substrates are available now. For example, 1200 failure-free thermal cycles (−55 to +125°C, 15 minute soak, <5 minute transition time between chambers) and 2000 power cycles on a 0.006 inch thick compliant substrate containing 20 to 52 pin leadless ceramic packages have been reported in military applications [10].

There are problems with compliant layer substrates, too, however. For example, the top epoxy laminate layer (nonreinforced laminate material) is susceptible to dissolution by some of the chemicals used in fabrication and assembly. The epoxy resin dissolves in methylene chloride, but alcohol and other solvents used in manufacturing do not cause problems.

There is also some concern about Z axis expansion which, depending on the thickness of the top layers, can be significantly higher than it is a reinforced laminate. In addition to causing cracking in the plated through holes, the "buttery" layer may be responsible for problems such as pad lifting during repair operations.

Also, air bubbles between the laminates may occur during fabrication, causing board delamination later. Hence, the process should be closely controlled, and operators must be instructed to watch out for air bubbles. The ductility of the copper plating on the top surface also must be controlled, to prevent possible cracking in the plating.

Either because of the foregoing problems or for other reasons, compliant layer substrates have not gained much acceptance in the industry despite their significantly lower cost. The least expensive and most widely used substrate, which we discuss next in detail.

4.7 GLASS EPOXY SUBSTRATES

When using leaded packages, especially plastic, the conventional glass glass epoxy substrate is the cheapest option. It is the most commonly used material for single, double, or multilayer printed circuit boards in both commercial and military applications. The substrate is basically a composite material consisting of epoxy (resin) and glass fibers (base material). The base material gives the laminate its structural stability and the resin provides ductility.

A composite structure is advantageous because it combines the favorable properties of its ingredients. For instance, glass fibers are strong but they break easily when bent. Forming a composite of the brittle glass fibers and a tough resin imparts better ductility to the composite because the glass fibers absorb most of the stress during bending.

The glass fibers are made essentially by the same process as any other glass: high temperature melting of sand. The molten glass is turned into glass fibers of different diameters, which are formed into bundles or yarns and assembled in the form of a cloth having the characteristic weave structure that adds strength to the laminate. There are many ways in which the glass fibers are tied together to form the yarns and many ways in which the yarns are woven together.

The epoxy resin dispersed within the weave structure of the glass cloth is the medium that holds together successive layers of the glass cloth weave. Besides the base epoxy, laminate resins contain other chemicals, specially added to impart certain properties. Curing agents promote cross-linking of the epoxy chains, thus hastening the cure. Stabilizing agents prevent the epoxy from decomposing at elevated temperatures during processing, assembly, and use. Flame retardants, such as bromine, make the laminate self-extinguishing, which is very important for safety. Still other additives called adhesion promoters enhance the adhesion of the glass cloth to the epoxy resin. This adhesion should be sufficient to avoid delamination of the glass cloth from the resin under mechanical or thermomechanical stresses.

The traces are made of copper used in the subtractive process. They are the unetched portions of the copper foil from the copper-clad base laminates. The critical characteristic for the traces is peel strength. A low peel strength will lead to trace delamination or pad lifting during assembly, component repair/rework, and handling. For FR-4 laminates, the peel strength should be at least 5 pounds per inch of trace width after a 1 hour conditioning treatment at 125°C.

Most copper foil used in multilayer laminates is electrodeposited copper. The typical purity of this copper is 99.8% and typical tensile strengths are 15,000 and 30,000 psi for ½ oz. and 1 oz. foil, respectively.

As we will see later on, various layers of laminates are processed (etched to form the required circuitry), laminated, drilled, and then plated to achieve the final interconnection.

4.7.1   Types of Glass Epoxy Substrates

Table 4.4, adapted from Reference 1, summarizes the types of glass epoxy laminate in use today in the printed circuit industry. We discuss now some of the commonly used glass epoxy substrates.

1. G-10 and G-11. These are glass fiber based laminates with an epoxy resin system; they can be drilled but not punched. G-10 lacks flame retardant; otherwise it is similar to FR-4, the most widely used laminate in the United States today.
2. FR-2, FR-3, FR-4, FR-5, and FR-6. These laminates are characterized by the presence of a flame retardant (hence the FR designation), which makes the laminates self-extinguishing if a component shorts out and causes a fire.

FR-2 is a paper-based material with a phenolic resin system.

FR-3 is paper based also but has an epoxy-based resin.

FR-4 laminate, the industry favorite, has glass fibers as the base material and epoxy as the resin.

FR-5 is similar to FR-4 but has better strength and electrical properties at higher temperatures.

FR-6 has a glass fiber cloth and polyester resin system; it is designed for low capacitance or high impact resistance.

The G-10 and FR-4 laminates are the most common laminates for multilayer boards. These laminates are cheap and easily processible into multilayers by using layers of partially cured material called "prepreg" with some glass and resin system. They are laminated using heat and pressure. The completed laminate can be drilled and plated for interconnection among the layers.

Other laminates that are not glass epoxy are commonly used in certain applications. Examples of some of them are given below.

1. Polyimide. This laminate has a glass fiber base and a polyimide resin system. It has better strength and stability at high temperatures than FR-4. Hence, it finds widespread use in the military.
2. GT and GX. These laminates have a glass fiber base too, but the resin is Teflon, which imparts to the materials a controlled dielectric property that permits their use when laminate dielectric constant is critical. GX has a better tolerance of dielectric properties than GT and finds use in high frequency applications.
3. XXXP and XXXPC. These paper-based laminates with a phenolic resin are punchable only; they cannot be drilled. The XXXPC

Table 4.4 Laminate materials used in printed circuit boards [1]

COMMON DESIG-NATION	RESIN SYSTEM	BASE MATERIAL	DESCRIPTION
XXXP	Phenolic	Paper	Punchable at room temperature.
XXXPC	Phenolic	Paper	Punchable at or above room temperature. XXXP and XXXPC are widely used in high volume *single*-sided consumer products.
G-10[a]	Epoxy	Glass fibers	General purpose material system.
G-11	Epoxy	Glass fibers	Same as G-10, but can be used to higher temperatures.
FR-2	Phenolic	Paper	Same as XXXPC, but has a flame retardant (FR) system that renders it self-extinguishing.
FR-3	Epoxy	Paper	Punchable at room temperature and has flame retardant.
FR-4[a]	Epoxy	Glass fibers	Same as G-10, but has a flame retardant.
FR-5	Epoxy	Glass fibers	Same as FR-4, but has better strength and electrical properties at higher temperatures.
FR-6	Polyester	Glass fibers	Designed for low capacitance or high impact resistance; has flame retardant.
Polyimide[a]	Polyimide	Glass fibers	Better strength and demonstrated stability to a higher temperature than FR-4.

Table 4.4   *(Continued)*

COMMON DESIG-NATION	RESIN SYSTEM	BASE MATERIAL	DESCRIPTION
GT or GX[a]	Teflon	Glass fibers	Controlled dielectric laminate. GX has better tolerance of dielectric properties than GT.

[a]Principal materials for multilayer PCBs.

laminate can be punched at or above room temperature, but XXXP can be punched only at room temperature. These laminates are widely used in Japan for consumer products.

G-10, FR-4, polyimide, GT, and GX can be used to make multilayer boards; XXXP and XXXPC are used only for single- and double-sided boards.

4.7.2   Operating Temperatures for Glass Epoxy Boards

Each laminate has a characteristic highest continuous operating temperature. That is, the laminate will break down and the assembly will not function properly if the board is used above this temperature. The highest continuous operating temperature is determined by either electrical properties or mechanical properties (see Table 4.5 which is adapted from Reference 1, page 2–22). As indicated in Table 4.5, FR-2, FR-3, and FR-6 can be used to only 105°C; XXXP and XXXPC can be used up to 125°C; FR-4 can be used up to 130°C for electrical and mechanical factors. G-11 and FR-5, on the other hand, can be used up to 170°C for electrical factors and 180°C for mechanical factors. Polyimide, which can be used to 260°C, is usually classified as a high temperature laminate. GT and GX can operate up to relatively high temperatures, too: about 220°C.

4.7.3   Fabrication of Glass Epoxy Substrates

Depending on the materials, processes, and equipment, there are many ways to fabricate a multilayer board. The major steps in the commonly used

Table 4.5 Highest continuous operating temperature for common printed circuit board materials [1].

	HIGHEST CONTINUOUS OPERATING TEMPERATURES			
	ELECTRICAL FACTORS		MECHANICAL FACTORS	
MATERIAL	°C	°F	°C	°F
XXXP	125	257	125	257
XXXPC	125	257	125	257
G-10	130	266	130	284
G-11	170	338	180	356
FR-2	105	221	105	221
FR-3	105	221	105	221
FR-4	130	266	130	248
FR-5	170	338	180	356
FR-6	105	221	105	221
Polyimide	260	500	260	500
GT	220	428	220	428
GX	220	428	220	428

processes are listed in Table 4.6 and Figure 4.8. There are two major processes: tin-lead plating and solder mask over bare copper (SMOBC). Steps 1 through 34 (Table 4.6) are the same for both processes; steps 35 through 45 apply to SMOBC only. Table 4.7 compares the final process steps in SMOBC and tin-lead boards. Let us discuss and illustrate some of the process steps in detail.

The final substrate is achieved in three stages. Compounding or mixing the epoxy is called the A stage. In the B stage, very large sheets of the glass fabric are impregnated with the resin and dried to an intermediate cure. In the C stage, copper sheets and the B stage material are pressed together. Handling and storage of the B stage material is very critical in determining its final properties; it must not absorb moisture, and it must be allowed to stabilize before being converted to the C stage (lamination of copper sheet).

A typical board vendor begins the process with a C stage (fully cured) laminate containing glass epoxy and copper. He also uses B stage material (no copper on outer layers) during lay-up for lamination (step 17) to achieve the dielectric separation between copper layers. In steps

Table 4.6 Process steps in fabrication of multi-layer boards.

1. Raw material
2. Shear C stage material into panel sizes
3. Identify material with markings
4. Extra material cure (bake) [optional]
5. Drill or punch registration tooling holes
6. Material scrub
7. Apply dry film photoresist (inner layers)
8. Artwork image transfer (inner layers)
9. Develop artwork image (inner layers)
10. Etch inner circuit patterns
11. Remove photoresist
12. Inner circuit inspection
13. Inner circuit test
14. Scrub for lamination/oxide treatment
15. Bake
16. Shear B stage and punch registration holes
17. Layup for lamination
18. Lamination
19. Remove laminate from fixtures
20. Laminate trim (shear epoxy flash)
21. Mark ID number on panels
22. Postlaminate bake
23. Drill plated through-holes
24. Deburr holes
25. Hole cleaning (etch back)
26. Plated through-hole sensitization (electroless copper)
27. Scrub for photoresist application
28. Apply dry film photoresist (outer layers)
29. Artwork image transfer (outer layers)
30. Develop artwork image (outer layers)
31. Electroplate copper
32. Electroplate tin-lead
33. Strip dry film photoresist
34. Etch copper circuits
35. Strip tin-lead
36. Apply solder mask and legend/cure
37. Hot air level tin-lead solder/clean
38. Tape board for edge connector plating
39. Electroplate nickel-gold on edge fingers
40. Remove tape/clean
41. Drill nonplated holes
42. Route boards to final contour and bevel leading edge
43. Final test
44. Final inspection
45. Packaging/shipping

2 and 3, the large sheets of C stage material are cut down to size and marked by the fabricator for processing. To improve the dimensional stability, the C stage material is baked (step 4). After tooling holes have been punched in all layers for accurate alignment, the laminate is ready for developing the image. Dry film resist is laminated on the copper

Inner Layer Process

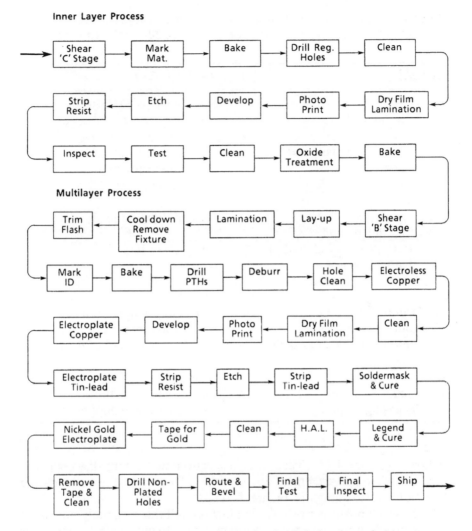

Figure 4.8 **Summary of process steps in the fabrication of glass epoxy board; see text for details.**

surface; then the area that will eventually have the copper traces is exposed to UV light to receive the artwork image transfer (see Figure 4.9). The dry film that was not exposed to UV light is removed, leaving behind copper covered with dry film and exposed copper.

The exposed copper is etched away in the commonly used subtractive process, with dry film serving as the etch resist. Figure 4.10 shows copper before and after etching. Now the dry film has served its purpose and is

Table 4.7 **A comparison of solder mask on bare copper (SMOBC) and tin-lead plating process steps. Process steps 1 to 34 (Table 4.6) are the same.**

SMOBC	TIN-LEAD
	1-34. Same as SMOBC
35. Strip tin-lead	35. Tape board for edge connector plating
36. Apply solder mask and legend/cure	36. Plate nickel-gold on edge connector fingers
37. Hot air level tin-lead solder/ clean	37. Remove tape/clean
38. Tape board for edge connector plating	38. Reflow tin lead
39. Electroplate nickel-gold on edge fingers	39. Clean boards
40. Remove tape/clean	40. Apply solder mask and legend
41. Drill nonplated holes	41. Drill nonplated holes
42. Route boards to final contour and bevel leading edge	42. Route boards to final contour and bevel leading edge
43. Final test	43. Final test
44. Final inspection	44. Final inspection
45. Packaging/shipping	45. Packaging/shipping

removed (Figure 4.11). Then after inspection and testing, the copper is given black oxide treatment to promote adhesion to the resin during lamination (also shown in Figure 4.11).

Now the board is ready for lamination. All the separately processed layers just described are put in the lamination fixture, with "prepreg" as the B stage material is called, between the layers (see Figure 4.12). After pressure and heat have been applied, with proper controls, the excess resin from the B stage material that has oozed out is sheared off.

The laminated board is drilled, and the burrs are removed using pressurized air or water. The holes are cleaned with sulfuric acid, potassium permanganate (most common), or plasma (see Figure 4.13). Electroless copper is then plated to a thickness of 10 to 20 microinches on the nonconductive glass epoxy in the plated through hole.

Now the outer layers are processed in the same fashion as the inner layers. In other words, steps 28, 29, and 30 are a repetition of steps 7, 8, and 9. Then the entire board except for the area covered with dry film

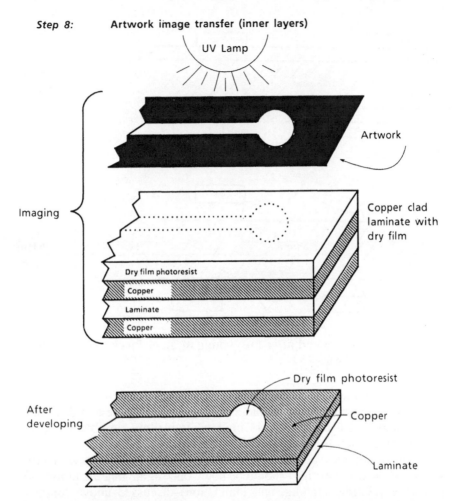

Step 8: **Artwork image transfer (inner layers)**

UV Lamp

Artwork

Imaging

Copper clad
laminate with
dry film

Dry film photoresist

Copper

Laminate

Copper

Dry film photoresist

After
developing

Copper

Laminate

Figure 4.9   Using dry film photoresist for the development of circuitry (steps 8 and 9, Table 4.6).

receives electrolytical plating (with copper followed by tin-lead). The dry film has served its purpose and is now removed. Next the tin-lead acts as an etch resist while copper is being removed. Then the tin-lead is removed to expose the bare copper underneath, and solder mask is applied over any copper that does not need to see solder later. The areas that will be soldered are not covered with mask. Solder is applied to these areas with the hot air leveling (HAL) or HASL (hot air solder leveling) process.

After the solder mask application, the legends are applied. The inks used in the legend on the outer layers of a multilayer board are made of a two-part epoxy. That is, the two components are mixed together and screened onto the board, which is then cured. The legend color should

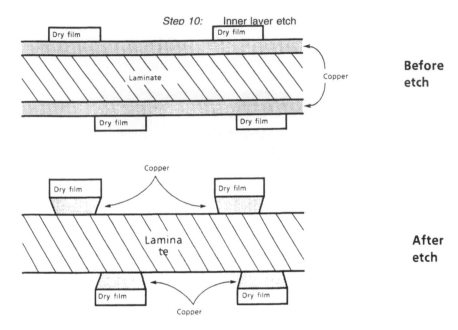

Figure 4.10   Before and after the etching of inner layer circuit patterns (step 10, Table 4.6).

be such that the legend is legible on the solder mask. A color that doesn't form a good contrast with the solder mask color should be avoided. Since solder masks are usually green or blue, white, yellow, and black inks are most often chosen for legends.

This completes all the process steps. Figure 4.14 shows a cross section of the final result. It should be noted once again that the fabrication processes (plating, etching, and lamination) discussed above for glass epoxy boards also apply to boards with compliant layers and constraining core.

4.8   PLATING PROCESSES

There are two main types of plating processes: electroplating and electroless plating. Electroplating is an electrolytic chemical deposition process that uses electricity as the driver to move molecules of metal to a target. Electroless plating is a chemical deposition process requiring a reducing agent that takes metal out of the chemical bath and deposits it on the target. Plating is a critical process, one of those that determine the quality of the board. There are many different types of metal plating in

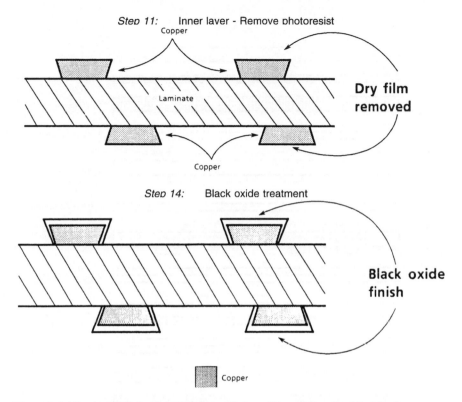

Figure 4.11   Removal of photoresist and treatment with black oxide (steps 11 and 14, Table 4.6).

a multilayer board, including copper, gold, nickel, tin, and solder, and they are found in many locations.

The bonds between the various platings differ markedly in strength. For example, the bond between the solder plating and the copper electrolytic plating is a metallurgical bond and is very strong because the two plating layers are held together by the compounds called intermetallics, which are formed between them. On the other hand, the bond between the electroless barrel plating and the inner trace is relatively weak because it is a physical bond of adhesion. Any process-related problem such as the presence of smeared epoxy on the edge of the inner trace prior to electroless copper plating is likely to cause a reduction in the contact area between the trace edge and the plating and consequently to weaken the bond even further.

The basic plating mechanism is similar for all metals. A solution containing the metal ions is the source of the metal, and plating occurs

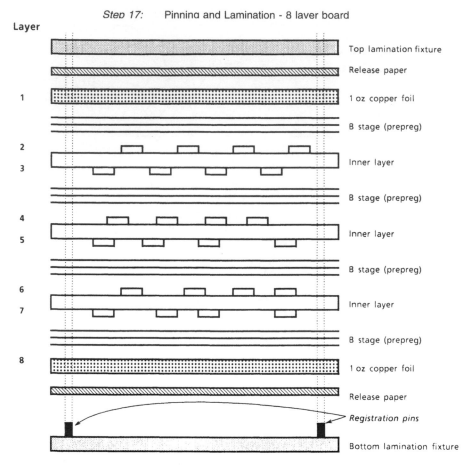

Figure 4.12 Layup for lamination of an eight-layer board (step 17, Table 4.6).

when the positively charged metal ions capture electrons from this specific source, depositing the metal on a substrate in the solution.

4.8.1 Copper Plating

There are two types of copper plating: electroless and electrolytic. The difference lies in the source of the electrons. For electroless copper plating, in which no electrical current is used, the source is a chemical reaction. In electrolytic copper plating, the current that is passed through the plating solution serves as the source of electrons.

Step 25:   Desmear and etchback (hole cleaning)

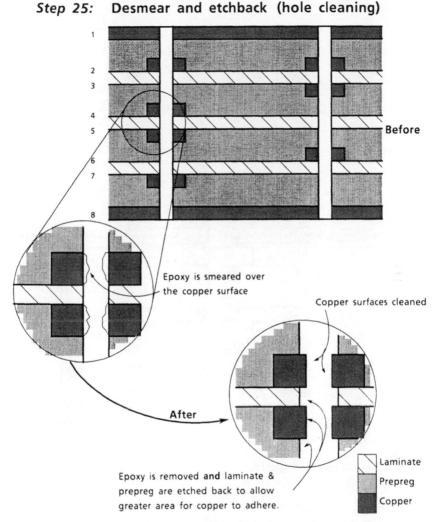

Figure 4.13   Hole cleaning after drilling (step 25, Table 4.6).

The purpose of electroless copper deposition is to make the nonconductive surfaces within the drilled hole conductive so that subsequent plating operations can take place electrolytically. The purposes of electrolytic copper plating are to provide a connection between the layers in a multilayer board and to strengthen the barrel of the plated through hole.

Many plating solutions are used for electrolytic copper deposition. These include copper sulfate, copper fluoroborate, and pyrophosphate solutions. The tensile strength (the maximum stress that the metal plating

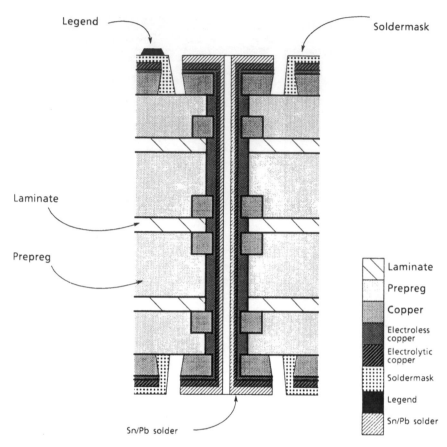

Legend

Soldermask

Laminate

Prepreg

Laminate

Prepreg

Copper

Electroless copper

Electrolytic copper

Soldermask

Legend

Sn/Pb solder

Sn/Pb solder

Figure 4.14 A finished board.

can withstand under tension) of the plating solutions is about 3 to 5 × 10^4 psi. The higher the tensile strength, the stronger the plated through hole. However, the ductility (the extent to which the copper plating can be "stretched" or elongated before fracture) differs markedly. A high value of ductility is preferred to allow "yielding" before failure instead of sudden failure.

Electroless plating offers very low ductility (<1% elongation) compared with electrolytic platings—a good reason for requiring electrolytic plating for plated through-hole strengthening. The fluoroborate and pyrophosphate plating solutions give a better ductility than the copper sulfate plating bath.

The residual stress in the platings also differs substantially. It is compressive in electroless copper plating and tensile in the electrolytic form. A compressive residual stress enhances the adhesion of the electroless plating to the nonmetal laminate and to the inner layer trace, comprising another reason for using electroless copper before electrolytic copper.

4.8.2 Gold Plating

Gold plating is deposited electrolytically as a contact surface because gold has a low contact resistance and a high resistance to corrosion. Gold is also used as a tarnish-resistant layer because it doesn't oxidize even at elevated temperatures. Gold sometimes is also used as a metal etch resist during board fabrication.

There are four different types of plating solution, which are classified according to their pH as follows: acid, neutral, alkaline cyanide, and alkaline noncyanide. The properties of gold deposits (such as hardness and porosity) depend on the plating solution. Acid gold is the most common plating solution for connector pad plating. Cobalt is used as a brightener in acid solutions.

4.8.3 Nickel Plating

Nickel plating, which is usually deposited electrolytically, forms an undercoat to precious metals such as gold to improve wear resistance during use. Nickel also forms an effective barrier layer to intermetallic formation. Intermetallic formation is essential but can be detrimental if thickness is excessive.

4.8.4 Lead-Tin Solder Plating

There are two methods by which a lead-tin solder coating is applied: electroplating and solder coating by dipping and hot air solder leveling.

In lead-tin plating, difficulties occur if the current density is not properly controlled. Uneven composition may result, for example, if the tin composition is decreased as the current density is lowered. This would cause formation of a lead-rich solder, which is less solderable. Even with good control, however, weak "knees" are encountered in the plated through holes. Also, since plated surfaces are porous, they need to be reflowed to fuse the solder. This tends to make the surface uneven with thinner solder on the edges and thicker solder in the middle of the pad. Control of plating thickness is also a problem, especially when thick coatings are required. Electroplated solder is notoriously porous, but it provides excellent control over the plating thicknesses. Fusing the solder after electroplating gives a solder coating without porosity.

The other option for applying solder to the board surface is a coating

Table 4.8   Solder thickness (in μ inches) on various types of lands on the same board from a hot air solder leveled (HASL) board

TYPES OF DEVICES	MIN SOLDER THICKNESS	MAX SOLDER THICKNESS	AVERAGE THICKNESS
Discrete devices	33	900	135
50 mil pitch devices	68	866	228
25 mil pitch devices	151	600	284

process commonly referred to as the HASL (hot air solder leveling). It is a physical deposition process using bulk liquid metal into which the board is dipped to apply coatings. A secondary step (hot air blow) is used to level or reduce the amount of material contained on the surface.

The advantage of HASL over electrodeposition is that HASL solder coatings are dense and have good adhesion (due to intermetallic formation). It also acts as an instant solderability tester in a production process because solder will not adhere to a poor surface in HASL.

There are some disadvantages with HASL as well. The nonuniformity of the coating (across the pad cross section) from HASL can lead to solderability problems, especially in surface mount assemblies. As shown in Table 4.8, 33 to 900 μin. (microinches or 0.9 mils) variation was found with hot air leveling. There are problems with both thin and thick surface coatings, especially in surface mounting.

Thinner surfaces (<80 μin.) are definitely not desirable because they are almost entirely consumed as intermetallic during reflow, especially if the assemblies are reflow soldered twice (primary and then secondary side, as in double-sided boards). Since intermetallic compounds passivate easily if exposed, they have poor solderability.

Thicker coatings, especially with wide variations within the same land pattern of the same package, are not good because they compound the coplanarity problem for fine pitch devices. Thicker coatings are good for solderability, but anything exceeding 300 μin. is really not necessary.

Solder thickness uniformity has the same effect on assembly as components lead coplanarity. Board warpage has a similar impact as solder finish variation and lead coplanarity. Warpage on standard 62 mil thick boards should be held at 0.5% as opposed to 1%, which is the industry standard.

Thermal shock to boards during hot air leveling is another disadvantage because of the potential for board warpage and hence paste printing problems. Be that as it may, HASL (hot air solder air leveling) is the most widely used process.

The other alternative to HASL for a solder surface is SSD (solid solder deposition). There are two types of SSD processes—chemical and physical deposition. The chemical process is used to apply additional tin lead material to the appropriate thickness necessary for component attachment. In the physical deposition process, bulk metal is forced into cavities of a mask to provide the solder needed for surface attachment. Because of inconsistent quality and high costs the SSD processes have not been widely accepted, even though they showed a lot of promise in their early days.

The use of nonsolder anti-tarnish finishes such OSP (organic solderability protection) and noble metals are some of the other techniques used to achieve the more uniform surface finish that is so critical for fine pitch and ball grid arrays. These techniques are discussed next.

4.9   ALTERNATIVE COATINGS FOR BOARD SURFACES

As the usage of fine pitch devices has increased, the need for a uniform and consistent soldering surface has become more critical. Finer pitch components such as TSOPs (thin small outline package), QFPs, and micro BGAs require a flatter surface. Because of the inherent problem of uneven solder thickness in the most widely used process (HASL), the industry is looking for alternative processes. Examples of these alternatives are organic solderability protection (OSP) and noble metal coatings such as palladium and gold coatings. See Table 4.9 for key attributes of various coatings.

OSP is an anti-tarnish coating of an organic compound (benzimidazole-based compound) over a copper surface to prevent oxidation. It is a water-based organic compound that selectively bonds with copper to provide an organometallic layer that protects the copper. Various chemistries of OSPs are available. Some common ones are benzotriazol, imidazol, and benzimidazol. They keep the copper surface solderable. The OSP surface has many advantages. For example, it keeps the surface flat and it is lead free and hence may be considered more environmentally friendly.

OSP is applied in thicker (8–20 μin. or 0.2–0.5 μm) or thinner (4–8 μin. or 0.1–0.2 μm) coatings on a copper surface. OSP provides flat pads compatible with the gold surface on the board (OSP bonds selectively with copper). So many companies are using or considering the use of OSPs as a replacement for HASL that OSP is the new industry trend. In addition to providing a flat surface, OSP-coated boards have the potential for lower overall cost (board and assembly cost). The results of one study

Table 4.9 Key attributes of various board surface finishes

FEATURE	COATING TYPE				
	HASL	OSP	NiAu	NiPd	
SMT pad surface topology	Dome	Flat	Flat	Flat	
Solderability (as received)	Good	Good	Good	Good	
Max reflow cycles	>3	<2	>3	>3	
Shelf life	>6 months	<1 week	>6 months	>6 months	
Handling	Normal	Critical (finger print concerns)	Normal	Normal	
Exposed copper	No	Yes	No	No	
Use on thin boards (PCMCIA)	No	Yes	Yes	Yes	
Waste treatment and safety	Poor	Good	Fair	Fair	
Process control	Poor (thickness control concerns)	Fair/Poor	Fair (bath control)	Fair (bath control)	
Coating thickness	33–1500 μin.	8–20 μin.	15–20 μin.	5–20 μin.	
Cost	1	0.4–0.6	3	1.13	

show that the shelf life of the OSP coating is equivalent to that of the HASL coating [11].

However, there are some issues with OSP as well. The OSP surface is not robust, especially when the surface is very thin. There is the potential for exposed copper on the surface, which leads to nonwetting. There are some potential process incompatibilities as well. For example, if paste or flux does not cover all pad surfaces during soldering, there may be insufficient hole fill in wave soldering and a dewetted appearance in reflow soldering (on the edges and corners of surface mount pads). This is why it is important for flux to get into the PTH during wave soldering to achieve top side fillet. And in the SMT process the paste must cover all of the pad surface to avoid a dewetted appearance at the pad edges. There is the potential for incompatibility with no-clean flux and terpene-based solvents. Also, OSPs may not withstand multiple heat cycles and the solderability may degrade since OSPs easily degrade at temperatures over 100°C in the presence of air, leaving the underlying copper surface susceptible to oxidation [12]. There is also some concern that test probes may not make adequate contact during electrical testing.

The other alternatives to HASL are noble metal coatings such as palladium and gold. The purpose of the gold or palladium coating is to prevent oxidation of the underlying nickel layer to achieve better solderability. Palladium is the poor man's noble metal. It costs about one third as much as a gold surface as indicated in Table 4.9. With Pd plating, there is no problem in hole filling during wave soldering as may be the case with OSP. A thin layer of electroless palladium (5 to 10 μin.) is applied over a barrier layer of 100–200 μin. of nickel-phosphorous (Ni-P) over the copper surface.

Dealing with an electroless nickel bath is very difficult because phosphorous is co-deposited. In an alternative process, a thicker layer of palladium (about 20 μin.) is deposited directly over the copper surface. The thickness of palladium is controlled by the time of plating. The rate of plating is generally 2.5 μm to 5 μm (100 μin. to 200 μin.) per hour. With palladium plating, the solder joint is formed with intermetallic between copper and tin and not with palladium.

The alternative to palladium is gold. Gold surface comes in various varieties. The electrolytic plating process can be used to apply up to 100 μin. (2.5 μm) of gold to achieve good corrosion resistance. But cost and the formation of brittle gold-tin intermetallics are of concern with the high thickness of gold. On the other hand, the electroless plating process can apply 15–20 μin. of gold.

Another method of applying gold is a chemical, self-limiting ion exchange process known as immersion gold. It is applied over a 100–200 μin. Ni-P barrier layer. One can achieve a deposit of 5 to 10 μin. of pure gold in the immersion gold process. Both immersion gold and electrolytic gold are also wire bondable. The concern with this process is solder joint reliability.

In general, metallic coatings such as gold and palladium tend to provide better solderability and thermal stability than OSPs. In one study [13] where the authors evaluated three types of noble metal coatings (5 μin. Au over 150 μin. Ni, 10 μin. Pd over 150 μin. Ni, and 20 μin. Pd over copper) for solderability, they found that the surface degraded after 8 hours of steam aging due to diffusion of Ni and its subsequent oxidation. However, they found that OSP (azole) was more susceptible to solderability degradation under similar 8 hour steam aging conditions.

HASL is much thicker in comparison to any of these coatings and hence provides better solderability even after steam aging. Nothing solders like solder. Despite its disadvantages, HASL is currently the most widely used surface finish. Its use will decline, however, as finer pitch surface mount devices become more widespread.

In summary, there are various options for PC surface finish—namely HASL, OSP, and noble metal coatings. All of these finishes have advantages and disadvantages. The selection of a specific coating should be based on the specific application and familiarity with all the issues related to that technology.

4.10   SOLDER MASK SELECTION

The solder mask is a polymer. However, unlike the laminate it is not a composite but a homogeneous material. As the name suggests, this material is used to mask off the outer areas of the board where solder is not required to prevent bridging between conductors. In the past not all boards required a solder mask because the traces and pads were spaced quite far apart. The solder bridges between adjacent traces during wave soldering were not critical. But with the advent of fine lines and spaces, the use of a solder mask has become almost mandatory for boards that are going to be wave soldered. Refer to Chapter 7, Section 7.8 for additional information on use of solder mask.

On a full SMT board (Type I SMT), where no wave soldering is required, tenting of via holes for testing may be more critical than the bridging of adjacent conductors. Tenting—the application of solder mask to block or plug a via—allows closer spacing between a via and the adjacent conductor lines. Tenting also helps in bed-of-nails testing, where almost all the via holes must be sealed to provide a good vacuum for the text fixture.

The two broad categories of solder masks are permanent and tempo-

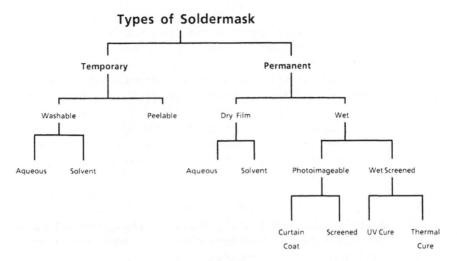

Figure 4.15   Types of solder mask.

rary, but there are subdivisions, as shown in Figure 4.15. The focus of
this section is on permanent solder masks, but let us briefly mention
the temporaries.

Temporary solder masks are washable or peelable. They are used
during wave soldering to prevent filling of holes that must be kept open
for the installation after soldering of such leaded items as unsealed parts
that might not withstand the cleaning or soldering environments. The
temporary masks that can be peeled off are also used to mask off gold
pads, which must not be soldered.

Washable masks are more convenient than peelable masks because
they come off in the cleaning operation after wave soldering and do not
need an extra step to remove them. The aqueous washable solder masks
require an aqueous cleaning system, and the solvent washable masks must
be washed in a solvent cleaning system. Let us now turn to the permanent
solder masks.

4.10.1   Wet Versus Dry Film Solder Masks

Permanent solder masks come in dry film and wet film. Dry film
masks can have an aqueous or a solvent base. In both cases, the mask starts
out as a polymer film, which is applied to the board by vacuum lamination.

Wet film solder masks, as the name implies, are liquid or pastelike.
They include photoimageable and wet screenable solder masks. The latter
are differentiated by the method of cure. Some wet screenable solder

masks can be cured by UV light and some can be cured thermally in convection or IR ovens. UV masks do not provide as good adhesion as thermal masks but require only seconds to cure as opposed to 30–60 minutes for thermal cure.

Each of the solder mask categories has advantages and disadvantages. They are inexpensive and highly durable. Being liquid, they flow between conductors and prevent the formation of air pockets. There is no trim waste, and the thickness of the mask can be controlled for each design.

Since the wet film solder masks are screened on (a mechanical process), they are difficult to register and have a tendency to skip over traces, especially on fine line boards. They also tend to bleed onto the pads and surface mount lands during cure. Wet screenable masks are difficult to use on boards with fine lines and spaces (<8 mils). They are also vulnerable to voids, bubbles and pin holes. The use of screenable mask has been on decline as photoimageable solder masks (discussed in Section 4.10.2) have gained in popularity.

The wet film solder mask cannot successfully tent via holes. Generally these materials fill the vias partially, which can prevent bridging but is ineffective in sealing holes to prevent vacuum leakage during testing. Partially filled vias trap process chemicals and are difficult to clean.

Dry film solder masks have some advantages over their wet screen counterparts. The former provide very accurate registration, which is critical in preventing solder bridging or bleeding on fine line boards, as well as sharper resolution. Tenting of via holes is also superior with dry film solder masks because they are never in a liquid state and do not drip into the via during vacuum lamination. However some problems arise when trying to laminate a semisolid dry film over an uneven board surface with traces and pads. Any warping and twisting of the board will compound the problem, possibly causing air pockets underneath the dry film near traces.

The curing of a dry film solder mask is extremely critical for achieving a reliable coating. Insufficient cure can cause cleaning problems because of diminished resistance to chemical attack by fluxes or cleaning solutions. Overcure results in a brittle mask, which can crack easily under thermal stress.

Most dry film solder masks are not resistant to thermal shock. Cracks develop in the cured masks within 100 cycles of thermal shock cycling from +100° to −40°C. This can be a problem especially in SMOBC boards because of exposed copper traces. However, Davignon and Gray [14] found a commercially available solder mask (Dynachem Conformask) that resists cracking during thermal shock.

Dry film masks are more costly than the wet film variety, and the vendor base is limited. Moreover, the application process for dry film solder masks is very difficult to control. Not many film thicknesses are commercially available, and this can limit flexibility and increase cost. Typically, most dry film solder masks are quite thick—3 to 4 mils. Trim waste also adds to cost.

Nonwetting of surface mount boards containing wave soldered chip components sometimes occurs because of the greatest thickness. In addition, the thicker mask surrounding the small vias may prevent them from filling with solder during wave soldering (crater effect).

The thicker masks can cause problems in reflow soldering as well. For example, a dry film mask applied between the pads of passive surface mount devices can cause tombstoning (standing up on edge) during reflow soldering because of the rocking effect of the mask. For this reason dry film solder mask should not be used between the pads of chip resistors and capacitors or in assemblies that have components glued to the bottom side for wave soldering. This property of dry film can be desirable in some applications, however. For example, as shown in Figure 4.16, the higher thickness of dry film over copper can be used to provide a solder dam. This can be very helpful for LCCC packages in achieving good solder fillets.

4.10.2   Photoimageable Solder Masks

Photoimageable solder masks combine the advantages of dry film and wet solder masks. Dry film is also a photoimageable mask. In this section, however, our discussion is focused on wet film photoimageable masks, which provide accurate registration, are easy to apply, encapsulate the circuit lines totally, have excellent durability, and are cheaper than dry film [15].

Photoimageable masks can be either screened on or applied by a process called curtain-coating, in which the board is passed at high speed through a curtain or waterfall of solder mask.

The photoimageable mask may contain solvent along with photopolymer liquid. If the solvent is added in the mask, the liquid mask is screened on, solvent is dried off in an oven, and then the mask is exposed to UV light by off-contact or on-contact methods. (If no solvents are used, the liquid is 100% reactive to UV light.) The off-contact method requires a collimated light system to minimize diffraction and scatter in liquid. This makes the system very expensive. The on-contact approach needs no collimated UV light source, and the system is relatively cheaper.

Photoimageable solder masks can tent only very small via holes. Most photoimageable wet film masks will not even tent 0.014 inch via holes because it is difficult to cure polymer in via holes. If tenting is

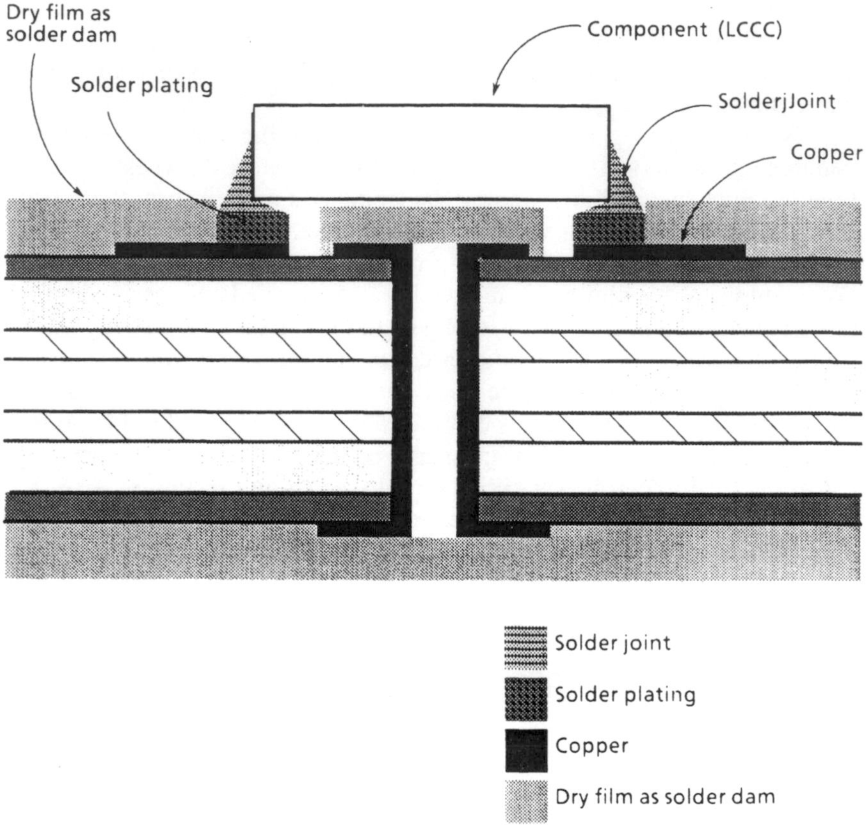
Dry film as solder dam
Component (LCCC)
Solder plating
SolderjJoint
Copper
Solder joint
Solder plating
Copper
Dry film as solder dam

Figure 4.16 Use of dry film solder mask for solder dam and tenting of via holes.

required, dry film is needed because only dry film can tent via holes effectively.

4.11 VIA HOLE CRACKING PROBLEMS IN SUBSTRATES

All the substrates discussed in this chapter have at least one problem in common: the cracking of via holes. A via hole is defined as a plated through hole that is used only for interconnection, not for the insertion of leads.

As packaging density on the board has increased, smaller and smaller via holes have become essential to save on real estate and to compensate for the impact on routing channels of higher density. As the via holes get smaller, their reliability becomes questionable. This is because thermomechanical fatigue generated by the laminate T_g causes cracking and creep effects. First, let us discuss the explanation for the cracking and then present some practical means to minimize it.

The most commonly seen failure modes are barrel plating cracks (Figure 4.17), which are all around the circumference of the hole.

The plating thickness of the barrel determines the stress it can withstand. As the aspect ratio (board thickness divided by drilled hole diameter) of the hole increases, the plating thickness in the middle of the barrel decreases. This is also described as the dog bone effect in plating, because plating thicknesses are higher near the top and bottom surfaces of the board but very thin in the middle of the hole. For any size hole, laminate expansion at elevated temperatures imposes a tensile force on the barrel plating. This force can be equated to a tensile force acting on the circular edges of a cylinder. However, the barrel plating cracks when stress (not force) exceeds the fatigue fracture stress of the plating. The holes with a large aspect ratio are weak in the middle and generally crack there.

The glass epoxy board is constrained by the glass in the X-Y direction. However, there is no constraint in the Z direction. The Z axis CTE of the laminate is driven by resin, its degree of cure, and the ratio of resin to glass. The CTEs of epoxy and polyimide are 60 and 40 ppm/°C, respectively, versus 13 ppm/°C, which is the CTE of copper. Thus the basic cause of this stress lies in the different coefficients of thermal expansion in the Z direction between the substrate and the copper barrel.

Reliability problems are encountered when the via hole diameter is reduced to 10 mils. At 18 mils and above, performance is satisfactory. Even 14 mil vias (finished via hole size) are generally considered reliable as long as the aspect ratio is kept under 3. (It should be noted that the drilled hole diameter, used for the calculation of aspect ratio, is larger than the finished hole diameter.) Via hole failure is also related to poor quality drilling, which creates ragged hole surfaces. The uneven surfaces plate nonuniformly and may result in barrel cracking due to localized thin copper in the plated through hole.

How do we minimize the via hole cracking problem? The easiest way to reduce barrel cracking is to use larger holes. This not only increases reliability but also reduces cost. But the reason for using smaller via holes is to increase the interconnection density on the board. Thus the trend has been to move toward smaller and smaller holes. The result is that board thickness must be decreased by using fewer layers. Again, reducing layer count is not always feasible, so the designer must make trade-offs.

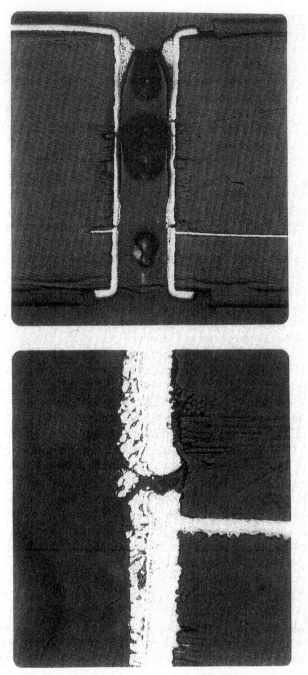

Figure 4.17   Barrel cracking in a 10 mil via hole at 50X (top figure) and 250X (bottom figure).

In addition to aspect ratio, the plating thickness must be uniform, with a 0.001 inch minimum. Achieving a uniform plating thickness is an industry-wide problem, especially with respect to small via holes.

If vias as small as 10 or 14 mils must be used, nickel has been found to improve reliability. It was the experience of at least one company that nickel underplating over electroless copper followed by the usual solder/ tin plating reduced the via hole cracking problem by a factor of 4 [16]. Nickel is a tougher metal and was reported to be a better defense against the stresses imposed during thermal cycling. It should be noted, though, that nickel plating of the plated through hole is a very expensive process and is used only as a last resort. Since nickel has poor "throw" into the hole, electroless nickel is used for plating. An electroless nickel bath is not an easy bath to control and hence it is expensive.

Generally nickel plating is deposited electrolytically to serve as an underplating to precious metals such as gold, to improve wear resistance during use. Nickel also forms an effective barrier layer to intermetallic formation and prevents leaching of precious metals during soldering. Here, the function is simply to strengthen the barrel or via in anticipation of higher stresses.

What is the best way to evaluate via hole quality? An IPC round robin test [17] found that the military thermal cycle test (−65° to +125°C, 30 minute dwell time at each extreme, 30 minute transition time) is a good way to weed out poor vendors and poor quality vias. An acceptable via hole must pass at least 300 cycles of this test, which allows one to discriminate among via holes of different quality. Good quality holes with uniform and sufficient copper thickness will not fail this test. The round robin test also established that the commonly used practice of microsectioning to determine via hole quality is not reliable.

The same round robin test also found that the commercial thermal cycle (0° to 100°C, 40 minute dwell at low temperature, 23 minute dwell at high temperature, 30 minute transition time) is not a good test. Almost all the 200,000 via holes (ranging in size from 10 to 20 mils in 30, 60, and 90 mil thick boards, built by 16 vendors) passed this test. Only the very poor quality via holes failed. This test can only determine infant mortality. The findings suggest that the aspect ratio (board thickness divided by drilled hole diameter) be maintained around 3.

4.12   SUMMARY

Despite the many considerations (such as CTE, cost, dielectric properties, and T_g) to be taken into account when designing surface mount boards, the selection of a substrate is basically determined by the types of components to be used. When leadless ceramic chip carriers are mounted on conventional glass epoxy substrates, solder joint cracking is generally seen in about 100 cycles. The cause of the excessive stress is the CTE differential between the ceramic package and the glass epoxy substrate.

There are three different approaches to solder joint cracking problems: using a substrate with a compatible CTE, using a compliant top layer substrate, and replacing leadless ceramic packages with leaded ones. The most widely used substrate—namely, glass epoxy—entails no CTE compatibility problems when used for plastic surface mount packages. This provides the solution for commercial applications only, however.

The most commonly used substrate for military application is one with a CTE value compatible with that of the ceramic package that has been specified. Each substrate option has its advantages and disadvantages. The designer needs to carefully balance the constraints of cost with reliability and performance needs. In addition, solder masks and via hole sizes should be selected carefully.

REFERENCES

1. Coombs, Clyde F. *Printed Circuit Handbook,* McGraw Hill, 2nd edition, 1979, pages 2–18, 2–22, 23–5.

2. Lee, L. C., et al. "Micromechanics of multilayer printed circuit board." *IBM Journal of Research and Development,* Vol. 28, No. 6, November 1984.

3. IPC-SM-782. Surface Mount Land Patterns: Configurations and Design Rules. IPC, Northbrook, IL.

4. Gray, F., Paper WCIV-38 presented at the Printed Circuit World Convention, Tokyo, 1987. Available from IPC, Northbrook, IL.

5. Hughes, E. W., and Beckman, E. C. Porcelain Enameled Metal Substrates for Surface Mount Components. Publication of Ferr-ECA Electronics Company, 3130 West 22nd Street, P.O. Box 8305, Erie, PA 16505.

6. Foster Gray. "Substrates for chip carrier interconnections." In *Surface Mount Technology,* International Society for Hybrid Microelectronics Technical Monograph Series 6984-002, 1984, Reston, VA, pp. 57–85.

7. IPC-CF-152. Metallic Foil Specification for Copper-Invar-Copper for Printed Wiring and Other Related Applications. IPC, Northbrook, IL.

8. Hanson, R., and Hauser, J. L. "New board overcomes TCE problem." *EP&P,* November 1986, pp. 48–51.

9. Jensen, W. M. U.S. patent number 4,318,954, March 9, 1982.

10. Chen, C. H., and Verville, J. M. "Direct soldering of ceramic chip carrier solves radio production problems." *Electronics,* February 1984, pp. 15–17.

11. Parker, J. L., Jr. "Shelf life and durability testing of OSP coated printed wiring boards." Proceedings of Technical Program SMI 95, San Jose, August 29–31, 1995, pp. 907–912.

12. Artaki, I.; Ray, U.; Vianco, P.; Gordon, H. M.; and Jackson, A. M. "Solderability Preservative Coatings: Electroless Tin Vs. Organic Azoles." *Proceedings of Technical Program SMI 93,* August 29–Sept. 2, 1993, pp. 414–420.

13. Ray, U.; Artaki, I.; Wenger, G. M.; and Machusak, D. A. "Printed wiring board surface finishes: Evaluation of electroless noble metal coatings." *Proceedings of Technical Program SMI 95,* San Jose, August 29–31, 1995, pp. 891–906.

14. Davignon, John, and Gray, Foster. "An evaluation of via hole tenting with solder mask designed to pass Mil-P-55110D thermal shock requirements." *Proceedings of Technical Program SMI 91,* San Jose, August 25–29, 1991, pp. 905–921.

15. Denkler, J. D. "The speed of liquid." *Circuits Manufacturing,* May 1986, pp. 21–24.

16. Manufacturing Technology Review of VHSIC Program by Martin Marietta Corp., August, 1987. Wright Patterson, Air Force Materials Lab, Dayton, OH.

17. IPC-TR-579 Round Robin Evaluation for Small Diameter Plated Through Holes, September 1988. Available from IPC, Northbrook, IL.

Chapter 5

Surface Mount Design Considerations

5.0 INTRODUCTION

The design engineer is responsible for designing a product that meets certain functional, performance, and environmental requirements. The functions, moreover, must fit within the form factor constraints for the product, which may be either cost driven or performance driven. This is not to imply that performance-driven products do not have to be cost effective. It means only that performance is the overriding issue. The designer must also ensure that the product will meet thermal and reliability requirements and can be designed and built in a timely manner to succeed in a given market window.

These requirements can be met by various packaging options: through hole, surface mount technology (SMT), application-specific or custom packages, or a combination of all these. Which packaging option will be chosen and why? Should the packages be ceramic or plastic? What kind of substrate should be used? What are the pros and cons of these various choices?

Once the packaging option has been selected, the product must be designed for manufacturability to ensure that the cost goals are met. To successfully accomplish these goals, the designer must be aware of the manufacturing processes and the equipment needed for building the product.

Before embarking on a particular course, the designer must take into consideration form, fit, function, cost, reliability, and time to market. This is a very broad undertaking indeed. In this chapter we explore the pertinent issues, with the objective of helping the designer to choose the appropriate packaging option.

There are a whole range of packaging options at the designer's disposal. It is important to understand the driving forces for various types of packages on board and assembly. In this chapter we focus on trade-offs of various packaging options as they relate to cost, performance and real estate constraints. The specific design for manufacturability issues related to these are discussed in Chapters 6 and 7, and the manufacturing and process issues are detailed in Chapters 8 through 14.

5.1   SYSTEM DESIGN CONSIDERATIONS

Systems analysis starts with determination of product needs in the marketplace. Therefore the designer must look at the proposed product from the systems point of view. Generally the design engineer is concerned only with the device or the board or the system. This is certainly a mistake, since customers do not buy a device or a board; rather, they buy function or product. Great devices or boards do not sell in the marketplace. Products do [1].

Since the definition of the product is the responsibility of the marketing department, the definition of the factors involved (form, fit, function, manufacturing cost, selling price, distribution channels, service, etc.) should start with the marketing department.

Many board designs are canceled after extensive development costs have been incurred, sometimes before and sometimes after the product has been introduced in the marketplace. There are as many reasons for cancellation as there are products. The cause in any given instance could be poor design or manufacturing, or inadequate marketing research. If a product is canceled during the prototype stage, losses may be relatively small. In extreme cases, however, a product is canceled after a complete SMT line has been set up just for that product, and the financial position of the company may be badly damaged. What it boils down to is this: failure to consider all aspects of product design can have serious adverse impacts on a business.

Volumes have been written on market analysis and research and strategies for attacking a particular market segment. It is not the objective of this book to discuss marketing strategies but simply to point out that no project should be initiated until the product (not the device, the board, or the box) has been defined. Product definition should be done by the product team, consisting of representatives from marketing, design, manufacturing, quality assurance, reliability, and other affected functional organizations.

After the product has been defined, the system requirements must be established. For example, what is the environment in which the product will function? Is it an office, home, or industrial environment? Or is the system intended, instead, for a high reliability military or critical life support environment? Answers to these questions will help determine whether the packages, either through-hole mount or surface mount, should be hermetically sealed (ceramic) or plastic. Selection of components, in turn, will influence selection of substrate material.

The form, fit, and function of the product must be defined next. This will further narrow the selection of packaging options because it will help determine the most cost-effective way to meet the functional requirements. There are many packaging options to consider. Section 5.2 deals with the driving forces for various packages. Their impacts on cost and performance are discussed later in applicable sections in this chapter.

Form, fit, and function constitute some of the most important considerations in determining whether surface mount technology or conventional through-hole technology is required. If in the evaluation of form, fit, and function it is determined that surface mount technology is needed, various questions must be answered before a particular type of SMT assembly is selected. The real estate requirement is not the only consideration, as is generally thought, although it is the primary one.

The final decision for surface mount packaging options should be made after cost and such technical issues as thermal management, interconnects, reliability, and design for manufacturability have also been considered. In the sections that follow, we discuss these items in some detail, because the designer must carefully consider them before committing to SMT. We will start with a discussion of generic driving forces for packaging in general and then discuss drivers for specific package types.

To make a product that is successful in the marketplace, we start with the idea of form, fit, and function. Once these variables have been defined, packaging technologies can be considered.

What do these often-used terms—form, fit, and function—mean? The *form* of a product (complete system or individual board) relates to the physical size, shape, and weight of the item. Another term for form is "factor constraints." Such constraints may be imposed by the various bus architectures (MULTIBUS I, MULTIBUS II, PC BUS, etc.). Different architectures have different slot spacings, which will establish the upper and lower height limits for component protrusions. The size of the board will determine the need for SMT, and the top and bottom height constraints will determine the component and related assembly process requirements.

The *fit* of the product refers to its relationship with other functions or products within the system. Fit will determine the need for a given type of connector or input/output (I/O) function.

Function refers to a product's basic mission in life. What must it do? Once we have established the functional requirements, we see that they can be met by devices of different types, each with different physical, thermal, cost, and availability characteristics.

5.2   PACKAGE DRIVERS

Continued emphasis on faster, smaller, and lighter electronics systems is making component, board, and system packaging more complex. With the rapid introduction of the increasingly faster microprocessors needed for more user friendly software, the complexity of computer systems is also increasing. In this section we look into the driving forces for various types of component packaging and discuss some of the common issues associated with them.

The driving forces for component packaging in computer systems are thermal and electrical performance, real estate constraint, and cost. These are described in the following paragraphs. The component packaging requirement varies for different types of systems. This is summarized in Figure 5.1. For example, high-end microprocessors run at higher frequencies and require thermally and electrically enhanced packages.

Examples of thermal enhancements are heat slugs, heat spreaders, heat sink, and fin-fan (fan mounted on heat sink), etc. Examples of electrical enhancements are multilayer packages and in-package capacitance. Generally hermetic multilayer ceramic packages are used for high-end applications such as high wattage microprocessors. For mid-range systems, performance is important, but so is cost (not that cost is not important for high-end systems). Thermally enhanced, multilayer packages (plastic

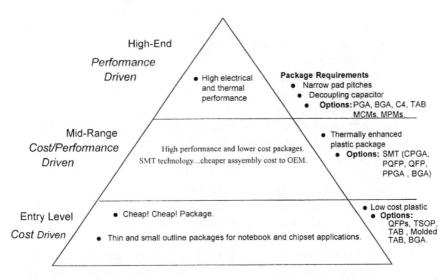

Figure 5.1   Component packaging requirements for different types of systems.

PGA or QFP, BGAs) may be appropriate for this application. And for low-end entry level and portable systems, cost and form factors are critical and generally surface mount packages such as QFP and TSOP are used.

Cost drivers. We generally associate higher price with a more complex system. However, even with increased complexity, the price of computer systems is dropping rather rapidly. The pyramid model for performance and cost shown in Figure 5.1 has turned into a pancake. In other words, most of the volume is in the cost-sensitive mid- and low-end systems. This means that we not only need performance but we need it at a very low price.

Thermal Drivers. Thermal enhancements have become essential for today's increasingly fast microprocessors. With the introduction of new generations of microprocessors, power dissipation has continually moved upward. And there is every indication that this will be true in the future. As we increase speed, power usage goes up. This problem is mitigated, fortunately, by the die shrink made possible by fine feature semiconductor processes. For the higher wattage packages, ceramic PGA (CPGA) is commonly used. There are also many other reasons for using PGAs as discussed later on.

Electrical Drivers. With increasing frequency, in addition to power, the number of bond pads on the die and the number of pins in packages also go up, as shown in Figure 5.2. Fortunately, even with an increase in the number of bond pads for high speed devices, the pin count is kept to a minimum by the use of decoupling capacitors in the package and on the die. In addition, by adding power and ground planes in the package (hence making the use of a multilayer package essential), the number of pins is reduced.

For example, if no enhancement is used, the number of package pins will equal the number of bond pads. By using package capacitance and multilayer packages, the pin count is kept to a minimum. As also shown by the lower line in Figure 5.2 (enhanced packages), the pin count is lower than that for low cost packages since the addition of too many layers and in-package capacitance is generally not feasible in a low cost package. For example, as a historical note, the Intel Pentium™ microprocessor was designed in a 296 pin enhanced ceramic PGA. But the same device had to be designed in a 0.25 mm 320 pin low cost TAB (tape-automated bonding) or TCP (tape carrier package) package in surface mount for notebook computers.

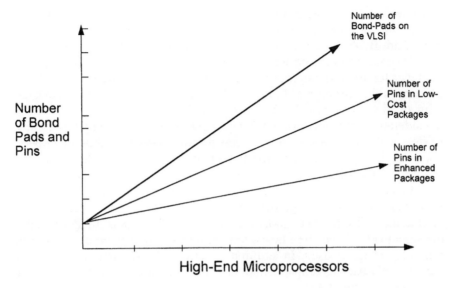

Figure 5.2 **Number of bond pads in silicon versus number of pins in low cost and enhanced packages.**

Real estate drivers. A major driving force for component packaging is real estate constraint. After certain pin counts, the pitch must drop to keep the body size within a practical range for manufacturing. For example, 84 pin PLCC is the largest 50 mil pitch package commonly used. When the pin count equipment increases beyond 84 pins, the pitches drop to 0.65 or 0.5 mm. These packages take us to slightly over 200 pins. If the need goes beyond that, finer pitches such as 0.4/.3 mm TAB or HD-QFP format become necessary.

These are the generic package drivers: cost, thermal and electrical enhancements, and real estate constraints. Now let us examine the driving factors for some specific packages.

5.2.1　PGA Drivers

As the processor frequency increases, so do demands on the electrical and thermal performance of the package. Increased electrical performance requires increased package layers and pin count. Such higher pin count devices can be packaged either in peripheral or array packages. Peripheral packages are available in surface mount fine or ultra fine pitch packages.

For array packages we have multiple options. In through hole, PGA is an example of an array package. Examples of array packages in surface mount are BGA, land grid array (LGA), and stud grid array (SGA).

Despite multiple options of surface mount packages, they have disadvantages for use with microprocessors (with expensive silicon). For example, one of the alternatives to PGA is the fine pitch surface mount package with pitches of 0.65 and 0.5 mm or lower. In these packages, there is considerable yield loss during package handling and test due to the fragility of fine pitch leads. Most of the defects during board assembly are experienced in fine pitch devices for the same reason. As the lead pitch is reduced, defects increase.

In comparison, the main disadvantage of PGA is its larger size. However, for high wattage devices, board or system space requirement is determined by the heat sink size. Hence the larger size of PGA really does not matter since large heat sinks are needed anyway no matter which type of package is used. In addition, a ceramic PGA easily incorporates electrical (in-package capacitance) and thermal enhancements needed for high performance devices.

The PGA package is a known commodity. Unlike ultrafine pitch and BGA packages, it has been used for a long time. And above all, it can be reliably socketed (and soldered). Socketing is preferred by original equipment manufacturers (OEMs) and users for various reasons. There is a strong need for socketing the most expensive component in a typical system. For example, socketing allows OEMs to design a common mother board that will accept all different frequencies of a given microprocessor with common pinout. (The Intel Pentium Pro™ is a good example.) It allows upgrades by end users who do not have to buy a whole new system in order to take advantage of the higher performance processors. For example, users can easily upgrade to a higher frequency processor if it is socketed. And sockets are also handy if the processor needs to be replaced for any other reason.

There is one more reason, perhaps the biggest one, for using PGAs. They enjoy the advantage of lower import duty on mother boards assembled overseas. Since import duty is generally based on the value of the products being imported, lower duty is levied on a board with sockets for processors, but without the processor. For this reason, many users even go to the trouble of converting fine pitch devices into PGAs by using interposer cards.

OEMs also don't like the idea of tying up millions of dollars worth of inventory for any longer period than they need to. Socketing allows them the freedom to purchase the processor just before shipping to the end customers.

All this adds up to a strong case for the continued use of PGAs instead of BGAs or any other surface mount packages for a long time to come.

5.2.2   Fine Pitch Drivers

The primary driver for fine pitch is real estate constraint, but cost is not too far behind. There is a significant cost saving, in some cases well above 50%, over PGA. And in some cases a package may be available only in fine pitch surface mount. An example would be the Intel Pentium™ microprocessor in ultra fine pitch (0.010 inch) TAB package for note-book computers.

Fine pitch plastic package is a common package for high pin count (over 100) ASIC (application specific integrated circuit) devices. In many cases, ASICs are used for reducing thermal load by combing 10 to 20 programmable logic array devices, which would consume 5 to 10 watts, into a 2 to 4 watts ASIC device. These packages allow thermal and electrical enhancements by the use of heat slugs and multilayer boards, while significantly reducing cost.

5.2.3   Ball Grid Array (BGA) Drivers

The most important reason for the potential proliferation of BGAs is the problems associated with the alternative of peripheral fine pitch and ultra fine pitch TAB packages. As mentioned earlier, the complexity of the board assembly increases dramatically with the reduction in lead pitch. Even at 0.5 mm pitch, considerable problems, generally associated with handling and lead coplanarity, are encountered.

The absence of fragile leads susceptible to handling damage such as lead bend, sweep, and coplanarity, make BGA a much more robust package. Therefore, it provides much better manufacturing yield. It is much easier to place, print, and solder a 1.5 mm pitch BGA package than a 0.4 mm pitch HD-QFP (high density quad flat pack) or TAB package since the placement and soldering processes of "conventional" surface mount apply for BGA. A BGA package with 1.5 mm ball pitch takes even less real estate than a 0.4 mm pitch HD-QFP above 300 pins. And it maintains a body size of around 35 mm even at 500 pins.

5.2.4   Issues in Component Packaging

We should note that there are issues with every package option. Let us briefly review them. Repair and rework of BGA is very difficult. Since the balls collapse (in plastic BGA) during rework and are difficult to reball by the user, one may have to throw away a perfectly functional device. The fact that BGA does not require as much rework as fine pitch does not necessarily help the argument for putting expensive silicon in BGA. We should also note that BGA will increase layer count and hence board cost.

Since BGA solder joints cannot be inspected, one must depend primarily on process control. To reap the benefits that BGA offers, one must have good process control in place, and at many companies, this is not as straightforward as it sounds.

However other packages have their own problems. For example, a quantum jump in surface mount manufacturing technology is required to establish a complete infrastructure for 0.4 mm and lower pitch packages. The major issues with ultra fine pitch packages with lead pitches below 20 mils (HD-QFP or TAB) are paste printing, placement, soldering, and rework. The availability of boards with an even thickness of solder may be a problem as well. Hence solder may have to be applied by the board supplier by paste printing or plating, or the use of a copper surface may be necessary.

Even the familiar PGA has some concerns. For example, high pin count PGAs used for high performance processors can least afford to have the high inductance of a socket. And if they are soldered, they lose their allure of socketing and are difficult to remove.

5.3   REAL ESTATE CONSIDERATIONS

When considering the use of surface mount technology, the most important factor is real estate. This is not to imply that other factors are not important. They are. But as a rule, SMT is considered only when a product does not meet the functional requirements within the specified form factors. Once it has been determined that SMT is necessary, the designer must consider the three types of SMT assembly: I, II, and III, which we introduced in Chapter 1.

To review briefly, Type I SMT refers to assemblies containing only surface mount components. In Type III, components are glued to the bottom side and then wave soldered. Type II SMT, in terms of both process and components, is simply a mixture of the other two types, I and III.

Different types of SMT assembly provide different degrees of real estate savings. Table 5.1 gives a rough estimate of real estate savings for the three types. The mixture of surface mount and through-hole mount

Table 5.1   Typical real estate savings in different types of SMT assemblies.

TYPE OF SMT ASSEMBLY	FUNCTIONAL DENSITY INCREASE
Type III SMT	1.05–1.10 X
Type II SMT	1.20–1.40 X
Type I SMT	2.00–4.6 X

components primarily determines the real estate savings, which also will vary for logic and memory boards.

Actual real estate savings will depend on component selection, inter-package spacing rules for design, and the skill of the board designer in taking advantage of every nook and cranny (packing efficiency) on the board. There is generally some packing inefficiency, since the available space is not optimally utilized because of proximity requirements (for some components to be near others), component orientation requirements, form factor restrictions, and most important of all, varying sizes and shapes of different components on the board.

Final real estate requirements can be determined only after actual board layout, but the designer needs a rough idea before beginning the layout. An estimate is possible by multiplying the area required by different devices (size of the land pattern plus interpackage spacing) by their total number on the board and adding a packing inefficiency factor. Typically, the packing inefficiencies vary from 10% for a memory board to 30% for a logic board. For logic boards, a minimum of 20% packing inefficiency should be allowed. Irregular sizes of different components cause some area to be wasted. An approximate estimate of real estate requirement can be represented by the following formula:

$$A = [(a_1 + I_1)n_1 + (a_2 + I_2)n_2 + (a_3 + I_3)n_3 + \cdots] \, K + M$$

where A = real estate area requirement

$a_1, a_2, \ldots$ = land pattern areas of different components (see Chapter 6 for actual values; Chapter 6 does not show the land pattern requirement for through-hole components)

$I_1, I_2, \ldots$ = interpackage spacing requirements (see Chapter 7 for actual values, including interpackage spacing between surface mount and through-hole components)

$n_1, n_2, \ldots$ = total number of each component

K = packing inefficiency constant: 1.10 for memory board and 1.30 for logic boards

M = reserved area required for miscellaneous purposes such as clearance for edge card guide, automated test equipment (ATE), card ejector brackets, front panel, etc.

Using the above formula, Table 5.2 gives an example of real estate area that should be allowed for some selected through-hole and surface mount components. The table shows the land pattern length and width and the interpackage spacing requirements for the package. The last column shows the area, which takes into account both the land pattern and interpackage spacing. A similar table can be generated for all the packages to be used on a given board. The required area *(A)* can be obtained by multiplying the area for the packages (last column in Table 5.2) by the number of each package on the board. To take into account the packing inefficiency, the total area must be multiplied by K (1.1 for memory, 1.2 or 1.3 for minimum and maximum limits, respectively, for logic boards). Now add to this number the reserved area. Remember that the final number is only an *estimate* for the real estate requirement. Such an estimate is intended to help determine if all the components will fit on one side of the board. The final determination can be made only after placement in CAD by using the land pattern and interpackage spacing requirements discussed in Chapters 6 and 7, respectively.

We should note here that in this discussion we have only considered the X and Y dimension requirements. The impact of various packages on layer count and board cost is discussed later in this chapter.

5.4   MANUFACTURING CONSIDERATIONS

Manufacturing, especially with new technologies, is a specialized discipline. The designer is not expected to understand the manufacturing processes thoroughly, but there must be a manufacturing engineer on the team. It is not in the best interest of the company to design a product without considering manufacturability. A design that cannot be built or tested in a cost-effect manner is a very poor design indeed, any theoretical excellence notwithstanding.

Most designers become very concerned about "wasting" space and tend to pack boards as tightly as possible. This is short-sighted. If you look at any well designed PC board, you will find plenty of room between packages for each inspection, repair, and test. Sufficient space should be set aside for routing and interpackage spacing requirements to ensure manufacturability. Chapters 6 and 7 discuss in detail such design-for-

Table 5.2 Land pattern/package area and interpackage spacing requirements for some selected surface mount and through-hole components

PACKAGE/PIN COUNT	WIDTH	LENGTH	W-SPACE	L-SPACE	AREA
SOIC8	0.292	0.197	0.050	0.050	0.084
SOIC14	0.292	0.344	0.050	0.050	0.135
SOIC16	0.292	0.394	0.050	0.050	0.152
R-PACK14	0.352	0.325	0.050	0.050	0.151
R-PACK16	0.352	0.390	0.050	0.050	0.177
SOICL16	0.452	0.413	0.050	0.050	0.232
SOICL20	0.452	0.512	0.050	0.050	0.282
SOICL24	0.452	0.614	0.050	0.050	0.333
SOICL28	0.452	0.713	0.050	0.050	0.383
SOJ26	0.382	0.680	0.060	0.050	0.323
PLCC16	0.370	0.370	0.050	0.050	0.176
PLCC20	0.420	0.420	0.050	0.050	0.221
PLCC24	0.470	0.470	0.050	0.050	0.270
PLCC28	0.420	0.620	0.050	0.050	0.315
PLCC32	0.520	0.620	0.050	0.050	0.382
PLCC44	0.720	0.720	0.050	0.050	0.593
PLCC68	1.020	1.020	0.050	0.050	1.145
PLCC84	1.220	1.220	0.050	0.050	1.613
PQFP84	0.810	0.810	0.100	0.100	0.828
PQFP100	0.905	0.905	0.100	0.100	1.010
PQFP132	1.105	1.105	0.100	0.100	1.452

0805(RES/CAP)	0.050	0.150	0.050	0.050	0.020
1206(RES/CAP)	0.060	0.210	0.050	0.050	0.029
1210(RES/CAP)	0.100	0.190	0.050	0.050	0.036
DIP16	0.300	0.800	0.100	0.100	0.360
DIP18	0.300	0.900	0.100	0.100	0.400
DIP20	0.300	1.000	0.100	0.100	0.440
DIP22	0.300	1.150	0.100	0.100	0.500
DIP24	0.300	1.250	0.100	0.100	0.540
DIP32	0.600	1.650	0.100	0.100	1.225
DIP40	0.600	2.050	0.100	0.100	1.505
PGA68	1.165	1.165	0.100	0.100	1.600
PGA96	1.165	1.165	0.100	0.100	1.600
PGA132	1.450	1.450	0.100	0.100	2.403
PGA148	1.560	1.560	0.100	0.100	2.756
PGASOCKET68	1.075	1.075	0.100	0.100	1.381
PGASOCKET132	1.470	1.570	0.100	0.100	2.622
PGASOCKET149	1.570	1.570	0.100	0.100	2.789
DIPSOCKET22	0.500	1.150	0.100	0.100	0.750
DIPSOCKET32	0.700	1.600	0.100	0.100	1.360
DIPSOCKET40	0.700	2.000	0.100	0.100	1.680
STAKEPIN	0.070	0.070	0.050	0.050	0.014
OSCILLATOR	0.520	0.835	0.100	0.100	0.580
CRYSTAL	0.435	0.550	0.100	0.100	0.348
AXIALCAP2.2MF	0.020	0.447	0.100	0.100	0.066
AXIACAP22MF	0.180	0.420	0.100	0.100	0.146
SIP6	0.200	0.600	0.000	0.000	0.120
SIP8	0.200	0.800	0.000	0.000	0.160
SIP10	0.200	1.000	0.000	0.000	0.200

manufacturability issues as land pattern, cleaning, repair, testability, and inspection.

In addition to easy manufacturability, the designer must consider the processes to be used. As discussed in Chapter 7, different soldering processes have different rules. Process selection is also tied to component selection, and the effects on space saving and cost must not be ignored.

Many through-hole mount components are being used as surface mount components by modifying the leads as gull wing or butt leads. Naturally these modifications do not save space; rather, they are done to save a process step in hand or wave soldering. As a general rule, however, through-hole mount components are NOT expected to withstand the higher temperatures of reflow soldering. Before selecting items for use as surface mount components, it should be verified that they will not be damaged at the soldering temperatures contemplated.

Surface mount active components are generally designed for reflow soldering and should not be wave soldered. Some surface mount components may not withstand the reflow temperature, either. Such a process incompatibility may be encountered if an industry standard does not exist for a particular device, as is not uncommon with an evolving technology. For example, we had a surface mount delay line that had its internal connections soldered with eutectic solder with a melting point of 183°C. When this component was reflow soldered at the usual reflow temperature of 220°C, the internal connections melted, and the solder came oozing out of the package.

Space and cost are also functions of the various possible combinations of components to go on the primary or secondary side of the board. If all components could be had in surface mount, this would not be a problem. But since all components are not available in surface mount, very creative combinations are needed to fit all the components on the board.

If the form factor must not change, no functions can be taken off, and the board must have active surface mount devices on top and bottom along with through-hole devices on the top, there are two options. Neither of them is really desirable because of adverse impact on cost or on product reliability. Refer to Chapter 7 (Section 7.4: Soldering Considerations) for details on this complex situation.

Finally, the designer must consider what manufacturing capability exists and what processes are preferred by the manufacturing organization. If the capability for a Type II or Type IIC SMT exists, all options including through-hole assembly are available. (As discussed in Chapter 1, Type IIC is defined as a mixture of through-hole and surface mount components, including fine pitch, ultra fine pitch (UFP), and BGA. In our definition UFP also includes TAB.) If the capability for only Type III or Type I processes exists, the manufacturing options are limited. Even when there is full SMT capability, only the most appropriate technology should be used, because a Type I or Type III assembly may be more cost effective than a Type II assembly.

The selection of a packaging option should also be based on the volume requirement. For example, if product volume is very high, any functions can be combined in an application-specific integrated circuit. An ASIC package, depending on the real estate requirement, can be used in addition to or in lieu of SMT. For example, an ASIC package may make it possible to use Type III assembly instead of Type II.

The designer should not commit to SMT yet. The cost impact of the packaging options, discussed in the next section, remains to be dealt with.

5.5   COST CONSIDERATIONS

The switch to SMT is made for various reasons, but cost must not be neglected. When comparing the cost of a through-hole mount assembly (THMA) to that of a surface mount assembly (SMA), a basic distinction should be made between cost per assembly and cost per function. If surface mounting is used because all the functions do not fit on a given form factor, the cost per function is a more appropriate index than the cost per assembly.

The cost per assembly of a densely packed SMA is going to be higher because of the higher component count and finer lines and spaces on the board. But the cost per function will be less because it would take two THMAs to provide the functional equivalent. It will be more expensive to procure, handle, assemble, and test two THMAs than one SMA. There are additional savings if the need for connectors or cabling is reduced because there are fewer assemblies.

There are three major contributors to the cost of an assembly: the printed circuit board (PCB), the components, and the assembly. These are discussed in turn.

5.5.1   Printed Circuit Board Cost

The requirements for surface mount PCBs are not significantly different from those for conventional through-hole assemblies. However, the 0.050 inch center surface mount components reduce the gap between the adjacent pads to only 0.025 inch. Generally 0.050 inch center devices require a pad width of 0.025 inch. Since even micro vias generally require pads larger than 0.025 inch, less than 0.025 inch is left between the pads. This tiny gap demands that finer features (trace width and space) be used for routing traces between pads and vias. Refer to Figure 5.3, which

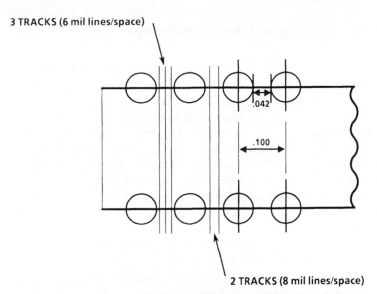

THROUGH HOLE PAD PATTERN

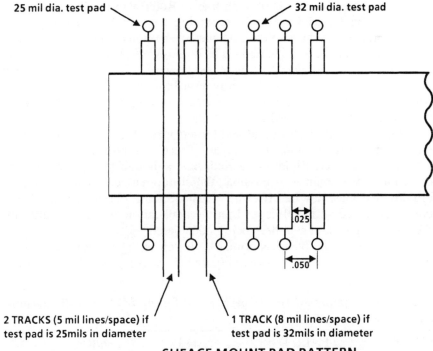

SUFACE MOUNT PAD PATTERN

Figure 5.3 Need for finer lines and spaces in surface mount boards. Note 3 six-mil lines (3 track) in the through-hole board (top figure) versus only 2 five-mil lines (2 track) in surface mount boards. This is due to the change in lead/pad patch from 100 mils in the through hole board to 50 mils in the surface mount board.

compares the line width and space requirements for through hole (top figure) at 100 mil center with 58 mil (typical) pad sizes and surface mount (bottom figure) at 50 mil center with 25 mil (typical) devices.

As shown in Figure 5.3, the line width can be 8 mils for routing two traces in through hole and only one trace in surface mount pad patterns. For routing two traces in surface mount, the line width must drop to 5 mils. If larger than 0.025 inch test pads are used for test, only one trace (6 to 8 mils) can be routed. For ATE, 0.032 inch test pad vias (square if possible) are generally used to provide a larger test target for the test probes (see Chapter 7). As is clear from Figure 5.3, SMT boards will require finer lines on the PCBs.

Table 5.3 shows the number of traces between BGA pads. A desirable pad size for a plastic BGA is 0.030 inch in diameter. This means that the pitch of the package needs to be 0.060 inch in order to allow routing of two 0.006 inch traces between BGA pads. If the pitch is reduced to 0.050 inch, then pad size will have to reduced to 0.025 inch to allow either one 0.006 inch trace or two 0.005 inch traces. Refer also to Chapters 6 and 7 for additional considerations in designing with the BGA package.

If surface mount active logic devices are mounted on both sides of the board, the use of blind and buried via technology may become necessary to achieve the greatest density. There are fewer PCB suppliers for such dense, double-sided, fine line boards with blind and/or buried vias than for conventional boards. Also, there is a cost premium associated with blind and/or buried via technology.

Figure 5.4 illustrates blind and buried vias. Blind vias extend from an outer layer to one or more inner layers. These vias can trap contaminants, affecting reliability. Blind and buried vias, although they increase cost, do reduce layer count. It is possible, therefore, that cost increases due to blind/buried vias may be offset by a cost decrease due to reduced layer count. This trade-off will depend on the application and the maturity of the technologies and suppliers being considered.

The cost of the board is primarily driven by layer count and board trace width and space, as shown in Figure 5.5. As the line width decreases,

Table 5.3   Number of traces between different BGA pad dimensions and pitches

PITCH	BALL D	PAD D	6/6	5/5
60	30	30	2	2
50	30	25	1	2
40	25	25	0	1
20	4	4	0	1

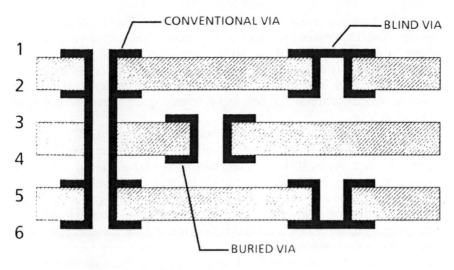

Figure 5.4 Blind and buried vias.

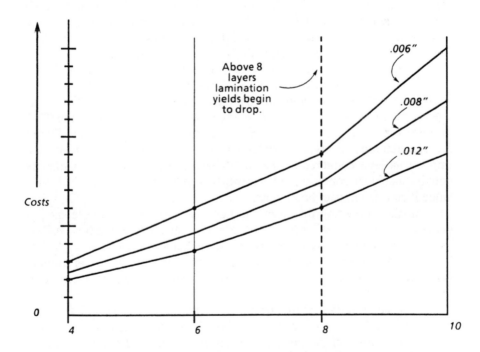

Number of Layers (.062" thick board)

Figure 5.5 Effect of layer count and trace width/space on PCB cost.

the yield decreases due to opens and neckdowns. This combination increases cost. The cost of the board also increases with an increase in layer count, rising significantly if the layer count goes beyond 8 layers for a given trace space and width. In such a case it is cost effective to switch over to finer lines and spaces. As the vendor base matures, significant cost differences between 5/5, 6/6, and 8/8 trace/space board technologies may disappear.

The PCB cost for surface mount boards, as compared with through-hole boards, will be lower if the component count is the same in both cases. In other words, if the functionality of the boards is the same, the packaging density on the surface mount board will allow enough space between packages for routing purposes to permit the layer count to be reduced. Reduction of layer count from 8 to 6 layers or from 6 to 4 layers can generally result in a PCB cost savings of about 25%. Looser component packaging is uncommon, however.

Generally the real estate freed up by smaller surface mount components is used to provide additional functions. Hence the component packaging generally becomes tighter in a surface mount board. Tighter packaging density requires boards with finer lines and spaces, thus increasing board cost. If, however, increased packaging density makes it possible to replace two boards with one, the overall PCB cost will be lower.

Different packages have different impacts on layer count and board area. For example, as shown in Figure 5.6, the peripheral fine pitch packages such as HD-QFP require more board area but lower layer count. BGA packages require less board area but more layer count. And ultrafine pitch COT (TAB) packages require more board area and layer count (last bar in Figure 5.6). In addition to the type of package and its pin counts, track density also impacts the layer count requirement as discussed earlier. In addition to track density and layer count, it is also important to keep in mind via hole size since it can also impact board cost.

Surface mounting increases the need for a smaller via hole size. A decrease here saves real estate, but again, the reliability of the hole suffers as a result of increased aspect ratio (see Chapter 4, Section 4.11: Via Hole Cracking). In any event, the vendor base for small vias is limited. As shown in Figure 5.7, there is generally a cost premium for via holes of 18 mils and less.

In general (either loose or tight packaging), the switch to SMT requires finer lines and spaces. Any board can be designed with wider lines and spaces, but the process usually is manual and takes more time in the CAD cycle. This may not be acceptable because of schedule constraints. Also, staggered interconnection or test vias must be used to allow routing with wider traces. Staggered vias require extra real estate, hence take up some of the real estate benefits of SMT. Thus there is seldom a

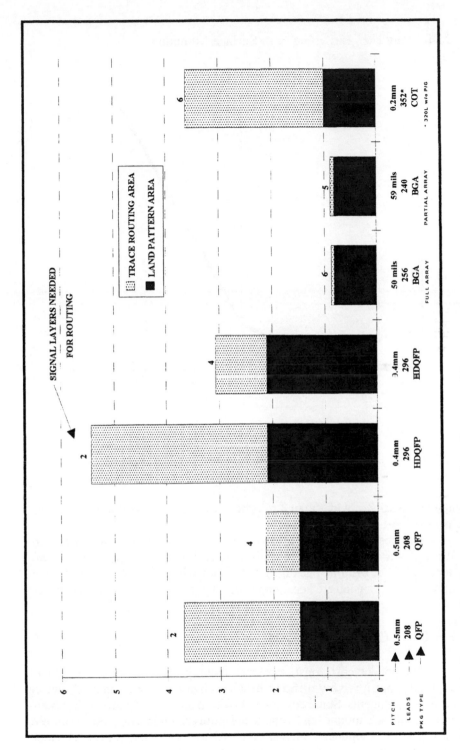

Figure 5.6 Impact of various packages on board area and layer count.

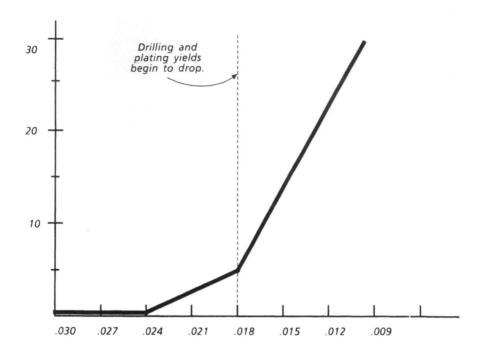

Finished hole diameter in inches

Figure 5.7 Effect of finished via hole size on PCB cost.

need to use traces wider than 8 mils because there is essentially no savings in PCB cost.

The cost of the PCB also depends on the type of substrate used. With plastic packages, the commonly used FR-4 glass epoxy boards are necessary. For ceramic packages, substrates with a compatible coefficient of thermal expansion are necessary. See Chapter 4 for details on substrates.

5.5.2 Component Cost

Component cost constitutes the major portion of the total cost of assembly, but it is very difficult to generalize the cost impact of surface mount components. Some components are cheaper, others are more expensive in surface mount, and others are equivalent in cost. For example, most passive components such as chip resistors and capacitors are cheaper. There is some premium on tantalum capacitors.

Some of the active devices may be more expensive, bearing a premium of up to 15%. This is slowly changing, however. Indeed, for many other components, depending on the supplier, there may not be any cost difference between surface mount and through-hole mount devices. For special or custom components, the cost depends heavily on the volume and on the business relationship between the supplier and the user. And for high pin count ASIC (application specific integrated circuits) components, through hole PGA (pin grid array) may be considerably more expensive than fine pitch PQFP (plastic quad flat pack) packages.

Also, surface mount components may be harder to purchase on a quick-turn around basis. Thus the schedule should allow a longer lead time for their procurement.

In general, cost may not be the only reason for selecting surface mount components. Through hole is a very old technology and is facing obsolescence (although some through hole will be with us for a long time—see Section 5.2.1: PGA Drivers). SMT will be needed to produce state-of-the-art technology and to be competitive.

5.5.3   Assembly Cost

Assembly cost includes component placement, soldering, cleaning, repair, and test costs. Soldering, cleaning, and test costs are essentially the same for both through-hole and surface mount assemblies. The cleaning and soldering operations for through hole and surface mount take roughly the same amount of time. Since the same equipment is used for testing, the test cost is the same, although the cost of the surface mount test fixture is generally higher.

The repair cost is lower for multileaded surface mount devices because they are easier to remove. In addition, the thermal damage during repair is almost nonexistent in surface mount devices, since the use of hot air eliminates the thermal damage caused by the pressure and high temperature of a soldering iron (see Chapter 14). It is difficult to quantify savings here because the amount of repair required is based on the process control, the adequacy of the design for manufacturability, and the quality of incoming materials and components.

For all practical purposes, among the various assembly processes that determine product cost, placement is the most significant. In general, the pick-and-place time for surface mounting is considerably less than that for the insertion of the through-hole mount devices. The basic reason for the higher insertion time and cost for through-hole components is that the insertion operation requires setup and operation of three or four

Table 5.4 Assembly cost comparisons for through-hole and surface mount technology

TYPE OF ASSEMBLY	SAVING IN PLACEMENT COST[a]
Through-hole assembly	(Used as standard) 0
Type III SMT	0
Type II SMT	19%
Type I SMT	30%

machines such as a component sequencer, an axial inserter, a radial inserter, and a DIP inserter.

On the other hand, all surface mount components can be placed by a single placement machine. This can reduce setup time considerably, as well as queue time, handling time, and work in progress. In some applications, if versatility is not as important as speed, more than one dedicated piece of surface mount equipment for different component types can be used for increased speed. Thus, since placement time depends on the speed of pick-and-place equipment and on whether the manufacturing line is set for versatile or dedicated placement machines, it is difficult to generalize savings in this area.

Dedicated placement equipment is much faster than versatile equipment. The savings also depend on the feeder input capacity of the machine and on whether components are fed by tube or by tape and reel feeders. No matter what system is used, about 25% savings can be expected in placement time. This is shown in Table 5.4 for versatile, hence relatively slow, placement equipment.

5.6   THERMAL CONSIDERATIONS

With increased levels of integration on silicon chips (i.e., more transistors per unit area of silicon), thermal management remains a concern even with through-hole devices. Surface mount technology only compounds the problem. The surface mount packages, being smaller, yet containing the same die used in the roomier DIP packages, have higher thermal resistance. In addition, because of increased packing density provided by smaller packages, the power density per unit area of the substrate increases.

Thermal management is necessary not just for the sake of reducing the internal temperature of the electronic equipment. It is also necessary to cool the devices on the board so that they can operate more reliably and fail less frequently. One does not have to radically change the basic thermal management approach (e.g., from forced air cooling to liquid cooling) just because of SMT. Instead, thermal management is undertaken mainly to achieve reliability at both package and system levels.

Thermal management is a complex subject, and here we simply highlight the thermal problems that are compounded by SMT. With the advent of SMT, such problems have become more critical for both suppliers and users of the packages. Since the problems exist at both the package and the assembly levels, suppliers and users must take certain steps to minimize these difficulties.

The existing thermal problem that is exacerbated by SMT involves an increase in junction temperature (T_j), which can adversely affect the long-term operating life of the device. Junction temperature is a function of three items: power dissipation, junction-to-ambient thermal resistance, and ambient temperature. It can be calculated by the following equation:

$$T_j = (\Theta_{ja} \times P_d) + T_a$$

where T_j = junction temperature ($^\circ$C)
$\quad \Theta_{ja}$ = thermal resistance, junction to ambient ($^\circ$C/W)
$\quad P_d$ = power dissipation at T_j (W)
$\quad T_a$ = ambient temperature ($^\circ$C)

The ability of the package to conduct heat from junction to ambient is expressed in terms of the thermal resistance (Θ_{ja}). It represents the total thermal resistance from junction to ambient. This property generally is separated into two parts: thermal resistance from junction to case (Θ_{jc}) and thermal resistance from case to ambient (Θ_{ca}). See Figure 5.8.

Several factors control the junction-to-ambient thermal resistance. By regulating those factors, the junction temperature can be controlled and the device kept cooler when powered. The suppliers of surface mount devices are reducing the thermal resistance by changing the lead frames from the Kovar (54% Fe, 29% Ni, and 17% Co) and alloy 42- (Fe 58% and Ni 42%) alloys to copper itself, which is a better thermal conductor. Thus by changing the lead frame material, package suppliers have been able to reduce the thermal resistance of surface mount packages: the packages can get hotter, but the heat is conducted away more efficiently by the copper leads.

Another change component suppliers are making consists of providing heat spreaders inside the package, to conduct heat away much more rapidly. Changes in the die attach materials, use of larger die pads, and specification of better molding compounds also help in this endeavor.

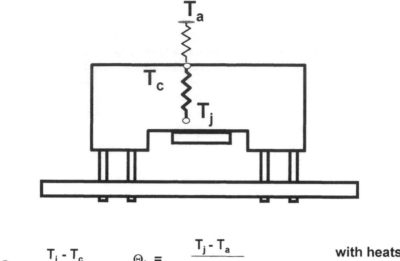

$$\Theta_{jc} = \frac{T_j - T_c}{P} \qquad \Theta_{ja} = \frac{T_j - T_a}{P} \qquad \text{with heatsink :}$$

$$\Theta_{ca} = \frac{T_c - T_a}{P} \qquad\qquad \Theta_{cs} = \frac{T_c - T_s}{P}$$

Figure 5.8   Thermal resistance formulas.

High wattage (more than 8 watts) microprocessors generally are not only packaged in ceramic packages, they may require a heat spreader or heat slug in the package. Figure 5.9 shows conventional, heat spreader, and heat slug ceramic packages. Both heat spreader and heat slug packages use copper-tungsten material to dissipate heat. The composition of Cu-W is maintained in such a proportion that it matches the CTE of ceramic. As is clear from Figure 5.9, the heat spreader is brazed to the top of the package, but in a heat slug, the silicon sits directly on the heat slug. Generally, heat slug is used for device wattage of more than 15 watts. Slug packages are considerably more expensive than the spreader packages. The reason for using heat spreader or slug is to reduce either the heat sink height/area, to reduce air flow rate, to lower junction temperature, or any combination thereof, as illustrated in Figures 5.10 and 5.11.

Despite the changes just described, surface mount packages have a junction-to-ambient thermal resistance somewhat higher than that of DIPs (Table 5.5). The problem is more critical for smaller and thinner SOIC packages than for PLCCs.

Users must also take appropriate steps to ensure that even the use of packages with greater thermal resistance does not increase the junction temperature to unacceptable levels, compromising reliability.

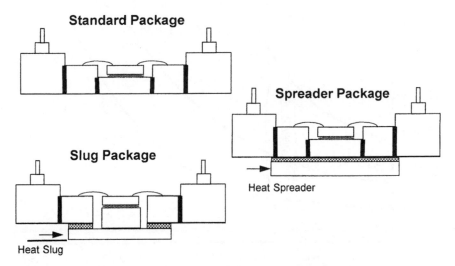

Figure 5.9   Standard, spreader, and slug packages.

The very first thing the user must do to analyze aboard thermal problem is to determine the types of devices that are the major contributors of heat on the board. For example, programmable logic array (PLA) devices, which account for a significant portion of the power on the board, are the culprit in most single-board computers. The second greatest contributors are the transistor-transistor logic (TTL) devices. This situation may not be applicable to every case, but for each product, one must determine the significant contributors to thermal problems in order to provide a cost-effective thermal solution.

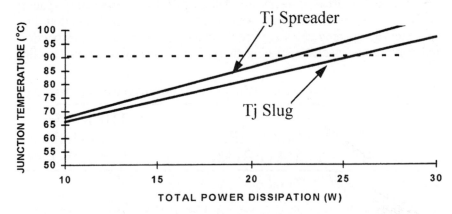

Figure 5.10   Junction temperature (T_j) for the slug and spreader with fan sink at different power levels. (Courtesy of Intel Corporation.)

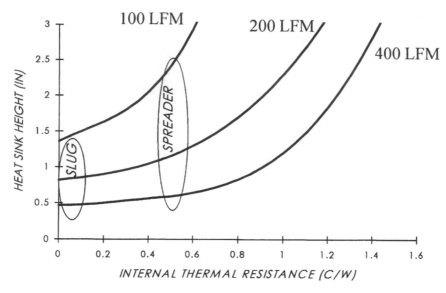

Figure 5.11   Heat sink sizes for slug and spreader for various air flow. (Courtesy of Intel Corporation.)

When feasible, 10 to 15 PLA or TTL devices can be converted into a single ASIC gate array package. This will not only reduce the power on the board but will provide plenty of room on the board for other packages. However, it is essential to avoid the temptation to fill up the space savings created by the ASIC device with additional functions. That will bring you back to where you were in the first place.

The user must also develop a design rule for interpackage spacing to minimize thermal problems. As discussed in Chapter 7 (Section 7.6), interpackage spacing rules are required for repair, cleaning, inspection, and testing. Interpackage spacing rules should also take thermal cooling

Table 5.5   Thermal resistance (Θ_{ja}) of DIP and surface mount packages

TYPE OF PACKAGE	LEAD FRAME MATERIAL	THERMAL RESISTANCE, Θ_{ja} (°C/W)
DIP 20 pin	Kovar	68
SOIC 20 pin	Copper	97
PLCC 20 pin	Copper	72

into consideration and, when developing thermal design rules, the user can establish requirements for placing hotter devices near cooler devices. In addition, the minimum acceptable interpackage spacing between components can be increased to maintain a maximum wattage per unit area of the board.

The thermal design rules should also establish the maximum power rating (P_d) for each type of package (SOIC, PLCC, etc.). The P_d can be derived from the equation given earlier. A maximum acceptable junction temperature (generally about 100 or 115°C) should be established to ensure the desired reliability. The next important variable is the ambient temperature, which depends on the market in which the product is designed to operate.

The final variable, junction-to-ambient thermal resistance, is supplied by the vendor. By plugging in the values for these parameters, an accept-

Table 5.6 **Maximum power rating (P_d) for surface mount packages for junction temperatures of 115° and 130°C and an ambient temperature of 70°C (Source: M. Kastner and P. Melville, eds.** Signetics SMD Thermal Considerations. **Signetics Publication 98-9800-010, 1986)**

PACKAGE	Θ_{ja} (°C/W)[a]	POWER (W) FOR JUNCTION TEMPERATURE	
		UNDER 115°C	UNDER 135°C
SO 14	110–130	0.340	—
SO 16	110–120	0.375	—
SOL 16	90–110	0.450	—
SO 20	80–90	0.500	0.720
SO 24	70–80	0.560	0.810
SO 28	70	0.640	0.930
PLCC 20	70–80	0.560	0.810
PLCC 28	59–73	0.620	0.890
PLCC 44	44–53	0.850	—
PLCC 52	38–42	1.070	—
PLCC 68	41–47	0.960	—
PLCC 84	32–38	1.180	—

able power rating for various packages can be established. Table 5.6 can be used in developing the maximum acceptable power rating for various surface mount packages for different values of Θ_{ja}, T_j, and T_a.

In addition, the user can take the following steps: (a) use conductive adhesive between the package and the substrate, (b) provide thermal pads for the packages such as LCCCs and mount them flush with the substrate, (c) connect devices with thermal vias and wider traces with the ground planes, and (d) use heat sinks or fan sinks on top of the package to conduct heat.

Another approach is to select different substrates with better heat dissipation characteristics. With ceramic packages, for example, the metallic constraining core used for matching the package and substrate coefficients of thermal expansion also helps in conducting heat.

Changing the airflow rate from, say, 200 to 400 linear feet per minute (lfm) (see Figures 5.10 and 5.11), derating the equipment for a lower ambient temperature, or changing the cooling medium entirely (from air to liquid cooling) are some of the more drastic steps that users can take. These steps are generally not warranted because of SMT alone.

5.7   PACKAGE RELIABILITY CONSIDERATIONS

Package reliability and solder joint reliability are different in surface mount and through-hole technologies. Solder joint reliability (discussed in Section 5.8) has been an issue for ceramic packages in military applications because of CTE mismatch between package and substrate. Although the plastic packages mounted on the commonly used glass epoxy substrates do not have solder joint reliability problems, the packages themselves may be susceptible to cracking.

When moisture-saturated surface mount packages are subjected to reflow soldering (vapor phase, IR, convection, hot plate) conditions, the moisture expands inside the package and may crack the plastic body [2]. Package damage can be either internal or external, and may be difficult to detect nondestructively. The device may eventually fail due to wire bond pad corrosion, which is considered a long-term reliability failure. Wire bond damage, such as broken wires, bond cratering, and bond shearing may also occur. In cratering, the ball lifts, taking with it a portion of the bond pad metallization and the underlying oxide or silicon.

If deleterious contaminants such as flux can seep in through such a crack, they may cause bond pad corrosion, seriously degrading the long-term reliability of the package. It should be emphasized here that the impact of cracking on the reliability of plastic packages is not fully understood. Corrosion failures have been seen only under extreme test conditions.

Cracks are difficult to visually inspect because they occur on the bottoms of the packages. The cracking problem is more widespread in larger plastic packages (over 68 pins). However, some cracking has been reported in smaller packages such as dynamic random access memories (DRAMs) and larger small outline integrated circuits (SOICs). The categories of packages that are considered susceptible to cracking include, but are not limited to, J-bend and gull-wing leaded packages such as plastic leaded chip carriers (PLCCs), SOICs, plastic quad flat packs (PQFPs), plastic ball grid arrays (PBGAs), small outline J-lead devices (SOJs), thin quad flat packs (TQFPs), and thin small outline packages (TSOPs).

Various long- and short-term solutions are being proposed by the industry for this problem. The Package Cracking Task Force of the Surface Mount Council published a white paper [3] summarizing the status of the problem with recommendations for the industry.

As mentioned in Chapter 2, the Council is only a management body chartered by the industry to coordinate the efforts of the Government and the industry to promote SMT. Since the Council itself does not write standards, it recommended that the industry organizations address this issue through the development of standards. As a result the IPC formed a committee to develop a standard on Package Cracking (IPC-SM-786A: Procedures for characterizing and handling of moisture/reflow sensitive ICs) [4] originally chaired by the author.

In addition to IPC, EIA also has standards on moisture sensitive package classification and handling. The names of the EIA standards which accomplish essentially the same objectives as IPC-SM-786A are JESD 112 and JESD 113. However, there are some differences between the IPC and EIA standards, so users are often confused as to which requirements to follow.

Again, at the direction of the Surface Mount Council, IPC and EIA are working together to develop two joint industry standards (J-standards). One of these is J-STD-020, Moisture/reflow sensitivity classification for plastic integrated circuit surface mount devices [5]. This standard is primarily for the use of package suppliers since they are the ones responsible for classifying a package as moisture sensitive or not and letting their users known about it. J-STD-020 replaces part of IPC-SM-786A and EIA standard JESD 112.

The second standard, which covers handling and shipping moisture sensitive packages for use by both suppliers and users, is still to be named and developed. In the interim, IPC-SM-786A and EIA JESD 113 will continue to be used for handling and shipping of moisture sensitive packages. Once the joint industry standard is released, the IPC and EIA standards will be phased out.

Now let us discuss factors that control package cracking and then focus on short- and long-term solutions.

5.7.1 Package Cracking Mechanism

The plastic DIP packages have never been thought to be moisture resistant, but no cracking has been reported in them. Why, then, since the industry is using the same molding compound in PLCCs as in DIPs, have the plastic DIPs not been found to be susceptible to cracking? In wave soldering, only the leads undergo high soldering temperatures. The package body rarely is heated beyond 150°C. However, in reflow soldering (vapor phase, infrared (IR), or convection), both the packages and the leads are subjected to temperatures of 215° to 230°C. If such a package is saturated with moisture, the moisture will expand during reflow and may crack the package.

Delamination between the mold compound and the die surface may result in sheared or cratered ball bonds, due to concentration of stress on the bonds. These sheared or cratered ball bonds can cause intermittent failures because, as the plastic body expands over a temperature excursion, electrical contact may be made or broken. The failures may show as high infant mortality failures, or increased failure rates during system life.

The CTE differences between different materials such as molding compound, lead frame, die, and die-attach material compound the problem when moisture turns to steam during reflow. Each of these components has its own CTE, which determines how great the expansion or contraction of the material will be in response to changes in temperature.

Large differences in CTE will increase internal stress and thus increase the susceptibility of a given device to package cracking. For example, the CTE mismatch between the lead frame and molding compound is less for copper lead frames than for alloy 42 lead frames. However, the CTE mismatch between the lead frame and the die is greater for copper than for alloy 42. Therefore, the choice of lead frame material by the component manufacturer involves a compromise in terms of CTE mismatch.

Packages with the same exterior configuration (lead count, thickness, etc.) might not have the same moisture sensitivity. The same number of days of ambient moisture can result in different amounts of moisture being absorbed by the package, depending upon package materials, internal geometries, design, etc. Furthermore, the same amount of moisture in different packages can cause different amounts of damage. The primary variables determining whether a given device will be susceptible to these failure mechanisms are as follows:

- Ratio of die paddle size to minimum plastic thickness
- Quantity and distribution of moisture absorbed by the package prior to surface mount
- Maximum package temperature during solder reflow/rework
- Adhesion of molding compound to die, lead frame, and other internal elements
- Mold or potting compound material properties
- CTE mismatches between different materials used in the package
- Component assembly mold process
- Die fabrication and wire bonding process (cratering only)

It has been pointed out that no cracking is seen if there is no die or a very small die in the package. Thus we can conclude that along with moisture, die size plays an important role. The thickness of the plastic body is also important: a thicker package is less susceptible to cracking, but the package will remain in a stressed condition because the moisture attempts to escape.

Steiner and Suhl [2] have determined that the threshold temperature for cracking is 180° to 200°C. This means that as far as package cracking is concerned, no reflow process (vapor phase, IR, or convection) is better than another. However, this problem will not occur if the soldering process (e.g., laser or hot bar soldering) does not subject the packages to higher soldering temperatures. This problem may be more serious if plastic surface mount devices are wave soldered because of much higher soldering temperatures (240°–260°C).

In addition to soldering processes, the ramp rate can also be a factor in package cracking [6]. Ramp rate is not as critical as the maximum package temperature but it should be kept at 6°C per second or less whenever possible, with such processes as IR or convection heating [5]. In addition, the package dwell time at maximum temperature has no significant effect. Multiple reflow cycles, however, will likely cause significant degradation in crack resistance. Therefore, the number of reflow cycles (including rework) should be kept to a minimum.

The specific mechanism of package cracking is depicted in Figure 5.12 [7]. The top diagram shows the package before moisture absorption, including the lead frame, the plastic encapsulant, and the plastic thickness below the lead frame. The package absorbs an amount of moisture that depends on the thickness of the plastic and the storage conditions. When the package is soldered by the reflow process, the entrapped moisture expands (middle diagram).

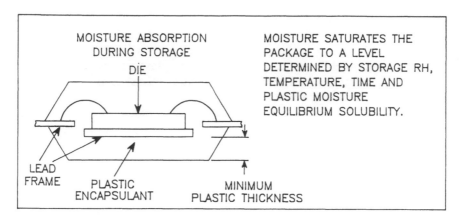

MOISTURE ABSORPTION DURING STORAGE

DIE

LEAD FRAME

PLASTIC ENCAPSULANT

MINIMUM PLASTIC THICKNESS

MOISTURE SATURATES THE PACKAGE TO A LEVEL DETERMINED BY STORAGE RH, TEMPERATURE, TIME AND PLASTIC MOISTURE EQUILIBRIUM SOLUBILITY.

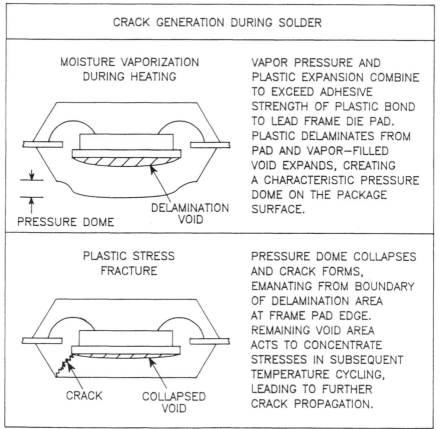

CRACK GENERATION DURING SOLDER

MOISTURE VAPORIZATION DURING HEATING

PRESSURE DOME

DELAMINATION VOID

VAPOR PRESSURE AND PLASTIC EXPANSION COMBINE TO EXCEED ADHESIVE STRENGTH OF PLASTIC BOND TO LEAD FRAME DIE PAD. PLASTIC DELAMINATES FROM PAD AND VAPOR–FILLED VOID EXPANDS, CREATING A CHARACTERISTIC PRESSURE DOME ON THE PACKAGE SURFACE.

PLASTIC STRESS FRACTURE

CRACK

COLLAPSED VOID

PRESSURE DOME COLLAPSES AND CRACK FORMS, EMANATING FROM BOUNDARY OF DELAMINATION AREA AT FRAME PAD EDGE. REMAINING VOID AREA ACTS TO CONCENTRATE STRESSES IN SUBSEQUENT TEMPERATURE CYCLING, LEADING TO FURTHER CRACK PROPAGATION.

Figure 5.12 Mechanism of cracking in large plastic surface mount package [7].

If the adhesive strength of the plastic bond with the die is exceeded, a delamination void results, such that the package cracks, usually at the edges (Figure 5.12, bottom), and the pressure dome collapses because of the escape of moisture. The crack generally emanates from the corner of the die paddle (lead frame directly below the die) as shown in the bottom diagram. Side cracking has also been observed, however. See Chapter 10, Figure 10.10 for a cross-sectional photograph of a die sitting on the die pad.

Cracks initiate from points of high stress at the edge of the lead frame die attach pad and propagate along the path of least resistance—downward through the thinnest portion of the plastic molding or radially out along the lead frame interface. This phenomenon was reported first by Oki Electric on flat packs but has recently been observed in small outline devices containing large static or dynamic RAMs and mostly high pin count (68 pin) PLCCs. Some cracking has also been observed in lower pin count PLCCs as well. As mentioned earlier, die size is more critical than package size or pin count.

Fukuzawa et al. [8] of Oki Electric observed cracks that developed in plastic surface mount packages (56 pin, plastic quad flat pack) after being subjected to temperatures from 215° to 270°C during the attachment process. They theorized that the package temperature experienced during reflow was far greater than the glass transition temperature (165°C) for the silica-filled novolac epoxy, resulting in localized delamination between the die paddle and the plastic. Vapor pressure from absorbed moisture in the package, in combination with CTE mismatch between the die pad and the surrounding molding compound, caused cracks to develop in the thinner bottom part of the package. Further evidence of this rupture was the dome that appeared on the underside of the package (Figure 5.12, bottom). Water condensation in the gap between the chip and the plastic led to aluminum pad corrosion.

Similarly, Steiner and Suhl [2] also reported that vapor pressure from absorbed moisture during reflow soldering causes package cracking. The packages were subjected to a temperature of 85°C and 85% relative humidity (RH) for 168 hours. Ito et al. [9] confirmed that moisture absorbed by the molding compounds coupled with the temperature rise experienced by the packages during reflow will promote delamination between the molding compound and the die pad interface.

Figure 5.13 shows the moisture absorption levels in a 68 pin PLCC at 85°C and three relative humidity conditions. After the preconditioning at various times, Alger et al. [7] subjected the packages to vapor phase soldering. Packages with moisture content levels above the "package crack jeopardy line" (Figure 5.13) may crack when subjected to reflow soldering environments.

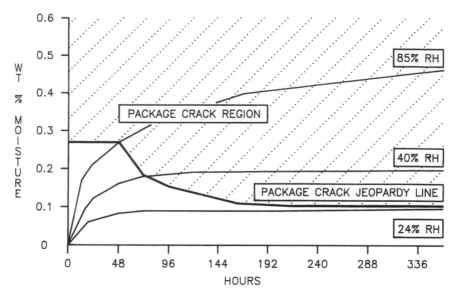

Figure 5.13 Moisture absorption in 68 pin PLCCs at 85°C and difference relative humidities [7].

Figure 5.13 shows moisture absorption in accelerated experimental conditions. Is this level of moisture absorption likely to be attained in normal storage conditions over time? The answer to this question, which is very critical in establishing whether the problem is real or simply a theoretical possibility, is contained in Figure 5.14 [7]. A Japanese Company (Shinko) collected the moisture absorption data shown in Figure 5.14 at 25°C for over 18 months. As the humidity in the environment changed, so did the moisture absorption level. Thus we see that moisture levels can exceed the threshold limit to cause package cracking.

In the beginning, the threshold limit for package cracking was considered to 0.11% (weight percent) of moisture, as shown in Figure 5.13. Bella et al. [10] showed that two layer plastic ball grid array (PBGA) packages with moisture content greater than 0.12 percent by weight exhibit serious delamination during reflow. They also found that four layer PBGAs absorb 20 to 25% less moisture than two layer PBGAs. We should note that all plastic packages including BGAs are susceptible to moisture. However, most of the surface mount packages other than BGAs need more time to achieve this percent of weight gain [10].

However, many package aspect ratios and geometries make the weight percent moisture less useful for standardization of handling procedures across the industry. Therefore, the moisture sensitivity classification levels in J-STD-020 [5] are based on representative useful floor exposure

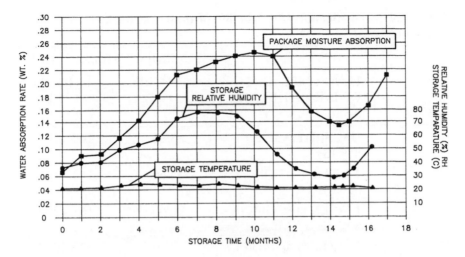

Figure 5.14   Effect of relative humidity on moisture absorption in PLCC at 25°C storage over a period of 18 months [7].

times used in manufacturing and not on the percentage of moisture gained after ambient exposure. There are six levels of classification per J-STD-020 as shown in Table 5.7.

Using the "derating" table shown in Table 5.8, one can determine the right exposure time for the relative humidity and temperature conditions applicable to a particular environment. For example, just by decreasing the factory temperature from 30°C to 20°C, the allowed exposure time can be more than doubled at any given RH condition. Table 5.8 also shows that thicker packages generally have less exposure time than the thinner packages for up to 60% RH, but the reverse is true for RH above 60%. In other words, thinner packages are less moisture sensitive than thicker packages at 60% or less RH but more moisture sensitive than thicker packages at RH conditions above 60%.

It was mentioned earlier but is worth repeating that the total moisture content alone is not the determining factor for package cracking. As packages are saturated with moisture, a moisture gradient occurs, and the location of the moisture is very important. For example, moisture absorption in the outer regions of the package is not as critical as absorption toward the center. Other important factors include reflow soldering temperature, lead frame design and surface preparation, strength of the plastic, die size, and plastic thickness. The die size and the thickness of the molding compound are particularly significant.

Because of greater CTE mismatch between the metallic lead frame paddle and the plastic molding compound, cracking severity increases

Table 5.7 **Classification and floor life of desiccant packed components [5]**

LEVEL	FLOOR LIFE (OUT OF BAG) AT BOARD ASSEMBLY SITE	MOISTURE EXPOSURE	
		°C/%RH	TIME (HRS)
1	UNLIMITED at </= 85%RH	85/85	168
2	1 YEAR at </=30°C/ 60%RH	85/60	168
3	1 WEEK at </=30°C/ 60%RH	30/60	168 + MET[a]
4	72 HOURS at </=30°C/ 60%RH	30/60	72 + MET
5	As specified on label: 24 or 48 hours at </= 30°C/60%RH	30/60	Time on label + MET
6	Mandatory bake: Bake before use and once baked must be reflowed within the time limit specified on the label	30/60	Time on label

[a]MET = Manufacture's exposure time. This compensation factor accounts for the time that the component manufacturer requires to process components prior to bag seal. It also includes a default amount of time to account for open bag time at a distributor before receipt by the customer.

The exposure times specified by this table relate to specific temperature and humidity conditions in the table; e.g. a Level 3 device must be baked before reflow if it has been exposed to 60% RH (relative humidity) at 30°C for 168 hours or more. Shook et al. developed equivalent exposure times or a "derating" procedure for varying temperature-humidity exposure conditions that a factory could maintain [11]. The findings of their study are shown in Table 5.8.

with die size. A rule of thumb is that die sizes greater than 4 mm (160 mils) on each side should be considered to be susceptible to cracking. As shown in Figure 5.15, die size and plastic body thickness (bottom side of package) can be used as guidelines in determining safe and unsafe regions. All packages that fall into unsafe regions should be declared susceptible to cracking. When using Figure 5.15, note that there will be some exceptions. The best way to correctly characterize a package is to subject it to the reliability tests discussed in Section 5.7.3.

The larger die sizes also generate more heat and hence require heat

Table 5.8 Recommended equivalent total floor life (hours) @ 20°C, 25°C, and 30°C for ICs with novolac, biphenol, and multifunctional epoxies (reflow at same temperature at which the component was qualified) [11]

MAX EPOXY THICKNESS ABOVE OR BELOW DIE	MOISTURE SENSITIVITY LEVEL	MAXIMUM PERCENT RELATIVE HUMIDITY								Hours @
		20%	30%	40%	50%	60%	70%	80%	90%	
Thickness = 1.6 mm	Level 3	476	282	220	188	168	154	144	136	30°C
PQFPS > 84 pins, PLCCs		666	394	310	266	238	218	204	192	25°C
(square) MQFPs		940	554	434	372	334	308	288	272	20°C
	Level 4	124	98	86	78	72	68	66	64	30°C
		174	140	122	112	106	100	96	92	25°C
		248	198	176	162	152	146	140	136	20°C
	Level 5	76	62	56	52	48	46	44	44	30°C
	(48 hrs)	110	92	84	80	74	70	68	66	25°C
		158	134	122	114	108	104	102	98	20°C
	Level 5	36	30	28	26	24	24	23	22	30°C
	(24 hrs)	56	50	46	44	42	40	38	38	25°C
		84	76	70	68	64	62	62	60	20°C
Thickness = 1.0 mm	Level 3	∞	∞	334	218	168	140	122	108	30°C
PLCCs (rectangular)		∞	∞	482	312	242	202	176	156	25°C
18–32 pins		∞	∞	704	450	346	288	252	226	20°C
SOICs (wide body)										
SOICs ≥ 20 pins, PQFPs	Level 4	354	152	106	86	72	64	58	54	30°C
≤ 80 pins		512	218	154	126	108	94	88	80	25°C
		750	314	224	182	162	140	128	120	20°C

Continued

Table 5.8 *(Continued)*

MAX EPOXY THICKNESS ABOVE OR BELOW DIE	MOISTURE SENSITIVITY LEVEL	MAXIMUM PERCENT RELATIVE HUMIDITY								Hours @
		20%	30%	40%	50%	60%	70%	80%	90%	
Thickness = 1.0 mm										
PLCCs (rectangular)	Level 5	154	90	68	56	48	44	40	36	30°C
18–32 pins	(48 hrs)	220	130	100	84	74	68	62	58	25°C
		318	188	146	124	110	100	94	88	20°C
SOICs (wide body)	Level 5	60	42	32	28	24	22	20	18	30°C
SOICs ≥ 20 pins, PQFPs	(24 hrs)	74	64	52	46	42	38	36	34	25°C
≤ 80 pins		132	96	80	72	66	60	58	54	20°C
Thickness = 0.5 mm	Level 3	∞	∞	∞	∞	168	82	56	44	30°C
TSOPs, SOICs < 18 pins		∞	∞	∞	∞	314	128	90	70	25°C
TQFPs		∞	∞	∞	∞	380	200	138	108	20°C
	Level 4	∞	∞	∞	156	72	48	36	28	30°C
		∞	∞	∞	278	114	78	62	48	25°C
		∞	∞	∞	400	174	120	92	76	20°C
	Level 5	∞	∞	∞	80	48	34	26	20	30°C
	(48 hrs)	∞	∞	∞	126	78	58	44	36	25°C
		∞	∞	∞	194	120	90	72	60	20°C
	Level 5	∞	∞	66	38	24	18	14	10	30°C
	(24 hrs)	∞	∞	104	60	44	34	26	22	25°C
		∞	∞	158	96	70	54	46	40	20°C

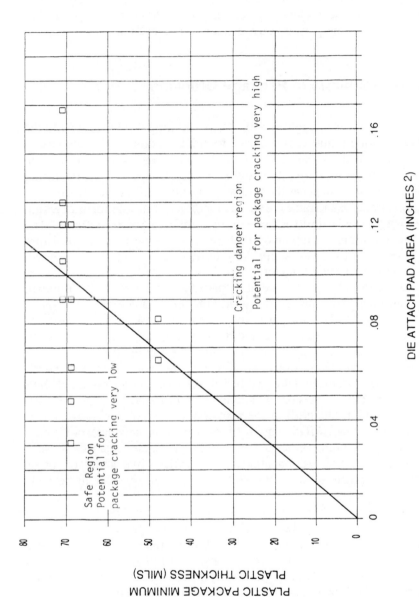

Figure 5.15 Guidelines for determining package susceptibility to cracking based on plastic thickness and die attach pad area [7].

DIE ATTACH PAD AREA (INCHES 2)

PLASTIC PACKAGE MINIMUM
PLASTIC THICKNESS (MILS)

Safe Region
Potential for
package cracking very low

Cracking danger region

Potential for package cracking very high

spreaders for heat dissipation. Although heat spreaders effectively reduce the thickness of the molding compound, they do increase the risk of cracking because the plastic body is made relatively thinner.

5.7.2   Solutions to Package Cracking

The industry is struggling with short- and long-term solutions to the package cracking problem. It has been found, for example, that increasing the thickness of the molding compound around and below the die can prevent cracking. Similarly, it is now known that package manufacturing variables that promote better lead frame adhesion, such as rougher internal leads, perforated lead frames, and anchor holes, can minimize cracking by allowing the plastic to adhere more tightly to the lead frame. The use of new molding compounds that are more impervious to moisture penetration would provide a true long-term solution, but it is unknown when or indeed if such a material can be developed. If success is ever achieved, the use of expensive ceramic packages will drastically diminish, along with their associated solder joint cracking problems.

Another potential solution is to prevent the absorption of moisture at the interface of the lead frame and the molding compound. If moisture vapor does not collect at this interface, cracking will not occur. Suhl [12] reported excellent results from applying hexamethyldisilazane (HMDS) vapor to the lead frame and die just before the molding operation. HMDS is used to promote adhesion of silicon wafers to novolac photoresist, hence is a familiar chemical to the semiconductor industry. However, it should be noted that HMDS is hygroscopic (absorbs moisture), odiferous, flammable (flash point of 13°C), and toxic. It causes headaches, sore throat, dizziness, and asthma attacks. So HMDS, like about 65 other hazardous chemicals used by the silicon chip industry, should be handled accordingly.

Suhl [12] treated one group of samples with HMDS and used the other half as a control. The samples in both groups were considered susceptible to cracking. All the samples were saturated with moisture and subjected to vapor phase reflow conditions. Suhl found that the packages treated with HMDS showed significantly lower cracking (5%) than the untreated standard packages (100% cracking) that served as controls.

Approaches based on use of a better molding compound or treatment with HMDS are still in the experimental stages, but baking is evolving as the short-term solution. Figure 5.16, which shows moisture desorption for packages saturated with moisture at various relative humidities [7], suggests that the absorbed moisture can be baked out. The "package crack

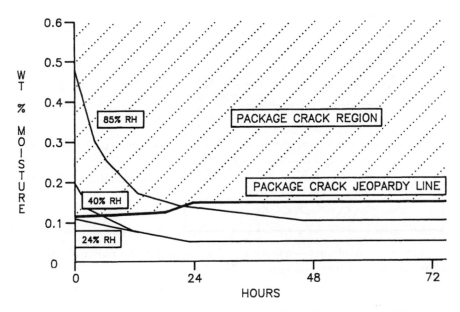

Figure 5.16 **Moisture desorption in 68 pin PLCCs baked at 125°C for different times [7].**

jeopardy line" in Figure 5.16 develops when the packages are baked at 125°C for various times.

To prevent cracking, packages should be baked and packaged in desiccant bags and classified per Table 5.7 for acceptable floor life out of bag. In addition, baked packages must be shipped in desiccants to prevent moisture absorption during shipping and storage at the user's facility.

A construction of a shipping container with desiccant is shown in Figure 5.17. A desiccant "dry" pack consists of desiccant material and humidity indicator card (HIC) sealed with the components inside a moisture barrier bag (MBB). A warning and information label, sufficient to ensure proper handling, is provided on the outside surface of the bag. The specific requirements for the bag and desiccant material are specified in IPC-SM-786A [4].

A sample humidity indicator card (HIC) with three levels of resolution is shown in Figure 5.18. This indicator has dots graded from 10% RH to 30% RH. If the 20% RH dot has turned pink and the 30% RH dot is not blue, the components have been exposed to a level of moisture beyond that recommended, and the usable dry pack life has expired. These units need to be baked dry prior to reflow. The HIC label also provides the user with the moisture sensitivity classification level as described in Table

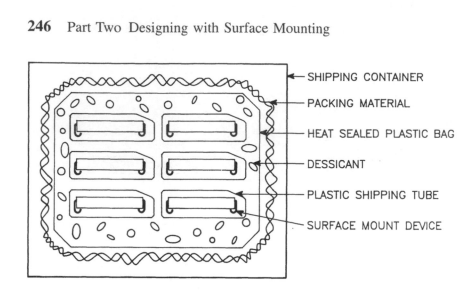

Figure 5.17 **Moisture-sensitive packing to minimize moisture absorption during shipping and storage [7].**

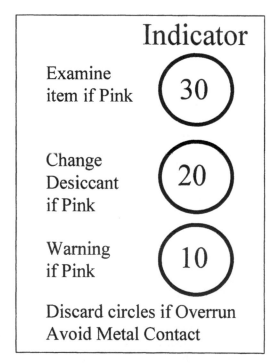

Figure 5.18 **Sample humidity indicator with three levels of resolution.**

5.7. In addition, the factory seal date, from which the remaining shelf life can be determined, may also be on this label.

Generally, the user has six months from the date of seal to use the package if the bag is kept sealed. However, once the color has changed to pink, the packages need to be rebaked. After baking, components can be resealed with fresh desiccant in the MBB. One possible way to avoid rebaking is to store parts in controlled environments such as desiccant cabinets. Desiccant cabinets must be able to provide an environment of 25+/−5°C and <20% RH. Nitrogen or dry air can also be used to store baked components.

Two baking options can be used to provide flexibility in manufacturing. These are the high and low temperature baking cycles. The process conditions for the high temperature bake are:

Temperature: 125(±5)°C
Duration: 24–25 hours
Chamber relative humidity: <50%

The high temperature baking process is not suitable for plastic tubes or tape and reel because they cannot withstand the baking temperature and the packages must be removed from the tubes or tapes and reels before being baked. They can be placed either in a metallic tube or in waffle packs, which can withstand the baking temperature.

Each package that was not received baked and bagged from the vendor should be subjected to a maximum of TWO high temperature bake cycles before being soldered to the board. Packages received in a baked and bagged condition should be subjected to only ONE high temperature bake cycle before being soldered unless solderability of the post baked component has been verified. This is mainly because repeated baking may cause solderability problems.

If repeated baking is to be used, any adverse impact on component solderability should be determined. Baking packages at 125°C for 24 hours (one baking cycle) will add about 15 to 20 microinches to the copper-tin intermetallic thickness and should not pose solderability problems. The tin-lead is more than 200 microinches thick.

Care should be taken to ensure that outgassing of shipping media, such as trays, does not impact lead solderability. The users should verify use temperatures of shipping media with component suppliers before using them in high temperature applications.

Low temperature baking, which provides much more flexibility for a manufacturing environment, may be especially helpful for packages that come in plastic tube or on tape and reel. Low temperature baking takes longer, however. The process conditions for the low temperature bake are:

Temperature: 40(+5–0)°C
Duration: 192 hours minimum (8 days)
Chamber relative humidity: <5%

The low temperature baking process is suitable for plastic tubes as well as tapes and reels. Packages can be subjected to as many low temperature bake cycles as required.

There are three separate conditions that require baking by the user:

1. The component's desiccant packaging has expired.
2. The components are classified as moisture sensitivity level 6 (see Table 5.7).
3. The component has been exposed longer than the extended floor life.

After baking, the post baking life is determined by the classification level of the package as determined originally by the component supplier. Upon receipt, "dry" pack shipments should be inspected for bag seal date located on the caution or bar code label. Bag integrity should be verified. There should be no holes, tears, or punctures of any kind that would expose either the contents or an inner layer of multilayer MBB.

Bags may be opened for component inspection by cutting at the top of the bag near the seal. However, component exposure is not allowed to exceed 0.5 hours at 30°C/60% RH to ensure that the floor life is not adversely affected. Meeting these conditions, and if the humidity indicator card (HIC) has not exceeded its limit, the bag may be resealed with the same desiccant and HIC. The bag should be hot sealed so as not to damage or cause delamination of the MBB.

After assembly, if components need to be reworked, boards need to be baked before reflow to prevent component damage. Since it is not desirable to bake a board at 125°C for a long time (24 hours), a lower temperature (90°C) may be used, but the time needs to be increased (28 to 85 hours). Alternatively, if the rework temperature is kept at 200°C maximum, baking of the board may not be necessary.

5.7.3   Moisture Sensitivity Classification for Package Cracking

As mentioned earlier, cracking is difficult to spot during routine manufacturing operations such as inspection and electrical testing. To prevent potential package cracking problems, moisture-sensitive packages should be qualified for use. To simulate a worst case, packages should be saturated with moisture in the required temperature and humidity environment and then subjected to process conditions. The environmental conditions to which packages are subjected are shown in Table 5.9. This table specifies a precondition sequence to simulate the surface mount process for determining a plastic package's susceptibility to cracking. All components would be subjected to this test sequence before reliability stress testing. The test conditions suggested in Table 5.9 are established by J-STD-020 [5].

Packages that have been subjected to the preconditioning test flow sequence of Table 5.9 should be cross-sectioned and examined under a microscope. Cracks are difficult to find, but they commonly occur along the die edges or in the center of the die, then proceed toward the leads. Complete cracking of packages on the bottom is not very common.

The reason for drying the package is to ensure that we know exactly the level of moisture in the package. Three cycles are chosen to simulate two reflow cycles and one rework cycle. For reflow, either vapor phase or convection reflow can be used.

Since the convection reflow process has essentially replaced vapor phase (see Chapter 12), the use of convection reflow for package classification was added to the revised specification J-STD-020.

The acceptance criteria are the same, regardless of moisture sensitivity classification level. The acceptance criteria are applied in a hierarchical manner.

If the components pass the electrical test, there is no visual evidence of external cracks, and there is no evidence of delamination or cracks observed by acoustic microscopy or other means, the component is considered to pass that level of moisture sensitivity. This is the ideal condition. However, if internal cracks are observed, components are required to be

Table 5.9   Generic flow for moisture sensitivity classification [4]

STEP 1: Drying, 125°C hours

STEP 2: Temperature/humidity exposure (as defined in Table 5.7 for specific classification level being assessed)

STEP 3: Three cycles exposure to vapor phase or convection reflow (220°C + 5°C/–0°C)

STEP 4: Acceptance criteria evaluation

evaluated using cross-sectioning or another method to assess crack length and depth using the following criteria:

1.   Cracks are not allowed to intersect the bond wire, ball bond, or wedge bond.
2.   Cracks are not allowed to extend from any lead finger to any other internal feature (lead finger, chip, die attach paddle).
3.   Cracks are not allowed to extend more than two-thirds the distance from any internal feature to the outside package.

It is also important that there be no measurable change in any of the above three conditions. If there is a measurable change, evaluation at the next sensitivity level will be required. For further details refer to IPC-SM-786A and J-STD-020 [4,5].

5.8   SOLDER JOINT RELIABILITY CONSIDERATIONS

In through-hole mount assemblies, solder joint reliability has not been an issue because the plated through hole gives added mechanical strength to the joint. However, solder joint reliability causes concern in surface mounting because the surface mount lands provide both mechanical and electrical connections. Are surface mount solder joints less reliable because of the absence of plated through holes? The answer is both yes and no. Solder joint reliability is of concern in military applications where the CTE of packages (ceramic) and glass epoxy board are not compatible. However, in commercial applications using plastic leaded packages or small ceramic chip components on a glass epoxy board, solder joint reliability is not of concern. Since the difference in CTE between plastic packages and glass epoxy boards is very minor, the leads of the packages take up the small expansion mismatch created by thermomechanical cycling.

Ceramic chip resistors and capacitors are also widely used on glass epoxy boards. But because of the small size of these components, the stresses they generate do not exceed the mechanical strength of the solder joint; hence there is no reliability concern. When using properly designed land patterns (see Chapter 6), the surface mount solder joints (plastic packages and ceramic chip components mounted on glass epoxy boards) pass all the environmental reliability tests, as shown in Table 5.10. The tests were conducted on functional assemblies using the test parameters given later in Table 5.11 (Section 5.8.1).

As the size of the ceramic package increases, stresses can increase

Table 5.8   Thermal cycling results on solder joint reliability for Type I, II and III SMT. See Table 5.9 for test conditions. (Intel Corporation.)

	TYPE III ASSEMBLY	TYPE I ASSEMBLY	TYPE II ASSEMBLY
Random vibration	Pass	Pass	Pass
Mechanical shock	N/A	Pass	Pass
Humidity storage	Pass	Pass	Pass
Temperature cycle (operating)	N/A	Pass	Pass
Temperature cycle (nonoperating)	Pass	Pass	Pass
Life test	Pass	Pass	Pass
Total solder joint cycles	210,000	580,000	1.3 million

substantially. This is generally the case for ceramic capacitors larger than 0.250 inch and for high pin count leadless ceramic packages. For example, it is well known that because of a mismatch in CTE values, LCCCs mounted on glass epoxy substrates cause solder joint cracks in fewer than 200 thermal cycles. There are three commonly used approaches to solder joint cracking problems related to CTE mismatch: using substrates with compatible or matched CTEs, using substrates with top compliant layers that can absorb the stress, and using leaded ceramic packages.

The compliant top layer approach is rarely used because the reliability problems may be shifted to the traces or the via holes. Using substrates with compatible CTEs is the way most commonly chosen, but even this approach has its problems. For example, cracking may be eliminated in the solder joints, but there are failures in the plated through holes. Even if the solder joints do not crack as a result of compatible CTE values, the cracking in the plated through holes has the same effect: unreliable assembly.

To prevent via hole cracking, either the board thickness is reduced or the via hole diameter is increased. Use of a substrate material with a lower modulus of elasticity has been tried to prevent via hole cracking. Since the lower modulus material is soft, it cannot also damage vias even under stress. See Chapter 4 (Section 4.3: Selection of Substrate Material) and sections dealing with substrates.

In the third approach, using leaded instead of leadless ceramic packages on substrates with unmatched CTEs, the stress on the solder joint can be taken up by the compliant leads. The use of leaded packages not only reduces the overall cost but almost assures the reliability of the solder joints. The leads can be either gull wing or J type. However, the use of leaded ceramic packages has not been a very successful approach.

The leaded ceramic packages typically have leads made from Kovar or alloy 42 instead of copper, to ensure that the lead frame material matches the ceramic body. These leads are not ductile like copper leads. They may, however, be adequate for smaller chip carrier sizes that do not generate excessive stress on the solder joints. In addition to having a ductile lead material, ceramic packages require proper lead design. For example, gull wing leads provide better compliancy than J leads. And even S-shaped leads are being used by some for best results. However, S-shaped leads are neither widely available nor commonly used. Using leaded ceramic packages on inexpensive glass epoxy substrates calls for additional work in the areas of lead frame material selection and lead configuration design to ensure compatibility with mass soldering.

The concept of using leaded ceramic packages for military applications is not new. As a matter of fact, the military originally wanted to use leaded packages but abandoned the idea for various reasons. The leadless ceramic packages were selected because they were easier to assemble. In addition, they can be placed closer together (giving a higher density assembly), they have superior thermal performance, and they are cheaper than leaded packages. As discussed earlier, the use of leadless packages requires ceramic substrates or exotic substrates. This adds significantly to the cost of the assembly. In the final analysis, the overall increase in assembly cost far exceeds any savings from the leadless packages. Now the trend is shifting back to leaded packages for the same reasons—reliability and cost concerns.

Achieving adequate solder joint reliability at reasonable cost for military applications is no easy matter. The leadless packages give the most problems. The U.S. Department of Defense has put LCCCs on a liability suspect list even though the devices are mounted on substrates with matched or tailored CTE values. This by no means implies that leadless packages do not have any application. They do, especially where it is critical to have the shortest possible signal path as well as efficient cooling through the substrate. The use of leadless packages can and should be reduced whenever possible, however. They should be selected only when leaded ceramic or plastic packages are not feasible at all. Moreover, when using leadless packages, via hole and solder joint reliability should be extensively evaluated before a substrate material is chosen.

5.8.1 Solder Joint Reliability Tests

What are the relevant solder joint reliability test methods? What constitutes a failure? How should failures be monitored? The answers are not as obvious as they might appear. This is basically because there are no hard data to correlate the accelerated reliability test results with what is known of field failures or intended service life. Among the various types of test methods available, which one should be chosen? The answer will depend on the purpose of the test. A test method that closely resembles the field application will give the most representative results.

The electronics industry uses many test methods for ascertaining solder joint reliability, but the most common are powered functional cycling, thermal cycling, and mechanical cycling. Table 5.11 shows an example of test parameters for functional, thermal, and mechanical cycling tests used on Type I, II, and III surface mount assemblies for commercial/industrial applications.

In addition to subjecting the assemblies to environmental test conditions such as shown in Table 5.11, for any application, the power functional cycling is also very desirable; it is a recommended test regardless of whether any other tests are conducted. Engelmaier has shown that powered functional and thermal cycling are more realistic tests to induce solder joint fatigue [13]. Functional cycling tests should be conducted in an environment that closely simulates the field environment.

The other commonly used test, thermal cycling, is appropriate for comparison purposes and for assemblies with large CTE mismatch, such

Table 5.11 Environmental test conditions to which surface mount assemblies were subjected. See Table 5.10 for test results. (Courtesy of Intel Corporation.)

RANDOM VIBRATION:
  (0.01 g^2/Hz at 10 Hz sloping to 0.02 g^2/Hz from 20 Hz to 1 kHz)
MECHANICAL SHOCK:
  (50 g sine for 11 ms)
HUMIDITY STORAGE:
  (5 days at 55°C, 95% RH)
TEMPERATURE CYCLING (OPERATIONAL)
  (-15 to $+70$°C for 5 days)
TEMPERATURE CYCLING (NONOPERATIONAL)
  (-40 to $+125$°C; 2500 cycles)
LIFE TEST:
  (3000 hours at 60°C)

as ceramic chip carriers or chip components on FR-4. For thermal cycling to produce useful results in a reasonable time frame, a CTE mismatch of at least 3 ppm/°C must exist between the component and the substrate.

The dwell temperatures and times and the ramp rates are the most critical parameters in functional and thermal cycling. It is important to note that dwell times at extreme temperatures are significantly shorter in accelerated testing than in the field, and thus the stress relaxation is less complete. Therefore the number of cycles to failure in the field is always less than the number of cycles to failure in an accelerated test. This discrepancy is explained by the stress-strain curves of Engelmaier [14] in Figures 5.19 and 5.20, which illustrate different cyclic stress-strain behaviors of metals. Figure 5.19 shows a hysteresis loop without the stress relaxation characteristic of fatigue cycling of copper or even solder with zero dwell times. It is more typical of engineering metals, such as copper, aluminum, or steel, that do not creep or relieve stress at room temperature. Figure 5.20, on the other hand, shows the hysteresis loop for a solder joint with cyclic dwells long enough for complete stress relaxation, as would occur in field use.

The area in the cyclic hysteresis loop is a measure of the fatigue damage per cycle. Since cyclic fatigue damage is cumulative with each fatigue cycle and exhausts the available fracture toughness, the larger the hysteresis loop, the lower the number of cycles to failure. This means that the difference in the hysteresis loops generated in accelerated testing

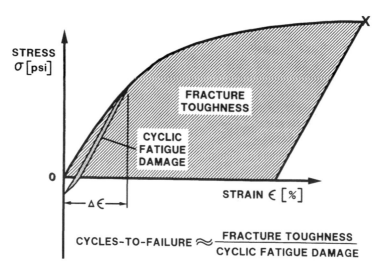

Figure 5.19   Illustration of cyclic stress-strain behavior of solder; stress relaxation in the hysteresis loop is not shown [14].

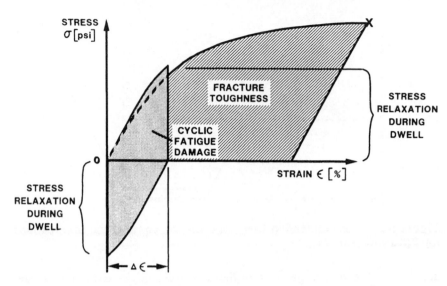

Figure 5.20   Illustration of cyclic stress-strain behavior of solder; stress relaxation is shown in the hysteresis loop, which indicates loss of ductility and premature fatigue [14].

and in the field needs to be accounted for when establishing product reliability. It also means that to achieve high reliability, the component/ substrate system needs to be designed to produce a small hysteresis loop. This can best be accomplished by a combination of component and substrate CTE tailoring (taking into account the temperature difference between component and substrate in a functional environment) and attachment (lead, solder, and substrate) compliancy.

Thermal cycles of various types with different dwell temperatures and times are used to test solder joint reliability. There is absolutely no consensus in the industry about the upper and lower temperatures and the ramp rate. In addition to thermal cycle (one-chamber) tests, thermal shock (two-chamber) tests with rapid change in temperature extremes are used. These conditions rarely are encountered by a product in service, however, so unless field conditions warrant, thermal shock test results may be misleading [15].

Generally, thermal cycling from −55° to +125°C for 500 cycles for military applications and 0° to +100°C for 1000 cycles for commercial applications is used [16]. The dwell time is 30 minutes at each temperature extreme and the transition time between temperature extremes is 30 minutes, as shown in Figure 5.21.

The third commonly used method of determining solder joint reliability, mechanical cycling, is generally preferred because it is a highly

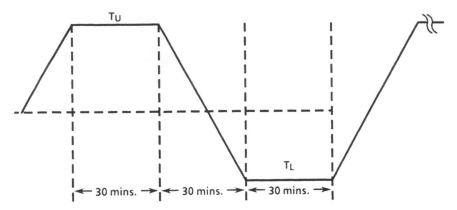

Figure 5.21 Recommended thermal cycle for commercial applications on FR-4 substrate [16].

accelerated test as compared with thermal cycling. For example, 20,000 mechanical cycles can be performed in the same amount of time (84 days) needed for only 1000 thermal cycles. Since both mechanical and functional cycles subject the solder joints to shear stress, the failure mechanism in either case is by low cycle fatigue (as long as the joint is not overstressed).

Mechanical cycling generally calls for 40 to 80 inch radius alternate convex-concave bending of the circuit board. A frequency of 240 cycles per day until failure is commonly used with mechanical testing [15]. Equipment for conducting mechanical cycling tests is commercially available [17]. The tester was originally designed for the Compliant Lead Task Force of the Institute of Electrical and Electronics Engineers (IEEE). One of the problems with both mechanical and thermal cycles has been that it takes around three months to see any failure. For mechanical cycling, only thin boards (40 mil thick) were used because the tester mentioned in reference 17 could not accommodate thicker (62 mils) boards. (There is no thickness limit for thermal cycling, but it still takes about 1000 cycles or 80 days to see the first failure.) A thinner board being very flexible, took 20,000–30,000 cycles (84–126 days) to show the first failure. Now, with some modification in this tester, 62 mil thick boards can be used and failure is seen within 8 days (under 2000 cycles). The existing testers can be modified to accommodate thicker boards. Also, caution should be exercised when using the tester mentioned in reference 17. Variations of as much as 15% in the convex and concave bending from the set values may be seen. Such a wide variation, if allowed to occur, will give questionable test results.

No matter which method (thermal or mechanical cycling) is used, the failure criteria should be the same. When evaluating solder joint reliability, daisy-chained packages should be *continuously monitored.* The majority of test monitors in the industry today, however, are ordinary data loggers that do not continuously monitor solder joints. Depending on the number of channels available and the number of resistance loops being monitored, every loop is monitored every 5 to 6 seconds. Since most (not all) solder joint failures occur during the transient phase (between the upper and lower test limits), the solder joint will be open for only a few microseconds. A multiplexer data logger could easily miss the first failure indication, causing the final results to be off by 10 to 30%. In other words, failures might not be seen until some time after they really happened. This is true of failures caused by shear stress, as is the case for solder joints.

When the solder joint failure mode is due to tensile stress instead of shear stress, the resistance loops do not open and close as quickly. Thus a multiplexed data recording instrument can easily catch the first open. For example, when monitoring plated through holes in a printed circuit board, failures are likely to be spotted early because the copper barrels fail due to tensile stress and are open for a longer period.

For failure, discontinuities lasting at least one microsecond need to be observed and recorded. A discontinuity, that is, loop resistance of 1000 ohms or more, should be confirmed by 10 such failures [16]. Equipment and software are commercially available to continuously monitor resistance, hence solder joint failures [18]. However, the number of suppliers for such equipment is very limited.

Failure also may be defined by the occurrence of a change in resistance of some percentage (50–200%), but this approach is less exacting. Visual observation can also be used, but it is highly unreliable and labor intensive. However, valid confirmation of a totally cracked joint can be achieved by way of visual examination.

Having completed the reliability tests, can we predict the field performance of the product accurately? No. But the accelerated tests results can be used to compare the results of several designs and to help us select superior land pattern designs and better package/substrate combinations.

5.9 INTERCONNECT CONSIDERATIONS

As mentioned earlier, surface mounting has essentially forced the industry into using fine line boards. As trace spacings become smaller and circuit speeds become higher, the potential for circuit malfunction increases as a result of cross-talk or coupling energies from adjacent traces. With the use of SMT and faster devices, therefore, the design must reflect consideration of cross coupling effects [19].

Present indications are that active lines (lines with impedance to ground of 100 ohms or less) can be spaced at 6/6 or even 5/5 without significant cross coupling. However, signal to ground plane spacing should be within 8 mils, and adjacent traces should run parallel for less than 3 inches.

With surface mounting, the integrity of the power and ground planes is increased because the number of component mounting holes is reduced. (However, there can be exceptions, especially when fancut patterns; i.e., vias for every pad, are used.) This enables the designer to route key signal traces on these planes without damaging their effectiveness as reference planes for the signal traces on adjacent layers. Prime candidates for these traces are high energy clock lines and high impedance lines, which require greater decoupling from the other signal traces.

When working on surface mount boards, the designer must develop rules and guidelines for acceptable trace width and spacing, maximum allowable lengths of parallel circuit lines, and the optimum dielectric thickness for the substrate material. Typical guidelines for the use of these traces on ground and power planes are shown in Figure 5.22 and are summarized as follows:

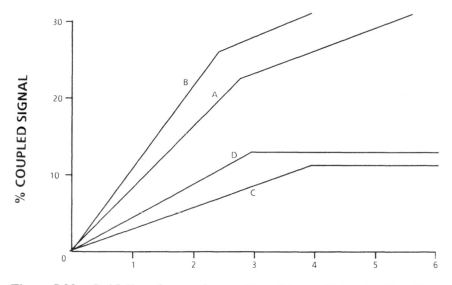

Figure 5.22 Guidelines for maximum allowable parallel active lines free of cross-talk. A and B: traces on external layers; D and C: traces on internal layers; B and D: lightly loaded; A and C: heavily loaded. (Courtesy of Intel Corporation.)

1.   Maximum trace length of 1.5 inches.
2.   All traces routed parallel with minimum spacing of 200 mils.
3.   Adjacent signal plane lines routed perpendicular to traces on ground and power planes.

There is some good news in SMT interconnect design. For example, surface mount decoupling capacitors are most effective than through-hole mounted capacitors because the former have less lead inductance to the ground and power planes. A rule of thumb would indicate that half as many capacitors are necessary in surface mount design versus through-hole board design. Also, because lead lengths are short, there is less inductance loss when using surface mount components.

Propagation delay is a function of the total lead length (outside lead length and inside wire bond length) of the package. As we move from through-hole technology to SMT, considerable improvement in propagation delay can be expected because of significant reductions of total lead length in SMT. The propagation delay gets even smaller in packages such as chip-on-board or ball grid arrays (BGAs) since they have smaller total lead lengths than "conventional" SMT packages. Let us clarify this point by an example.

The propagation delay is defined as the time required for travel of the electrical signal through wire bond (package body) and outside lead (air). To calculate the delay time, the speed of the signal is divided by the square root of the dielectric constant. The speed of travel for an electrical signal is 3×10^{10} cm/second or 12 inches/nanosecond (same as the speed of light) when the dielectric constant of the medium is 1 (air). In other words, the propagation delay will be 1 nanosecond for a 12 inch long lead length in air (12 divided by the square root of 1). Since the signal travels in the package body (wire bond) and air (outside package lead), the propagation delay is a function of the effective dielectric constant of these two mediums. To keep our calculations simple, let us assume a dielectric constant of 9 for the medium (for ceramic package and air combination). The propagation delay will be 1 nanosecond for every 4 inches of lead length (12 divided by the square root of 9).

Let us refer back to Chapter 1, Figure 1.15 which shows a propagation delay of 0.3 nanosecond for a 64 pin DIP package. This means that the authors in Reference 1, Chapter 1 have assumed a total lead length of 1.2 (4 × 0.3) inches in a DIP package (the corner leads could have this lead length). By comparison the propagation delay is shown to be 0.05 nanosecond in a 64 pin LCCC. This means that the total lead length is 0.2 inch (0.05 × 4) in a 64 pin LCCC.

The propagation delay for a plastic package would be even lower than the numbers just discussed because of the lower dielectric constant of plastic material. If we carry this analysis further, the propagation delay in a flip chip (not shown in Figure 1.15) will be practically nonexistent since a flip chip has no lead and is directly mounted on the substrate.

5.10   CAD LAYOUT CONSIDERATIONS

Most computer-aided design (CAD) systems were designed for through-hole boards. Various requirements imposed by SMT call for additional capability in the CAD software. It is safe to generalize that more work is needed by CAD system suppliers to completely meet the CAD requirements of SMT boards. The designer must take this into account and should be prepared for the impact of the CAD tool on the product schedule as well as on cost.

There are many types of CAD systems available for SMT applications, and all have their pros and cons. In particular, the product designer must be aware of the implications for PCB features due to SMT. For example, changing to SMT design generally also means using boards with narrow trace width and space. Certain design requirements are different for SMT boards, and the CAD system must address them successfully. For example, in through-hole boards, the location of via holes is not an issue. But in surface mounting, vias are required for interconnection and testing, and specific guidelines for their location must be developed. SMT also requires specific guidelines for via placement, since errors in this respect can have a negative impact on real estate, testability, and manufacturability.

To preserve as much space as possible for routing traces, the number of vias must be minimized. This means that the CAD system must be able to route from the surface pads. Some CAD systems, however, look for familiar through-hole vias for autorouting, and hence must be manually routed, which takes more time in CAD and also increases the potential for errors.

Typically, SMT may require about 20% additional time in the CAD cycle because surface mount calls for such postprocessing steps as generating a separate layer of solder paste artwork and generating test maps for assembly testing. Additional time may also be required if the CAD software does not allow complete autorouting and design rule checking.

Although most CAD systems allow placement of components on both sides, very few allow for "flip-flop screens," which permit the CAD designer to view the opposite side of the board. Also, placement of components on both sides, especially in logic boards, may require the use of blind and buried vias. Not all CAD systems have this capability.

Some CAD systems are limited in the number of pad sizes they can accommodate in their library. This may necessitate a compromise on the optimum pad size for best reliability and manufacturability results.

5.11 SUMMARY

There are many design issues that must be settled before a particular packaging direction is selected. The designer must consider market needs, time to market, package drivers and their limitations, real estate, cost per function, and package moisture sensitivity. This is not enough, however. The product must also satisfy thermal and solder joint reliability requirements. As the packaging density increases, moreover, thermal problems are compounded, with a potential adverse impact on overall product reliability.

Solder joint reliability for surface mount is a source of concern because of CTE mismatch between ceramic packages and FR-4 glass epoxy substrates. This problem, which is confined to military applications, is being addressed by using expensive substrates with matched CTE. Success, however, has been less than complete.

Because the plastic packages used in commercial applications have compliant leads, they do not experience problems related to CTE mismatch. The large plastic packages, especially plastic ball grid arrays, may be prone to cracking at reflow soldering temperatures, however, and this is an industry-wide problem. Long-term solutions are still evolving, but baking before reflow offers one answer.

The increase in package density has necessitated the use of fine lines at closer spacings. This can increase cross-talk between the lines, especially if they carry a high speed signal. The product design also is influenced by the type of CAD system that is available. The cost or the schedule or both may be affected by the type of CAD system used for SMT board design.

Thus, for every board, the designer must consider all the pros and cons of designing the board in SMT. If the decision is to go ahead with SMT, then it is important to follow the specific guidelines and rules discussed in Chapters 6 and 7.

REFERENCES

1. Davidow, W. H. *Marketing High Technology.* New York: Free Press, 1986.

2. Steiner, T. O., and Suhl, D. "Investigation of large PLCC package cracking during surface mount exposure." *IEEE Transactions on Component, Hybrid, and Manufacturing Technology,* Vol. CHMT-10, No. 2, June 1987, pp. 209–216.

3. Prasad, Ray, Intel; Spalik, Joe, IBM; Fehr, Gerald, LSI Logic; and Nixon, David, Aerospace Corporation. Moisture induced plastic IC package cracking, Surface Mount Council white paper, May 1988, Available from IPC, Northbrook, IL.

4. IPC-SM-786 "Procedures for characterizing and handling of moisture/reflow sensitive ICs," Available from IPC, Northbrook, IL.

5. J-STD-020 "Moisture/reflow sensitivity classification for plastic integrated circuit surface mount devices," Available from IPC, Northbrook, IL.

6. Yalamachili, P.; Gannamani, R.; Munamarty, R.; McClusky, P.; and Christou, A. "Popcorning." Optimum processing prevents PQFP, SMT May 1995 pp. 39–42.

7. Alger, C.; Huffman, W. A.; Gordon, S.; Prough, S.; Sandkuhle, R.; and Yee, K. "Moisture effects on susceptibility to package cracking in plastic surface mount components." *IPC Technical Review,* February 1988, pp. 21–28.

8. Fukuzawa, I., et al. "Moisture resistance degradation of plastic LSIs by reflow soldering." *Proceedings of the 23rd Reliability Physics Symposium,* 1985, pp. 192–197.

9. Ito, S., et al. "Special properties of molding compounds for small outline (SO) packaged devices." *Proceedings of the 36th Electronic Components Conference,* 1986, p. 360.

10. Poborets, Bella; Ilyas, Qazi S. M.; Potter, Mark; and Arggyle, John. "Reliability and moisture sensitivity evaluation of 225 pin, 2 layered overmolded (OMAPC) ball grid arrays, *IEEE, ECTC 1995 Conference Proceedings,* pp. 434–439.

11. Shook, Richard L.; and Conrad, Timothy R. "Diffusion model to derate moisture sensitive ICs for factory use conditions." *IEEE, ECTC 1995 Conference Proceedings,* pp. 440–449 and additional data in a private communication.

12. Suhl, D. "The mechanism of plastic package cracking in SMT and two solutions." IPC-TP-683 Technical paper presented at the IPC spring meeting, Hollywood, FL, April 1988.

13. Engelmaier, W. "Functional cycles and surface mount attachment reliability." In *Surface Mount Technology,* International Society

for Hybrid Microelectronics, Reston, VA, Technical Monograph Series 6984-002, 1984, pp. 87–114.

14. Engelmaier, W., "Test method considerations for SMT solder joint reliability." *Proceedings of IEPS,* October 1984, pp. 360–369.

15. Engelmaier, W. "Is present-day accelerated cycling adequate for surface mount attachment reliability evaluation?" Technical paper IPC-TP-653, presented at the IPC fall meeting, Chicago, October 1987.

16. IPC-785. Guidelines for accelerated reliability testing for surface mount solder attachment. Available from IPC, Northbrook, IL.

17. IEEE Compliant Lead Task Force Flex Tester, Penn Tech Services, Inc., Southampton, PA.

18. Solder Joint Monitor. STD Series, Models 32, 128, and 256, Anatek, Wakefield, MA.

19. DeFalco, J. A., "Reflection and crosstalk in logic circuit interconnections." *IEEE Spectrum* 1970, pp. 44–50.

Chapter   6
Surface Mount Land Pattern Design

6.0   INTRODUCTION

Surface mount land patterns, also referred to as footprints or pads, define the sites at which components are to be soldered to a printed circuit board. The design of land patterns is critical because it not only determines solder joint strength, hence reliability, but it also influences the areas of solder defects, cleanability, testability, and repair/rework. In other words, the very producibility of surface mount assemblies depends on land pattern design. However, the producibility of a surface mount assembly (SMA) is not determined by pad design alone. Materials, processes, components, and board solderability also play very important roles. These issues are covered elsewhere in this book.

Poor component tolerances and the lack of standardization of surface mount packages have compounded the problem of standardizing the land pattern design. Numerous package types are being offered by the industry, and the variations in a given package type can be numerous, as well. More important, the tolerance on components varies significantly, causing real problems for land pattern design and adding to the manufacturing problems for SMT users.

Consider the component tolerance found in "standard" 1206 resistors and capacitors. These items are anything but standard: the tolerance of the most important dimensions—the termination width—varies from a minimum of 10 mils to a maximum of 30 mils (nominal dimension is 20 mil). With a tolerance variation of this magnitude, the design of land patterns is challenging indeed.

The technical societies, such as the Institute of Interconnecting and Packaging Electronic Circuits (IPC), the EIA, and the Surface Mount Council, are doing everything they can to improve the current situation as far as component standardization and tolerances are concerned. For example, the IPC chartered the Surface Mount Land Pattern Committee

(IPC-SM-782) to standardize the land patterns for surface mount components [1].

Through the efforts of the Surface Mount Council, there is now a significant amount of coordination between the EIA and the IPC to ensure compatibility between component outlines (established by the EIA) and their land patterns (established by the IPC). As chairman of IPC-SM-782, I had the privilege of working with many individuals from EIA and IPC member companies to develop land pattern designs for surface mount components. For example, any new component outline considered by the EIA parts committee is first reviewed by the IPC land pattern committee before a final decision is made [2]. The IPC-SM-782 land pattern guidelines offer a good example of the industry's effort to promote standardization in surface mount technology.

IPC-SM-782 is a good source for the latest information on land pattern design. It is revised on a regular basis to reflect the changes in component availability. It should be noted, however, that IPC-SM-782 is primarily a land pattern document and not a design for manufacturability (DFM) document. We should also keep in mind that neither this book nor any other book is a substitute for an in-house DFM document, although they may be good places to start. Why? An in-house DFM document is critical for fully optimizing one's design for one's specific manufacturing processes and equipment. A DFM document should encompass issues discussed in Chapters 3, 4, 5, 6, and 7 of this book and should provide a specific number for land pattern design, via size, interpackage spacing, and a host of other variables as discussed in Chapters 5, 6, and 7.

In this chapter we discuss the basic concepts of land pattern design for different package types, including the formulas that serve as the basis for the land patterns. This information can also be used for developing land patterns for newer components as they become available or when the dimensions of the existing components change.

The formulas were developed after extensive testing for reliability and manufacturability of assemblies soldered by reflow and wave soldering (reflow soldering of active components and reflow and wave soldering of passive components). The test boards were tested for reliability by thermal and mechanical cycling for solder joint failures. Some of the results of the test program have been published [3].

To simplify the land pattern design guidelines discussed in this chapter, let us divide surface mount components into categories. For each category, we first discuss the basic concept and then the formulas for land pattern design. Finally, the specific land pattern dimensions are given.

6.1 GENERAL CONSIDERATIONS FOR LAND PATTERN DESIGN

New surface mount packages are being introduced constantly, and their dimensions are being standardized. In such a situation, it is important to keep in mind certain design considerations for land pattern design. For example, it is not advisable to use industry standards such as JEDEC/ EIA standards on component outline as procurement specifications. Users must develop their own in-house specifications and qualify a limited number of vendors that meet those specifications. (Limiting the number of vendors reduces the tolerances the land pattern design will have to support.)

The land pattern design is also process dependent; that is, the same design may give different defects depending on the process variables. For example, for different solder paste thicknesses, the degree of solder joint reliability and the number of manufacturing defects will be different. One should also expect varying results depending on the type of solder mask and the reflow processes used. Not surprisingly, therefore, we find some variations in land pattern dimensions used by various companies. Yet most of these different land pattern dimensions may be correct, since companies do not use the same process variables. In addition, all companies have different expectations of manufacturability and reliability. Thus it is very important that the land pattern design accommodate reasonable tolerances in component packages; also, it should take into account the processes and equipment used in manufacturing. Hence it is recommended that the basic concepts discussed in this chapter or in any other publication serve only as guidelines.

As noted earlier, the user must develop an in-house DFM document including land pattern design based on in-house materials, processes, equipment, and reliability requirements. Final land patterns should be validated for manufacturability and reliability before being implemented on products. The IPC-SM-785 [4] can be used as a guideline to validate any new land pattern design for reliability.

6.2 LAND PATTERNS FOR PASSIVE COMPONENTS

Passive components are soldered by wave, reflow (vapor phase, convection, and infrared), and other processes, each with its own unique thermal profile characteristics and requirements. For example, in wave soldering there is an unlimited supply of solder from the solder pot. In reflow soldering, however, the supply of solder is limited and can be controlled to a limited degree by regulating the metal content and the thickness of the solder paste. Also since components in wave soldering are held in place by adhesive, solder defects such as part movement and tombstoning (component standing on its end) do not occur and need not be addressed in designing. Even in reflow soldering, the incidence of part movement in vapor phase and convection/infrared (IR) soldering is different because of the dissimilar thermal profile characteristics of these two processes. (See Chapter 12 for a discussion of thermal profiles.)

Given the process and thermal profile differences among soldering methods, should the land pattern designs for each be different? From the standpoint of optimizing the land pattern design, it is advantageous to have different land patterns for different soldering processes. But from a systems standpoint, having different land patterns for different processes poses some problems. For example, some CAD systems limit the total number of land pattern sizes that can be supported.

Even if the CAD capability is not a concern, it would be very confusing for a CAD designer to have to maintain different sets of land patterns for different soldering processes for the same part in the CAD library. In many cases this approach could be very restrictive if the company uses different soldering processes for various applications because no one process meets all requirements.

The primary reason for having different land patterns for reflow and wave soldering is the susceptibility of components to moving or standing up during soldering. During wave soldering, since the components are glued, part movement is not a concern. Thus a land pattern design that is optimum for reflow soldering will work also for wave soldering. Obviously, then, it is desirable for the land pattern design for components to be transparent to both wave and reflow processes. It should be kept in mind that this issue of common land patterns is applicable only to discrete devices, such as resistors, capacitors, small outline transistors (SOTs), and sometimes small outline integrated circuits (SOICs). Active devices such as plastic leaded chip carriers (PLCCs) and fine pitch are not wave soldered, so for them this issue does not arise.

Whether using the same or different land pattern designs for different soldering processes, the land patterns must be validated for reliability and manufacturability before being used on products. In this connection, some additional do's and don'ts for land pattern design of passive components should be kept in mind. For example, component misalignment can be caused both by improper land pattern design and by excessive component misplacement. Components that have poor land pattern design but are accurately placed can misalign after reflow soldering. Nor do components self-align when they have a good land pattern design but are misplaced. Ball grid arrays (BGAs) may be an exception to this statement because they tend to self-align even if misplaced almost by 50%. However, the smudging caused by misplaced BGAs will result in an unacceptable level of solder balls. For this reason, one should not count on self-alignment to correct the sloppy performance of pick-and-place machines or the machine operators and programmers.

There are many causes of tombstoning. But they all have one thing in common—uneven downward forces on each side of the pad. It is the downward force generated by the surface tension of solder on the pad that keeps the chip in place. This uneven surface tension force can be generated by uneven solderability of component terminations on each end, uneven termination width (poor tolerance, a common industry problem), uneven solder deposit during screen printing, misplacement of components, rapid rate of heating (in vapor phase soldering), and shadowing of a chip component by a large neighboring component (in IR soldering). These factors make it possible for the solder joint on one end to reflow slightly before the other end, causing unequal surface tension force. Depending on the degree of force differential, tombstoning, minor part movement, or something in between will occur. The user should be aware of these causes in taking appropriate corrective action.

In addition to noting the commonly known causes and solutions mentioned above, Giordano and Khoe suggest that vendors should supply chip components with wider bottom termination widths and maintain tight tolerances at least on the bottom side to allow a greater margin of manufacturing flexibility, eliminating expensive rework [5].

To prevent misplacement and tombstoning, the component should be placed accurately so that both terminations fall symmetrically on solder paste. Whether components self-align during reflow depends on the degree of misplacement. Minor misplacement in passive devices and smaller active devices does get corrected, but as noted earlier, one cannot count on it.

The incidence of misalignment and tombstoning is also minimized by the elimination of solder mask, especially a dry film solder mask, between pads and proper preheat before reflow. This is discussed in Chapter 7. In addition to symmetric placement of terminations, the gap between pads plays a very important role.

It is generally recommended that the gap between the pads be maintained such that both terminations fall squarely on their pads. This means that the gap between the chip component lands is slightly reduced to prevent tombstoning. There is one exception to this guideline, however. When wave soldering 0805 and smaller chip components, the gap should be kept as large as possible. The smaller chip resistors and capacitors are susceptible to bridging of their terminations. Such bridges are hidden by the chip component and cannot be easily detected during visual inspection. The land patterns of resistors and capacitors discussed later on take these factors into account.

6.2.1 Land Pattern Design for Rectangular Passive Components

Rectangular passive components such as ceramic capacitors and resistors are the most widely used, but some companies select a land width narrower than the component width and others use lands wider than the components for the same soldering process. To determine the effect of land geometry on reliability and manufacturability (solder defects), a study was conducted by the author. Two different test boards with various land pattern configurations were designed. One board was used for wave soldering and the other for vapor phase soldering.

An analysis of the test results showed that the solder joint failures were dependent on land lengths, not land widths [3]. Therefore, although the width of the land and the width of the component should be roughly equal, a minor variation in land width is not important. However, failure in solder joints was noticeable for land lengths of less than 50 mils for 1206 resistors. This study also found that resistors are more susceptible to part movement and tombstoning if larger land geometries are used.

To accommodate these findings, I developed formulas for resistor and capacitor land patterns using minimum and maximum component dimensions instead of nominal dimensions. The nominal dimension is meaningless because of poor tolerances in components. When using minimum and maximum dimensions, a distinction is made between resistors and capacitors because resistors are about half the thickness of capacitors. If this distinction is not made, the land patterns for resistors and capacitors of a given component size will be the same, causing part movement in resistors because of their smaller thickness, hence lower mass. This is why the lands for resistors and capacitors, for a given component size, are of different lengths. However the widths of the land and of the gaps between the lands are the same for both resistors and capacitors, as shown in Figure 6.1, and as summarized below.

Land patterns for resistors and capacitors:

Pad width
$$A = W_{max} - K$$

Pad length
$$B \text{ (Resistor)} = H_{max} + T_{max} + K$$
$$B \text{ (Capacitor)} = H_{max} + T_{min} - K$$

Gap between lands
$$G = L_{max} - 2T_{max} - K$$

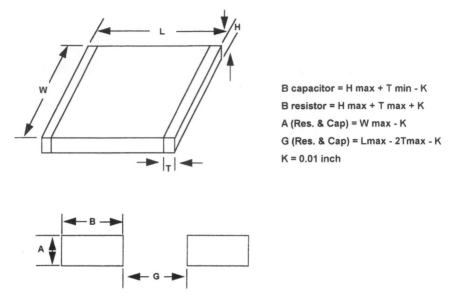

B capacitor = H max + T min - K

B resistor = H max + T max + K

A (Res. & Cap) = W max - K

G (Res. & Cap) = Lmax - 2Tmax - K

K = 0.01 inch

Figure 6.1   Formula for land pattern design of rectangular resistors and capacitors.

where L = length, W = width, T = solderable termination width, H = height, and K = constant (recommended to be 0.01 inch). Note that the formulas for A and G are the same for both resistors and capacitors, but the formulas for B are different.

The land dimensions of commonly used ceramic resistors and capacitors shown in Figure 6.2 are based on these formulas.

6.2.2   Land Pattern Design for Tantalum Capacitors

For most passive components, the solderable termination width and the component width are equal, as are the component height and the solderable termination height. However, molded plastic tantalum capacitors are an exception. For these components, the solderable termination width and height, depending on the package style and vendor, may be less than the component body width and height. Hence although the general concept is the same as in Figure 6.1, the exact formulas shown in this figure cannot be used for tantalum capacitors.

When using the formulas shown in Figure 6.1 for molded plastic tantalum capacitors, use w instead of W and h instead of H to indicate the land pattern dimensions. The applicable formulas and land pattern

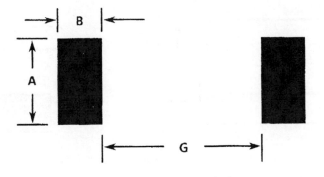

DIMENSIONS IN INCHES			
RESISTOR DESIGNATION	A (WIDTH)	B (LENGTH)	G (GAP)
R0403	0.030	0.030	0.015[a]
R0603	0.030	0.050	0.020[a]
R0805	0.050	0.060	0.030[a]
R1206	0.060	0.060	0.060
R1210	0.100	0.060	0.060
R2010	0.100	0.075	0.120
R2512	0.120	0.120	0.150
CAPACITOR DESIGNATION	A (WIDTH)	B (LENGTH)	G (GAP)
C0403	0.030	0.030	0.015[a]
C0603	0.030	0.050	0.020[a]
C0805	0.050	0.060	0.030[a]
C1206	0.060	0.070	0.060
C1210	0.100	0.070	0.060
C1812	0.120	0.090	0.090
C1825	0.250	0.120	0.090

[a]The gap between the lands of 0805 and smaller resistors and capacitors should be increased by 10 mils and should be covered with liquid photoimageable solder mask when wave soldering. This is necessary to prevent potential bridging of terminations underneath the components during wave soldering.

Figure 6.2   Land pattern dimensions (inches) for rectangular ceramic resistors and capacitors. See Figures 3.2 and 3.5, respectively, for component dimensions.

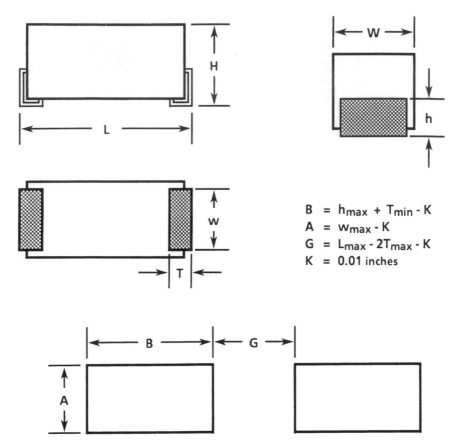

$$B = h_{max} + T_{min} - K$$
$$A = w_{max} - K$$
$$G = L_{max} - 2T_{max} - K$$
$$K = 0.01 \text{ inches}$$

Figure 6.3   Formulas for land pattern design of tantalum capacitors.

dimensions for tantalum capacitors are shown in Figures 6.3 and 6.4, respectively.

As discussed in Chapter 3, there are two types of tantalum capacitor: the molded plastic type and the kind with a welded stub on one side. Since the weld stub type capacitors are not used much, we will not discuss them.

6.3   LAND PATTERNS FOR CYLINDRICAL PASSIVE (MELF) DEVICES

Metal electrode leadless face (MELF) devices are used for capacitors, resistors, and diodes. See Chapter 3 for a description and component

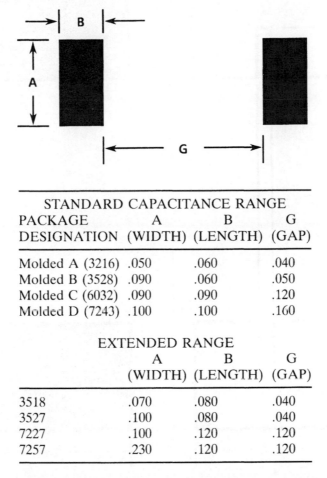

STANDARD CAPACITANCE RANGE			
PACKAGE	A	B	G
DESIGNATION	(WIDTH)	(LENGTH)	(GAP)
Molded A (3216)	.050	.060	.040
Molded B (3528)	.090	.060	.050
Molded C (6032)	.090	.090	.120
Molded D (7243)	.100	.100	.160

EXTENDED RANGE			
	A	B	G
	(WIDTH)	(LENGTH)	(GAP)
3518	.070	.080	.040
3527	.100	.080	.040
7227	.100	.120	.120
7257	.230	.120	.120

Figure 6.4 Land pattern dimensions (inches) for molded tantalum capacitors. See Figure 3.7 for component dimensions.

dimensions. MELF devices are soldered by both wave and reflow processes. This is one component that may have different land patterns for reflow and wave soldering. The land pattern for reflow soldering will also work for wave soldering, but the reverse is not true, because the MELF components, being cylindrical, have a tendency to roll either during handling or during reflow soldering.

The land pattern dimensions for MELFs are shown in Figure 6.5. Some companies use a notch in the land pattern to keep the device in place during reflow soldering. The notch (dimensions *D* and *E* in Figure 6.5) can be omitted for wave soldering, since the adhesive will prevent

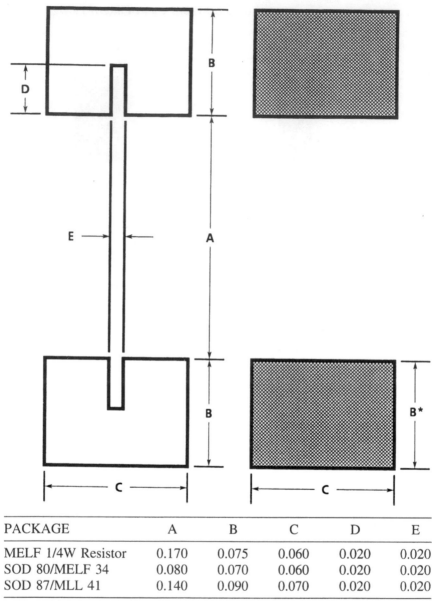

PACKAGE	A	B	C	D	E
MELF 1/4W Resistor	0.170	0.075	0.060	0.020	0.020
SOD 80/MELF 34	0.080	0.070	0.060	0.020	0.020
SOD 87/MLL 41	0.140	0.090	0.070	0.020	0.020

Rectangular pads are now more common for both wave and reflow. Caution: If a MELF is misplaced along its longitudinal direction, one of the MELF ends may not have an end fillet when using notch in the pad. This could be a rejectable condition.

Figure 6.5   Land pattern design for tubular MELF devices: resistors, capacitors, inductors, and diodes (dimensions in inches). See Figure 3.9 for MELF component dimensions.

the component from rolling sideways. The depth of the notch is determined by the following formula:

$$D = \frac{B - (2B+A-L_{max})}{2}$$

where L_{max} is the maximum body length of the tubular device (not shown in Figure 6.5. See Chapter 3, Figure 3.9 for MELF dimensions).

The formulas for *A, B,* and *C* are similar to those discussed earlier for ceramic resistors and capacitors (Figure 6.1).

There is one important point that should be kept in mind when using a notch. The notch will cause lack of end fillet if the component is not properly placed. For this reason, the use of the notch is not recommended.

6.4 LAND PATTERNS FOR TRANSISTORS

As discussed in Chapter 3, the commonly used types of small outline transistor are SOT 23, SOT 89, SOD 123, SOT 143, SOT 223, and TO 252 (DPAK). There is no formula given for the land pattern design of SOTs.

Land pattern design for SOTs is simple: maintain the distance between the center of lands equal to the center distance between the leads, and provide at least 15 mils for internal and external pad extension. This concept is very similar to the concepts for PLCC and SOIC land pattern designs, which are discussed next. Based on these two simple concepts, the land pattern dimensions for SOT 23, SOT 89, SOD 123, SOT 143, SOT 223, and TO 252 (DPAK) are shown in Figures 6.6, 6.7, 6.8, 6.9, 6.10, and 6.11, respectively.

In many applications, MELFs are being replaced by SOD 123 because they are easier to assemble than the MELFs.

6.5 LAND PATTERNS FOR PLASTIC LEADED CHIP CARRIERS

Plastic leaded chip carriers are the most commonly used packages for commercial applications for pin counts above 20. Various pad sizes for PLCCs have been found to provide reliable solder joints [3]. This is due to the compliancy (ductility) of J leads and the good CTE compatibility between the plastic package and the glass epoxy board.

For PLCCs, the total pad length is not sufficient. The land pattern

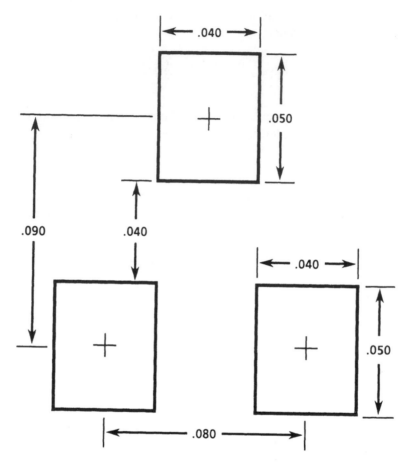

Figure 6.6 Land pattern for SOT 23 transistor/diode (dimensions in inches). See Figure 3.17 for component dimensions.

design must also provide enough pad extension beyond the package lead to accommodate a good fillet. In PLCCs, the outside fillet is the major fillet and absorbs most of the stresses. Thus there must be enough outside pad extension for outside fillets to ensure solder joint reliability. A 0.015 inch pad extension is considered to be satisfactory for good outside fillets.

The land pattern design should also provide sufficient space for routing between the pads and should accommodate reasonable variation in component tolerances. As mentioned earlier, the JEDEC standards generally have loose component tolerances. PLCCs are no exception.

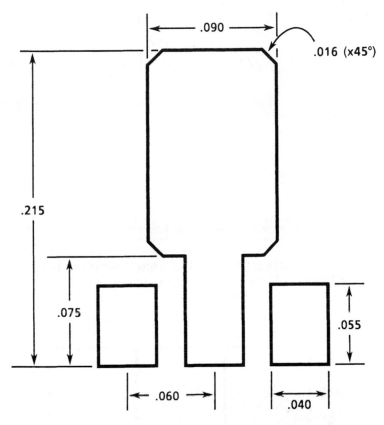

Figure 6.7 Land pattern for SOT 89 transistor/diode (dimensions in inches). See Figure 3.18 for component dimensions.

The formulas for the land pattern for PLCCs (see Figure 6.12) take into account the considerations discussed above and result in the following dimensions:

$$\text{pad width} = 0.025 \pm 0.005 \text{ inch}$$
$$\text{pad length} = 0.075 \pm 0.005 \text{ inch}$$
$$A \text{ or } B = C + K$$

The 25 mil by 75 mil PLCC pad size has been selected to meet both reliability and manufacturability requirements. A wider pad tends to reduce the gap between the pads and to restrict the routing space and may cause

PACKAGE	A	B	G
SOD 123	0.025	0.060	0.070
SMB	0.090	0.100	0.080

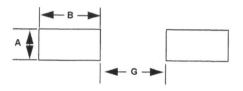

Figure 6.8   Land pattern for SOD 123 transistor/diode (dimensions in inches). See Figure 3.20 for component dimensions.

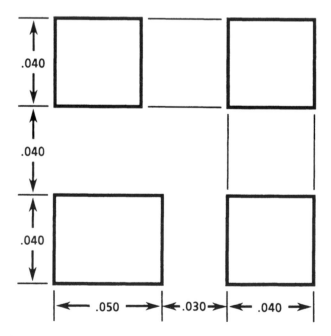

Figure 6.9   Land pattern for SOT 143 transistor/diode (dimensions in inches). See Figure 3.21 for component dimensions.

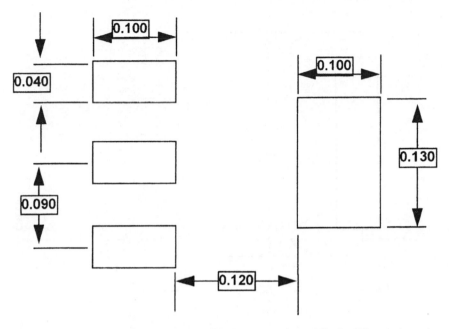

Figure 6.10   Land pattern for SOT 223 transistor/diode (dimensions in inches). See Figure 3.22 for component dimensions.

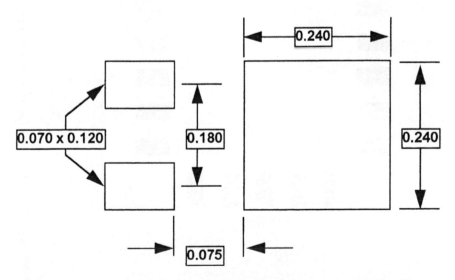

Figure 6.11   Land pattern for TO 252 (DPAK) transistor/diode (dimensions in inches). See Figure 3.23 for component dimensions.

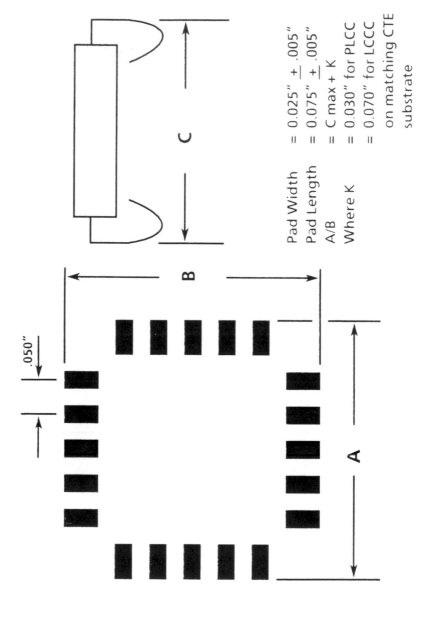

Pad Width   = 0.025" $\pm$ .005"

Pad Length   = 0.075" $\pm$ .005"

A/B   = C max + K

Where K   = 0.030" for PLCC

              = 0.070" for LCCC

               on matching CTE

               substrate

Figure 6.12   Formulas for land pattern design of plastic leaded chip carriers.

solder bridging. For this reason, surface mount packages that have a 50 mil center lead pitch (such as PLCC, SOIC, and SOJ) use a 0.025 inch pad width.

The value of K in Figure 6.12 is 0.030 inch for PLCC; this will allow a minimum pad extension of 0.015 inch, since dimensions A and B are based on the maximum values of C and K. If the C dimension is lower than the maximum allowed in the JEDEC standard, the pad extension will be greater than 0.015 inch. This extension will be reduced on one side and increased on the opposite side by placement tolerance of the pick-and-place machine.

The land pattern dimensions for PLCCs shown in Figure 6.13 are based on the formulas in Figure 6.12. Figure 6.13 also shows the land pattern dimensions for leadless ceramic chip carriers. The reasons for adopting the same land pattern dimensions for PLCCs and LCCCs are discussed next.

6.6   LAND PATTERNS FOR LEADLESS CERAMIC CHIP CARRIERS

The land pattern dimensions for LCCCs in Figure 6.13 are the same as those for PLCCs. If we refer back to Chapter 3, Figure 3.11, we will find that the LCCC packages, for a given pin count, are about 0.040 inch smaller than the PLCC packages. To achieve the same land pattern for both LCCCs and PLCCs, the value of K has to be different for LCCCs. Hence in Figure 6.12 we have $K = 0.070$ inch for LCCCs and $K = 0.030$ inch for PLCCs.

Why should we have the same land pattern for PLCCs and LCCCs? Using the same land pattern for both allows a 0.015 inch outside pad extension for PLCCs but a 0.035 inch outside pad extension for LCCCs. The larger pad extension for LCCCs is needed for better reliability because the solder fillet in these chip carriers is formed only on the outside of the package.

In PLCCs, solder fillets are formed both inside and outside the package, and also the PLCC leads take up some strain because they are compliant. Hence we can tolerate smaller outside pad extensions for PLCCs. Having the same land pattern dimensions for both PLCCs and LCCCs simplifies the land pattern design and provides reliable solder joints when LCCCs are soldered to a substrate with a compatible coefficient of thermal expansion.

Larger pad extension for LCCCs may be needed for packages larger than 44 pins. Some companies use a larger pad size (25 mils × 100 mils)

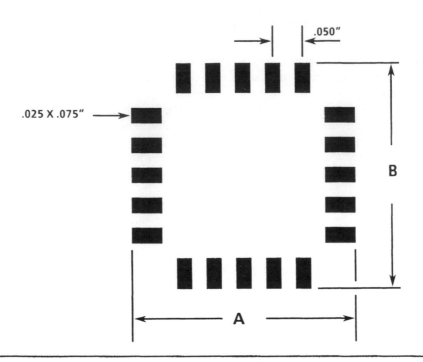

PIN COUNT OF PLCC AND LCCC	LAND PATTERN DIMENSIONS	
	WIDTH (A)	LENGTH (B)
RECTANGULAR PACKAGES:		
18-short	.350	.490
18-long	.350	.550
22	.350	.550
28	.420	.620
32	.520	.620
SQUARE PACKAGES:		
16	.370	.370
20	.420	.420
24	.470	.470
28	.520	.520
44	.720	.720
52	.820	.820
68	1.020	1.020
84	1.220	1.220
100	1.420	1.420
124	1.720	1.720
156	2.120	2.120

◄ **Figure 6.13   Land pattern design for plastic leaded chip carriers and leadless ceramic chip carriers (dimensions in inches). See Figure 3.11 for component dimensions.**

and a larger pad extension when the CTEs of package and substrates are not compatible, such as when using glass epoxy substrates for LCCCs. Use of glass epoxy boards for LCCCs is not recommended however, for reasons detailed in Chapter 4.

6.7   LAND PATTERNS FOR SMALL OUTLINE INTEGRATED CIRCUITS AND R-PACKS

There are some differences between gull wing packages (SOICs or R-packs) and J-lead packages (PLCCs, SOJs). For example, the gull wing leads of SOICs are more compliant than the J leads of PLCCs because they are not as work hardened during lead forming. Also, the SOIC packages are relatively small in comparison to PLCC packages, hence induce less stress on solder joints. For these reasons, the reliability of solder joints for SOICs is not of serious concern. However, manufacturability is a concern for SOICs because the leads of SOICs and R-packs are prone to damage during handling.

There are other differences as well. For example, as discussed earlier, in PLCCs the major fillet is on the outside of the package; but in the gull wing packages such as SOICs, the major fillet is on the inside.

The formulas for SOICs take into account the location of the major fillets to ensure proper fillet formation, as summarized below and shown in Figure 6.14:

$$\text{pad width} = 0.025 \pm 0.005 \text{ inch}$$
$$\text{pad length} = 0.075 \pm 0.005 \text{ inch}$$
$$D = F - K$$

where K is a constant and is recommended to be 0.01 inch.

The F dimension is used for the land pattern design to ensure formation of major fillets and solder joint reliability. Also, the F dimension for SOICs (Figure 6.14) has as tight a tolerance as the dimension C (Figure 6.12) in PLCCs.

SOICs, SOPs, and R-packs appear physically the same, although they may differ in body width. For example, as discussed in Chapter 3, the body widths of SOICs are 150 (narrow body) and 300 mils (wide or

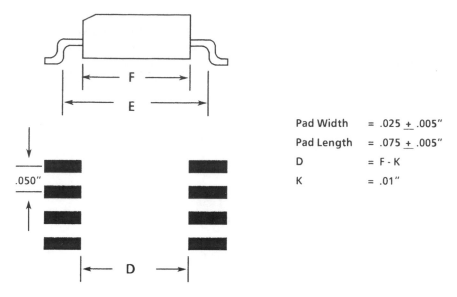

Pad Width	= .025 ± .005"
Pad Length	= .075 ± .005"
D	= F - K
K	= .01"

Figure 6.14   Formulas for land pattern design of SOIC and R-packs.

large body) and the body width of an R-pack is 220 mils. The SOPs are slightly wider than the narrow body SOICs but much narrower than the wide body SOICs.

EIAJ has SOPs (small outline packages) at 50 mil pitch that are similar to SOICs with some minor variations. For example, they are slightly wider in body than the narrow body SOICs but much narrower than the wide body SOICs. SOPs are classified based on their package height, which varies from 1.5 mm (Type I) to 4 mm (Type VI). Refer to Chapter 3, Section 3.4.2.

Since SOICs, SOPs, and R-packs physically look alike and have the same lead configuration (gull wing), the same formulas (Figure 6.14) apply for all three types of packages.

The land pattern dimensions for SOICs and R-packs are shown in Figure 6.15, and the land pattern dimensions for SOPs are shown in Figure 6.16.

When SOICs, SOPs, or R-packs are wave soldered, the trailing leads tend to bridge. To avoid this, a dummy "robber" pad is added to the trailing side. The purpose of the "robber" pads is to create a dummy bridge site that is not connected to any functional circuit. The dummy robber pads are needed only on the trailing end. They can be added on each end of the SOIC or SOP if one is not sure which end is the trailing

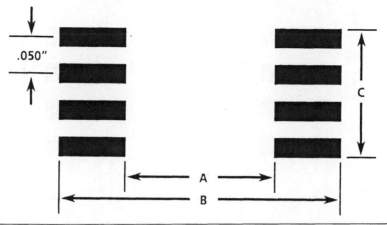

DIMENSIONS IN INCHES			
LEAD COUNT	A	B	C
SO-08	0.140	0.290	0.175
SO-14	0.140	0.290	0.325
SO-16	0.140	0.290	0.375
SOL-14	0.300	0.450	0.325
SOL-16	0.300	0.450	0.375
SOL-20	0.300	0.450	0.475
SOL-24	0.300	0.450	0.575
SOL-28	0.300	0.450	0.675
SOL-32	0.360	0.510	0.775
R-PACK-14	0.200	0.350	0.325
R-PACK-16	0.200	0.350	0.375

Figure 6.15   Land pattern for SOIC and R-pack (dimensions in inches).

end. These pads reduce functional bridges (not total bridges seen on the board) but do take up additional real estate. Also, if a dummy pad is added, it should be made clear to the shop floor personnel that those bridges are acceptable so that a functional board is not rejected by mistake.

It should be noted that SOICs, SOPs, and R-packs should be wave soldered only after the process has been fully characterized. Because of CTE mismatch between the lead and the molding compound, the flux from the wave soldering process may seep inside the package and cause corrosion of the bond wire or silicon.

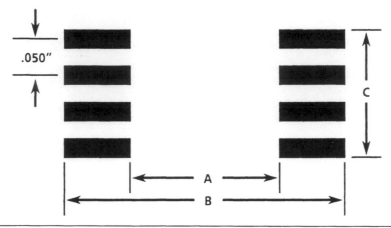

DIMENSIONS IN INCHES			
LEAD COUNT	A	B	C
SOP-6	0.130	0.280	0.125
SOP-8	0.130	0.280	0.175
SOP-10	0.130	0.280	0.225
SOP-12	0.130	0.280	0.275
SOP-14	0.130	0.280	0.325
SOP-16	0.210	0.360	0.375
SOP-18	0.210	0.360	0.425
SOP-20	0.210	0.360	0.475
SOP-22	0.280	0.430	0.525
SOP-24	0.280	0.430	0.575
SOP-28	0.360	0.510	0.675
SOP-30	0.360	0.510	0.725
SOP-32	0.430	0.580	0.775
SOP-36	0.430	0.580	0.825
SOP-40	0.510	0.660	0.875
SOP-42	0.510	0.660	0.925

Figure 6.16   Land pattern for SOP (dimensions in inches).

6.8   LAND PATTERNS FOR SOJ (MEMORY) PACKAGES

The small outline, J-lead devices are similar in design to PLCCs except that the leads are on only two sides of the package. SOJ packages are primarily used for memory devices. Static RAMs, are also available in SOJ packages. The land pattern dimensions for these devices are shown in Figure 6.17.

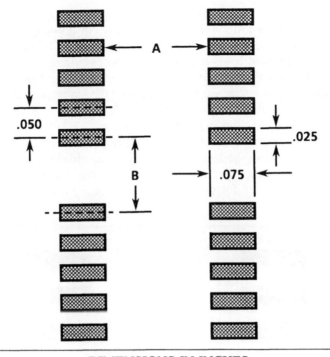

	DIMENSIONS IN INCHES		
	BODY WIDTH		
DEVICE		A	B
20 PIN SOJ	0.300	0.230	N/A
26 POSITION/20 PIN	0.300	0.230	0.200
26 POSITION/20 PIN	0.350	0.280	0.200
24 PIN SOJ	0.300	0.230	N/A
28 PIN SOJ	0.300	0.230	N/A
32 PIN SOJ	0.300	0.230	N/A
32 PIN SOJ	0.400	0.330	N/A
40 PIN SOJ	0.400	0.330	N/A

Figure 6.17   Land pattern for SOJ (dimensions in inches).

6.9   LAND PATTERNS FOR DIP (BUTT MOUNT) PACKAGES

Since not all packages are available in surface mount, it is sometimes desirable to shear DIP leads for butt mounting. For example, in an assembly having active surface mount devices on both sides but only a few DIPs on the top side, using DIPs with butt leads may cut down on the number of soldering and cleaning steps. Chapter 7 (Section 7.4) discusses cases in which butt-mounted leads may save on process steps to cut cost.

Use of DIPs with butt leads is not always desirable, however. Even when it is cost effective to use butt-mounted DIP leads, the reliability of a butt solder joint may be questionable. Refer to the pros and cons of different lead configurations in Chapter 3, Section 3.7.2. Also, care should be taken when shearing leads to prevent damage to components.

Figure 6.18 shows the land pattern dimensions for butt-mounted DIP leads. Some users prefer bending DIP leads into the gull wing form. The land pattern dimensions for gull wing DIP leads are not shown, but the

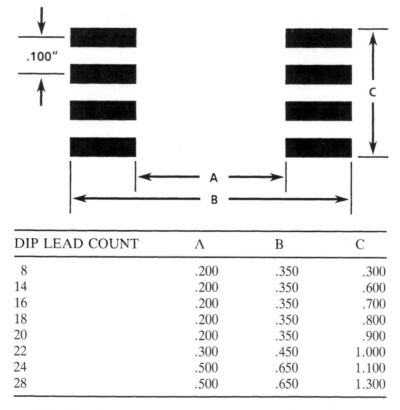

DIP LEAD COUNT	A	B	C
8	.200	.350	.300
14	.200	.350	.600
16	.200	.350	.700
18	.200	.350	.800
20	.200	.350	.900
22	.300	.450	1.000
24	.500	.650	1.100
28	.500	.650	1.300

Figure 6.18   Land pattern design for butt lead DIP packages (in inches). The body widths for DIPs are 300 mils (14–20 pins), 400 mils (22 pins), and 600 mils (24 pins and above). The lead pitch is 100 mils for all pin counts.

concept discussed in connection with SOIC can be used to derive the land pattern dimensions.

6.10   LAND PATTERNS FOR FINE PITCH, GULL WING PACKAGES

Land patterns for gull wing fine pitch (0.8 mm, 0.65 mm, 25 mil, and 0.5 mm center) packages are based on the formula shown in Figure 6.19. The basic idea behind the formula is very simple. Ensure that there is at least 0.015 inch extension under the heels of each lead for heel fillet. This is opposite of the concept discussed earlier for PLCCs. Since the heel fillet in gull wing packages is the inner fillet, it is critical for reliability that pad extension under the package be provided. The pad width is slightly more than half the pitch to allow for any undercutting of pad during board fabrication. The length of the pad should be about 0.070 inch.

Based on these concepts, the land pattern for a 0.025 inch pitch PQFP is shown in Figure 6.20. This is the U.S. version of the bumpered fine pitch package.

The land pattern for the EIAJ fine pitch package is not shown because, even for the same pin count, the land pattern design may have to be different for packages from different suppliers. This may appear strange but it is true. There is also one very important point that should be kept in mind when designing the land pattern for EIAJ fine pitch PQFPs. It is highly recommended that you look at the data sheet furnished by the component supplier before designing the land pattern. The concept just used for the fine pitch is just as valid for the EIAJ packages. Also, the land pattern for EIAJ fine pitch packages (0.65 mm and 0.5 mm) should

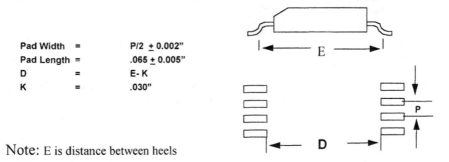

Pad Width =	P/2 $\pm$ 0.002"
Pad Length =	.065 $\pm$ 0.005"
D =	E- K
K =	.030"

Note: E is distance between heels

Figure 6.19 Formula for land pattern design of fine pitch packages with gull wing leads.

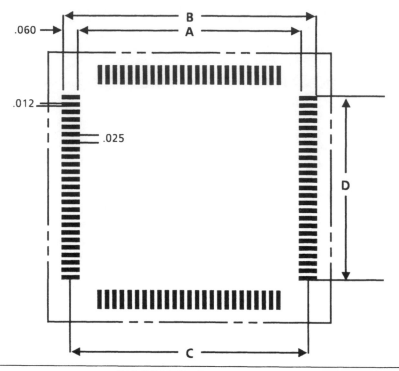

DIMENSIONS IN INCHES

LEAD COUNT (Pads per side)	A	B	C	D
84 (21 Pads)	0.665	0.805	0.735	0.500
100 (25 Pads)	0.765	0.905	0.835	0.600
132 (33 Pads)	0.965	1.105	1.035	0.800
164 (41 Pads)	1.165	1.305	1.235	1.000
196 (49 Pads)	1.365	1.505	1.435	1.200
220 (55 Pads)	1.540	1.680	1.615	1.350
244 (61 Pads)	1.545	1.685	1.615	1.500

Figure 6.20 **Land pattern for U.S. version of 0.025 inch pitch gull wing PQFP with corner bumpers (dimensions in inches).**

be designed in hard metric or by conversion to at least three decimal places to ensure correct design.

There are also other lead pitches such as 0.4 mm and 0.3 mm, etc., that are used. The concept just discussed can be used for them as well. Again, make sure that you look at the data sheet before designing the land pattern.

6.11   LAND PATTERN DESIGN FOR BALL GRID ARRAYS (BGAs)

The commonly used pitches for ball grid arrays (BGAs) are 50 and 60 mils. 40 mil pitch BGAs and even finer pitch micro BGAs are also used. The narrower the pitch, the more difficult the routing will be and hence the higher the layer count will be. Refer to Chapter 5, Section 5.5.1 for a discussion of PCB routing and related costs.

Twenty-five mil diameter pad size is commonly used for plastic BGAs. Brown and Bromley [6] found that 20 mil diameter pads provided slight improvement in reliability when compared to 24 mil pads with no impact on manufacturing process. Use of smaller pads improves board routing as well. Figure 6.21 shows BGA pad design with two different via connection options. The junction with fillet (right) is preferred because it distributes the stress over a larger pad surface area. As shown in Figure 6.21, the via is connected with the BGA pad by copper plane (not just a trace) and is covered with solder mask. This provides a strong connection and prevents lifting of the BGA pad.

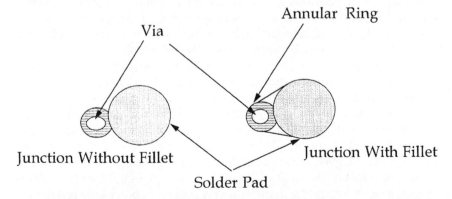

Figure 6.21   BGA pad design with two pad to via pad connection options. The junction with fillet (right) is preferred because it distributes the stress over a larger pad surface area.

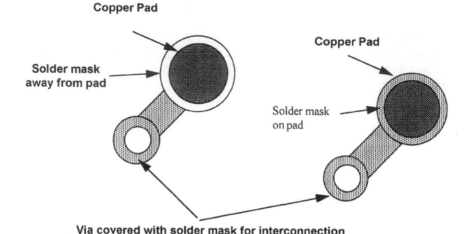

Copper Pad

Copper Pad

Solder mask
away from pad

Solder mask
on pad

Via covered with solder mask for interconnection

Figure 6.22   Copper defined (left) and solder mask defined (right) BGA pad designs.

There are some advocates of vias in pad design because they increase routing efficiency and provide a rivet-like, strong structure. However, vias in pad design are not recommended because of the potential for insufficient solder fillet.

There are two major approaches to BGA pad design: solder mask defined and copper defined pad. These are shown in Figure 6.22. In a solder mask defined pad, the mask defines the pad size where the copper pad is larger than the mask opening. The mask holds down the pad since it overlaps it. It performs better in rework and multiple reflow cycles. Also more copper provides better heat spreading to the mother board. In this approach, the solder mask helps control part movement and paste flow and controls the ball shape to be more spherical.

However, there are some negatives in the solder mask defined approach. The mask misregistration affects the pad and hence solder joint quality. Also the solder mask can act as a crack initiation site.

As shown in Figure 6.22, in the copper defined pad approach, the mask opening is larger than the copper pad. The copper defined pad is commonly used on all other SMT pads. Mask misregistration is not an issue because the mask is kept away from the pad. The copper undercutting and etching defines the final pad size. In this approach, there is no CTE mismatch issue with solder mask, which is better for solder joint reliability. Mawer et. al [7] found three fold improvement in reliability in copper defined pads over solder mask defined pads. Also, this approach allows more space for routing, and the traces connecting to the pad show wetting that can be x-rayed.

6.12   LAND PATTERN DESIGN FOR TAB

The basic concept for designing land patterns for TAB (tape-automated bonding) is the same as for fine and ultra fine pitch packages. (TAB is also referred to as TCP—tape carrier package.) Since TAB is a metric package, one should use metric dimensions for accurate land pattern design. As with other packages, the land width should be about half the package pitch (for 0.25 mm pitch, use 0.1 to 0.12 mm +/– 0.005 mm land width).

There is one difference for TAB—the land length is generally much longer. It should be long enough to provide both toe and heel fillet (extensions beyond the toe of 0.360 to 0.600 mm and beyond the heel of 0.170 to 0.550 mm). A typical land length for TAB is 2 to 3 mm. The longer land length is preferred because of the use of the hot bar process.

It should also be noted that unlike other packages, TAB is first bonded to the substrate using an adhesive for heat transfer. So the land pattern design must include a central pad, called the die attach pad, as shown in Figure 6-23. It is intended not for electrical connection but for heat transfer. The die attach pads should be 3.75 mm larger than the die size on all sides for placement tolerance and die attach fillet (fillet with die attach adhesive).

In TAB land pattern design, 1 mm diameter fiducials are used to provide good contrast on all four corners of the TAB site. The land pattern for TAB is shown in Figure 6.23.

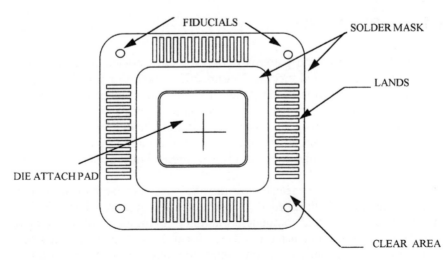

Figure 6.23   Land pattern for TAB (TCP).

6.13   LAND PATTERNS FOR SOLDER PASTE AND SOLDER MASK SCREENS

The relationship between the pieces of artwork used for solder paste film, solder mask film, and the top layers of PCB film should be closely controlled. Generally the supplier modifies the film to achieve the user's requirements. As a general guideline, the width and length of the film pad for the solder paste screen should be 0.002 inch smaller than the solder pads on the PCB. Similarly, the width and length of the film pad for the solder mask screen should be 0.002 inch bigger than the solder pads on the PCB. The 0.002 inch dimension is valid for solder masks such as dry film or photoimageable solder masks that provide good resolution. For wet film solder masks, however, 0.005 inch should be used. The latter guideline is adequate to prevent the solder mask from getting on the pad and to prevent bridging of paste before reflow.

The area around the fine pitch land pattern may be free of solder mask if a solder mask is not available to achieve the required resolution between the fine pitch pads. This may be especially true when using ultra fine pitch (0.4 mm pitch and below) packages. However, this does increase the potential for solder bridging. The TAB land pattern should be clear of solder mask because it is an ultra fine pitch package. If soldering is done by the hot bar process, bridging is not a serious concern.

The same concept that is used for fine pitch is generally also valid for BGAs; i.e., there should be about 0.002 inch clearance between the solder mask and BGA lands.

6.14   SUMMARY

Surface mount land patterns define the sites at which components are soldered for needed electrical connection and mechanical strength. They play a critical role in the reliability of surface mount solder joints. Certain basic land pattern design parameters are applicable to all components for achieving this goal; these are solderable lead/termination width and height, lead contact points or areas on the printed circuit board, and extension of pads beyond contact points or areas (both inside and outside the package).

The foregoing parameters determine the size and shape of the solder fillets, which provide the needed strength for solder joint reliability. Keeping these parameters and component tolerances in mind, formulas for land pattern design should be used. For every formula, a constant K is defined to accommodate variations in component tolerances. The basic variables in the land pattern design guidelines can be summarized as follows:

1. The center-to-center distance between adjacent pads should be equal to the center-to-center distance between adjacent component leads.
2. The pad width should be equal to half the lead pitch for standard 0.050 inch pitch surface mount components.
3. The pad width for fine pitch packages should be slightly greater than half of the lead pitch. This means that the pad width is slightly bigger than the spacing between the pads. This compensates for some undercutting of pads during board fabrication and provides some additional tolerance for paste misregistration and component placement in manufacturing.
4. The pad length is determined by the height and width of solderable terminations or leads and the contact points or areas of components. Pad length plays a more critical role than pad width in solder joint reliability.
5. The gap between opposite pads of the land pattern depends on the component body width and dimensional tolerances and is most critical in properly locating the components on the pads. This gap has an impact on both reliability and solder defects.

REFERENCES

1. ANSI/IPC-SM-782. Surface Mount Land Patterns: Configurations and Design Rules available from IPC, Northbrook, IL.
2. News, IPC and EIA cooperate on outlines and land patterns, *IPC Technical Review,* July 1988, p. 9.
3. Prasad, R. P. "Surface mount land patterns and design rules. Surface mount land patterns play a key role in the reliability of SMT." *Printed Circuit Design,* December 1986, pp. 13–20.
4. IPC-SM-785. Guidelines for accelerated reliability testing for surface mount solder attachment. Available from IPC, Northbrook, IL.
5. Giordano, Jerry, and Khoe, David, Bourns, Inc., "Chip resistor design helps prevent tombstoning," *Surface Mount Technology,* August 1988, pp. 23–25.
6. Brown, Jason, and Bromley, Brenda "PBGA solder joint reliability/manufacturability as a function of PCB pad size" *Proceedings of Surface Mount International*, September 10–12, 1996, pp. 161–165.
7. Mawer, Andrew, Cho, Dennis, and Darveaux, Robert "The effect of PBGA solder pad geometry on solder joint reliability" *Proceedings of Surface Mount International*, September 10–12, 1996, pp. 127–135.

Chapter 7

Design for Manufacturability, Testing, and Repair

7.0 INTRODUCTION

Design for manufacturability is the practice of designing board products that can be produced in a cost-effective manner using existing manufacturing processes and equipment. The benefits of a manufacturable design are better quality, quicker time to market, lower labor and material costs, shorter throughput time, and fewer design iterations. The quickest way to waste thousands of dollars is to produce an SMT design that cannot be assembled, repaired, or tested using the existing equipment. Design for manufacturability is essentially a yield issue, hence a cost issue. It plays a critical role in reducing defects in printed circuit assemblies.

Design for manufacturability (DFM) is gaining more recognition as it becomes clear that cost reduction of printed circuit assemblies cannot be controlled by manufacturing engineers alone. The printed circuit board designer also plays a critical role in cost reduction. The days of throwing the design over the fence to the manufacturing engineer are gone—if indeed they ever really existed.

Companies planning surface mount products generally work out manufacturable designs by trial and error and at considerable expense. Getting decent yields can be a source of constant irritation and often total frustration.

How should one go about establishing design guidelines for manufacturability? In this chapter we aim to minimize the frustration of producing a manufacturable SMT board by presenting some major DFM guidelines, suitable for use in developing in-house design rules.

These guidelines have been validated on thousands of products built in production environments. Certain guidelines are equipment and process dependent—they may not apply equally to all manufacturing facilities. Other guidelines are generic, hence will apply to every company regardless of the equipment or processes used.

Any DFM guidelines, including the ones presented in this chapter, must be verified on prototype assemblies before an item is released for high volume production. Validation of DFM should not pose a major problem because most designs go through one or two revisions, at least in the beginning stages, to fine-tune the electrical performance. During those revisions, one should also be on the lookout for manufacturing defects.

We must keep in mind that design for manufacturability alone cannot eliminate all defects in surface mount assemblies. Defects fall into three major categories: design-related problems, incoming material related problems such as PCB, adhesive, solder paste, etc., and problems related to manufacturing processes and equipment. Each defect should be analyzed for its source, to permit appropriate corrective action to be taken.

Simply put, all major issues in design, materials, and process must be fine-tuned to achieved zero defects. Materials and process issues are covered in Chapters 8 through 14 of this book. In this chapter we discuss DFM guidelines to prevent design-related problems.

There are two major design-related issues: land pattern design and design for manufacturability. The first is very critical because it determines the strength of the solder joint, hence its reliability; it also has an impact on solder defects. Land pattern design, which was covered in Chapter 6, should be considered to be an integral part of design for manufacturability.

The design trade-offs, which were covered in Chapter 5, are important in determining whether SMT meets cost, real estate, and reliability constraints. They should also be considered an integral part of DFM. In this chapter we concentrate on DFM issues that affect manufacturability from the manufacturing yield viewpoint, in order to minimize board assembly, testing, and rework costs.

7.1 DFM ORGANIZATIONAL STRUCTURE

The traditional approach of doing design serially, where the design proceeds from the logic or circuit designer to physical designer (CAD layout) to manufacturing and finally to the test engineer for review, is not appropriate because each engineer is making decisions independently in evaluating and selecting alternatives. This results in suboptimized designs [1]. What is worse is a situation where the manufacturing engineer sees the design only in a physical form on a PCB. This is generally the case when products are built by subcontract assembly houses.

So how should a product be designed? There is almost universal agreement on the parallel or concurrent design approach, where the logic designer, CAD layout designer, manufacturing, test, and process engineers, and purchasing and marketing representatives sit in one room to review the design and discuss the alternatives to meet thermal, electrical, real estate, cost, and time to market requirements. This should be done in the early phases of the design to evaluate various alternatives within the boundaries of the company's in-house self-created DFM document. Various trade-off issues for team study are covered in Chapter 5. The DFM team should consist of members from various organizations such as logic design, CAD layout, manufacturing, test, and component engineering, and purchasing and marketing. This team should be headed by a program manager with good technical and people skills who has the full support of the team members and their managers.

For the design team to be able to design a manufacturable product, it is important to establish guidelines and rules. The distinction between guidelines and rules is very important. Rules are necessary for compatibility with the planned manufacturing equipment and processes. If the design rules are not followed, the product either cannot be manufactured or must be built manually. In the case of SMT, having to build manually is tantamount to not being able to build at all. Guidelines, on the other hand, are nice to have and may make life easier for the manufacturing engineer, but they are not critical in getting the product out the door. This means that rules cannot be violated, but guidelines may be overruled for good and sufficient reason.

Depending on the product and on marketing constraints, even the rules can be violated as long as the members of the product team (manufacturing engineer, test engineer, CAD designer, electrical designer, component engineer, reliability engineer, and purchasing and marketing representatives responsible for the product) understand and agree on the consequences of each violation. In general, the guidelines can be violated by the DFM team, but violating the rules requires, in addition to the team's approval, the signature of a higher level manager who has been designated by the company as its DFM authority. In order to lend some weight to DFM guidelines and rules, it is important that such a policy be in place. In addition, there should also be a procedure in the DFM requiring the team to provide reasons for each violation, alternatives attempted, and the impact of the violation on product cost and schedule. Also, violating either the guidelines or rules once must not set a precedence for subsequent violation on other products. Such a policy will generally not be viewed as a bureaucratic hurdle since it is created by the team members who have to live by it. This implies that they are diligent in creating the DFM document in the first place, and since it is their policy, they can abide by it.

It is not enough, however, to establish design rules and guidelines for manufacturability. A corporate culture must exist to support adherence to guidelines and rules. In this book we have called everything "guidelines" because a book cannot be a substitute for an in-house specification, whereas it can be used as a basis for developing in-house guidelines and rules.

It is not an exaggeration to say that the design team plays a critical role in the success and financial viability of any product using SMT. This has been true in the through-hole mount design as well, but not to the same degree. There is only one way to assemble a through-hole mount board: stuff and wave solder it. In SMT, however, the design has many options, depending on the type of assembly. For example, in SMT, the same component can be placed on either the top side or the bottom side, with different manufacturing consequences in each case. These issues are discussed later on. Clearly the design team needs to be thoroughly familiar with the surface mount manufacturing processes.

This is not to belittle the importance of DFM in through-hole assemblies, because DFM for autoinsertion is important. But unlike through-hole mount, SMT does not have the option of manual placement, and the rules for testability are very different because in SMT we no longer have the ends of the devices that also served as the test nodes. In a nutshell, DFM and the role of the design team in the success of SMT cannot be overemphasized.

We now discuss various issues the design team must confront in designing a manufacturable SMT assembly.

7.2   GENERAL DESIGN CONSIDERATIONS

To reduce manufacturing costs, products should fit into a standard form factor, that is, a standard board shape and size and a standard tooling hole location and size. (See Chapter 11, Figure 11.4.) This is the first item of importance in DFM, and it means that even with the variety of new products available, it is possible to have similarities that allow the use of standard manufacturing processes.

If the products do not fit into a standard form factor, manual setups are required, which leads to longer throughput time and higher tooling and manufacturing costs. Even when cost is not a major factor, especially for prototype volumes, the schedule may be important. The lead time required for fixtures and tooling for the screen printer, placement, and soldering machines may be unusually long if a nonstandard form factor is used. This issue must be taken into account before committing to a nonstandard form factor.

Board size should not be confused with standard form factor. Even if it is not possible to have a standard board size (which is generally the case in industry because of multiple product designs for different applications), a standard form factory strategy can be implemented in manufacturing. If it is not possible to have one standard form factor, it is desirable to have a limited number rather than many form factors in a manufacturing line. Limitation can be achieved by selecting only a few standard panel sizes that accommodate various small boards in a multipack (panelized) design. The individual boards within the multipack panel must be on a standard grid with respect to each other and with the tooling holes on the panel. The grid chosen must be the same as the one used to create the board. Any multipack design must allow for easy depaneling in individual boards.

The maximum size of the standard panel or board should be selected with the capabilities of the machines in mind, as well as potential warp and twist problems in the board. Limiting the size of the panel to a maximum of 12 inches by 12 inches is a good idea, even though panels of 12 inches by 18 inches are not uncommon. Any board or panel size larger than the 12 inch square may require either permanent or temporary stiffeners to prevent warpage when it goes through the soldering processes. To prevent board warpage and machine jams, it is generally a good idea to ensure that the panel width does not exceed 1.5 times the panel length.

The panels should be designed for routing with little manual intervention. Use of solid panel designs is preferred over routed slots for structural stability. But solid panel designs require depaneling technologies such as laser, water jets, or machine routing. In some cases grooves may be used for depaneling for manual separation. Figure 7.1 shows the thickness of

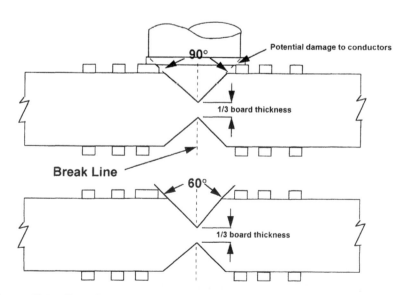

Figure 7.1   Board scoring for break-off.

the web in the groove to be about one-third of board thickness. In a typical 0.062 inch thick board, the web should be between 0.012 inch to 0.018 inch to ensure structural stability of the board during the assembly and reflow process. The grooves should be parallel to the solder wave (perpendicular to board flow direction). Instead of grooves, unplated holes are also used between pre-routed slots to separate these boards. No matter which depaneling technology is used, the total cost of depaneling should be taken into account.

There are a few other items that should be considered in panel designs. The panels should have standard tooling holes to ensure proper accuracy and location. The leading edge of the panels should be free of slots if possible.

Fiducials or alignment targets should also be provided on the panels and on the individual boards. If fine pitch components are being used, even a local fiducial is desirable. Figure 7.2 provides examples of global and local fiducials. The local fiducials for fine pitch can either be at the center of the package or at the corner as shown in Figure 7.2 (bottom right). It is also very helpful to provide local alignment marks for BGA placement and inspection as shown in Figure 7.2 (top). There are three alignment marks shown on two corners of a BGA package land pattern.

BGA Alignment marks

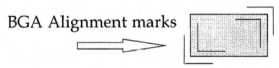

Target Position for BGA (shaded area): Center alignment mark

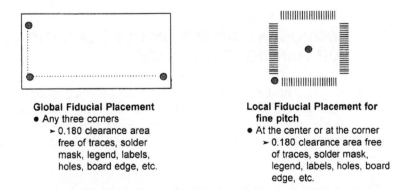

Global Fiducial Placement
- Any three corners
  - ➤ 0.180 clearance area free of traces, solder mask, legend, labels, holes, board edge, etc.

Local Fiducial Placement for fine pitch
- At the center or at the corner
  - ➤ 0.180 clearance area free of traces, solder mask, legend, labels, holes, board edge, etc.

Figure 7.2 Global fiducials (bottom left) and local fiducials for fine pitch (bottom right) and alignment marks for BGA (top) for placement accuracy.

Note: Solid round dot typically 0.060 inch diameter preferred

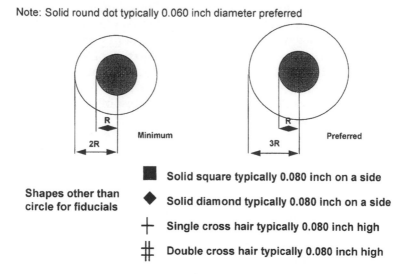

Figure 7.3   Recommended sizes and shapes of fiducials for vision system.

If the BGA is placed correctly, the corners of the package should overlap the center alignment marks, as in Figure 7.2. These alignment marks can be used as a guide for judging the degree of misplacement from the target position.

Figure 7.3 shows the recommended sizes and shapes for global and local fiducials. The design requirements for fiducials are generally governed by the pick-and-place machines. However, they all require some clearance around the fiducials.

7.3   COMPONENT SELECTION CONSIDERATIONS FOR MANUFACTURABILITY

The objective of this section is not to discuss surface mount component issues in general but only to point out some do's and don'ts of component selection for easy manufacturing. Surface mount components were covered in detail in Chapter 3. For easier manufacturing, the following guidelines should be used.

1.   Use only surface mount components that solve the real estate constraints or reduce the number of manufacturing steps (from Type II SMT to full surface mount Type I SMT, for example) and bear no cost premium. (Also, bear in mind that some surface

mount components may have a cost premium and longer lead time depending on supply and demand.) Unless surface mount components meet at least one of these requirements, their use will not be cost effective.

2. All surface mount components should be autoplaceable. It would seem that all surface mount components must be autoplaceable, but this is not true. For example, tantalum capacitors with welded stubs cause problems during placement. Instead, use encapsulated tantalum capacitors, which are more reliable and easy to place. Connectors, heat sinks, and sockets also fall into a difficult to autoplace category.

3. Component terminations with solder plating are better than solder-dipped terminations because of the "dog bone" effect that occurs during dipping. The uneven surfaces can prevent the placement jaws from centering components.

4. Select only components that can withstand the soldering temperature anticipated. Many through-hole components have been converted to surface mount by cutting or bending the leads, but not all of them can withstand the reflow temperatures. For example, delay lines that have internal connections soldered with eutectic solder (63% Sn and 37% Pb, mp = 183°C) will melt during reflow soldering at 215 to 220°C.

5. Use 1206 and 0805 size passive components (resistors and capacitors) whenever possible. Smaller sizes such as 0603 and 0402 should be used only when the required value or dielectric is unavailable in another size because they are harder to place than the larger sizes.

   For wave soldering avoid the use of Z5U capacitors. Instead, use X7R because they are less susceptible to cracking. Also, do not use capacitors larger than 1812 for wave soldering, as they are susceptible to cracking due to thermal shock. Finally, do not use any ceramic part larger than 0.25 inch on a side on an FR-4 board to avoid CTE mismatch and hence potential solder joint cracks.

6. Use marked components if there is no cost premium. Parts without marking do create problems if they get mixed up or have to be repaired in the field, since the documentation (schematic or parts list) must be referred to when making replacements. Use of components on tape and reel eliminates part mix-up.

7. For secondary side attachment in Type II or Type III SMT, low profile SOT devices are preferred over the high profile SOTs because the former are easier to bond with adhesive. For reflow soldering, there is no difference. However, the higher profile SOTs are better for cleaning, for both wave and reflow soldering.

8. Avoid using metal electrode leadless face (MELF) passive components unless the product is targeted for a very cost-sensitive market. MELFs are cheaper than their rectangular counterparts. However, they tend to be less reliable and harder to place. Also they tend to roll off during reflow soldering. Rolling off, which can be prevented by using the land pattern featuring a notch in the pad (see Chapter 6), is not an issue in wave soldering, since the components are glued. For diodes, MELFs can also be replaced with SOTs or two lead gull wing small outline diodes (SODs).

9. To prevent leaching (dissolution of the solderable termination), use nickel barriers on resistors and capacitors. Leaching exposes nonsolderable ceramic dielectric, resulting in poor solder joints. This requirement for a nickel barrier underplating is more critical for wave soldering than for reflow soldering, where the limited amount of solder paste does not cause leaching. Now it is almost an industry standard to have nickel barriers in chip components. See Chapter 10 for a detailed discussion of this subject.

10. Avoid using surface mount connectors and sockets unless there is good reason. They do not always provide real estate savings. Also, the relatively weaker surface mount solder joints of sockets may not withstand the mechanical stress of repeated removal and insertion of components. When surface mount connectors are used, they should have tooling holes for mechanical fastening to the board to prevent solder joint cracking.

11. Use only qualified components. Some components appear to be surface mountable but are not, as discussed in item 4 above. Qualification requirements are defined by their application and the manufacturing processes (placement, soldering, cleaning, repair, etc.) that the components must withstand. By defining the qualification requirements and using only qualified components, many problems in manufacturing and service can be avoided. For example, during the qualification stage it can be determined whether or not the component requires baking before reflow to prevent package cracking. See Chapter 5 (Section 5.7) for details on package cracking. Another example would be using small outline integrated circuits (SOICs) for wave soldering, especially when using aggressive soldering fluxes. The potential damage of using SOICs for wave soldering is that the flux may seep between the lead frame and molding compound due to their CTE mismatch and cause corrosion of silicon or wire bond inside the package.

12. The last but most important guideline is to minimize the different types of components and to prevent component proliferation. The benefits of component commonality or standardization are many.

It is easier, for example, to establish standard land patterns and to limit dealings to a few vendors.

Dealing with the smallest possible number of vendors provides improved purchasing leverage and reduced material overhead. Also, depending on the input feeder slot capability of the pick-and-place machine, most of the commonly used component types can be line loaded and made available right on the placement machine. This certainly provides quick response or throughput time in manufacturing by eliminating setup time. This all adds up to improved quality and reliability and reduced cost.

Using application specific integrated circuit (ASIC) devices whenever possible is the ultimate in reducing the usage of different part types. Substituting one ASIC for various programmable logic arrays (PLAs) is an example. One will not only save considerably in placement time, but also in programming, inventory, and feeder requirements. An added benefit is that thermal problems will be alleviated, since PLAs are the biggest culprit in compounding the thermal problems on a board. Because of nonrecurring development costs, however, ASICs are feasible only in high volume applications. Also one must allow 4 to 6 months of development time for the ASIC in the product schedule. Nevertheless, when the ASIC route is feasible, the savings in both component and manufacturing costs can be substantial.

7.4   SOLDERING CONSIDERATIONS

When converting to SMT, one of the first questions generally asked is: What type of SMT should be used? Should it be Type III, Type II, or Type I? Sometimes the answer is very clear—a function of component availability and real estate constraints. At other times one must weigh various factors before making a decision. The initial trade-off must include options other than SMT. Refer to Chapter 5 for the considerations to be balanced in making the final decision for or against SMT. Here we assume that the board must be SMT. Thus the first order of business is to complete a preliminary part placement to determine the real estate constraint, hence the type of SMT assembly. Refer to Section 5.3 in Chapter 5 for a rough estimate of real estate requirements.

For Type III and Type I SMT, the soldering issues are clear-cut. But for Type II, the same components, depending on whether they are placed on the top side or the bottom side, have different soldering implications. For example, in Type II assemblies, the discrete devices such as resistors, capacitors, transistors, and diodes can be mounted either on the secondary (wave soldering) or the primary (reflow soldering) side.

Other surface-mounted devices such as R-packs and SOICs should be mounted on the primary side for reflow soldering. Many companies place them on either side, but this is not always preferable because of the potential problems in solder defects (shadowing, bridging, etc.) and adverse impact on package reliability. The risks of wave soldering any leaded surface mount devices are not worth taking. Since surface mount packages are susceptible to package cracking due to moisture absorption, wave soldering of plastic surface mount devices should be avoided unless the reliability data indicate otherwise.

Many companies wave solder SOICs and butt mount and reflow solder DIPs. Now we examine the pros and cons of this approach to determine when and if it is wise.

Let us discuss the soldering options for an assembly like that of Figure 7.4, which has only a few through-hole devices on top and active surface mount devices on both sides, due to real estate constraints. (We assume that the board cannot grow any larger and that we need all the functions.) This puts the designer in a quandary. If both sides of the boards are reflow soldered, the through-hole components must be hand soldered and this will increase cost. If the leaded surface mount devices on the bottom side (along with the through-hole devices on the top side) are wave soldered, package reliability may be degraded. However, if we butt mount the through-hole devices and reflow both sides, the butt solder joints may not be reliable. Some studies have indicated that butt joints may actually be satisfactory, but the results require further evaluation. We should also note that butt joints are not allowed in Class III (military and high reliability) applications.

DIP leads can be bent to form either a J or a gull wing, but care must be taken to prevent package cracking. Such an approach will eliminate the reliability concerns associated with wave soldering of leaded surface mount devices and will reduce soldering steps, hence cost. Appropriate tooling is available.

One of the very desirable methods for soldering an assembly such as the one in Figure 7.4 is called paste in hole process. It is discussed in detail in Chapter 9, Section 9.6.3 (Printing for Through Hole in Mixed Assembly). Another option for soldering such an assembly is to use a cover plate over the surface mount devices on the bottom side (See Figure 7.5.) If such an option is used, the process sequence is as follows. Components are reflow soldered on the secondary (bottom) side and then cleaned. Then the primary (top) side of the assembly is reflow soldered. Note that adhesive is not used to hold the secondary side components in place. They do not fall off because they are being held in place by the surface tension of solder. After a second cleaning, the through-hole devices are inserted on the primary side, and the surface mount components on

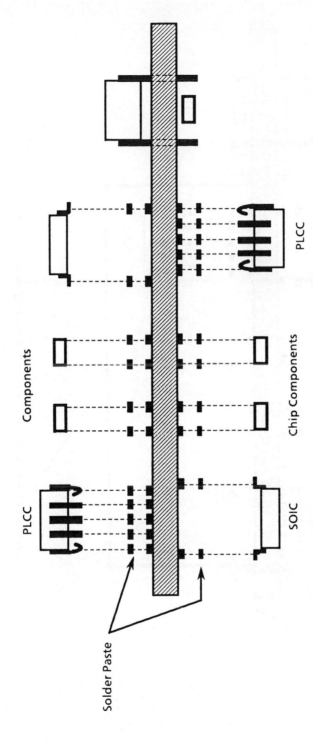

Figure 7.4 A Type II SMT packaging case in which it is definitely desirable to convert the last few through-hole components to surface mount to achieve Type I SMT assembly.

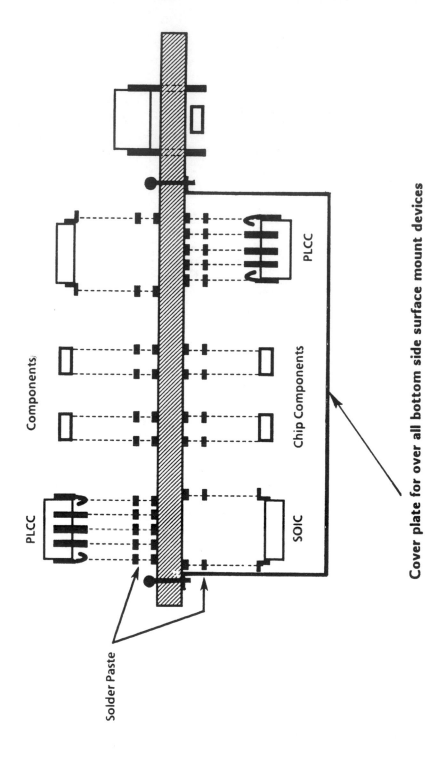

Cover plate for over all bottom side surface mount devices

Components

PLCC

Chip Components

PLCC

SOIC

Solder Paste

Figure 7.5 Use of cover plate to shield surface mount devices from the solder wave during wave soldering of through-hole devices.

the secondary side are covered using a cover plate as shown in Figure 7.5. Since the secondary side surface mount devices are covered with the cover plate, no flux or solder touches these devices during wave solder. This prevents any reliability problem associated with wave soldering of active surface mount devices.

The cover plate material can be made of either stainless steel or titanium. Titanium is an expensive and difficult material to work with, but will hold its shape better after repeated use. If the cover plate is made of some other metals, such as copper or aluminum, the solder pot will get contaminated. High temperature plastic may also be an option. In addition to material selection, one must also be aware that the cover plate must be custom made for each product, and the through-hole components must be segregated in one corner or end of the board. Hence, this may not be a feasible option if certain through-hole components must be located next to certain surface mount components. Also, the wave solder pot may have to be lowered to allow sufficient clearance for the cover plate. Despite these technical challenges, many well known companies are using cover plates for boards that must have surface mount active devices on both sides and many through-hole devices on one side.

The case illustrated in Figure 7.6 may also justify paste in hole process (Section 9.6.3 in Chapter 9) or modification of DIP leads. Since this assembly does not have any components on the secondary side, by converting the few through-hole devices to surface mount, the wave soldering step can be eliminated, reducing cost. However, the through-hole devices must withstand reflow soldering temperatures.

If passive devices are added to the secondary side (Figure 7.7), the soldering cost advantage disappears even if the through-hole devices are converted, because the number of soldering steps is not reduced. At least two soldering operations are necessary no matter what happens to the

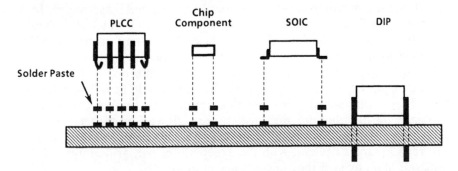

Figure 7.6   A Type II SMT packaging case in which full conversion to SMT will save soldering and cleaning process steps.

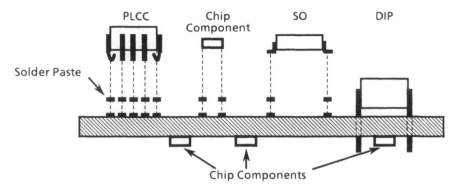

Figure 7.7 A Type II SMT packaging case in which full conversion to SMT will not save soldering and cleaning process steps.

through-hole devices. In such a case it is better to begin by reflow soldering the top side surface mount devices and then wave soldering the through-hole devices on top and surface mount passive devices on the bottom, in a single step.

The foregoing discussion, as it relates to Figure 7.4 through 7.7, suggests that paste in hole process may be desirable at times where butt mounting of through-hole devices may not be desirable because of reliability concerns. It is also not desirable to wave solder surface mount devices such as SOICs and PLCCs because the reliability of the package is compromised. We should note that packages with leads on all four sides should never be wave soldered because of excessive bridging and shadowing problems. And as discussed earlier, when PLCCs or SOICs are wave soldered, there is always the potential for flux seeping inside the package during wave soldering and causing corrosion of silicon and wire bonds.

Let us take another case to illustrate solder considerations in design: a full surface mount, double-sided board (Type I SMT assembly). In a double-sided Type I assembly, it is important to decide which side should bear the larger components (PLCCs above 84 pins or smaller PLCCs attached to heat sink), which may fall off if the assembly is inverted. Obviously larger or heavier components (those attached to heat sinks) should be placed on one side only, and this side should be reflow soldered last. There is an added benefit to this option. The smaller components mounted on the opposite side, which are reflowed twice, tend to self-align more during the second reflow cycle. Self-alignment is generally not seen in larger devices, with the exception of ball grid arrays. BGAs will self-align even if they are misplaced up to 50%.

Here is a third design involving soldering considerations. Some double-sided Type I assemblies containing only passive devices on the secondary side have both wave and reflow soldering options for the secondary side. Wave soldering the secondary side instead of reflow soldering is not any cheaper, but the former may be desirable for many reasons. For example, via holes can be fully filled only during wave soldering. Plugging of vias is necessary to achieve the vacuum required for automated test equipment (ATE) work. If wave soldering is not used, one can fill the vias during reflow by printing solder paste over them. The screen or stencils can be ordered with this requirement in mind. Either option is acceptable, but a choice must be made at the design stage to permit time for the artwork for screens or stencils to be generated if the reflow option is taken. Also, if the wave soldering option for the secondary side is chosen, only wet film solder mask should be used, as discussed later in Section 7.8.

And finally, as mentioned earlier, we should keep in mind that most large active devices, especially PLCCs, quad flat packs (QFPs) and BGAs, are susceptible to cracking due to moisture during reflow or wave soldering. This is also generally referred to as "popcorning." See Chapter 5, Section 5.7 for a detailed discussion on the package cracking phenomenon.

7.5   COMPONENT ORIENTATION CONSIDERATIONS

One of the important considerations for manufacturing is proper alignment of components on the PCB. The guidelines for component orientation vary depending on type; that is, components of similar types should be aligned in the same orientation for ease of component placement, inspection, and soldering.

Orienting similar components in the same direction is very desirable for ease of programming pick-and-place equipment. It would be nice to orient pin 1 of all devices in the same direction, say, north. But this is not always feasible due to routing constraints. However, pin 1 orientation should be limited to a maximum of two directions on the board. Uniform orientation also facilitates the inspection of components for misplacement and is very helpful for troubleshooting during testing. Figure 7.8 shows the pin 1 ends of polarized tantalum capacitors and active devices in north and west directions. (Instead of referring to the orientation as north or west, one can also use degrees of rotation from the reference, for example, 0° and 270°.)

For wave soldering, proper alignment is necessary to prevent uneven fillets or solder skips (see Figure 7.9). The recommended component orientation for wave soldering to prevent these defects is shown in Figure 7.10.

It is incorrectly thought that any orientation for passive components

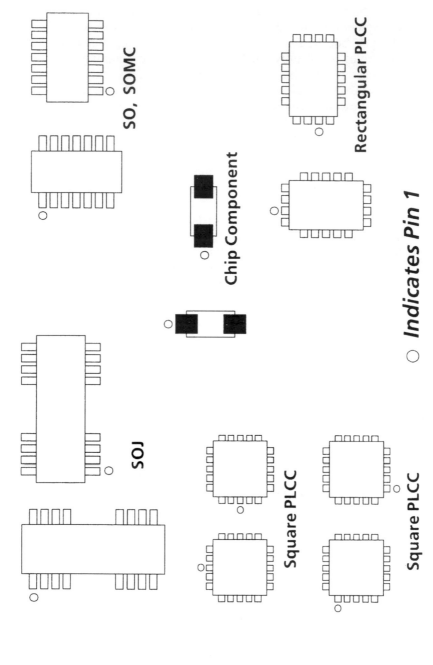

Figure 7.8 Recommended component orientation for ease of placement, inspection, and testing.

- ## Uneven fillets

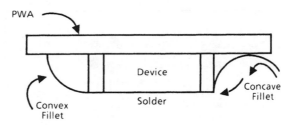

- ## Solder skips can occur on trailing termination

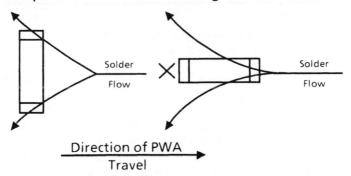

Figure 7.9 **Uneven fillets or solder skips will result if both component terminations are not soldered simultaneously.**

is acceptable if dual-wave or one of the various other "chip wave" geometries is used. A dual-wave solder pot is certainly very effective for minimizing solder skips, but it does nothing to prevent uneven fillets. The orientation shown in Figure 7.10 is the only one that will yield uniform solder fillets on both terminations. Otherwise, trailing terminations generally have excess solder fillets (Figure 7.9, top), hence unnecessary excess stress on the solder joint, which may crack ceramic chip capacitors. Cracking of chip components also may be caused by thermal shock, as discussed in Chapter 3 (Section 3.2.3) and Chapter 14 (Section 14.3.2).

It should be also noted from Figure 7.10 that when small and large components are adjacent to each other and the spacing between them is less than 0.100 inch, the smaller component must be on the leading edge during wave soldering. Otherwise the larger component may shadow the trailing smaller component. This effect is similar to the case of staggered

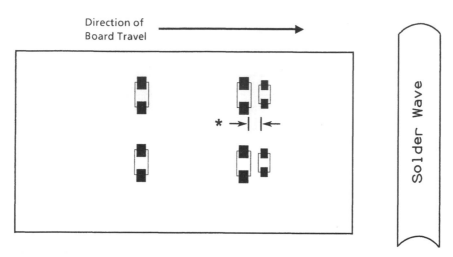

* When pad-to-pad dimension is less than .100, smaller component must enter wave first, because of shadowing

Figure 7.10 Recommended component orientation for wave soldering.

components, shown in Figure 7.11, where the leading component may shadow the trailing component. In other words, both unequal and staggered components may shadow the trailing component during wave soldering if the spacing between them is less than 0.100 inch. Other cases of interpackage spacing are discussed in Section 7.6.

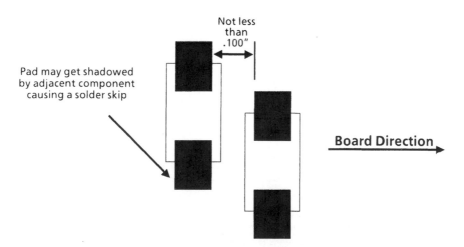

Figure 7.11 Minimum distance between staggered components to prevent shadowing of trailing component in wave soldering.

Preferred IC orientation

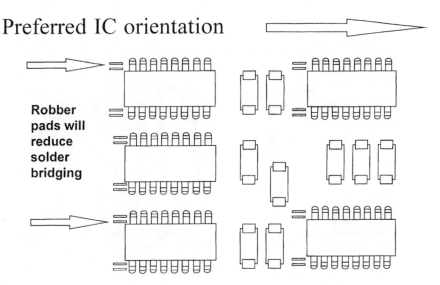

Figure 7.12   Preferred SOIC orientation with "robber pads" on the trailing sides.

When wave soldering SOICs, it is important to keep two things in mind: the orientation of the SOICs with respect to the wave and the use of dummy or "robber" pads on the trailing terminations as shown in Figure 7.12. The orientation is important to prevent shadowing (solder skips) of leads. For example, if the SOICs are at 90° of the direction shown in Figure 7.12, the trailing leads may not get soldered.

However, when SOICs are oriented in the direction shown in Figure 7.12, the trailing pins instead of experiencing solder skips, tend to bridge. Since it is very difficult to avoid bridging, we can create a bridge by design at the location of our choosing by adding dummy or "robber" pads that are not connected to any circuits. These bridges are harmless since they are not connected to any functional circuits.

When more than one package type of SOIC is wave soldered, only one type should be laid out in a given row. This is shown in Figure 7.13. Otherwise the larger packages tend to shadow the smaller packages as is the case with passive devices discussed earlier (Figure 7.10).

Whenever SOICs are being wave soldered, it generally means that there are through-hole components on the top side of that board. If there are no through-hole components on the primary side, it may not make sense to wave solder SOICs on the secondary side. It may be better to reflow solder them like the primary side surface mount components.

In any event, it is the reality that most likely there will be some through-hole components on the primary side mixed with surface mount components which may be on the primary and secondary sides. In such

SOIC's in rows. Only one
package type in each row.

SOIC's staggered. Mixed
package types in each row.

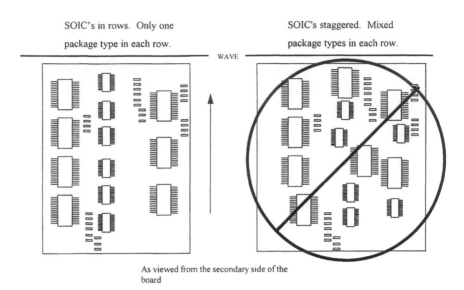

As viewed from the secondary side of the
board

**Figure 7.13   Poor and good layout patterns of SOICs for wave soldering.
Use only one package type in a given row.**

cases, the protruding pins of through-hole DIPs or stake pins should not
be inserted within the SOIC rows on the secondary side (wave solder
side). The reason for this is that the protruding through-hole pins tend to
shadow SOICs and may cause solder skips.

The acceptable and unacceptable conditions of mixing through-hole
pins in SOIC rows as seen from the secondary side are shown in Figure
7.14. Also, as shown in Figure 7.14, if the through-hole pins need to be
in the same row as the SOICs, they must be at the trailing edges. There
is no such restriction on the primary side where SOICs and other surface
mount components are reflow soldered.

7.6   INTERPACKAGE SPACING CONSIDERATIONS

Interpackage spacing is one of the most important issues in design
for manufacturability; it controls the cost effectiveness of placement,
soldering, testing, inspection, and repair. A minimum interpackage spacing
is required to satisfy the various manufacturing requirements, but there
is no maximum limit—the more the better. We all know that designers
like to pack surface mount components in as tightly as possible. After
all, isn't that one of the reasons for going to SMT? Therefore, to prevent
excessively tight packing with a view to keeping the assembly manufactur-

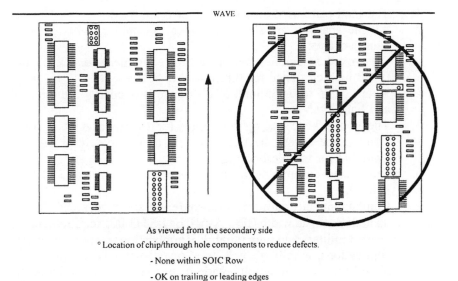

As viewed from the secondary side

° Location of chip/through hole components to reduce defects.

- None within SOIC Row

- OK on trailing or leading edges

Figure 7.14   Acceptable and unacceptable locations of through-hole pins in SOIC rows for wave soldering.

able in a cost-effective manner, we must develop parameters for interpackage spacing. This is not the place to make compromises, and every company should establish rules (not guidelines) for this purpose. (See Section 7.1 for the distinction between design guidelines and rules.)

There are three types of spacing: pad to pad, lead to lead, and body to body. For manufacturing purposes, the lead-to-lead or termination-to-termination spacing is what matters. But for the CAD designer, the pad-to-pad spacing is the important rule for placement. Thus interpackage spacings should be specified between adjacent land patterns of different packages for ease of CAD layout. It might be more appropriate to use the term "interpad spacings," but this could cause confusion with the interpad spacings of the same package; hence we will continue to refer to interpackage spacings.

7.6.1   Assumptions in Interpackage Spacing Requirements

The interpackage spacing guidelines presented in this chapter make the following assumptions:

1. The land patterns shown in Chapter 6 are used.
2. All the spacings are shown between the land patterns of adjacent packages.
3. For mixed assemblies, autoinsertion equipment requires about 0.100 inch clearance for through-hole devices.
4. About 0.075 to 0.085 inch of clearance is needed for automated and hand probe testing.
5. The interpackage spacing must allow for test probe access at ATE. Hand probe access at ATE and test clip access must also be available for troubleshooting.
6. Automated cleaning with adequate pressure nozzles (>30 PSI) is used.
7. Component shift during reflow soldering and placement may be up to 25% of pad width (IPC-A-610 or J-STD-001 requirement for Class 3 military and high reliability applications). For commercial applications, a shift of 50% is allowed, but this would be too liberal for the assumptions made in this section. In addition, the spacing between two isolated conductors should not be less than 0.005 inch. Since there is no shift or float during wave soldering, the component shift requirement is not considered for this process. For this and other reasons, interpackage spacing is process dependent, being different for reflow and for wave soldering. The requirements specified later are identified as such.
8. Solder joint melting of adjacent components is allowed; hence the number of times a part can be replaced and repaired is minimized (two times maximum). The assumption is that solder joints can be reflowed four times (twice in reflow for a double-sided assembly) and twice in rework. This is a very important point. Much larger (>0.5 inch) interpackage spacing is required with hot air reflow if melting of the solder joints of the adjacent components is to be completely prevented.
9. Inspection is visual, and the assemblies can be tilted at various angles for inspection of solder joints.

7.6.2   Interpackage Spacing Requirements

Before mentioning specific numbers for interpackage spacing, we need to compare the requirement for through-hole assemblies. It should be pointed out that the interpackage spacing specified in the literature [2] can be as high as 0.150 to 0.200 inch. There is no problem with these numbers, because as discussed earlier, the more the better.

The interpackage spacing used in conventional through-hole assemblies is 0.100 inch. The spacing for surface mount must be less than 0.100 inch, otherwise the advantages of SMT are lost to a great degree. Therefore all the spacings specified in this section are less than 0.100 inch between the leads (and lands) of adjacent packages. Thousands of actual boards, meeting all the manufacturability requirements in a production environment, have been built using the interpackage spacing specified in this section. One of them has been built by three different manufacturers in production quantities in this country and overseas. However, this figure (<0.100 inch for SMT) is not the sole basis for the guidelines presented here. The important thing to keep in mind is that the actual interpackage spacing (lead to lead or termination to termination) will be 25 to 30 mils greater than what is shown in the figures in this section. For actual distance, refer to the component package dimensions shown in Chapter 3 and the land pattern dimensions shown in Chapter 6.

As a general rule of thumb, the pad-to-pad spacing of adjacent components should be 0.050 to 0.070 inch minimum, depending on component type and orientation [3]. For PLCCs, this means 0.085 inch between leads of adjacent components. Since, as discussed in chapter 6, at least a 0.015 inch pad extension beyond the lead is required for outside fillets, in such a case, the interpackage spacings between through-hole and surface mount components can be reduced below 0.100 inch. For components such as SOJ and passives, where leads or terminations are on only two sides of the packages, the interpackage spacing should be 0.040 to 0.050 inch, as indicated in Figure 7.15.

For mix-and-match boards, the spacing between the pads of through-hole mount devices and the pads of surface mount components should be at least 60 mils (see Figure 7.16). The idea is to ensure 100 mils clearance for through-hole components for the autoinsertion equipment. With continuing decrease in the use of DIPs, the use of autoinsertion equipment is declining. Since only a few DIPs per board are being used, most companies insert them manually.

The interpackage dimensions shown in Figures 7.15 and 7.16 apply only to reflow soldering for surface mount components and wave soldering of through-hole components. These dimensions can be about 10 mils smaller for memory devices because room for test accessories such as test clips and hand probing may not be needed for memory arrays.

If the secondary side passive components are wave soldered, the requirements are different. Those for staggered and unequal adjacent components were discussed earlier (Figures 7.10 and 7.11). When the adjacent chip components for wave soldering are of the same size (height and length), however, the possibility of shadowing does not arise and the interpackage distance can be reduced to 0.040 inch (Figure 7.17).

If the heights of adjacent components as shown in Figures 7.10 and

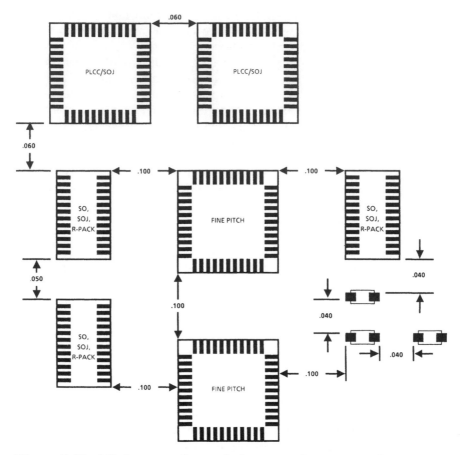

Figure 7.15 Minimum pad-to-pad clearances between surface mount pads of active devices for reflow soldering. At least 100 mils of clearance must be provided all around the fine pitch land patterns to allow multi-level paste printing. (See Figure 9.21 in Chapter 9.)

7.11 are different but their lengths are equal (e.g., the 1206 chip resistor and capacitor), the interpackage spacing needs to be only 0.040 inch. Either component can be on the trailing side.

Note also from Figure 7.17 that the external distance between chip component and DIP pad is 0.060 inch but the internal distance is only 0.040, because it is assumed that the DIPs are clinched outward. The internal distance between a DIP and a passive device is only 0.040 inch, but the internal distance between a pin grid array (PGA) and a passive device is 0.060 (Figure 7.18). Greater distance for PGAs is required because the numerous pins make these devices susceptible to bridging.

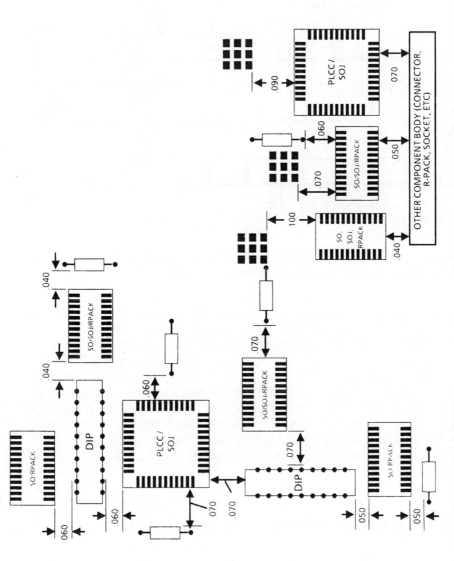

Figure 7.16 Minimum pad-to-pad clearances between surface mount pads of active devices for reflow soldering and through-hole devices for wave soldering.

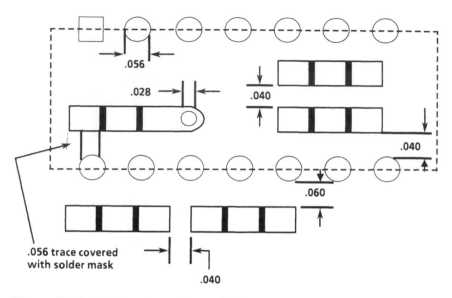

Figure 7.17 Minimum pad-to-pad clearances between surface mount pads of passive devices and DIPs for wave soldering.

Figure 7.18 also shows a 0.100 inch internal distance to pads of axial devices, and it is assumed that these leads are clinched inward.

Table 7.1 summarizes the *minimum* interpackage dimensions among various types of surface mount and through-hole components. The 70 mil dimensions in Table 7.1 are intended to allow sufficient space for autoinsertion and for PLCC/DIP clips for testing. There should also be a minimum of 0.050 inch of clear space all around the edges of the PCB if boards are to be tested from the connector. This requirement is increased to 0.100 inch minimum if a vacuum-actuated, bed-of-nails fixture is used.

It is worth repeating that the requirements discussed above are *absolute* minima.

7.7   VIA HOLE CONSIDERATIONS

It is mistakenly thought that holes simply disappear when using SMT. This is obviously untrue, however, since via holes are needed for interconnection. The vias are relatively quite narrow in diameter (14 to 18 mils), versus about 38 mils for the plated through holes used for component insertion.

The total number of via holes can be considerable. Table 7.2 compares

INSIDE A PGA PATTERN:

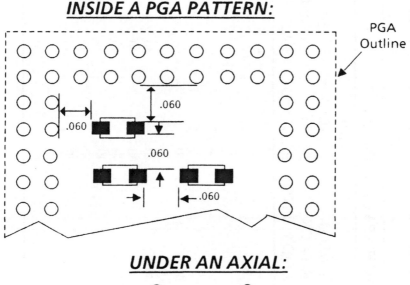

UNDER AN AXIAL:

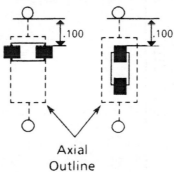

Axial
Outline

Figure 7.18 Minimum pad-to-pad clearances for a surface mount and through-hole PGA and axial component for wave soldering.

plated through holes and via holes in a board before and after its conversion to Type II SMT (80% surface mount components).

As shown in Table 7.2, even after conversion to SMT, a large number of via holes is required. In fact, the number of via holes in the through-hole version is significantly lower than the surface mount version because DIP lead ends also serve as test points. In the surface mount version the vias also serve as test points and must be filled to achieve adequate vacuum for the bed-of-nails (ATE) test. The solder mask requirements for via holes are discussed in Section 7.8.

Via holes within the surface mount pads should be avoided because if solder flows inside such vias during reflow soldering, insufficient solder fillets may result. Figure 7.19 indicates that the recommended minimum

Table 7.1 A summary of MINIMUM interpackage spacings among various types of surface mount and through-hole components.

	MINIMUM SPACING (MILS)				
	PLCC/SOJ PAD, SIDE	SO/R-PACK PAD, SIDE	SO/SOJ/ R-PACK BODY, END	CHIP COMPONENT PAD, SIDE	CHIP COMPONENT PAD, END
PLCC/SOJ side, pad to	60[a]	—	—	—	—
SO/R-Pack side, pad to	60[a]	50	—	—	—
SO/SOJ/R-pack end, body to	70[a]	50	50	—	—
Chip component side, pad to	70[a]	50[a]	40	40	—
Chip component end, pad to	40	40	40	40	40
DIP side, pad to	60	60	70	60	40
DIP end, body to	70[a]	50	40	40	40
Axial side, body to	70[a]	50	40	40	40
Axial end, pad to	60	60	70	70	50
Stake pin, pad to	90	70	100	70	40
Other component, body to	70[a]	50	40	40	40

[a] A 10 mils smaller gap may be used between components in a memory array.
NOTE: Allow at least 100 mils of clearance all around the land pattern of a fine pitch (under 25 mil pitch) device.

324

Table 7.2 Via hole and plated through-hole comparisons in an Intel single-sided computer board before and after its conversion to SMT.

	THROUGH-HOLE MOUNT VERSION	SURFACE MOUNT VERSION
Component holes	3150	1300
Via holes	800	2200
Total holes	3950	3500

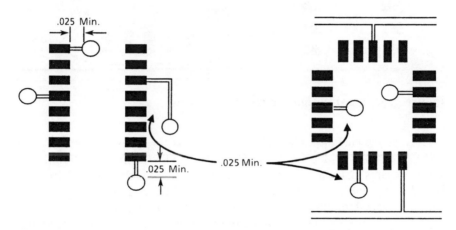

Figure 7.19 Good design practice for via hole location in surface mount boards.

distance between via holes and solder pads for reflow soldered assemblies is 0.025 inch. This distance can be reduced to 0.015 inch if solder mask is used between the via and the pad. Any smaller gap, however, will increase the risk of the solder mask getting on the lands, causing insufficient solder fillet. Insufficient solder fillets also may result when flush vias are connected to surface mount lands. Figure 7.20 illustrates bad design practices for reflow soldering, including flush vias and wider conductors. Designing wide traces connected to solder pads has the same effect as flush vias on solder pads (i.e., insufficient solder fillets).

Wide traces should either be covered with solder mask (preferably solder mask over bare copper to prevent cracking of solder mask) or necked down to 0.012 inch maximum.

As a general rule, vias within a pad should be avoided unless they can be filled by the board fabricator. That not only takes care of any potential flux entrapment problem but is highly desirable for attaining a good vacuum seal during in-circuit, bed-of-nails testing.

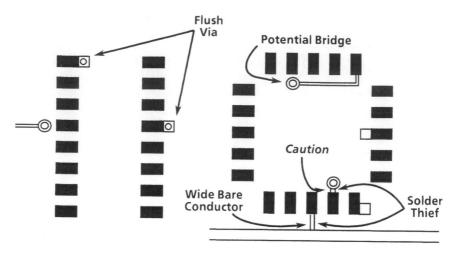

Figure 7.20   Poor design for reflow soldering.

Note that via hole location is a concern only for reflow soldering (Type I SMT). In wave soldering (Types II and III SMT), the unlimited supply of solder ensures that solder fillets are not robbed of solder by the vias. In Type II and III assemblies all the vias are filled and there is no problem with the vacuum needed to hold the board down during ATE testing.

Vias within pads of passive components glued to the underside (Type III SMT) are acceptable or even preferred because they may provide escape holes for outgassing. This feature may prove to be especially helpful if the flux used in wave soldering fails to dry sufficiently during preheat.

Whenever possible, via holes should be located underneath surface mount packages in Type I SMT assemblies. However, in Type II and Type III SMT, via holes underneath surface mount packages should be minimized or avoided because during wave soldering of these assemblies, flux may be trapped underneath the packages. As a practical matter, vias are placed both inside and outside components because of real estate constraints. This is not a real problem if the cleaning process is properly controlled or if no-clean flux is used.

7.8   SOLDER MASK CONSIDERATIONS

Solder mask is used for tenting (covering) vias, as shown in Figure 7.21. Tenting is good for Type I SMT boards because it can cover the via holes and prevent vacuum leakage during testing. Of course the test

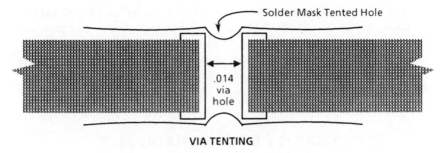

Figure 7.21   Tented via hole with solder mask.

pads should not be tented, since solder mask, being nonconductive, will prevent electrical contact between the test probe and the test nodes.

Generally dry film solder mask is used for fine line boards, but it should not be used in the gap between the pads of a chip component. The thick dry film (0.003–0.004 inch) solder mask acts as a pivot between the pads and compounds the tombstoning problem. However, some very thin dry film solder masks are becoming available. An example of a commercially available very thin dry film solder mask is Dynachem Conformask, which is only 0.0015 inch thick. Such dry film masks can be used between the pads of chip components as is the case with LPI (liquid photo imageable) and screenable solder masks. Examples of commercially available LPI and screenable solder masks are Probimer 52, Probimer 65, and Paiyo PSR 4000. Like other wet film masks, these masks are about 0.001 inch thick.

Wet film solder mask is not a problem because its coating is relatively thin (0.001–0.002 inch). In wave soldering, moreover, a wet film mask allows routing of traces between pads of chip components, increases adhesive board strength, and minimizes formation of voids during adhesive cure. Since it cannot tent vias, its use is limited to Type II and III assemblies, where tenting is not an issue.

Because of the very high resolution requirements, not many vendors have the capability to mask the small spaces (10–12 mils) between adjacent pads when fine pitch components are used. Therefore, it may be necessary to require no solder mask around the land patterns of fine pitch packages. Unfortunately this is where one needs solder mask, especially in surface mount components that are to be reflow soldered. During solder paste application and reflow soldering, the fine pitch leads (<25 mils pitch) have a tendency to bridge. Fortunately solder mask technology has improved and many vendors are able to provide solder mask where it is needed most in surface mounting.

If solder mask is used over tin-lead, the solder melts and crinkles the mask, which may crack and cause entrapment of flux and moisture. This damage may eventually lead to circuit failure due to corrosion.

To prevent crinkling of solder masks, all masks should be applied over bare copper. Also refer to Chapter 4, Section 4.10 for additional discussion on solder mask.

7.9   REPAIRABILITY CONSIDERATIONS

The interpackage spacing guidelines presented earlier may or may not cause problems depending on the type of equipment needed for repairs. For example, if the repair tool does not use a vacuum tip for component removal and replacement, there may be access problems. If overly thick conductive tips (Figure 7.22 left) or mechanical tips in conjunction with hot air (Figure 7.22 right) are used for repair, they may not fit into the right interpackage spacing discussed earlier. This problem is disappearing as the repair/rework tool manufacturers are making hot air tools with vacuum pickup (removal and replacement) capability for passive components.

Access is not a problem for active components, because the hot air

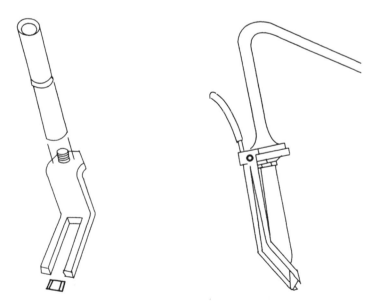

Figure 7.22   Equipment for the repair of passive devices: (left) conductive and (right) hot air with mechanical tweezers.

tools used for their removal and replacement blow hot air from the top (see Chapter 14). However, the nozzles should have tight tolerances to minimize the reflow of solder joints of adjacent components.

With the best of tools available at this time, it is very difficult to prevent totally the reflow of adjacent solder joints unless the interpackage spacing is more than 0.5 inch. This dimension, which is not feasible due to real estate constraints, is not absolutely necessary either, from a metallurgical standpoint.

As long as the number of times a component can be removed and replaced is controlled, the intermetallic thickness due to repair will not grow to the point of causing failure. For this reason, a maximum of two repairs should be allowed, assuming that the solder joint has already seen two reflows if it is on a double-sided board. As we said earlier, the manufacturability requirements are process and equipment dependent, and one should be mindful of the manufacturing situation before setting requirements.

Repairability may be a serious concern when a chip capacitor is used for decoupling under SOJ memory packages commonly used in memory modules (Figure 7.23). If the chip capacitor shown in Figure 7.23 needs to be replaced, the SOJ will have to be removed first. But defects in such devices are generally not seen since the solder joint cannot be inspected. As is the case with BGA, process control rather than inspection and repair is the method for ensuring yield in such applications.

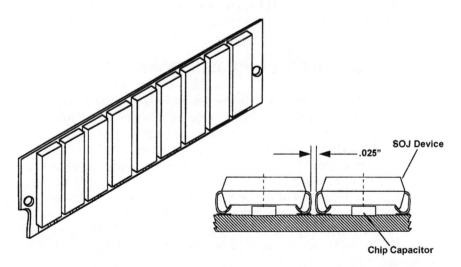

Figure 7.23   Chip capacitor under SOJ package used in memory modules.

7.10   CLEANLINESS CONSIDERATIONS

Given the smaller stand-off heights between surface mount components and the board, the cleaning of surface mount assemblies is of some concern. When a relatively active flux is used, which may be the case with wave soldering, cleaning problems are compounded. The elements in designing for cleaning are component orientation and solder mask use.

The guidelines recommended for component orientation and inter-package spacing in Sections 7.5 and 7.6 apply for cleaning as well. Use of solder mask is recommended whenever possible. It promotes cleanliness and helps in other areas, as discussed earlier. It is important to use components with adequate stand-off heights (0.015 inch minimum for packages one inch or larger on a side). The fine pitch packages made per JEDEC (U.S.) standards and most BGAs have sufficient stand-off for cleaning. But the EIAJ (Japanese) fine pitch packages may have zero stand-off height (EIAJ standard is 0-0.010 inch). The flux trapped under such packages, as is the case if the board goes through wave soldering after reflow, cannot be removed by any cleaning process no matter how many times the board is cleaned. In such cases, temporary solder masks (peelable if no-clean flux is used or solvent soluble in the cleaning process) should be used during wave soldering to prevent flux from getting under these low stand-off packages. This is an effective method but does add a manual operation in the process.

The most important element of designing for cleaning is the use of surface insulation resistance (SIR) patterns to monitor the process. Refer to Chapter 13 (Section 13.9: Designing for Cleaning) for details.

7.11   TESTABILITY CONSIDERATIONS

In this section we discuss only the design issues in testing that must be taken into account when the board is being designed. The guidelines for procurement of test fixtures, test probes, and accessories are discussed in Chapter 14. When boards are designed for through-hole technology, board testability comes "free," at least insofar as testability can be defined as the ability to probe for in-circuit testing. This is because in through-hole boards, the lead ends of the devices also serve as test nodes and can be accessed by test probes from the bottom side. This is not always the case in Type I and Type II SMT boards. Type III SMT boards are essentially the same as through-hole boards, since all the active devices are through hole.

Design for testability of SMT boards boils down to whether or not to provide a test pad for each test node. In addition to taking up real estate, furnishing test pads raises cost. Not all test pads, however, demand their own real estate on the board. Most of the vias required for interconnection also serve as test pads.

The number of test nodes needed depends on the level of defects in manufacturing. If the defect rate is really low, we can tolerate a lower number of test nodes and accept a poor diagnostic capability. However, if the defect rate is high, we need adequate diagnostic capability, which means that we must provide access to all test nodes.

The alternative to ATE testing is functional testing, which can be much more expensive [4]. Also the solder joints in SMT boards are hard to inspect (inside fillet of SOICs and fine pitch, for example). The higher density of the SMT boards puts an additional constraint on visual inspection. For example, it becomes difficult to visually detect even obvious defects such as opens and shorts (bridges) in the fine pitch components. And if BGA is used, visual inspection of solder joints is not feasible since they are hidden under the package.

In addition, many of the components do not have part markings, which makes visual checking of the values of the devices impossible. Improperly trained operators can easily make mistakes in identifying polarity and pin 1.

This is not to suggest that ATE is the only right answer for SMT assemblies. There is no right or wrong approach in designing for testability. Rather, one must determine the appropriate approach. With process under control, one can tolerate access to less than all components or even depend on some form of self-test or functional test.

Companies just starting out in SMT, however, still need to get process under control. Until they do, access to all test nodes must be provided, to fully utilize the troubleshooting capability of ATE.

Thus in the short term, we need to depend on the full diagnostic capability of ATE and should provide full nodal access. In the long term, we need to get the processes under control and not waste the space and money on conventional (ATE) testing. Instead, we should strive for a combination of appropriate process control and test approaches for a cost-effective solution.

7.11.1   Guidelines for ATE Testing

Designing for testability revolves around whether all the test nodes must be accessible by automated test equipment (ATE). In any event, the test pads must be accessible from both sides: the bottom access is needed for ATE test probes, and the top access is needed for hand probing while the board is under test in ATE. In such a situation the bottom test pads cannot be accessed manually. Providing the access on both sides requires additional real estate because the design must include adequate interpackage spacing.

Here are some guidelines for using test points (TPs):

- Use on TP for Ground and one TP for Power for every 10 ICs to reduce risk of ringing.
- Connect all unused pins to Ground or Power via resistor to improve quality of testing.
- Whenever possible, avoid double-sided test probe approach to reduce fixturing and test cost.
- One TP is needed for every test node unless 100% diagnostic capability is needed for individual component.
- Test point spacings should be 100 mils whenever possible. Reason: Probes for smaller spacings are fragile and prone to damage.
- For smaller TPs (under 30 mils dia), deposit solder paste on TPs to form bumps for ease of probing with serrated test probe. As is discussed later, use of TP smaller than 0.032 inch increases the potential for missing the test pad target.

Interpackage spacing also has an impact on the selection of test probes and the size of the test pads (outer diameter of via pads). The two types of test probe generally used for in-circuit or bed-of-nails testing in ATE are 50 and 100 mil probes. The smaller probes allow the use of smaller (25–40 mils) test pads.

The smaller the test pad, the higher the need for fixture accuracy to guarantee good probe contact. Use of 25 mil test pads is rare, and a 32 mil test pad size may be a good compromise between the 25 and 40 mil sizes to meet the conflicting requirements of real estate constraints and fixture accuracy for good probe contact. Pads of 0.032 inch diameter allow routing of one line between the test pads at 0.050 inch pitch. If one can use 0.005 inch trace, the test pad diameter can be increased to 0.035 inch without compromising routing efficiency. The 100 mil probes require larger test pads, but this is seldom troublesome because the lead ends of conventional devices serve the purpose.

Figure 7.24, which should be used as a guideline for establishing the spacing between test pads and the adjacent package, is based on the diameter of the test probes. Since most boards contain both surface mount and through-hole components, both types of test probe are used. One should keep the probe density per unit area to a minimum to make sure that the fixture can be pulled down with the available vacuum.

Since the test pads are very small, there is no room for overlap of solder mask. Additional space for probe contact may be provided by

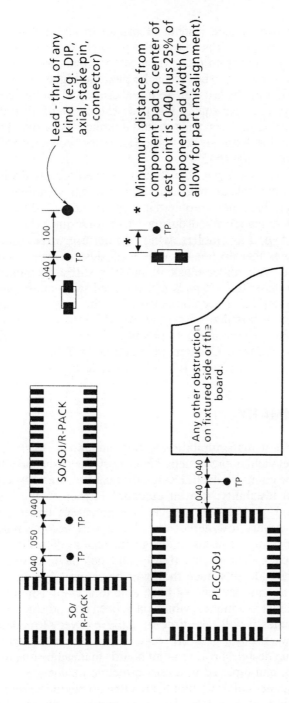

Figure 7.24 Interpackage spacing for ATE testing for 50 and 100 mil test probes.

333

making the test pads square instead of round. In addition to providing additional space for probe contact on the pad corners, a square test pad is readily distinguished from other (nontest) via pads. However, square test pads constrain routing. The larger the pad, the higher the probability of the test probe hitting the test pad. So if real estate and routing constraints are not very severe, larger square test pads should be used. But the use of smaller than 0.032 inch test pads should definitely be avoided to ensure that the expensive test fixture is not wasted since the probability of missing the test pads will be higher [5].

As many via holes as possible should be filled or sealed with solder mask to prevent air leakage when the fixture is being pulled down. As mentioned earlier, this is not a problem in Type II and III SMT assemblies, since all the via holes are filled during wave soldering. The problem is encountered in Type I assemblies, which do not use wave soldering.

The via holes that do not serve as test pads in Type I assemblies should be tented with solder mask. If there is still a problem with air leakage, the via holes for test pads can be filled with solder paste. This is not a separate step but is accomplished when the paste is screened onto other solder pads. If possible, screening of paste into vias under devices should be avoided, to minimize entrapment of flux and solder balls under the components. Refer to Chapter 14, Section 14.7 for details on test fixtures, test probes and various other issues in test.

7.12   SUMMARY

Designing for manufacturability, testing, and repair is very important for time to market yield improvement, hence cost reduction and ultimately, the survival of the company itself. Design for manufacturability, especially repairability and testability, has an essential impact on the real estate savings that SMT provides. If we get our processes under control to achieve a zero or very low defect rate, DFM will become much more critical with respect to the needs to design for repair and testing. In other words we can pack components tightly and not have to worry about finding or repairing defects since they will be essentially nonexistent.

However, the industry has not been able to accomplish zero defects even in conventional assemblies, which have been around since the 1950s. For most companies it will take time to achieve a very low defect rate. In the interim, we need to give up some of the real estate benefits that SMT provides and design surface mount boards that can be manufactured, inspected, tested, and repaired in a cost-effective manner.

The consequences of designing boards that do not meet these requirements can be serious to the extent of exerting adverse impacts on revenues or schedules or both. This means that a product may not succeed in the marketplace, and that could be a serious problem indeed.

REFERENCES

1. Garrison, Tucker. "Manufacturing PCB densities and packaging alternatives." *Circuits Assembly,* January 1995, pp. 28–32.

2. Solberg, Vern. "SMT design guidelines for automation—The robotic assembly process." *Proceedings of the Technical Program,* August 1988, pp. 271–280, available from Surface Mount Technology Association, Edina, MN.

3. Prasad, R. P. "Designing for high speed, high yield SMT." *Surface Mount Technology,* January 1994, pp. 49–53.

4. Kunin, D., and Desai, N. "Design put to test." *Circuits Manufacturing,* March 1987, pp. 52–58.

5. Parry, Mike. "A contract manufacturer talks about designing for testability." *Surface Mount Technology,* March 1995, pp. 46–50.

Part Three
Manufacturing with Surface Mounting

Chapter 8
Adhesive and Its Application

8.0 INTRODUCTION

An adhesive in surface mounting is used to hold passive components on the bottom side of the board during wave soldering. This is necessary to avoid the displacement of these components under the action of the wave. When soldering is complete, the adhesive no longer has a useful function.

The types of components most commonly glued to the bottom side of Type III and Type II SMT boards are rectangular chip capacitors and resistors, the cylindrical transistors known as metal electrode leadless face (MELFs), small outline transistors (SOTs), and small outline integrated circuits (SOICs). These components are wave soldered together with the through-hole mount (THM) devices.

Rarely, adhesive is also used to hold multileaded active devices such as plastic leaded chip carriers (PLCCs) and plastic quad flat packs (PQFPs) on the bottom side for wave soldering. Wave soldering of active devices with leads on all four sides is generally not recommended for reasons of reliability and excessive bridging. The problem of bridging can be minimized by passing the board over the solder wave at 45° to board flow, but reliability (e.g., popcorning and flux seepage into the package) remains a concern. If active devices, including SOICs, are to be wave soldered, they must be properly qualified.

In another uncommon application, an adhesive holds both active and passive components placed on solder paste on the bottom or secondary side of the board during reflow soldering, to allow the simultaneous reflow soldering of surface mount components on both sides.

This chapter focuses on adhesive types, and on dispensing methods and curing mechanisms for adhesives. Some important technical considerations in the selection and qualification of such materials for surface mounting are also discussed.

8.1   IDEAL ADHESIVE FOR SURFACE MOUNTING

Reference 1 provides some general guidelines for selecting nonconductive and conductive adhesives. In this chapter, our focus will be on nonconductive adhesives, which are the most widely used. Electrically conductive adhesives are used for solder replacement, and thermally conductive adhesives are used for heat sink attachment. We will review them briefly in Section 8.4.

Many factors must be considered in the selection of an adhesive for surface mounting. In particular, it is important to keep in mind three main areas: precure properties, cure properties, and postcure properties.

8.1.1   Precure Properties

One-part adhesives are preferred over two-part adhesives for surface mounting because it is a nuisance to have to mix two-part adhesives in the right proportions for the right amount of time. One-part adhesives, eliminating one process variable in manufacturing, are easier to apply, and one does not have to worry about the short working life (pot life) of the mixture. The single-part adhesives have a shorter shelf life, however. The terms "shelf life" and "pot life" can be confusing. "Shelf life" refers to the usable life of the adhesive as it sits in the container, whereas "pot life," as indicated above, refers to the usable life of the adhesive after the two main components (catalyst and resin) have been mixed and catalysis has begun.

Two-part adhesives start to harden almost as soon as the two components are mixed, and hence have a short pot life even though each component has a long shelf life. Elaborate metering and dispensing systems are generally required for automated metering, mixing in the right proportions, and then dispensing the two-part adhesives.

Colored adhesives are very desirable because they are easy to spot if applied in excess amount, such that they contact the pads. Adhesive on pads prevents the soldering of terminations, hence it is not allowed. For most adhesives it is a simple matter to generate color by the addition of pigments. In certain formulations, however, pigments are not allowed because they would act as catalysts for side reactions with the polymers, perhaps drastically altering the cure properties. Typical colors for surface mount adhesives are red or yellow, but any color that allows easy detection can be used.

The uncured adhesive must have sufficient green strength to hold components in place during handling and placement before curing. This property is similar to the tackiness requirement of solder paste, which must secure components in their places before reflow. It should not cure at room temperature, in order to allow enough time between dispensing of the adhesive and component placement. The adhesive should have sufficient volume to fill the gap without spreading onto the solderable pads. It must be nontoxic, odorless, environmentally safe, and nonflammable.

Some consideration must also be given to storage conditions and shelf life. Most adhesives will have a longer shelf life if refrigerated. Finally, the adhesive must be compatible with the dispensing or stenciling method to be used in manufacturing. This means that it must have the proper viscosity. Adhesives that require refrigeration must be allowed to equilibrate to ambient temperature before use to assure accurate dispensing. The issue of changes in viscosity with temperature is discussed in Section 8.5.3.

8.1.2   Cure Properties

The cure properties relate to the time and temperature of cure needed to accomplish the desired bond strength. The shorter the time and the lower the temperature to achieve the desired result, the better the adhesive. The specific times and temperatures for some adhesives are discussed in Section 8.6.

The surface mount adhesive must have a short cure time at low temperature, and it must provide adequate bond strength after curing to hold the part in the wave. If there is too much bond strength, reworking may be difficult; too little bond strength may cause loss of components in the wave. However, high bond strength at room temperature does not mean poor reworkability, as discussed in Section 8.1.3.

The adhesive should cure at a temperature low enough to prevent warpage in the substrate and damage to components. In other words, it is preferable that the adhesive be curable below the glass transition temperature of the substrate (126°C for FR-4). However, a very short cure time above the glass transition temperature is generally acceptable. The cured adhesive should neither increase strength too much nor degrade strength during wave soldering.

To ensure sufficient throughput, a short cure time is desired and the adhesive cure property should be more dependent on the cure temperature than the cure time. Low shrinkage during cure, to minimize the stress on attached components, is another cure property. Finally, there should be no outgassing in adhesives, because this phenomenon will entrap flux and cause serious cleaning problems. Voids also may result from rapid curing of adhesive. (See Section 8.6.1.2.)

8.1.3   Postcure Properties

Although the adhesive loses its function after wave soldering, it still must not degrade the reliability of the assembly during subsequent manufacturing processes such as cleaning and repair/rework. Among the important postcure properties for adhesives is reworkability. To ensure reworkability, the adhesive should have a relatively low glass transition temperature. The cured adhesives soften (i.e., reach their T_g) as they are heated during rework. For fully cured adhesives, T_g in a range of 75° to 95°C is considered to accommodate reworkability.

Temperatures under the components often exceed 100°C during rework because the terminations must reach much higher temperatures (>183°C) for eutectic tin-lead solder to melt. As long as the T_g of the cured adhesive is below 100°C, and the amount of adhesive is not excessive, reworkability should not be a problem. As we will discuss in Section 8.7, differential scanning calorimetry (DSC) can be used to determine the T_g of a cured adhesive.

Another useful indicator of reworkability is the location of the shear line after rework. If the shear line exists in the adhesive bulk, as shown in Figure 8.1, it means that the weakest link is in the adhesive at the reworking temperature. The failure would occur as shown in Figure 8.1 because one would not want to lift the solder mask or pad during rework. In the failure mechanism shown in Figure 8.2, on the other hand, there is hardly any bond between the substrate and adhesive. This can happen due to contamination or undercuring of the adhesive.

Other important postcure properties for adhesives include nonconductivity, moisture resistance, and noncorrosivity. The adhesive should also have adequate insulation resistance and should remain inert to cleaning solvents. Insulation resistance is generally not a problem because the building blocks of most adhesives are insulative in nature, but insulation resistance under humidity should be checked before final selection of an adhesive is made.

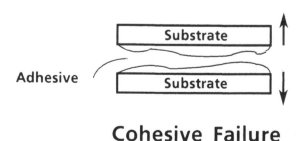

Cohesive Failure

Figure 8.1   Schematic representation of a "cohesive" failure in adhesive.

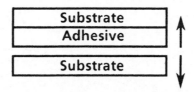

Adhesive Failure

Figure 8.2 Schematic representation of an "adhesive" failure at the adhesive-substrate bond line.

The test conditions for checking surface insulation resistance (SIR) are discussed in Chapter 13 (Cleaning). It should be noted that SIR test results can flag not only poor insulative characteristics but also an adhesive's voiding characteristics. See Section 8.6.1.2.

8.2 GENERAL CLASSIFICATION OF ADHESIVES

Adhesives can be based on electrical properties (insulative or conductive), chemical properties (acrylic or epoxy), curing properties (thermal or UV/thermal cure), or physical characteristics after cure (thermoplastic or thermosetting). Of course these electrical, chemical, curing, and physical properties are interrelated.

Based on conductivity, adhesives can be classified as insulative or conductive. They must have high insulation resistance, since electrical interconnection is provided by the solder. However the use of conductive adhesives as a replacement for solder for interconnection purposes is also being suggested. Silver fillers are usually added to adhesives to impart electrical conduction. We discuss conductive adhesives briefly in Section 8.4, but this chapter focuses on nonconductive (insulative) adhesives because they are most widely used in the wave soldering of surface mount components.

Surface mount adhesives can be classified as elastomeric, thermoplastic, or thermosetting. Elastomeric adhesives, as the name implies, are materials having great elasticity. These adhesives may be formulated in solvents from synthetic or naturally occurring polymers. They are noted for high peel strength and flexibility, but they are not used generally in surface mounting.

Thermoplastic adhesives do not harden by cross-linking of polymers. Instead, they harden by evaporation of solvents or by cooling from high temperature to room temperature. They can soften and harden any number

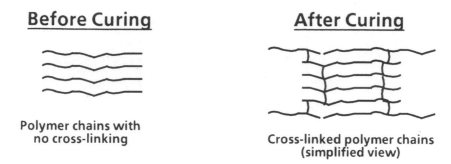

Figure 8.3 **A schematic representation of the curing mechanism in a thermosetting adhesive.**

of times as the temperature is raised or lowered. If the softening takes place to such an extent that they cannot withstand the rough action of the wave and may be displaced by the wave, however, thermoplastic adhesives cannot be used for surface mounting.

Thermosetting adhesives cure by cross-linking polymer molecules, a process that strengthens the bulk adhesive and transforms it from a rubbery state (Figure 8.3, left) to a rigid state (Figure 8.3, right). Once such a material has cured, subsequent heating does not add new bonds. Thermosetting adhesives are available as either one- or two-part systems.

8.3   ADHESIVES FOR SURFACE MOUNTING

Both conductive and nonconductive adhesives are used in surface mounting. Conductive adhesives are discussed in Section 8.4. The most commonly used nonconductive adhesives are epoxies and acrylics. Sometimes urethanes and cyanoacrylates are chosen.

8.3.1   Epoxy Adhesives

Epoxies are available in one- and two-part systems. Two-part adhesives cure at room temperature but require careful mixing in the proper proportions. This makes the two-part systems less desirable for production situations. Single-component adhesives cure at elevated temperatures, with the time for cure depending on the temperature. It is difficult to formulate a one-part epoxy adhesive having a long shelf life that does not need high curing temperature and long cure time. Epoxies in general are cured thermally and are suitable for all different methods of application. The catalysts for heat curing adhesives are epoxides. An epoxide ring contains an oxygen atom bonded to two carbon atoms which are bonded to each other. The thermal energy breaks this bond to start the curing process.

The shelf life of adhesives in opened packages is short, but this may not be an important issue for most syringing applications since small (5 gram) packages of adhesives are available. Any unused adhesive can be discarded without much cost impact if its shelf life has expired. For most one-part epoxies, the shelf life at 25°C is about 2 months. The shelf life can be prolonged, generally up to 3 months, by storage in a refrigerator at low temperature (0°C). Epoxy adhesives, like almost all adhesives used in surface mounting, should be handled with care because they may cause skin irritation. Good ventilation is also essential.

8.3.2   Acrylic Adhesives

Like epoxies, acrylics are thermosetting adhesives that come as either single- or two-part systems, but with a unique chemistry for quick curing. Acrylic adhesives harden by a polymerization reaction like that of the epoxies, but the mechanism of cure is different. The curing of adhesive is accomplished by using long wavelength ultraviolet (UV) light or heat. The UV light causes the decomposition of peroxides in the adhesive and generates a radical or odd-electron species. These radicals cause a chain reaction to cure the adhesive by forming a high molecular weight polymer (cured adhesive).

The acrylic adhesive must extend past the components to allow initiation of polymerization by the UV light. Since all the adhesive cannot be exposed to UV light, there may be uncured adhesive under the component. Not surprisingly, the presence of such uncured adhesive will pose reliability problems during subsequent processing or in the field in the event of any chemical activity in the uncured portion. In addition, uncured adhesives cause outgassing during solder and will form voids. Such voids may entrap flux.

Total cure of acrylic adhesives is generally accomplished by both UV light and heat to ensure cure and to also reduce the cure time. These adhesives have been widely used for surface mounting because of faster throughput and because they allow in-line curing between placement and wave soldering of components. Like the epoxies, the acrylics are amenable to dispensing by various methods.

The acrylic adhesives differ from the epoxy types in one major way. Most but not all acrylic adhesives are anaerobic (i.e., they can cure in the absence of air). To prevent natural curing, therefore, they should not be wrapped in airtight containers. To avoid cure in the bag, the adhesive must be able to breathe. The acrylic adhesives are nontoxic but are irritating to the skin. Venting and skin protection are required, although automated dispensing often can eliminate skin contact problems. Acrylic adhesives can be stored for up to 6 months when refrigerated at 5°C and for up to 2 months at 30°C.

8.3.3   Other Adhesives for Surface Mounting

Two other kinds of nonconductive adhesives are available: urethanes and cyanoacrylates. The urethane adhesives are typically moisture sensitive and require dry storage to prevent premature polymerization. Nor are they resistant to water and solvents after polymerization. These materials seldom are used in surface mounting.

The cyanoacrylates are fast bonding, single-component adhesives generally known by their commercial names (Instant Glue, Super Glue, Crazy Glue, etc.). They cure by moisture absorption without application of heat. The cyanoacrylates are considered too fast bonding for SMT, and they require good surface fit. An adhesive that cures too quickly is not suitable for surface mounting because some time lapse between adhesive placement and component placement is necessary. Also, cyanoacrylates are thermoplastic and may not withstand the heat of wave soldering.

8.4   CONDUCTIVE ADHESIVES FOR SURFACE MOUNTING

8.4.1   Electrically Conductive Adhesives

Electrically conductive adhesives have been proposed for surface mounting as a replacement for solder to correct the problem of solder joint cracking [1]. It is generally accepted that leadless ceramic chip carriers (LCCCs) soldered to a glass epoxy substrate are prone to solder joint cracking problems due to mismatch in the coefficients of thermal expansion (CTEs) between the carriers and the substrate. This problem exists in military applications, which require hermetically sealed ceramic packages.

Most electrically conductive adhesives are epoxy-based thermosetting resins that are hardened by applying heat. They cannot attain a flowable state again, but they will soften at their glass transition temperature. Nonconductive epoxy resin serves as a matrix, and conductivity is provided by filler metals. The metal particles must be present in large percentage so that they are touching each other, or they must be in close proximity, to allow electron tunneling to the next conductive particle through the nonconductive epoxy matrix. Typically it takes 60 to 80% filler metals, generally precious metals such as gold or silver, to make an adhesive electrically conductive. This is why these adhesives are very expensive. To reduce cost, nickel-filled adhesives are used. Copper is also used as a filler metal, but oxidation causes this metal to lose its conductivity.

The process for applying conductive adhesive is very simple. The adhesive is screen printed onto circuit boards to a thickness of about 2 mils. Nonconductive adhesive is also used to provide the mechanical strength for larger components. For smaller chip components, nonconductive adhesive is not needed for mechanical strength.

After the adhesive has been applied and the components placed, both conductive and nonconductive adhesives are cured. Depending on the adhesive, the heat for curing is provided by heating in a convection oven, by exposing to infrared or ultraviolet radiation, or by vapor phase condensation. Curing times can vary from a few minutes to an hour, depending on the adhesive and the curing equipment. High strength materials cure at 150° to 180°C; lower strength, highly elastomeric materials cure at 80° to 160°C [2].

Since flux is not used, the use of conductive adhesive is truly a no-clean process. In addition, the conductive adhesives work well with all types of board finishes such as tin-lead, OSP (organic solderability protection), and gold or palladium.

Electrically conductive adhesives have been ignored by the surface mounting industry for many reasons. The military market is still struggling with the solder joint cracking problem, and the exotic substrates selected thus far instead of adhesives have not been completely successful. However, there is not much reliability data for adhesives either, and since plastic packages with leads are used in commercial applications, the cracking problem does not exist in that market sector. The higher cost of conductive adhesives is also a factor.

The process step savings provided by conductive adhesive may be exaggerated. The only meaningful process step saving between solder reflow and conductive epoxy is elimination of cleaning when adhesive is used.

There are other reasons for not using conductive adhesives for surface mounting. As we indicate throughout this book, very few assemblies are entirely surface mount (Type I); most are a mixture of surface mount and through-hole mount (Type II). Electrically conductive adhesives, however, do not work for interconnection of through holes and therefore cannot be used for mixed assemblies. In any event, since the electrical conductivity of these adhesives is lower than that of solder, they cannot replace solder if high electrical conductivity is critical. When conductive adhesives can be used, component placement must be precise, for twisted components cannot be corrected without the risk that they will smudge and cause shorts.

Repairs may also be difficult with electrically conductive adhesives. A conductive adhesive is hard to remove from a conductive pad, yet complete removal is critical because the new component must sit flush. Also, it is not as convenient to just touch up leads with conductive adhesive, as can be done with solder. Probably the biggest reason for the very limited use of conductive adhesives, however, is unfamiliarity for mass application. The conductive adhesives do have their place in applications such as hybrids and semiconductors. It is unlikely, however, that they will ever replace solder, a familiar and relatively inexpensive material.

8.4.1.1   *Anisotropic Electrically Conductive Adhesives*

Anisotropic adhesives are also electrically conductive and are intended for applications where soldering is difficult or cannot be used. This process for their use was developed by IBM to provide electrical connections between tungsten and copper. Now there are many companies such as Alpha Metal, Amp and Sheldahl, and many Japanese companies who supply these adhesives. Examples of anisotropic adhesive applications include ultra fine pitch tape-automated bonding (TAB) and flip chips.

This type of adhesive is also referred to as Z-axis adhesive because it conducts electricity only in the Z axis (in the vertical direction between the pad on the board and the lead) and remains insulative in the X-Y horizontal plane. Since it is insulative in the X-Y direction, bridging with adjacent conductors is not a concern.

The anisotropic adhesives use a low concentration of conductive filler metals in a nonconductive polymer matrix. The filler metals do not touch each other in the matrix in order to prevent conductivity in the X-Y direction. The filler metals are either completely metallic or nickel-plated spherical elastomer plastics. The nickel-plated plastic spheres are considered more resilient.

As shown in Figure 8.4, the electricity is conducted between the lead and pad through each particle. This means that anisotropic adhesives cannot carry high current and hence their application is limited to low power devices.

The electrical contact is made between the lead and the pad by curing the adhesive under pressure. The conductive surfaces must be in contact during the entire cure cycle. The adhesives come in both thermoplastic and thermosetting epoxy matrix. The thermosetting adhesives require much higher pressure (several kilograms) during the cure cycle than the thermoplastic adhesives, which require about one-tenth as much pressure [3].

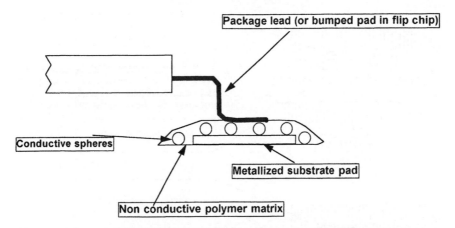

Figure 8.4   Lead attachment with anisotropic conductive adhesive.

8.4.2   Thermally Conductive Adhesives

In addition to nonconductive and electrically conductive adhesives, thermally conductive adhesives are also used. High performance devices, particularly microprocessors, typically generate a large amount of heat. Since the power in electronic devices continues to increase over time, removal of this heat is necessary for the microprocessor to deliver peak performance.

Most microcomputer system designs employ forced air cooling to remove heat from high power packages. Such designs require a heat sink to be attached to these packages. The heat sinks are attached by various interface materials such as thermally conductive tapes and adhesives (Figure 8.5, top) and thermal grease (Figure 8.5, bottom).

The thermally conductive adhesives include epoxies, silicones, acrylics, and urethanes. These are generally filled with metallic or ceramic filler to enhance their thermal conductivity. These fillers increase the thermal conductivity of these adhesives from about 0.2 W/m-K to more than 1.5 W/m-K.

The upper diagram in Figure 8.5 illustrates the structure of these materials. Like the nonconductive and electrically conductive adhesives discussed earlier, the thermally conductive adhesives also need to be cured to develop the necessary bond strength between the heat sink and the package.

Curing can be done thermally by baking in an oven or chemically by the use of an accelerator. A single-part adhesive typically needs thermally activated curing. Dual-part adhesives, where one part is the resin and the other part is the hardener, have to be mixed according to a pre-designed

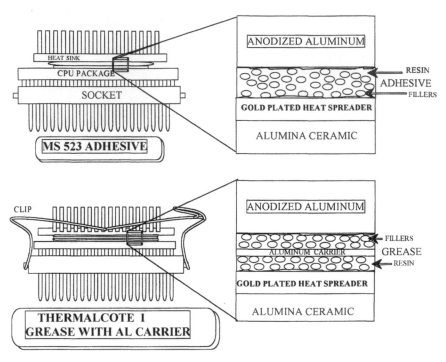

Figure 8.5 **Heat sink attachment to high power packages by thermally conductive adhesive and tapes (top) and thermal grease (bottom). (Courtesy Dr. Raiyomand Aspandiar, Intel Corporation).**

ratio. Curing is initiated as soon as mixing occurs, but to develop greater bond strength elevated temperature curing is generally required.

When heat sinks are attached to component packages with adhesives as opposed to thermal grease and thermal tapes, they do not require a secondary mechanical attachment. The adhesive serves as both the heat transfer interface as well as the mechanical bonding interface. A critical property of the adhesive to consider during evaluation is the modulus of rigidity. The adhesive should absorb the stresses generated by the expansion mismatches of the two bonded surfaces without debonding or cracking.

8.5   ADHESIVE APPLICATION METHODS

The commonly used methods of applying adhesives in surface mounting are pin transfer and syringing or pressure transfer. Proper selection of a method depends on a great number of considerations such as type of adhesive, volume or dot size, and speed of application. No matter which method is used, the following guidelines should be followed when dispensing adhesives.

1. Adhesives that are kept refrigerated should be removed from the refrigerator and allowed to come to room temperature before their containers are opened.
2. The adhesive should not extend onto the circuit pads. Adhesive that is placed on a part should not extend onto the component termination.
3. Sufficient adhesive should be applied to ensure that when the component is placed, most of the space between the substrate and the component is filled with adhesive. For large components, more than one dot may be required.
4. It is very important that the proper amount of adhesive be placed. As mentioned earlier, too little will cause loss of components in the solder wave and too much will either cause repair problems or flow onto the pad under component pressure, preventing proper soldering.
5. Figure 8.6 can be used as a general guideline for dot size requirements.
6. Unused adhesives should be discarded.
7. If two-part adhesives are used, it will be necessary to properly proportion the "A" (resin) and "B" (catalyst) materials and mix them thoroughly before dispensing. This can be done either manually or automatically. Two-part adhesives are not commonly used for surface mounting because they introduce additional process and equipment variables.

8.5.1 Stencil Printing

Like solder paste application, either screens or stencils can be used to print adhesive at the locations desired. Stenciling is more common than screening for adhesive printing, just as it is for solder paste printing. Stencils can deposit different heights of adhesive but screens cannot. (See dual-thickness printing in Chapter 9).

Stencil printing uses a squeegee to push adhesive through the holes in the stencil onto the substrate where adhesive is required. The stencils are made using an artwork film of the outer layer showing the locations at which adhesive needs to be deposited. Chapter 9 covers the stencil printing process and lists equipment for paste application; this process applies to the screening of adhesives as well. Stencil printing is a very

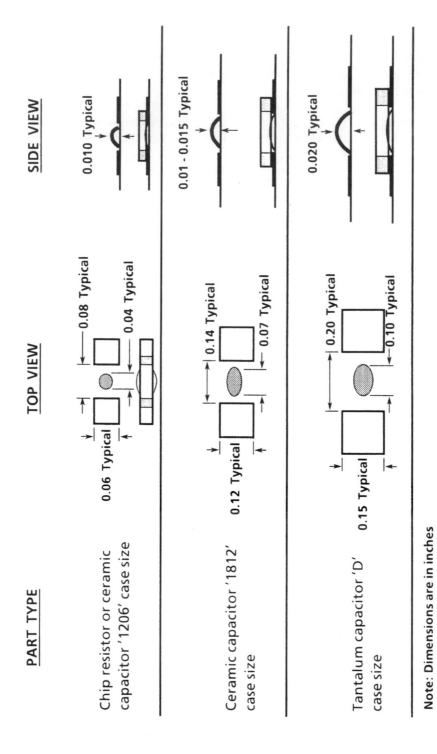

Figure 8.6 Guidelines for adhesive dot sizes for surface mount components. Note: Diameter of dot size is typically about half the gap between lands.

Note: Dimensions are in inches

PART TYPE

Chip resistor or ceramic capacitor '1206' case size

Ceramic capacitor '1812' case size

Tantalum capacitor 'D' case size

TOP VIEW

0.08 Typical

0.04 Typical

0.06 Typical

0.14 Typical

0.07 Typical

0.12 Typical

0.20 Typical

0.10 Typical

0.15 Typical

SIDE VIEW

0.010 Typical

0.01 - 0.015 Typical

0.020 Typical

fast process. It allows the deposition of adhesives on all locations in one stroke. Thickness and size of adhesive dots are determined by the thickness of the stencil or of the wire mesh and the emulsion on the screen.

Stenciling of adhesive is cumbersome, hence it is not a very common production process. Cleaning the stencils after printing is a difficult task. Also, care is necessary to prevent smudging of adhesives onto adjacent pads, to preserve solderability.

8.5.2 Pin Transfer

Pin transfer, like stenciling, is a very fast dispensing method because it applies adhesive en masse. Viscosity control is very critical in pin transfer to prevent tailing, just as it is for stenciling. The pin transfer system can be controlled by hardware or by software.

In hardware-controlled systems, a grid of pins, which is installed on a plate on locations corresponding to adhesive locations on the substrate, is lowered into a shallow adhesive tray to pick up adhesive. Then the grid is lowered onto the substrate. When the grid is raised again, a fixed amount of adhesive sticks to the substrate because the adhesive has greater affinity for the nonmetallic substrate surface than for the metallic pins. Gravity ensures that an almost uniform amount of adhesive is carried by the pins each time. Hardware-controlled systems are much faster than their software-controlled counterparts, but not as flexible. Some control in the size of dots can be exercised by changing the pin sizes, but this is very difficult.

Software-controlled systems offer greater flexibility at a slower speed, but there are some variations. For example, in some Japanese equipment, a jaw picks up the part, the adhesive is applied to the part (not the substrate) with a knife rather than with a pin, and then the part is placed on the substrate. This software-controlled system applies adhesive on one part at a time in fast succession. Thus when there is a great variety of boards to be assembled, software-controlled systems are preferable.

For prototyping, pin transfer can be achieved manually, using a stylus, as shown in Figure 8.7. Manual pin transfer provides great flexibility. For example, as shown in Figure 8.8, a dot of adhesive may be placed on the body of a large component, if this is necessary to ensure an adequate adhesive bond.

8.5.3 Syringing

Syringing, the most commonly used method for dispensing adhesive, is not as fast as other methods, but it allows the most flexibility. Adhesive

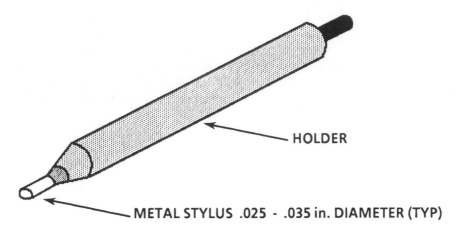

HOLDER

METAL STYLUS .025 - .035 in. DIAMETER (TYP)

Figure 8.7 A stylus for manual pin transfer of adhesive.

is placed inside a syringe and dispensed through a hollow needle by means of pressure applied pneumatically, hydraulically, or by an electric drive. In all cases, the system needs to control the flow rate of the adhesive to ensure uniformity.

As discussed in Chapter 11, adhesive dispensing systems that can place two to three dots per second are generally an integral part of the pick-and-place equipment. Either dedicated adhesive dispensing heads with their own X-Y table can be used, or the component placement X-Y table can be shared for adhesive placement. The latter option is cheaper but slower because the placement head can be either dispensing adhesive or placing components, not both.

The adhesive dispensing head as an integral part of the pick-and-place system is shown in Figure 8.9, which shows the syringe out of its

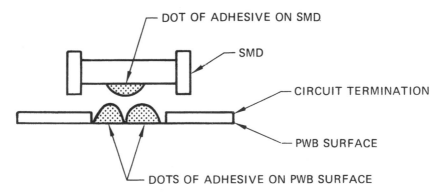

DOT OF ADHESIVE ON SMD

SMD

CIRCUIT TERMINATION

PWB SURFACE

DOTS OF ADHESIVE ON PWB SURFACE

Figure 8.8 Guideline for adhesive dots on large components.

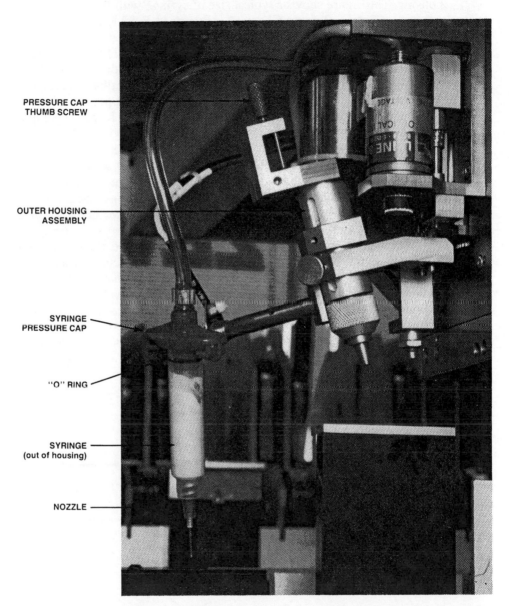

PRESSURE CAP
THUMB SCREW

OUTER HOUSING
ASSEMBLY

SYRINGE
PRESSURE CAP

"O" RING

SYRINGE
(out of housing)

NOZZLE

Figure 8.9   An automatic adhesive dispenser as an integral part of the pick-and-place equipment.

housing. This figure shows a typical size of syringe with 5 to 10 grams of adhesive capacity. Refer to Chapter 11, Figure 11.6, which shows the syringe inside its housing, ready for dispensing adhesive onto the substrate. The dispensing head can be programmed to dispense adhesive on desired locations. The coordinates of adhesive dots are generally downloaded from the CAD systems not only to save time in programming the placement equipment but also to provide better accuracy. Some pick-and-place equipment automatically generate an adhesive dispensing program from the component placement program. This is a very effective method for varying dot size or even for placing two dots, as is sometimes required.

  If the adhesive requires ultraviolet cure, the dot of adhesive should be placed or formed so that a small amount extends out from under the edges of the component, but away from the terminations and the leads, as shown in Figure 8.10. The exposed adhesive is necessary to initiate the ultraviolet cure.

  The important dispensing parameters, pressure and timing, control the dot size and to some extent tailing, which is a function of adhesive viscosity. By varying the pressure, the dot size can be changed. Stringing or tailing (dragging of the adhesive's "tail" to the next location over components and substrate surface) can cause serious problems of solder skips on the pads. Stringing can be reduced by making some adjustments to the dispensing system. For example, smaller distance between the board and the nozzle, larger diameter nozzle tips, and lower air pressure help

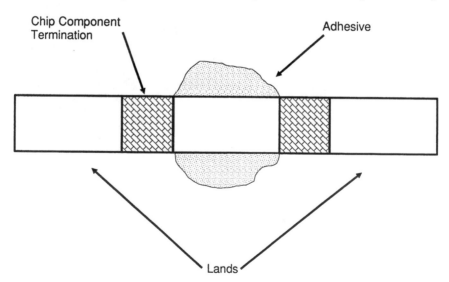

Figure 8.10   The adhesive extension required on the sides of components using UV curing adhesive.

reduce the incidence of stringing. If pressure is used for dispensing, which is commonly the case, any change in viscosity and restriction to flow rate will cause the pressure to drop off, resulting in a decrease in flow rate and a change in dot size.

The viscosity of adhesive also plays a role in stringing. For example, higher viscosity adhesives are more prone to stringing than lower viscosity adhesives. However, a very low viscosity may cause dispensing of excessive amounts of adhesive. Since viscosity changes with temperature, a change in ambient temperature can have a significant impact on the amount of adhesive that is dispensed. From a study conducted at IBM by Meeks [4] (see Figure 8.11) it was shown that only a 5°C change in ambient temperature could influence the amount of adhesive dispensed by almost 50% (from 0.13 gram to 0.190 gram). All other dispensing variables, such as nozzle size, pressure, and time remained the same. Temperature

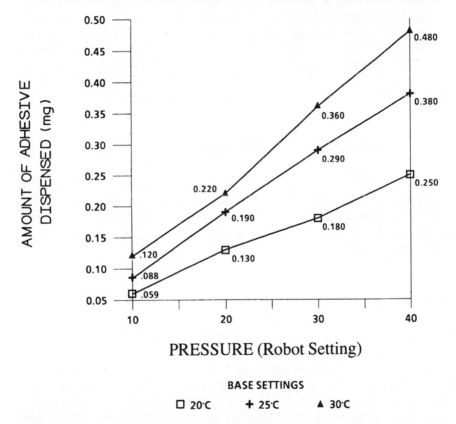

Figure 8.11   The impact of temperature on the amount of adhesive dispensed [4].

controlled housing or the positive displacement method (to dispense a known amount of adhesive) should be used to prevent variation in dot size due to change in ambient temperature.

Skipping of adhesive is another common problem in adhesive dispensing. The likely causes of skipping are clogged nozzles, worn dispenser tips, and circuit boards that are not flat [5]. The nozzles generally clog if adhesive is left unused for a long time (from a few hours to a few days, depending on the adhesive). To avoid clogging of nozzles, either the syringe should be discarded after every use or a wire can be put inside the nozzle tip. A very high viscosity can also cause skipping. When using automatic dispensers in pick-and-place systems, care should be exercised to keep gripper tips clean of adhesive. If contact with adhesive occurs during component placement, the gripper tips should be cleaned with isopropyl alcohol. Also, when using the pick-and-place machine for dispensing adhesive, a minimum amount of pressure should be used to bring the component leads or terminations down onto the pads. Once the components have been placed on the adhesive, lateral movement should be avoided. These precautions are necessary to ensure that no adhesive gets onto the pads. Dispensing can also be accomplished manually for prototyping applications using a semi-automatic dispenser (Figure 8.12). Controlling dot size uniformity is difficult with semi-automated dispensers.

8.6   CURING OF ADHESIVES

Once adhesive has been applied, components are placed. Now the adhesive must be cured to hold the part through the soldering process. There are two commonly used methods of cure: thermal cure and a combination of UV light and thermal cure. We discuss these curing processes in turn.

8.6.1   Thermal Cure

Most epoxy adhesives are designed for thermal cure, which is the most prevalent method of cure. Thermal cure can be accomplished simply in a convection oven or an infrared (IR) oven, without added investment in a UV system. The IR or convection ovens can also be used for reflow soldering. This is one of the reasons for the popularity of thermal cure, especially in IR ovens. (See Chapter 12 for details on the heat transfer mechanism in IR and convection-dominant ovens.) The single-part epoxy adhesives require a relatively longer cure time and higher temperatures.

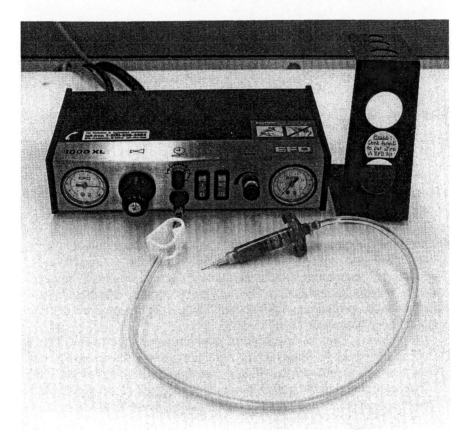

Figure 8.12 A semi-automatic dispenser for adhesive application.

When using higher temperatures, care should be taken that boards do not warp and are properly held.

8.6.1.1   Thermal Cure Profile and Bond Strength

As shown in Figure 8.13, the adhesive cure profile depends on the equipment. Batch convection ovens require a longer time, but the temperature is lower. In-line convection (and IR) ovens provide the same results in less time, since the curing is done at higher temperatures in multiple heating zones.

Different adhesives give different cure strengths for the same cure profile (time and temperature of cure), as shown in Figure 8.14, where strength is depicted as the force required to shear off a chip capacitor, cured in a batch convection oven, from the board at room temperature. For

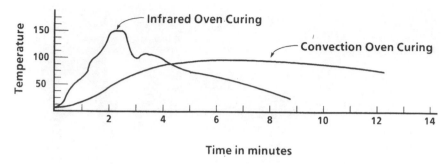

Figure 8.13 **Adhesive cure profiles in in-line IR and batch convection oven.**

each adhesive, the cure time is fixed at 15 minutes. The cure temperature in a batch convection oven was varied, and strength was measured with a Chatillon pull test gauge. Graphs similar to Figure 8.14 should be developed for evaluating the cure profile and corresponding bond strength of an adhesive for production applications. The curing can be done in either an in-line convection or an IR oven.

In adhesive cure, temperature is more important than time. This is shown in Figure 8.15. At any given cure temperature, the shear strength shows a minor increase as the time of cure is increased. However, when the cure temperature is increased, the shear strength increases significantly at the same cure time. For all practical purposes, the recommended mini-

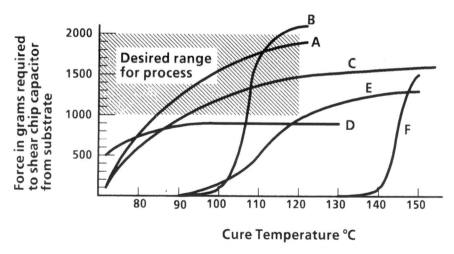

Figure 8.14 **Cure strength for adhesives A through F at different temperatures for 15 minutes in a batch convection oven.**

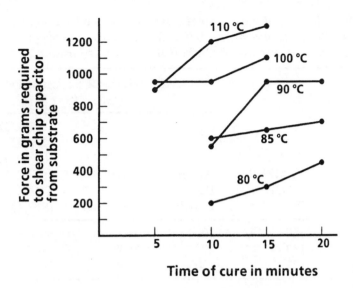

Figure 8.15 **Cure strength of adhesive C, an epoxy adhesive, at different times and temperatures.**

mum and maximum shear strengths for cured adhesive are 1000 and 2000 grams, respectively. However, higher bond strengths (up to 4000 grams) have been found to not cause rework problems, since adhesive softens at rework temperatures.

This point about cure temperature being more important than time is also true for in-line convection or infrared curing, as shown in Table 8.1. As the peak temperature in the IR oven is raised by raising panel temperatures, the average shear strength increases drastically. Table 8.1 also shows that additional curing takes place during soldering. Thus an adhesive that is partially cured during its cure cycle will be fully cured during wave soldering.

Most of the adhesive gets its final cure during the preheat phase of wave soldering. Hence it is not absolutely essential to accomplish the full cure during the curing cycle. Adequate cure is necessary to hold the component during wave soldering, however, and if one waits for the needed cure until the actual soldering, it may be too late. Chips could fall off in the solder wave.

It should be kept in mind that the surface of the substrate plays an important role in determining the bond strength of a cured adhesive. This is to be expected because bonding is a surface phenomenon. For example, a glass epoxy substrate surface has more bonding sites in the polymer structure than a surface covered by a solder mask.

Table 8.1 Impact of cure temperature on the bond strength of epoxy adhesive. The table also shows that additional curing takes place during wave soldering. (Courtesy of Chris Ruff, Intel Corporation.)

BELT SPEED (FT/MIN)[a]	PEAK TEMPER- ATURE (°C)	MEANSHEAR STRENGTH (GRAMS)		APPROXI- MATE REWORK (REMOVAL) TIME (SECONDS)
		AFTER IR CURE	AFTER WAVE SOL- DERING	
4.0	150	3000	3900	4–6
5.0	137	2000	3900	4–5
6.0	127	1000	3700	3–5

As illustrated in Figure 8.16, different solder masks give different bond strengths with the same adhesive using identical curing profiles. Hence an adhesive should be evaluated on the substrate surface that is to be used. If a new kind of solder mask is used, or if the solder mask is changed, an adhesive that was acceptable before may have only marginal shear strength under the new conditions and may cause the loss of many chip components in the wave.

8.6.1.2 Adhesive Cure Profile and Flux Entrapment

One additional requirement for adhesive thermal cure is also very important. The cure profile should be such that voids are not formed in the adhesive—this is an unacceptable condition. The voids will entrap flux, which is almost impossible to remove during cleaning. Such a condition is a serious cause for concern especially if the flux is aggressive in nature. The circuit may corrode and cause failure if the flux is not completely removed. Most often the cause of voids is fast belt speed to meet production throughput requirements. A rapid ramp rate during cure has also been found to cause voiding in the adhesive.

Voiding may not be caused by rapid ramp rate alone, however. Some adhesives are more susceptible to voiding characteristics than others. For example, entrapped air in the adhesive may cause voiding during cure. Voids in adhesive may also be caused by moisture absorption in the bare circuit boards during storage. Similarly, susceptibility to moisture

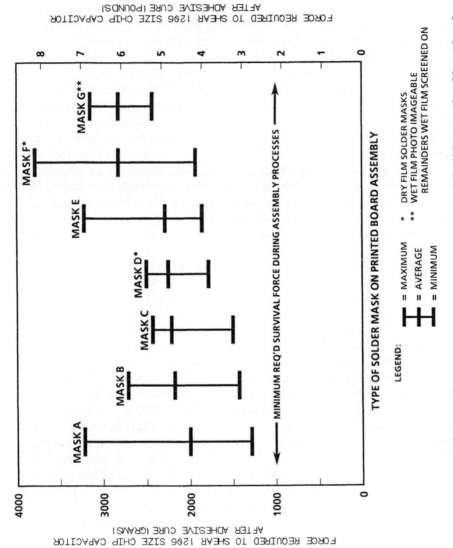

Figure 8.16 Bond strength of an epoxy adhesive with different types of solder masks. Note that letters A–G do not represent the same materials as in Figure 8.14 and 8.15.

absorption increases if a board is not covered with solder mask. During adhesive cure, the evolution of water vapor may cause voids. Whatever the cause, voids in adhesive are generally formed during the cure cycle.

Baking boards before adhesive application and using an adhesive that has been centrifuged to remove any air after packing of the adhesive in the syringe, certainly helps to prevent the formation of voids, as does the use of solder mask. Nevertheless, the most important way to prevent flux entrapment due to formation of voids in the adhesive during cure is to characterize the adhesive and the cure profile. This must be done before an adhesive with a given cure profile is used on products. How should an adhesive be characterized? We discussed earlier the precure, cure, and postcure properties of an ideal adhesive. Characterization of the adhesive cure profile to avoid void formation should also be added to the list.

There are two important elements in the cure profile for an adhesive, namely initial ramp rate (the rate at which temperature is raised) and peak temperature. The ramp rate determines its susceptibility to voiding, whereas the peak temperature determines the percentage cure and the bond strength after cure. Both are important, but controlling the ramp rate during cure is more critical.

Figure 8.17 shows the recommended cure profile for an epoxy adhesive in a 4-zone IR oven (Figure 8.17a), in a 10-zone IR oven (Figure 8.17b) and in a 10-zone convection oven (Figure 8.17c). A zone is defined as having both top and bottom heaters, thus a 10-zone oven has 10 top heaters and 10 bottom heaters. The cure profile will vary from oven to oven, but the profiles shown in Figure 8.17 can be used as a general guideline.

As shown in the figure, the average ramp rate for the in-line IR or in-line convection oven for epoxy adhesive cure is about 0.5°C/second. (This figure is obtained by dividing the rise in temperature by the time during the heating cycle: temperature of 160°C –30°C = 130°C divided by 300 seconds in the 4-zone IR oven (Figure 8.17a); temperature of 160°C–30°C = 130°C divided by 270 seconds in the 10-zone IR oven (Figure 8.17b); temperature of 130°C –30°C = 100°C divided by 200 seconds in the 10-zone convection oven (Figure 8.17c). Looking closely at Figure 8.17c, the initial ramp rate in the convection oven in the beginning is higher (temperature of 120°C –30°C = 90°C divided by 120 = 0.75°C/second), but the rate is only 0.5°C if we take into account the peak temperature.

In addition to the ramp rate, it is important to note that there is no significant difference in the total cure time between the 4-zone (Figure 8.17a) and 10-zone (Figures 8.17b and 8.17c) ovens. It is about six minutes for each oven. So it is important to keep in mind that switching to ovens with more zones should be done for reasons other than increasing

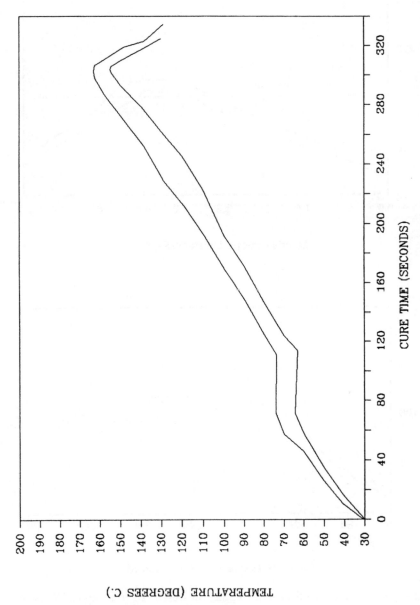

Figure 8.17   (a) Cure profile for an epoxy adhesive in a 4-zone IR oven (courtesy **Chris Ruff, Intel Corporation**)

365

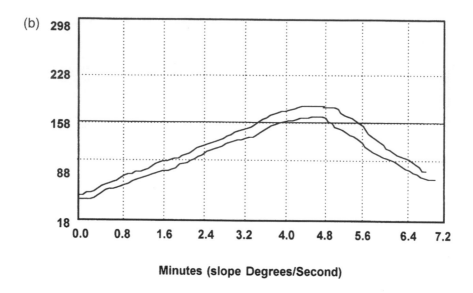

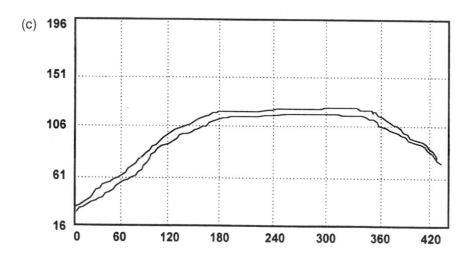

Figure 8.17 Cure profile for an epoxy adhesive in a 10-zone IR oven (b) and in a 10-zone convection oven (c). (Courtesy Dudi Amir, Intel Corporation).

throughput. A higher number of zones does make it easier to develop the thermal profile, however.

For comparison purposes, the ramp rate in a typical batch convection oven is only 0.1°C/second for the peak cure temperature of 100°C for 15 minutes. Such a low ramp rate in a batch oven may be ideal for preventing void formation, but it is not acceptable for production.

For the ovens discussed above, the ramp rate of 0.5°C/second translates into a belt speed of about 30 inches/minute. All the profiles shown in Figure 8.17 were developed at 30 inches/minute. Increasing the belt speed will increase the ramp rate but it will also increase the potential for void formation. For example, we found that a belt speed of 42 inches/minute pushed the average ramp rate to 0.8°C/second, but it caused voids in the adhesive. Increasing the ramp rate above 0.5°C/second may be possible in some ovens, but the adhesive must be conclusively characterized to show that higher ramp rates do not cause voids.

Cleanliness tests, including the surface insulation resistance (SIR) test, should be an integral part of the adhesive characterization process. Cleanliness tests other than SIR tests are necessary because SIR tests are generally valid only for water-soluble fluxes and may not flag rosin flux entrapment problems. Another way to determine the voiding characteristics of an adhesive for a given cure profile is to look for voids visually during the initial cure profile development and adhesive qualification phases.

Figure 8.18 offers accept/reject criteria for voids after cure; again, these are only guidelines. The SIR or other applicable cleanliness tests should be the determining factors for the accept/reject criteria applied to voids, ramp rate, and belt speed. In addition, data on acceptable bond strength to prevent loss of chips in the wave without compromising reworkability must be collected for the final profile. The bond strength requirement was discussed in the preceding section. It should be pointed out that the cure profile and the voiding characteristics of thermal and UV adhesive will differ, since the UV adhesive is intended for a faster cure.

8.6.2   UV/Thermal Cure

The UV/thermal cure system uses, as the name implies, both UV light and heat. The very fast cure that is provided may be ideal for the high throughput required in an in-line manufacturing situation. The adhesives used for this system (i.e., acrylics) require both UV light and heat for full cure and have two cure "peaks," as discussed later in connection with differential scanning calorimetry (DSC).

The UV/thermal adhesive must extend past the components to allow initiation of polymerization by the UV light, which is essentially used to

 Gross Voiding: (40% to 70%)

(unacceptable)

 Moderate Voiding: (25% to 40%)

(unacceptable)

 Gross Porosity: (5% to 20%)

(unacceptable)

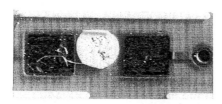 Moderate Porosity: (2% to 5%)

(minimum)

(acceptable)

 Minor Porosity: (0% to 2%)

(acceptable)

Figure 8.18 Accept/reject criteria for voids in adhesive (to prevent flux entrapment during wave soldering). (Courtesy of Chris Ruff, Intel Corporation.)

"tack" components in place and to partially cure the adhesive. Final cure is accomplished by heat energy from IR or convection or a combination of both.

It is not absolutely necessary to cure the UV/thermal adhesives by both UV light and heat. If a higher temperature is used, the UV cure step can be skipped. A higher cure temperature may also be necessary when the adhesive cannot extend past the component body (as required for UV cure; see Figure 8.10, discussed earlier) because of lead hindrance of components such as SOTs or SOICs.

For UV/thermal systems, it is important to have the right wattage, intensity, and ventilation. A lamp power of 200 watts/inch at a 4 inch distance using a 2 kW lamp generally requires about 15 seconds of UV cure. Depending on the maximum temperature in an IR or convection oven, a full cure can be accomplished in less than 2 minutes.

8.7   EVALUATION OF ADHESIVES WITH DIFFERENTIAL SCANNING CALORIMETRY

Adhesives are received in batches from the manufacturer, and batch-to-batch variations in composition are to be expected even though the chemical ingredients have not been changed. Some of these differences, however minute they may be, may affect the cure properties of an adhesive. For example, one particular batch may not cure to a strength that is adequate to withstand forces during wave soldering after it has been exposed to the standard cure profile. This can have a damaging impact on product yield.

Adhesives can be characterized by the supplier or by the user, as mutually agreed, to monitor the quality of adhesive from batch to batch. This is done by determining whether the adhesive can be fully cured when subjected to the curing profile. The equipment can be programmed to simulate any curing profile. It also can measure the glass transition temperature of the adhesive after cure, to determine whether the rework-ability properties have changed.

In the sections that follow, we discuss the results of adhesive characterization of epoxy and acrylic adhesives based on evaluations done at Intel [6]. The thermal events of interest for surface mount adhesives are the glass transition temperature and the curing peak.

8.7.1   Basic Properties of DSC Analysis

Differential scanning calorimetry is a thermal analysis technique used for material characterization, such as ascertaining the curing properties

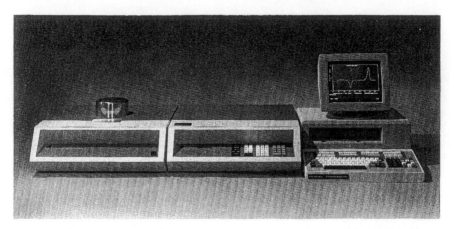

Figure 8.19 A differential scanning calorimeter. (Photograph courtesy of Perkin-Elmer.)

of adhesives. Typical DSC equipment is shown in Figure 8.19. The output from a DSC analysis is either an isothermal DSC curve (heat flow versus time at a fixed temperature) or an isochronal DSC curve (heat flow versus temperature at a fixed heating rate).

Basically, in the DSC method of analysis, the heat generated by the sample material (e.g., an adhesive) is compared to the heat generated in a reference material as the temperature of both materials is raised at a predetermined rate.

The DSC controller aims to maintain the same temperature for both the sample and the reference material. If heat is generated by the sample material at a particular temperature, the DSC controller will reduce the heat input to the sample in comparison to the heat input to the reference material, and vice versa. This difference of heat input is plotted as the heat flow to or from the sample (Y axis) as a function of temperature (X axis).

The reference material should undergo no change in either its physical or chemical properties in the temperature range of study. If the sample material does not evolve heat to or absorb heat from the ambient, the plot will be a straight line. If there is some heat-evolving (exothermic) or heat-absorbing (endothermic) event, the DSC plot of heat flow versus temperature will exhibit a discontinuity.

As seen in Figure 8.20, the glass transition temperature is represented as a change in the value of the baseline heat flow. This occurs because the heat capacity (the quantity of heat needed to raise the temperature of the adhesive by 1°C) of the adhesive below the glass transition temperature is different from the heat capacity above T_g.

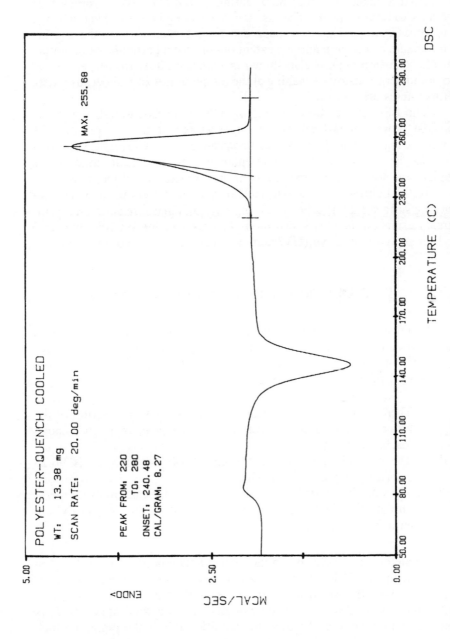

Figure 8.20   A typical DSC curve illustrating the various thermal events occurring in the sample under investigation.

As also seen in Figure 8.20, curing of the adhesive is represented by an exothermic peak. That is, the adhesive gives off heat when it undergoes curing. This cure peak can be analyzed to give the starting temperature of the cure and the extent of cure for a particular temperature profile. A fusion peak is also shown in Figure 8.20. However, adhesives do not undergo fusion or melting at the temperatures of use because these temperatures are too low.

The DSC curve shown in Figure 8.20 is an isochronal curve; that is, heat flow is measured as the temperatures of the sample and the reference material are increased at a constant rate. Isothermal DSC curves can also be generated at any particular temperature. Isothermal curves depict heat flow versus time at a constant predetermined temperature.

Both isochronal and isothermal DSC curves are very useful in characterizing surface mount adhesive curing rates and glass transition temperatures. Results on the characterization of an epoxy adhesive (thermal cure) and an analytic adhesive (UV/thermal cure) are presented next.

8.7.2   DSC Characterization of an Epoxy Adhesive

Figure 8.21 shows an isochronal curve for an uncured epoxy adhesive (adhesive A, Figure 8.14) subjected to a heating profile in the DSC furnace from 25°C to 270°C at a heating rate of 75°C/minute. The results from Figure 8.21 indicate the following.

1. There is an exothermic peak corresponding to the curing of the adhesive. The onset temperature of this peak is 122°C, and it reaches a minimum at 154°C.
2. The heat liberated during the curing of the adhesive is 96.25 calories per gram of adhesive.
3. The adhesive starts curing before it reaches the maximum temperature of 155°C (the peak temperature for most epoxy adhesives in an infrared oven).

When the same adhesive in the uncured state is cured in an IR oven and analyzed with DSC, the results are different, as shown in Figure 8.22, which compares the isochronal DSC curves of the adhesive in the uncured and cured states. Before cure, the adhesive is in a fluid state. Its T_g is below room temperature because by definition this is the point at which the adhesive transforms from a rigid state to a glassy or fluid state. After cure, the T_g of the adhesive will increase because of cross-linking in the carbon chains as discussed earlier (see Figure 8.3).

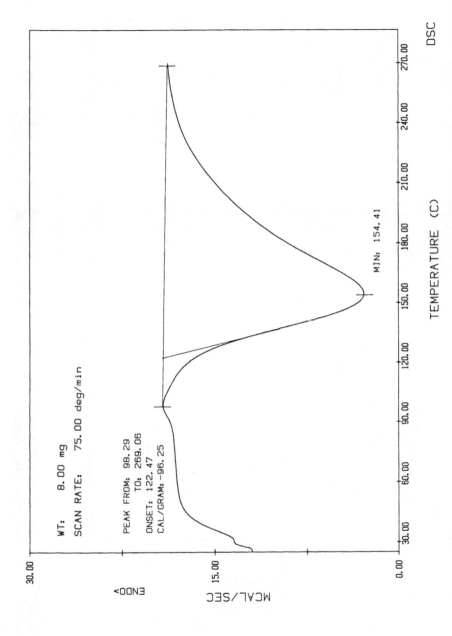

Figure 8.21 The isochronal DSC curve for an uncured epoxy adhesive. The curing exothermic peak is clearly visible.

373

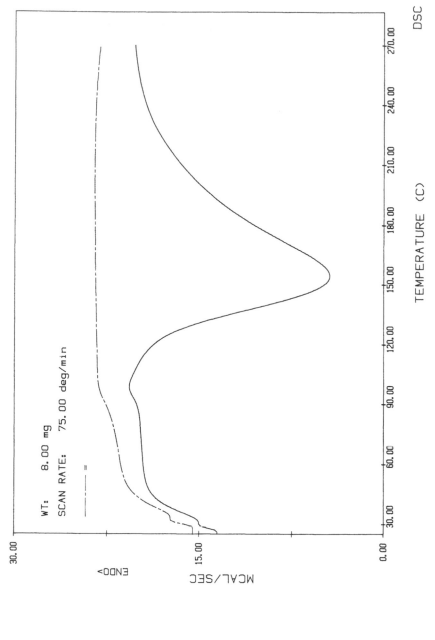

Figure 8.22 A comparison of the isochronal DSC curves for the epoxy adhesive in the cured (broken line) and uncured (solid line) states. The exothermic curing peak is absent for the cured adhesive.

An increase in heat flow occurs during the glass transition of an adhesive. Since the magnitude of this change is quite small, the Y axis in the DSC curves shown in Figure 8.22 must be expanded, as shown in Figure 8.23, to reveal the T_g effect. Figure 8.23 is a portion of the DSC curve from Figure 8.22 (broken curve) replotted with the Y axis scaled up. From Figure 8.23, it is apparent that the onset of the T_g occurs at 73°C and the midpoint of the T_g occurs at 80°C. It is appropriate to characterize T_g from the DSC curves as the temperature value at the midpoint of the transition range. (The onset value depends on the baseline slope construction, whereas the midpoint value is based on the inflection point on the DSC curve and is therefore independent of any geometric construction.)

Since the T_g in the cured state is quite low, it will be very easy to rework the adhesive after IR cure. After IR cure, however, the adhesive and the SMT assembly must be subjected to another heat treatment, namely, wave soldering. This further heating will increase the T_g of the adhesive, as evident from Figure 8.24, which shows a higher T_g (85°C).

The DSC curve in Figure 8.24 is for the same epoxy adhesive characterized in Figure 8.23 but after wave soldering. The 5°C increase in the T_g after wave soldering implies that additional physical curing of the adhesive occurs during the wave soldering step. A small increase in the T_g after wave soldering is not a real problem because the adhesive still can be reworked. The requirement for SMT adhesives is that after all soldering steps the T_g be below the melting point of the solder (i.e., 183°C).

Figure 8.24 also shows that the adhesive starts decomposing at about 260°C, as evidenced by the small wiggles at that temperature. The DSC curve shown in Figure 8.24 was run with the encapsulated sample placed in a nitrogen atmosphere. In other atmospheres, the decomposition temperature might be different.

8.7.3 DSC Characterization of an Acrylic Adhesive

As mentioned earlier, acrylic adhesives require UV light and heat, but they can be cured by heat alone if the temperature is high enough. However, as discussed in Section 8.3.2, the catalysts for UV curing are the peroxides (photoinitiators). Figure 8.25 shows the isochronal DSC curve of an acrylic adhesive (adhesive G) that had been cured in DSC equipment with a simulated IR oven curing profile. Also shown in Figure

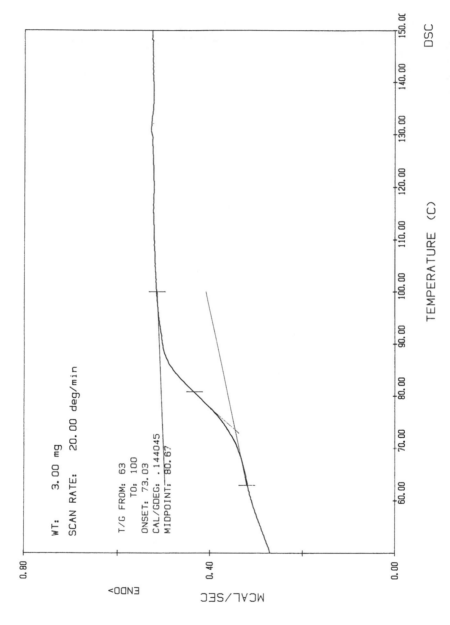

Figure 8.23 Part of the isochronal DSC curve showing the temperature range at which the T_g occurs for a epoxy adhesive first cured in an infrared oven. The T_g onset is at 73°C, and the midpoint is at 80°C.

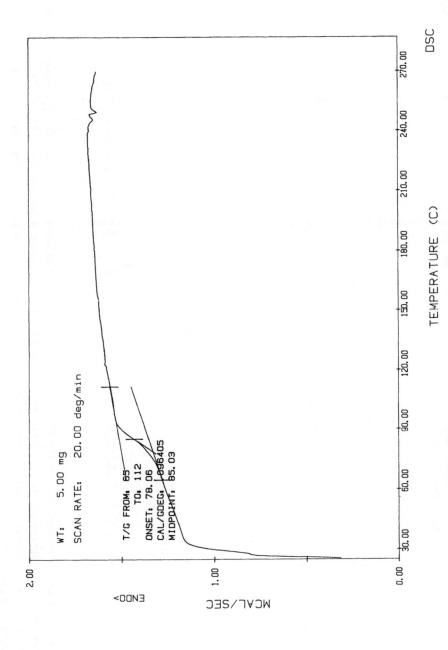

Figure 8.24 Part of the isochronal DSC curve showing the temperature range at which the T_g occurs for cured epoxy adhesive in an IR oven. The T_g onset is at 78°C, and the midpoint is at 85°C. The adhesive decomposes at 260°C.

377

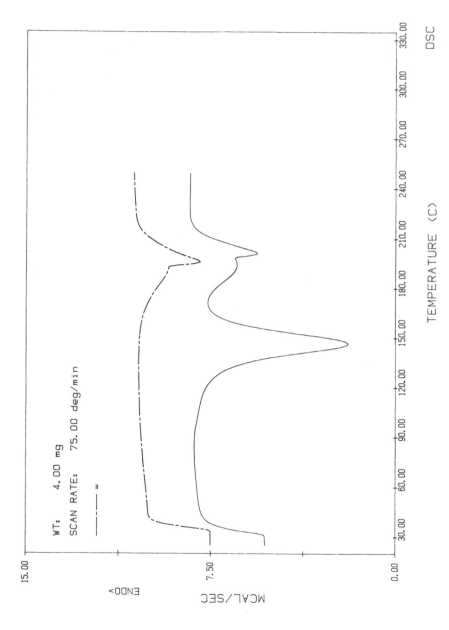

Figure 8.25 A comparison of the DSC isochronal curves for adhesive G, an acrylic (UV cure) adhesive. The broken curve represents already thermally cured adhesive at a peak temperature of 155°C. The presence of the peak in the broken curve means that the adhesive is only partially cured. The solid curve represents the uncured state of adhesive G.

8.25 for comparison is the isochronal DSC curve for the same adhesive in an uncured state. From Figure 8.25 we can draw two conclusions.

1. The uncured adhesive shows two cure peaks, one at 150°C and another at 190°C (solid line in Figure 8.25). The lower temperature peak is caused by the curing reaction induced by the photoinitiator catalyst in the UV adhesive. In other words, the photoinitiators added to the UV adhesive start the curing reaction at ambient temperature. They reduce the curing cycle time in the UV adhesives by "tacking" components in place. For complete cure, thermal energy is required.

2. DSC analysis offers two ways to distinguish a UV adhesive from a thermal adhesive. An uncured UV adhesive will show two curing peaks (solid line in Figure 8.25). If however, it is first thermally cured and then analyzed by DSC, it will show only one peak (broken line in Figure 8.25). This characteristic is similar to the curing characteristics of an uncured thermal adhesive as discussed earlier. In other words, a low temperature thermally cured UV adhesive will look like an uncured thermal adhesive.

Can an analytic UV adhesive be fully cured without UV? Yes, but as shown in Figure 8.26, the curing temperature must be significantly higher than that used for other adhesive cures. Such a high temperature may damage the temperature-sensitive through-hole components used in mixed assemblies.

Figure 8.26 compares the isochronal DSC curves for an adhesive in the uncured state and in the fully cured state. The adhesive was fully cured by heating it to 240°C, which is much higher than the 155°C maximum curing temperature in the IR oven. Figure 8.26 shows that the adhesive sample cured at 240°C does not exhibit any peak (broken line). This indicates that to achieve full cure of this adhesive, the material must be heated above the peak temperature of about 190°C. Again, as in Figure 8.25, the solid line in Figure 8.26 is for the same adhesive in an uncured state.

When UV acrylic adhesives are subjected to heat only, they do not fully cure. The partial cure can meet the bond strength requirement for wave soldering, however. Does this mean that UV adhesives can be thermally cured and one does not need to worry about the undercure as long as the bond strength requirements are met? Not necessarily. A partial cured adhesive is more susceptible to absorption of deleterious chemicals during soldering and cleaning than a fully cured adhesive, and unless the partially cured adhesive meets all other requirements, such as insulation resistance, it should not be used.

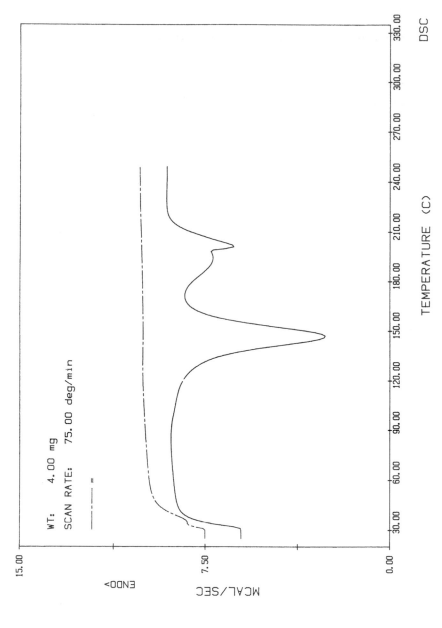

Figure 8.26 A comparison of the DSC isochronal curves for a UV-cured acrylic (adhesive G). The broken line curve represents the already thermally cured adhesive at a peak temperature of 240°C. The absence of peak in the broken curve means that the adhesive is fully cured. The solid curve represents the uncured state of adhesive G.

8.8   SUMMARY

Adhesive plays a critical role in the soldering of mixed surface mount assemblies. Among the considerations that should be taken into account in the selection of an adhesive are desired precure, cure, and postcure properties. Dispensing the right amount of adhesive is important, as well. Too little may cause loss of devices in the wave, and too much may be too hard to rework or may spread on the pad, resulting in solder defects. Adhesive may be dispensed in various ways, but syringing is the most widely used method.

Epoxies and acrylics are the most widely used types of adhesive. Adhesives for both thermal and UV/thermal cures are available, but the former are more prevalent. The cure profile selected for an adhesive should be at the lowest temperature and shortest time that will produce the required bond strength. Bond strength depends on the cure profile and the substrate surface, but the impact of temperature on bond strength is predominant.

Bond strength should not be the sole selection criterion, however. Consideration of voiding characteristics may be even more significant to ensure that flux entrapment and cleaning problems after wave soldering are not encountered. One way to prevent voiding is to control the ramp rate during the cure cycle of the adhesive and to fully characterize the adhesive cure profile before the adhesive is used on products. An integral part of that characterization process should be to visually look for gross voids and to confirm their impact by conducting cleanliness tests such as the test for surface insulation resistance. In addition, the cure profile should not affect the temperature-sensitive through-hole components used in mixed assemblies.

Minor variations in an adhesive can change its curing characteristics. Thus, it is important that a consistent quality be maintained from lot to lot. One of the simplest ways to monitor the curing characteristics of an adhesive is to analyze samples using differential scanning calorimetry (DSC). This can be performed either by the user or the supplier, per mutual agreement. DSC also serves other purposes. For example, it can be used to distinguish a thermal cure adhesive from a UV/thermal cure adhesive and a fully cured adhesive from a partially cured or uncured adhesive. DSC is used for characterizing other materials needed in the electronics industry, as well.

REFERENCES

1. Specifications on Adhesives, available from IPC, Northbrook, IL
   - IPC-SM-817. General Requirements for Dielectric Surface Mount Adhesives.
   - IPC-3406. Guidelines for Conductive Adhesives.
   - IPC-3407. General Requirements for Isotropically Conductive Adhesives.
   - IPC-3408. General Requirements for Anisotropically Conductive Adhesives.
2. Pound, Ronald. "Conductive epoxy is tested for SMT solder replacement." *Electronic Packaging and Production,* February 1985, pp. 86–90.
3. Zarrow, Phil, and Kopp, Debra. "Conductive Adhesives." *Circuit Assembly,* April 1996, pp. 22–25.
4. Meeks, S. "Application of surface mount adhesive to hot air leveled solder (HASL) circuit board evaluation of the bottom side adhesive dispense." Paper IPC-TR-664, presented at the IPC fall meeting, Chicago, 1987.
5. Kropp, Philip, and Eales, S. Kyle. "Trouble shooting guide for surface mount adhesive." *Surface Mount Technology,* August 1988, pp. 50–51.
6. Aspandiar, R., Intel Corporation, internal report, February, 1987.

Chapter 9

Solder Paste and Its Application

9.0 INTRODUCTION

In the reflow soldering of surface mount assemblies, solder paste is used for the connection between the leads or terminations of surface mount components and the lands. Solder paste is applied to the surface mount lands by screening, stenciling, or dispensing. Each process has its pros and cons.

Although the imperfections found after soldering are generally termed "soldering defects," this can be misleading, since the soldering process alone does not control product quality. Rather, the quality of a finished surface mount assembly is determined by many process and design factors. Screen printing, for example, is one of the processes that play a very important role in the quality of the final assembly.

In this chapter we cover the three major elements of the paste printing process—namely, solder paste quality, equipment variables, and printing process variables—and discuss their impact on print quality.

We begin with the basics of solder paste and the important considerations in its selection. We then proceed to the selection of solder paste printing equipment and conclude the chapter with the pros and cons of various solder paste application methods.

9.1 SOLDER PASTE PROPERTIES

Solder paste is essentially comprised of metal powder particles in a thickened flux vehicle, as shown in Figure 9.1. The actual process steps involved in making the metal powder and the paste formulations are mostly proprietary, but in general, molten solder of desired composition is rapidly chilled on a rotating wheel to form fine solder particles. The process takes place in an inert environment to minimize the oxidation of the particles.

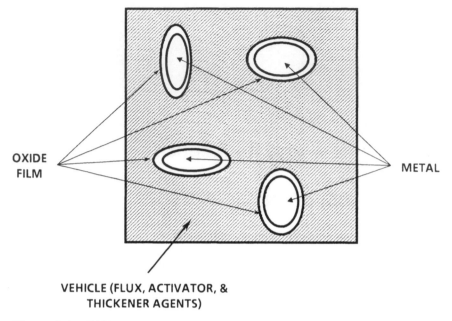

OXIDE FILM

METAL

VEHICLE (FLUX, ACTIVATOR, & THICKENER AGENTS)

Figure 9.1 Different constituents of solder paste.

Even though the solder bars used in wave soldering and the solder pastes used in surface mount have essentially the same chemical compositions, solder pastes cost about 20 times more. The reason is that the solder paste has to meet much more stringent quality requirements than solder bars for wave soldering. Also, there is added scrap in making solder powder of the required shape and size, and there is extra cost in adding solvents and flux to the paste.

Solder paste serves multiple critical purposes in the soldering of Type I (full surface mount) and Type II (mixed) surface mount assemblies. Since it also contains the flux necessary for effecting soldering, there is no need to add flux separately and to worry about controlling flux activity and density as in through-hole mount assemblies. The flux also acts as a temporary adhesive for holding surface mount components in place during placement and handling before reflow soldering. Clearly proper selection of solder paste is very important for producing defect-free, reliable surface mount assemblies.

The properties of solder paste that should be carefully controlled and evaluated before selecting a paste for surface mount assembly are discussed in the sections that follow.

There are various ways to evaluate solder paste. Vendor data and some industry specifications, such as Joint Industry Standard J-STD-005,

Requirements for Soldering Pastes [1], along with the recommendations discussed below, can be used to establish an in-house solder paste specification for vendor qualification.

Certain simple tests should be conducted by the user regularly to monitor the quality of solder paste. These tests, such as the solder ball test and viscosity measurement, can prevent the building of a large number of assemblies that may have to be reworked, recleaned, or even scrapped.

9.1.1   Metal Composition

Table 9.1 [2] shows some of the commonly used solder compositions that are available for various applications in the electronics industry. The critical factors that determine the composition of the solder in the solder paste are as follows.

1. Materials of the substrate and the components of the surface mount assemblies: high temperature solders (>225°C) can be used to attach ceramic packages to ceramic substrates. However, for the attachment of plastic surface mount packages to FR-4 (epoxy glass) substrates, lower melting solders are used to avoid degradation of the substrate and package materials during soldering. For soldering plastic surface mount components to FR-4 substrates, eutectic tin-lead (63% Sn-37% Pb, mp = 183°C) or tin-lead-silver (62% Sn-36% pb-2% Ag, mp = 179°C) solders are widely used. However, other solder alloys are available in paste form [3].

2. Compatibility of the solder with the metallizations on the substrate and the component leads: indium-lead alloy solders are recommended for soldering to gold terminations, since the popular tin-lead alloy solders form deleterious intermetallic compounds with gold. However, indium solders are rarely used. In order to form a brittle tin-gold intermetallic, there needs to be a sufficient amount of gold in the solder joint (about 4%). There is rarely enough gold either on the component lead or board surface to cause gold embrittlement.

3. Strength of solder: if high tensile and shear strength at elevated temperatures is a requirement, tin-antimony alloy solders are recommended.

4. Cost: silver is usually added to tin-lead solders to diminish the dissolution of the silver from component terminations (e.g., capacitors). A tin-lead solder containing 2% silver is recommended over a tin-solder containing 4% silver because the former is cheaper and

Table 9.1 Melting point of different compositions of solder for the electronics industry [2].

COMPOSITION	RANGE[a]	DIFFERENCE	PROPERTIES AND USES
75 Pb/25 In	250S-264L	14	Less gold
50 Pb/50 In	180S-209L	29	leaching, more
25 Pb/75 In	156S-165L	9	ductile than Sn/ Pb alloys; die attachment, closures, and general circuit assembly
37.5 Sn/37.5 Pb/ 25 In	134S-181L	47	Good wettability; not recommended for gold
80 Au/20 Sn	280E	0	Highest quality for gold surfaces; die attachment and closures
63 Sn/37 Pb[b]	183E	0	Widely used tin-
60 Sn/40 Pb[b]	183S-188L	5	lead solders for
50 Sn/50 Pb	183S-216L	33	surface
10 Sn/90 Pb	268S-302L	34	mounting and
5 Sn/95 Pb	308S-312L	4	general circuit assembly; low cost and good bonding properties; not recommended for silver and gold soldering because of high leaching rate

Table 9.1 *(Continued)*

COMPOSITION	RANGE[a]	DIFFERENCE	PROPERTIES AND USES
62 Sn/36 Pb/2 Ag[b]	179E	0	Tin-lead solders
10 Sn/88 Pb/2 Ag	268S-290L	22	containing small
1 Sn/97.5 Pb/1.5 Ag	309E	0	amounts of silver to minimize leaching of silver conductors and leads; not recommended for gold; Sn/Pb/Ag (62/36/2) is strongest tin lead solder
96.5 Sn/3.5 Ag	221E	0	Widely used tin-silver solders
95 Sn/5 Ag	221S-240L	19	providing very strong, lead-free joints; minimizes silver leaching; not recommended for gold
42 Sn/58 Bi	138E	0	Low temperature eutectic with high strength

[a]S = solidus, L= liquidus, E = eutectic.
[b]Most commonly used soldering alloys.

has an equivalent effect in diminishing silver dissolution from component terminations.

5. Environmental concerns regarding lead alloys are causing the industry to begin looking for lead-free solders. There are various candidates being tried, but the Sn-Pb solder remains the most widely used composition for the electronics industry. See Chapter 10 for a discussion on lead-free solders.

Table 9.2   For a given solder paste thickness, impact of metal content on reflowed solder thickness [4].

METAL CONTENT (%)	WET SOLDER PASTE (INCH)	REFLOWED SOLDER (INCH)
90	0.009	0.0045
85	0.009	0.0035
80	0.009	0.0025
75	0.009	0.0020

9.1.2   Metal Content

The metal content in solder paste determines the solder fillet size. Fillet size increases with an increase in the percentage of metal, but the tendency for solder bridging also increases with increase in metal content at a given viscosity. A higher metal content will result in higher thickness of the reflowed solder, as shown in Table 9.2 [4].

As Table 9.2 indicates, the final thickness of reflowed solder can vary from 50% of the paste thickness for 90% metal content to as low as 25% of the paste thickness for 75% metal content. Thus only a minor variation in the metal content of the paste from lot to lot can have a significant impact on the quality of solder joints. For example, a 10% variation in metal content can change an excess solder joint to an insufficient solder joint for the same paste thickness.

Typically, solder pastes for surface mount assembly contain 88 to 90% metal by weight (about 50% by volume). The metal percent should be within +/−1% of the nominal value specified on the user's purchase order [1].

9.1.3   Particle Size and Shape

Powder particle shape determines the oxide content of the powder as well as the paste's printability. Spherical powders are preferred over elliptical powders. The lower the surface area, the lower the oxidation. Hence the finer and irregularly shaped powders, with their larger surface area, have a higher metal oxide content than spherical powders.

The reason for this is as follows: as particle size becomes smaller, the ratio of surface area to volume increases.

$$\text{ratio} = \text{surface area/volume}$$
$$= \Pi R^2 / \tfrac{4}{3} \Pi R^3$$
$$= 3/4R$$
$$= 1.5/D$$

where D is diameter of solder particle. With smaller D, the ratio increases and the paste is more susceptible to oxidation and solder balling.

Solder pastes containing powders of irregular shape tend to clog screens and stencils. Figure 9.2 shows acceptable and unacceptable solder particles for use in solder paste. It should be noted that the pastes do not have to be as perfectly spherical as shown in Figure 9.2a. Some variation in particle sizes and shapes is only to be expected.

Printing problems are encountered with solder pastes containing powder particles that are large in diameter, which readily clog screen and stencils. On the other hand, a solder paste with powder particles that are too small is prone to form solder balls during reflow.

As shown in Table 9.3, the solder powder or particle size is classified in six different categories by the Joint Industry Standard J-STD 005 [1]. Solder powder is also given mesh designations. The mesh designation is based on ASTM B-214 (Test Method for Sieve Analysis of Granular Metal Powders). Many in the industry refer to powder size with mesh designation rather than in microns. Table 9.4 shows powder size designations both in mesh size and in microns for comparison. It also shows surface area ratios for various particle sizes [5]. As the size is decreased, the surface area increases and makes the powder more susceptible to oxidation.

The commonly used powder size for standard 50 mil pitch surface

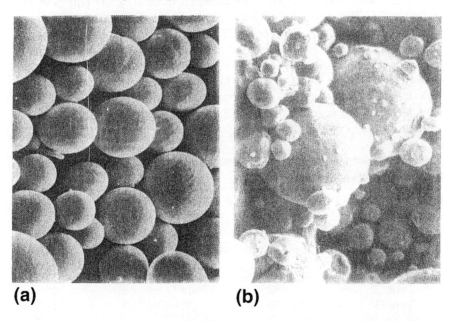

(a) **(b)**

Figure 9.2 **(a) Perfectly spherical and (b) unacceptable solder particles in solder paste.**

Table 9.3   Metal powder size classifications in microns (μm) [1].

TYPE	NONE LARGER THAN	LESS THAN 1% LARGER THAN	80% MINIMUM BETWEEN	10% MAXIMUM LESS THAN
1	160	150	150–75	20
2	80	75	75–45	20
3	50	45	45–25	20
4	40	38	38–20	20
5	30	25	25–15	15
6	20	15	15–5	5

mount components is −200/+325 mesh (Type 2 or 75–45 microns); that is, at least 99% by weight of the powder particles will pass through a 200 (holes/square inch) mesh and less than 20% of the powder particles by weight will pass through a 325 mesh. However, if fine pitch (25 mil pitch or lower) components are also on the board, Type 3 paste (25–45 microns or −325/+500) must be used. This is required even if there is only one fine pitch component on the board. Type 4 and Type 5 solder pastes are used for ultra fine pitch components (0.4 mm pitch and below).

For finer pitches, the powder size needs to be smaller. This is shown in Table 9.5. The general guideline to avoid stencil clogging is to ensure that the ratio of stencil aperture width to powder size is greater than 4.2. For example, as shown in Table 9.5, Type 3 powder (20–45 microns or 2 mil diameter) is needed for 10 mil stencil aperture (for 20 mil pitch printing) to achieve stencil aperture to powder size ratio of 5—an acceptable target. If, however, a Type 2 paste were to be used (about 3.2 mils powder size), the ratio would be 10/3.2 or 3.1, an unacceptable condition indicating the potential for stencil clogging.

Table 9.4   Solder powder mesh designation versus particle size and J-STD type classification [1, 5]

TYPE DESIGNATION PER J-STD-005 FOR CROSS-REFERENCE	MESH DESIGNATION PER ASTM-B214	PARTICLE SIZE RANGE MICRONS (μm)	PARTICLE SIZE AVERAGE (μm)	SURFACE AREA RATIO
Type 2	−200/+325	75–45	60	1
Type 3	−325/+500	45–25	35	1.71
Type 4	−400/+500	38–20	31	1.93
Type 5	−500	25–15	18	3.33

Table 9.5 Selection of appropriate powder size for different component pitches to achieve stencil aperture width to powder size ratio greater than 4.2

COMPONENT PITCH (INCHES)	STENCIL APERTURE WIDTH (INCHES)	MAXIMUM POWDER SIZE (MICRONS)	STENCIL APERTURE TO MAX POWDER SIZE	DESIRED PASTE TYPE
0.050	0.025	80	8	Type 2
0.025	0.014	50	4.3 or 7	Type 2 or Type 3
0.020	0.010	50	5	Type 3
0.015	0.008	40	5 or 6.6	Type 4 or Type 5
0.010	0.006	30	5	Type 5

There is one very important consideration in selecting powder size. The powder size should be the maximum size acceptable. This will not only reduce the incidence of solder balls but will also be cheaper.

9.1.4   Flux Activators and Wetting Action

The subjects of flux and cleaning are closely related and we discuss them in detail in Chapter 13. Flux is one of the main constituents of the solder paste vehicle. The activators in the flux promote wetting of the molten solder to the surface mount lands and component terminations or leads by removing oxides and other surface contaminants. In surface mounting, because of the lower stand-off between component and substrate, flux may be trapped, leading to reliability problems. Activators used in solder paste range from rosin to halide-containing compounds and organic acids (OAs).

The industry standard, J-STD-004, Requirements for Soldering Fluxes, classifies solder fluxes in three major categories based on their flux activity [6]. Since the flux in the solder paste determines its activity, solder paste classification is also based on the flux used in the paste. Solder paste is commonly available with three different flux types: rosin-based pastes, water-soluble pastes, and low residue or no-clean pastes. Let us review them briefly:

The rosin flux is primarily composed of natural resin extracted from the oleoresin of pine trees and refined. In the rosin category there are three subcategories: rosin (R), rosin mildly activated (RMA), and rosin activated (RA). The RA pastes are rarely used because of their high activity level. The RMA pastes have sufficient activator to clean the solder-coated or plated lands and component terminations or leads, thereby enabling the molten solder to wet these areas. The rosin (R) fluxes have the least activity. Since solvents containing chlorinated fluorocarbon compounds (CFCs) have been determined to cause depletion of the ozone layer, alternative solvents are now available to clean rosin fluxes. The 1987 Montreal Protocol signed by various nations including the United States banned the use of CFCs at the end of 1995.

Water-soluble fluxes are also referred to as organic acid (OA) fluxes. They are primarily composed of organic materials other than rosin or resin. Most water solubles are formulated with glycol bases. Glycols tend to be polar solvents, and the polyethylenes and polypropylenes, with heavy molecular weights, can become insoluble. The result is that they tend to bond to the board and become difficult to remove from low clearance areas under components [7]. However, OA fluxes can be cleaned and are widely used. See Chapter 13 for details.

The water-soluble solder pastes provide good soldering results because they have good flux activity. In the early days, poor tackiness and cleanability after reflow were among the main problems associated with water-soluble pastes. However, these technical problems have been solved and water-soluble solder pastes are widely used in the industry. They certainly have good wetting action (ability to remove oxides and prepare a clean surface for soldering). However, they may be too aggressive, calling for extra precautions during cleaning of the assemblies to avoid flux residue contamination.

The no-clean fluxes are primarily composed of natural resins other than rosin types, and/or synthetic resins. Typically RMAs have a resin content of 55–65% and no cleans have a resin content of 35–45%. So the no cleans have less body than RMAs and depend more on binders and other materials for their rheological characteristics. Some paste suppliers coat the solder powder with Parylene. When coating, a certain percentage of powder surface area is left open to ensure release of solder during reflow. The leftover Parylene floats off in the flux residue. By protecting the powder from reaction with the flux, the Parylene coating has been found to increase shelf life and improve tack time (from 1–4 hours to 6–18 hours) [7].

The no-clean fluxes are formulated with various levels of solids. Some of the no-clean fluxes have high levels of residue. Even if they pass electrical tests, they may leave excessive residue on the board and may be unacceptable cosmetically. Such fluxes provide good solderability and do not require the use of inert environments for reflow. At the other extreme, there are no-clean fluxes that will not leave any visible residue. No-clean pastes save not only cleaning costs but also capital expenditure and floor space. However, they are not without their problems. The no-clean pastes require a very clean assembly environment and good board and component solderability. In addition, they may also require use of an inert environment during reflow. No clean flux and paste issues are discussed in detail in Chapter 13 (Section 13.4.4).

No matter which category of flux is used, it needs to provide good solderability, meet board cleanliness requirements, and be cost effective. References 1 and 6 discuss various tests for qualification of different types of pastes and fluxes.

9.1.5   Solvent and Void Formation

The solvent dissolves the flux and imparts the pasty characteristics to the metal powder in the solder paste. It controls the tackiness of the paste by its evaporation rate under ambient conditions. The solvent should not be hygroscopic (i.e., should not absorb moisture). It should have a high flash point and should be compatible with the activator and the rheological modifiers.

The two most important factors that control the formation of voids in solder fillets are the solvent in the solder paste and the reflow profile. O'Hara and Lee [8] found higher incidences of voids in ball grid arrays (BGAs) as the boiling points of solvents in the paste decreased. They postulate that the other causes of voids may be poor board or component solderability, smaller particle size (and hence increasing oxide content), and less active flux. Not only do voids formed during reflow lower the strength of the fillet, but, because they represent an obstacle to the efficient transfer of heat from the device to the substrate, they could result in overheating of the product. Voids in the solder fillets of reflowed assemblies can be determined by x-ray analysis of the assemblies. The total area of voids calculated as a percentage of the total solder fillet area should not exceed 10%.

9.1.6   Rheological Properties

Rheology is defined as the study of the change in form and the flow of matter, generally characterized by elasticity, viscosity, and plasticity. It is affected by composition, shape and size of solder powder, chemical composition of the paste matrix or suspending agent such as the flux and solvent, and the chemical reaction between the powder and flux. The rheological properties of solder paste such as viscosity, slump, tackiness, and working life are controlled by the addition of rheological modifiers, also called thickening agents or secondary solvents. Rheological modifiers are generally high boiling solvents, since they have to function at temperatures up to the melting point. However, some amount of the high boiling solvents is trapped in the solder joint after solidification, since there is insufficient time for these modifiers to boil away completely. This leads to the formation of voids. Now let us review some of the important rheological properties of solder paste and their importance.

9.1.6.1   Viscosity

Viscosity is defined in Newton's law as the coefficient of shear stress versus shear rate:

$$\text{Viscosity} = \text{Shear stress/Shear rate.}$$

Viscosity can also be defined as the internal friction of a fluid, caused by molecular attraction, which makes it resist a tendency to flow. It is expressed in dyne-seconds per cm^2 (poise) or centipoise. Expressed more simply, viscosity is merely the degree of thickness of the fluid.

As mentioned earlier, viscosity is one aspect of rheology (and a very important one for paste printing). It is a measure of solder paste consistency. The viscosity value is dependent upon the equipment and temperature at which it is measured. Later we will discuss two different methods of measuring viscosity.

Solder pastes are thixotropic fluids. Thixotropic refers to the quality of certain materials that are paste or gel-like at rest but fluid when stressed. In other words, their viscosity changes with stress. When deformed at a constant rate of shear stress or strain rate, solder paste viscosity will decrease over time, thus implying that its structure breaks down progressively [9]. Furthermore, the viscosity of the solder paste decreases as the shear stress on the solder paste increases. Explained simply, the paste is thin when a shear stress is applied (as with a squeegee), but thick when no stress is applied. This property is highly desirable for printing since, as suggested by Figure 9.3, the solder paste will stay on top of the open areas of the screen or stencil, flowing only when the squeegee induces a stress on it [10].

Once the paste has been deposited on the lands, the squeegee-induced shear stress is removed and the paste returns to its highly viscous form, thus staying on the lands and not flowing onto the nonmetallized surface of the board. In addition to shear force, particle size and ambient temperature have an effect on solder paste viscosity [10]. Figure 9.4 represents the decrease in viscosity that occurs as shear force or temperature is

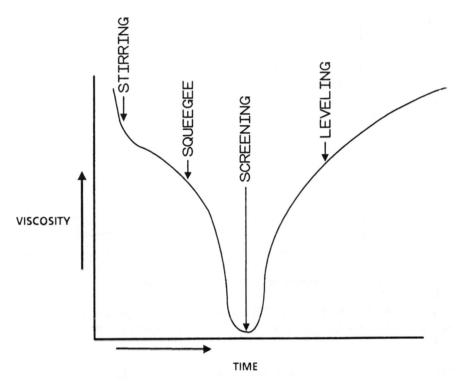

Figure 9.3   Typical solder paste viscosity change during screen printing operation.

increased. Viscosity is also decreased with larger particle size (finer particles increase viscosity).

It should be noted that Figure 9.4 is intended to show the effect on viscosity (on the Y axis) of only one variable (on the X axis) at a time. The reason for combining them is to simply illustrate a similar effect of these three variables on viscosity.

There are two common types of tools that are used for measuring solder paste viscosity. One of the earlier tools is the Brookfield viscometer shown in Figure 9.5. The viscosity should be measured at a given temperature, generally 25°C. There are various sizes of spindles (TA, TB, TC, TD, TE, TF, etc.); the TF spindle is the industry standard for solder paste viscosity measurement at 5 RPM when using model RTV.

The other method of measuring solder paste viscosity is by Malcom viscometer, which is shown in Figure 9.6. The measuring head geometry of the Malcom viscometer is based on a spiral pump design as schematically shown in Figure 9.7. The measuring head has an inner threaded

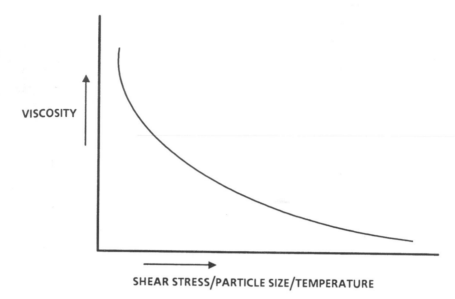

Figure 9.4 Impact of increasing shear force and temperature and particle size on solder paste viscosity with identical metal content and flux vehicle.

cylinder surrounded by an outer rotating cylinder. The combination of thread and rotating cylinder causes the paste to be continually pumped up through this mechanism. The shear rates corresponding to various rotational speeds are shown in Table 9.6 [11]. Generally the sample is measured at 5, 10, 15, 20, and 30 RPM. A stable reading must be achieved at each speed before changing to the next level. A stable display reading indicates the actual viscosity of the sample at the selected shear rate.

The key difference between the Brookfield and Malcom viscometers is that in the Malcom viscometer the material is continually pumped and the viscosity is measured under steady state. Hence the measurement is not affected by handling of paste prior to measurement. In the Brookfield viscometer, the viscosity can vary depending on the time of mixing the paste before measurement. It appears that the Malcom viscometer shows more stability in its results, although a 7–20% variation has been reported even by the Malcom viscometer.

Be that as it may, the Brookfield is widely used, perhaps because of its familiarity. Since there is no correlation between the viscosity numbers achieved using Brookfield and Malcom, one needs to develop an empirical set of data to establish viscosity requirements for different paste applications such as dispensing, screening, and stenciling.

No matter which viscometer is used, the solder paste should be

Figure 9.5   Brookfield viscometer. (Photograph courtesy of Intel Corporation.)

inspected visually for appearance: a homogeneous, light to medium gray paste is desirable. A thin layer of flux that makes the contour of solder particles still perceptible is acceptable. It should not exhibit a dry crusty appearance. To inspect for foreign substances, lumpiness, or a crusted surface, the solder paste may be stirred with a clean spatula. See Section

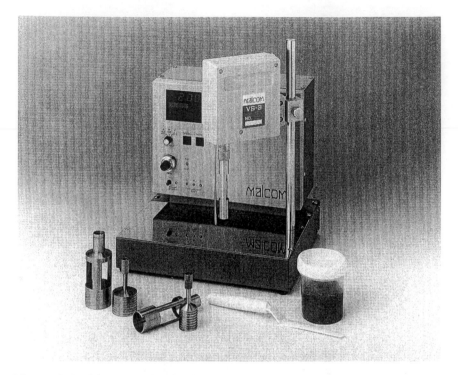

Figure 9.6 Photograph of spiral (Malcom) viscometer. (Courtesy Malcolm Instruments Corporation.)

9.5.1 for desired viscosity ranges and the impact of viscosity on print quality.

9.1.6.2   Slump

Slump is the ability of the paste to spread out after being deposited on the lands. For good soldering results, solder paste slump must be minimized. Slump also depends on the percentage of metal in the solder paste. The most reliable way to control slump is to establish bounds within which slumping is not a problem and ensure that the paste remains within those bounds. Excessive slump of a paste can cause bridging.

It is easy to conduct a slump test. A board should be designed with land patterns for varying pitches. The stencil thickness is selected to print the desired paste height. The paste is considered to fail the slump test for the pitch and paste height when adjacent lands are found to bridge after

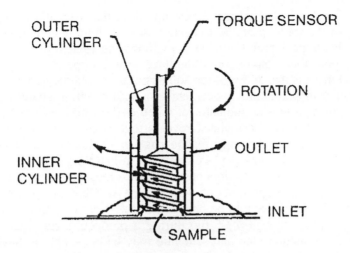

Figure 9.7   Schematic of spiral (Malcom) viscometer.

printing. A general procedure for slump test using sample land patterns is given in J-STD-005 [1].

9.1.6.3   Working Life and Tackiness

There is some confusion about the definition of "working life." The working life of a solder paste is generally defined as the length of time the paste can be left without degradation of its rheological properties on the screen/stencil before printing, or on the substrate after printing. Both parts of the definition may be correct, but they are not equally useful. For example, the length of time the paste can be left on the screen before printing does not help us much.

However, the length of time the paste can be left on the substrate after printing is useful information. It defines the maximum time the placement machine has to place the parts. Therefore a more useful defini-

Table 9.6   Shear rates corresponding to various rotational speeds for Malcom viscometer [11].

ROTATION SPEED (RPM)	SHEAR RATE (SEC⁻¹)
5	3
10	6
15	9
20	12
30	18

tion of "working life" is: the maximum time that can elapse between opening of the solder paste jar to paste reflow without degradation of the paste's rheological properties. This includes the total time needed for printing, placement, baking, and handling between operations.

Tackiness is the ability of the solder paste to hold the surface mount components in place after placement but before reflow soldering. The tackiness of a paste is an indicator of whether the working life of a paste has elapsed. If a tackiness check reveals that a paste has been pressed into service past its working life, the paste is no longer useful to hold components in place during placement and in handling before reflow.

Morris [12] discusses solder paste tackiness measurements. The tackiness of a paste is important for determining whether the paste on the screen or board will have to be changed in the event of a time delay for any reason (e.g., change of shifts) during a production run. Devising a controlled production plan and schedule may be preferable to having to conduct the tackiness test. Different manufacturers, depending on their operation, will have different requirements for tackiness.

There are commercially available tack testers that measure the force required to withdraw a standard weight from a standard size solder deposit. The test instrument measures peak force and total energy. The time that it takes to reduce the total tack energy by 20% is defined as the tack life. Whether that is acceptable or unacceptable is determined between the supplier and the user [1]. The tack time for different pastes can vary from 1/2 hour to 8 hours or more. Good pastes should easily meet 4 to 8 hours of tack time. In general (but not always), no-clean and rosin pastes have longer tack times than water-soluble pastes.

9.1.7   Solder Balls

Solder balls are small spherical particles of solder, usually 2 to 5 mils in diameter, which reside on the nonmetallic surfaces of the board. Solder balls, especially the mobile ones, are a reliability hazard because potentially they can short metallic conductors at any time during the life of the substrate. There are two mechanisms by which solder balls are formed.

1.   Solder balls are caused by very fine powder particles in the solder paste. They are carried away from the main solder deposit as the flux melts and flows before the solder itself melts. This happens especially when the paste is deposited outside the land area either by design (screen/stencil opening is larger than the land area) or by misregistration. These small powder particles then lose contact

with the larger solder paste deposit and when the solder melts, each particle becomes a small solder ball at the periphery of the original paste deposit. A collection of small solder balls around the main solder deposit is called a "halo."

2. Solder balls are also formed when the oxide layer on the surface of the solder powder particles is so thick that the rosin flux and any activator in the paste are not sufficient to remove it. Since oxides cannot melt at soldering temperatures, they are pushed aside as a solder ball by the surrounding oxide-free molten solder. Solder balls formed in this manner are usually larger than those formed by the first mechanism because of the presence of surface oxide which is less dense than the metal.

Oxidation of the solder powder particles is accelerated by improper storage, or by baking the paste at an excessively high temperature before reflow. Solder particle oxidation is also promoted by fretting, or the production of oxides by the mutual abrasion of powder particle surfaces during sieving to separate the particles into various size fractions [13].

When balls rub against each other during sieving (to separate balls of varying sizes) or ball attachment to the BGA package, there is the potential for the "black ball" phenomenon, where the balls take on dark colors. This happens due to severe rubbing of the balls against each other, and the balls tend to lose their solderability (hence there is no ball attachment during reflow). This problem has essentially been solved by the industry by ensuring that balls do not rub against each other for an extended time.

The tendency of a solder paste to form solder balls can be determined with a simple test. A few 6 to 10 mil thick solder paste patterns are deposited by stencil or screen printing on a nonmetallic substrate, such as frosted glass, ceramic, or even FR-4, and then reflowed, preferably using a reflow profile close to the one used in production.

The resulting solder deposit should be examined visually. If the deposit is one large shiny ball, with a smooth surface, the solder paste is acceptable. When using Type 1 through Type 4 solder pastes (Table 9.3), individual solder balls of greater than 3 mils (75 microns) should not form on more than one of the three test patterns used in the evaluation. When using even finer powders (Types 5 and 6), the maximum size of the acceptable solder ball is only 2 mils (50 microns) at not more than one of the three test patterns used in the evaluation [1].

Figure 9.8 shows examples of acceptable and unacceptable solder balls [1]. Naturally an absence of solder balls is preferred (top left). A minor occurrence of very fine solder balls (top right) is also acceptable.

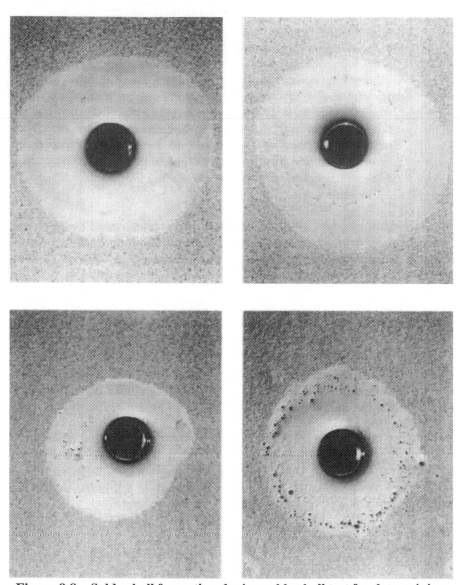

Figure 9.8 Solder ball formation during solder ball test for determining the suitability of solder paste. No solder ball (top left) is preferred, but a minor occurrence of very fine solder balls (top right) is also acceptable. Clustered solder balls (bottom left) and a solder ball halo ring with numerous solder balls (bottom right) are unacceptable [1].

Clustered solder balls are unacceptable (bottom left), as is a halo ring consisting of numerous solder balls (bottom right).

Oxidation of the solder paste metal alloy leads to the formation of solder balls and the loss of metal as oxide. Even though oxidation of the metal takes place on the surface of the metal particles, a good gauge of the tendency of a paste to form solder balls is the bulk oxide content.

In a manufacturing environment, it is not feasible to conduct the bulk oxide content test. It is more common and useful to test for solder balls, as described above. As a matter of fact, the solder ball test should be conducted regularly to monitor the quality of the solder paste and its tendency to ball.

It should be kept in mind that no-clean solder pastes may be more susceptible to solder balling because no-clean flux is relatively weak and not as effective in removing oxides from the powder or the board surface. This is one of the reasons for the widespread use of inert environments such as nitrogen, especially when the solids content in the flux is very low.

9.1.8   Printability

The printability of a solder paste is reflected in the accuracy and reproducibility of the screened or stenciled solder paste pattern onto the land patterns. The printability of the solder paste, which is one of the most critical tests, should be determined according to the following experiments. Dummy boards can be used for test substrates.

1. Weigh five cleaned dummy board substrates before (W_1) and after (W_2) the paste is screened or stenciled on. Use five substrates for each method of application, to get average data.
2. Determine the weight of the solder paste applied to each board by the formula $W_2 - W_1$.
3. Measure and record the paste height at four predetermined points on each substrate, using a depth scope.
4. Perform steps 1 through 3 for freshly removed solder paste and for solder paste exposed to the atmosphere for 4 hours.

The acceptance criteria for use of that particular solder paste should be as follows:

*   The solder paste weight should not vary more than 10% among the average measurements taken on one substrate.
*   The paste height should not vary more than ±1 mil among the average measurements taken on one substrate.

- The solder paste pattern should have uniform coverage, without stringing and without separation of flux and solder, and it should print without forming a peak.

Even though it is important, printability evaluation is rarely done in the industry. With the widespread use of BGAs, it is even more important to ensure uniform solder paste prints. For ceramic BGAs (CBGAs), uniform paste height is critical for good yield.

9.2   SOLDER PASTE PRINTING EQUIPMENT

The solder paste printing process involves a series of interrelated variables, but the printer is crucial for the achievement of the desired quality of print. The solder paste screen printers available on the market today fall into two main categories: laboratory and production. Each category has further subdivisions, since companies expect different levels of performance from laboratory and production printers. For example, a laboratory application that is R&D for one company could be prototype or production for another. Moreover, production requirements can vary widely depending on volume. Since a clear-cut classification of equipment is not possible, the best thing to do is to select a screen printer to match the desired application.

If the goal is to use the equipment entirely for R&D, the equipment controls do not have to match what will be used in manufacturing. However, if the goal is to use the equipment for process development and improvement, the basic features of the machine should closely match the ones to be used in manufacturing. The only difference may be in the level of automation to control printing parameters. For laboratory use or in the hybrid industry, smaller benchtop models generally are used, like the screen printer shown in Figure 9.9. Production applications usually call for larger equipment with semi-automatic or automatic control features.

Automatic printers require minimal operator intervention. For example, automatic printers can be programmed for squeegee travel speed, squeegee pressure, amount and frequency of a constant bead of paste application, and the frequency of cleaning the underside of the stencil. They may also include automatic board loading and unloading and an optical sensor to indicate transfer of the board from printer to pick-and-place machine. A vision system may also be included to accurately match the stencil with the board. And certain pads may be preprogrammed as targets for evaluating the performance of the printing process, giving the operator the option to either accept or reject the board. An example of an automatic printer is shown in Figure 9.10.

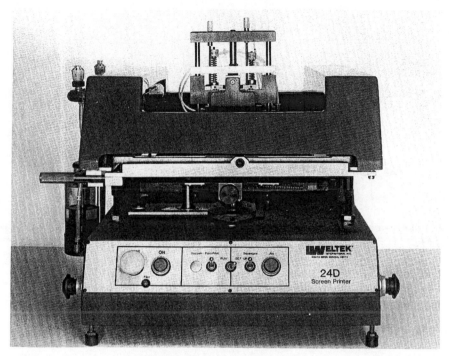

Figure 9.9 Laboratory model solder paste screen printer. (Photograph courtesy of Weltek International.)

Semi-automatic printers require full operator interaction during production. This includes loading and aligning circuit boards, setting printing parameters, inspecting the printed boards, unloading the boards, and periodically cleaning the stencil. Many of the print parameters entered into the computer for automatic printers will have to be set manually in semi-automatic printers. These include the rails on the transport tracks, squeegee pressure, stroke length, stroke speed and down stop, and snap-off distance. In addition, alignment, spreading the paste with a spatula, inspecting the board, and cleaning the stencil are often done manually. An example of a semi-automatic printer is shown in Figure 9.11.

All printers generally require the operator to replenish solder paste and wiper paper. The paste is either spread onto the stencil manually with a spatula or dispensed automatically from a cartridge.

9.2.1 Selecting a Printer

No matter what the application, the main features of screen printing equipment are mechanical rigidity of structure, registration method, squee-

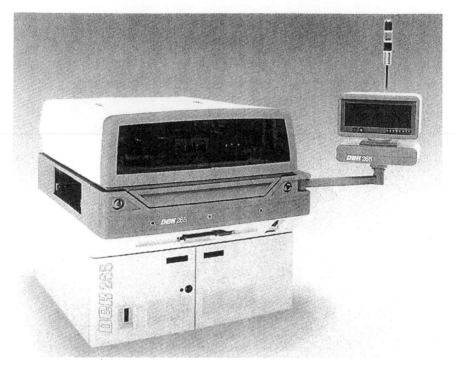

Figure 9.10   An example of an automatic screen printer. (Courtesy of DEK Corporation.)

gee velocity control, and squeegee pressure control [14]. Mechanical rigidity is important in maintaining parallelism between the squeegee travel path and the substrate, to achieve good print uniformity. Squeegee pressure and speed control must be repeatable, for the sake of uniformity of print. A good registration control with repeatable X, Y, and Θ (theta) adjustments expedites setup work.

Having considered these generic variables, how should one select a screen printer? Refer to Appendix B2 for a detailed questionnaire on selecting screen printers. Some of the general considerations in selecting a printer are throughput requirements, ease of use, and print quality and consistency. Other features that may be important in some applications are: vision capability, underside stencil wipe, and automatic board loading and unloading capability. Cost, service, and support are also important considerations.

The first item to consider is the maximum size of the substrate that will be used. The screen or stencil frame that will handle the substrate must be larger than the substrate, since in addition to the print area for

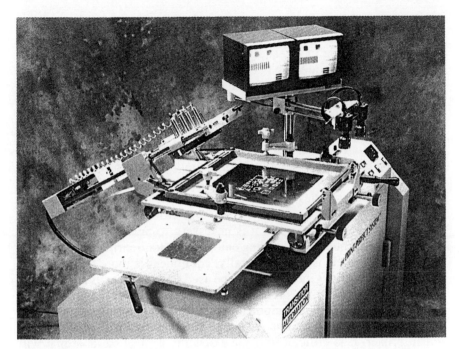

Figure 9.11 An example of a semi-automatic printer. (Courtesy of Transition Automation.)

the substrate, the frame must accommodate the area needed for squeegee travel. The equipment also must accommodate the outer frame of the screen/stencil. As a rough general rule, about 3 to 4 inches should be added on each side of the maximum print area to determine the maximum size of the machine. For example, to accommodate the printing on 7 inch × 12 inch and 9 inch × 9 inch boards in different orientations, a machine would have to be able to handle an outer frame size of 15 × 15 inches.

The maximum number of boards that can be processed per hour is another consideration. It should be kept in mind that once the machine is set up, the screen printer will almost never be a bottleneck in an SMT manufacturing line. Throughput for a screen printer is a function of squeegee travel speed and board size.

Since the squeegee travels at about 0.5 to 4 inches per second, a board can be printed almost every 20–30 seconds. Compare this throughput to 0.5 to 4 seconds per component during placement. This means that virtually no pick-and-place equipment available can keep up with a screen printer unless there are very few components on the board. Therefore, unless the manufacturing line is totally automated, there is no need to go overboard on functions that increase the throughput of the screen printer.

However, it is important to get a screen printer that gives accurate and repeatable print quality.

The print mode of the equipment is another consideration. Printers are designed with a squeegee bar and a flood bar. In flood mode the flood bar spreads the paste, and in print mode the squeegee applies pressure, such that the actual printing takes place. When the flood bar is spreading the paste, the squeegee is up (riding), and when the squeegee is printing the flood bar is up (riding). Since the two bars are identical, if the design allows, their functions can alternate. Some equipment can perform only in the flood/print mode, and others offer both print/print and flood/print modes. Some of the older machines are not suitable for stencil application because they are not flexible and cannot provide the print/print mode.

Only the print/print mode can be used with stencils, because the holes are completely open and it is necessary to prevent the paste from oozing out of the holes and dripping onto the substrate surface. For screens, the flood/print mode is used. Naturally, equipment is more desirable if it can function in various modes and can be used for both stencils and screens.

As mentioned earlier, print quality depends on the machine variables. Table 9.7 summarizes the equipment variables that determine print quality and repeatability [15]. These are generic variables and will apply to every screen printer. It is difficult to name a specific control variable that will be applicable to different types of equipment. However, Table 9.8 suggests a range for printer settings that can be used as a general guideline for fine-tuning the settings of different machines [4].

9.3 SOLDER PASTE PRINTING PROCESSES

In printing solder paste, the substrate is placed on the work holder, held firmly by vacuum or mechanically, and aligned with the aid of tooling pins or vision. Either a screen or stencil is used to apply solder paste. Figures 9.12 and 9.13 present cross-sectional and top views, respectively, that show the major differences between screens and stencils. The frames of the screens and stencils are similar; the differences lie in the construction of individual openings used for depositing the solder paste.

Figure 9.13 shows schematically a screen and three different types of stencil. Figure 9.14 gives close-up photographic views corresponding to the diagrams of Figure 9.13. Figure 9.13 shows only one PLCC on a screen/stencil and Figure 9.14 shows openings for only a few components. In reality there are many components on the screens or stencils. The number of components on the screen/stencil matches the number of surface mount components on the board. Figure 9.15 is a photograph of a stencil that shows the openings for all the components on the board. Since each

Table 9.7 Solder paste printing equipment variables [15]

- Structural
  - Stiffness
  - Parallelism
  - Precision of mechanical parts (fit and movement)
- Squeegee
  - Velocity
  - Acceleration
  - Deceleration
  - Pressure (down force)
  - Stroke parallelism
  - Parallelism in substrate
  - Down stop
- Modes of operation
  - Contact/Off-contact
  - Bi-directional/unidirectional printing (flood/print or print/print modes)
  - Flood bar
  - Multiple wet pass
- Screen holder
  - X axis
  - Y axis
  - Z axis
- Rotation (Theta)
- Peel off
- Snap off

board is unique, stencils and screens are unique to particular boards and cannot be used for others. However, the frames can be used over and over for different board designs to save cost.

To lower cost even further, many companies use stencils with universal frames. In this case the stencil is purchased without frame and the universal frame is used with all the stencils (within predetermined size limits). The stencil is stretched with the screws on the frame either only on two sides or on all four sides. See Figure 9.16.

Stretching stencil only on two sides does not provide as much tension as stretching the stencil on all four sides. The frames that allow stretching only on two sides are much cheaper than the frames that allow stretching on all four sides.

The universal frame does add some additional cost in the beginning but there is considerable saving over using stencils with dedicated frames. Mounting the stencil upside down may provide further savings by using only one stencil for a two-sided board with similar design (but with reverse

Table 9.8   Solder paste printing parameters (will vary with equipment)

MACHINE VARIABLES	RECOMMENDED SETTINGS
Squeegee material	Rubber or metal
Squeegee Pressure[a,b]	
• Screens with rubber squeegee	15–25 lb.
• Stencil with metal squeegee	10–20 lb.
• Stencil with rubber squeegee	30–40 lb. (1.5–2.5 lb./inch of squeegee length)
Squeegee Speed[a]	
• Stencil for 50 mil pitch components	2.0–4.0 inches/second
• Stencil for fine pitch components	0.5–1.5 inches/second
Snap-off distance[a]	0.020–0.040 inch (zero for on-contact printing)
Angle of attack	45° or 60°
Leveling of screen/stencil: front/back and side to side alignment	Adjust so it is parallel
	Repeated passes will bring the pad openings of the screen or stencil in line with the solder pads on the substrate; align with micrometer or screws or visually (manual printers) or with vision (automatic printers)

[a] Critical printing parameters.

[b] Start with low pressure and then gradually increase pressure until clean sweep is achieved on the stencil.

image). This approach also saves storage space since the flat stencils without dedicated frames can be hung like file folders (Figure 9.17). The negative side of this approach is that the screws in the universal frame may strip over time and may not provide the needed tension on the stencil. This can be prevented by using washers under the screws and by using thicker hold-down bars that clamp the stencil to the frame.

Despite being more expensive, stencils with dedicated frames are more common than flat stencils. If the volume of production is high, the cost of dedicated frames is negligible. When using flat stencils, there is the potential for damaging them during storage and even the possibility of mounting them upside down.

Typically, a screen will contain open wire mesh around which solder paste must flow to reach the substrate surface. A stencil opening is fully etched and does not obstruct paste flow. The screens or stencils have

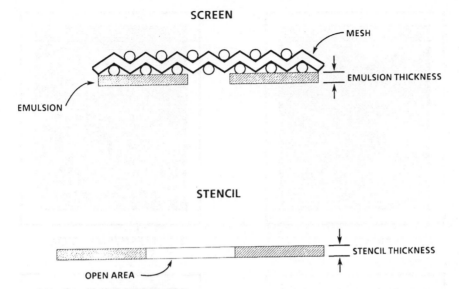

Figure 9.12 **Cross-sectional views of stencils and screens for solder paste printing.**

openings to match the land patterns on the substrate, where solder paste must be deposited for the electrical interconnections.

The screen/stencil is stretched in a metallic frame and is aligned above the substrate. The distance between the top of the board and the bottom of the stencil/screen is called the snap-off (see Figure 9.18). This gap or snap-off distance is a function of the equipment design and is about 0.020–0.040 inch.

9.3.1   Paste Printer Setup

In this section we describe the setup of manual or semi-automatic screen printers. Some of these functions can be performed automatically if the screen printer is equipped with those features. Initially, as shown in Figure 9.19, the solder paste is manually placed on the stencil/screen with the print squeegee at one end of the stencil. (In automatic printers, paste is automatically dispensed.) During the printing process, the print squeegee presses down on the stencil to the extent that the bottom of the stencil touches the top surface of the board. The solder paste is printed on the lands through the openings in the stencil/screen when the squeegee traverses the entire length of the image area etched in the metal mask.

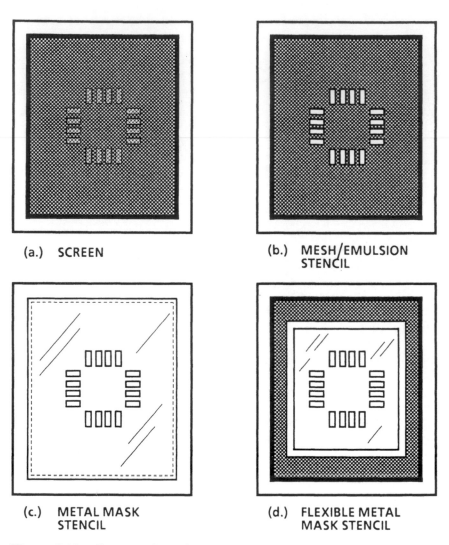

(a.) SCREEN

(b.) MESH/EMULSION STENCIL

(c.) METAL MASK STENCIL

(d.) FLEXIBLE METAL MASK STENCIL

Figure 9.13 Construction of screen (a), mesh/emulsion stencil (b), all metal mask stencil (c), and flexible metal mask (d).

When screens are used, there is also the flood squeegee close to the print squeegee, whose function is to flood the screen with solder paste before the printing operation begins. The flood and print squeegees were discussed previously.

After the paste has been deposited, the screen peels away or snaps off immediately behind the squeegee and returns to its original position. (See Figure 9.18.) Thus, snap-off distance and squeegee pressure are two

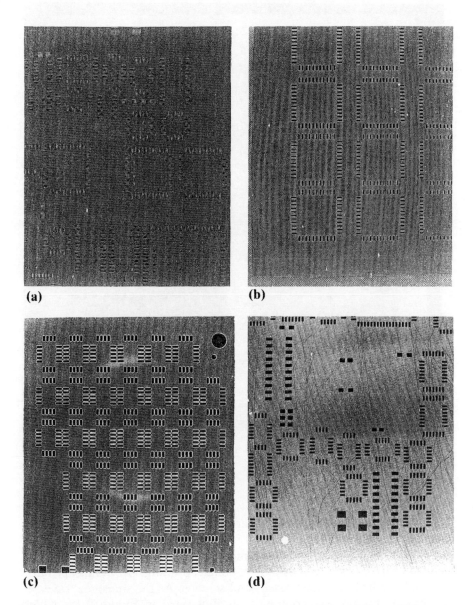

(a)

(b)

(c)

(d)

Figure 9.14 Close up views of openings in screen (a) and three different stencil types: mesh/emulsion stencil (b), all metal mask stencil (c), and flexible metal mask stencil (d), which correspond to the diagrams of Figure 9.13. (Photographs courtesy of Intel Corporation.)

Figure 9.15   A flexible metal mask stencil. (Photograph courtesy of Intel Corporation.)

important equipment-dependent variables for good quality printing. If there is no snap-off, the operation is called on-contact printing. This is used when an all-metal stencil or metal squeegee blade is used. If there is a snap-off, the process is called off-contact printing. Off-contact printing is used with flexible metal masks and screens. As the squeegee traverses the board, this process continues, as illustrated in Figure 9.18, which shows the location of paste, screen, squeegee, substrate, and snap-off distance. The process sequence for applying solder paste is as follows:

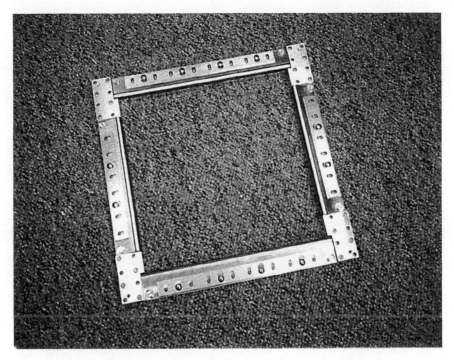

Figure 9.16   Frame for stretching stencils without frames. (Courtesy of SMT Division of Alpha Metals, Inc.)

- Prepare solder paste
- Set up and register screen/stencil
- Print solder paste
- Clean screen/stencil after printing.

When setting up the screen printer, the following guidelines should be observed for good registration of solder paste on the land patterns. (As mentioned earlier, the machines with vision capability may not require many of the steps described below.)

1. No matter which method of application is used, be sure that the solder paste has been stored properly. A tightly sealed, unopened container of solder paste generally can be stored for 6 months at 4 to 29°C (40–84°F). The shelf life is flux dependent. It is better

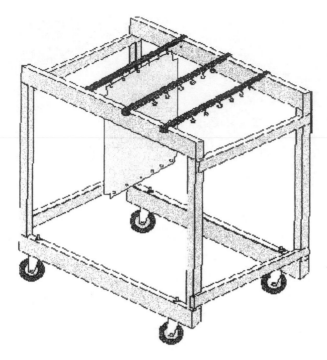

Figure 9.17   Storage rack for stencils without dedicated frames. (Courtesy of SMT Division of Alpha Metals, Inc.)

      to use the freshest paste possible. If opened, store in a refrigerated environment.

2.   Use only fresh paste every day. To accomplish this, use small jars that contain only one day's worth of paste or transfer paste from large jars as needed for the day (and put the rest back in the refrigerator). This helps improves paste-related yield. This simple step can also save thousands of dollars every month in a typical medium-volume manufacturing environment.

3.   Allow a refrigerated container to reach room temperature before use. It may be advisable to take the paste out of the refrigerator the night before for the next day's use to avoid any wait.

4.   Check the solder paste for solder ball characteristics and viscosity.

5.   Place a clean sheet of Mylar over the board.

6.   Print solder paste onto the Mylar and inspect the printed pattern for registration with the board and for uniform coverage. If registration is poor, realign, using X-Y-Θ adjustments on the printer, and print onto a clean Mylar sheet again. If coverage is not uniform, examine squeegee, solder paste, and screen for evidence

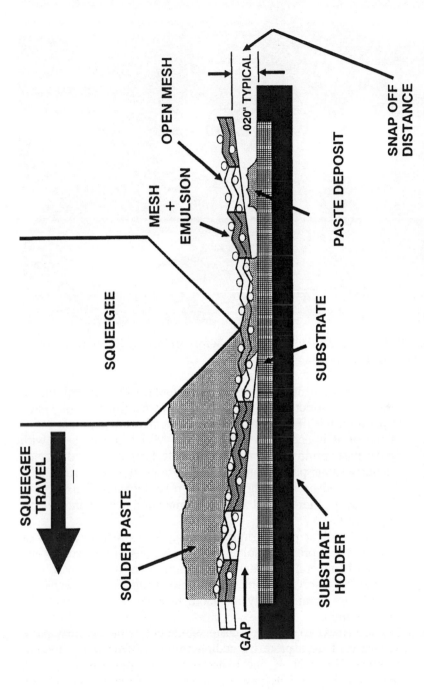

Figure 9.18 Applying solder paste on a substrate by squeegee in a screen printing process: schematic view.

Figure 9.19 Solder paste application before printing. (Photograph courtesy of Intel Corporation.)

of areas with insufficient solder paste, and readjust the equipment controls to correct the deficiency. When using printers with vision capability, use of a Mylar sheet is generally not required.

7.   If the print is acceptable, remove the Mylar and proceed with solder paste printing onto the board itself. Printing should provide adequate solder paste quantities for subsequent reflow processing. There should be no bridging between solder pads. This item is discussed in more detail later in connection with print defects (Section 9.4).

8.   When all solder paste printing is complete, wash the screen or stencil with appropriate solvents. The cleaning of the screens and stencils is generally a manual operation done in a batch solvent cleaner. Now solvent cleaners especially designed for screens and stencils are commercially available to make the job easier for the operators.

9.   Discard (recycle) any used paste. Some companies modify paste viscosity of used paste by adding thinners. This is not recommended. However, if one believes that the paste is still good, make sure it passes at least the solder ball test before using it on production boards.

The discussion above applies to solder application by both screening and stenciling. Now let us focus in more detail on the important process variables in each method of solder paste application and their pros and cons.

9.3.2   Screen Printing

The basic principle of applying paste is essentially the same whether screens or stencils are used. As illustrated in Figure 9.12, a screen made of a woven wire mesh is stretched over a frame with a glued-on photosensitive emulsion. The wire mesh supports the emulsion, which is etched where solder paste is needed. The screen is attached to an aluminum frame, which imparts rigidity and tension, hence a flat but flexible screen surface.

Screen construction was illustrated in Figure 9.13a. A screen contains open wire mesh around which solder paste must flow to reach the substrate surface. The wire mesh is usually made of monofilament polyester or stainless steel. Polyester materials are more resilient than stainless steel and last longer. The diameter of the wire and the size of the opening determine the mesh number, that is, the number of holes per square inch. Typically, screens used in surface mount assembly are of mesh number 80. (This is not the same mesh used in particle size classification as discussed earlier.) The diameter of the wire and the thickness of the emulsion primarily determine the solder paste thickness that can be deposited, but the snap-off distance (the distance between top of the substrate and the bottom of the screen) also plays a role.

Screen printing has been widely used in the hybrid industry for printing solder paste and in the printed circuit industry for printing wet film solder mask. Screen printing is also used for printing polymer thick films such as carbon inks for key pads which are widely used in the telecommunications industry. It has been adopted for surface mounting because of its many advantages, but chiefly its lower cost. The main disadvantage of screens is the difficulty of aligning them with the land patterns of the substrate. This is a difficult and time-consuming process because the lands on the board cannot be readily seen through the wire mesh.

Screens are restricted to off-contact printing because some snap-off distance is necessary to deposit a sufficient thickness of paste. The variables that control print quality and print thickness include snap-off height, squeegee pressure, wire diameter and mesh number, and emulsion thickness. When using screens, the following guidelines are helpful:

1.   Usually screen printing requires a slower speed and a greater snap-off distance than stencil printing. Also, lower viscosity pastes should be used for easy printing.

2. A feeler gauge or other suitable means can be used to adjust the Z-stops so that screen and work holder are parallel within 0.002 inch.
3. The snap-off distance is adjusted to 0.030 (+0.020/−0.005) inch, while maintaining the screen-to-board parallelism within 0.002 inch.

9.3.3   Stencil Printing

Screens and stencils are functionally the same, but there is a major difference, as mentioned earlier. In stencils, instead of screen mesh, the desired opening is chemically etched out either in metal sheets or in wire mesh covered with emulsion, so that the stencil opening does not obstruct solder paste flow. Instead of chemical etching, stencils are also cut with laser or built by additive processes. See Section 9.5.7 for a discussion on the pros and cons of preparing stencils in various ways. A stencil provides 100% open area for the paste to be printed through, whereas the screen provides only about 50% (the actual percentage depends on the mesh wire diameter and mesh number used for the screen).

Depending on the material in which the opening is etched, as shown in Figures 9.13b, 9.13c, and 9.13d (and the corresponding photographs in Figures 9.14b, 9.14c, and 9.14d), stencils can be classified in three major categories:

1. The selectively etched mesh/emulsion screen (Figures 9.13b and 9.14b)
2. The metal (brass, stainless steel, beryllium-copper, or nickel) mask (Figures 9.13c and 9.14c)
3. The flexible metal mask, a combination of categories 1 and 2 (Figures 9.13d and 9.14d).

Each type of stencil has its advantages and disadvantages. The selectively etched mesh/emulsion type (Figure 9.13b) is made the same way as the screen, but the openings are completely etched out. The main disadvantage of this type of stencil is high cost. Also, because the wire mesh is etched, the screen is less stable. This approach is rarely used now.

All metal mask stencils (Figure 9.13c) are not under constant tension and must be printed on-contact only. As discussed earlier, there is a variation of all metal masks used on universal frames (Figure 9.16). This approach provides the needed tension and is widely used because it saves on the cost of dedicated frames. These stencils in stretched universal frames can be used for either on-contact or off-contact printing.

The flexible metal mask (Figure 9.13d) combines the advantages of the all metal mask and the flexibility of screens. The metal portion of the mask is held in tension by the border of the flexible mesh to allow off-contact printing. This is the most widely used type of stencil.

The following guidelines should prove useful when using flexible metal mask stencils. First, polyester mesh is preferred over stainless steel mesh as a border for mounting because it allows better tension. However, polyester does not clean as well as a stainless mesh border. Second, the border area should be about 3 inches all around. A smaller border area will force the polyester fibers to stretch beyond the elastic point, and they will become damaged during printing [16]. Finally, the squeegee should stay on the metal mask; it should not be allowed to print across the polyester border.

9.3.4   Screen Printing Versus Stencil Printing

Stencils and screens are functionally the same, but they are used differently on the machine. For example, printing in both directions (print/print mode) is used for stencils to avoid the seeping of paste during the flood mode. The print/print mode requires the squeegee to jump over the paste at the end of each print stroke.

By contrast, in screen printing, both the flood and print modes are used. Also, stencils are easier to align on the substrate surface than screens because the openings provide clear visibility. Moreover, the holes generally do not become plugged, and it is easy to get consistently good print. Stencils are much easier to clean than screens. They are also sturdier than screens and hence last much longer. The viscosities required for stencils are much higher than those for screening.

Stencils can be used for selective printing but screens cannot. Selective printing can be very beneficial if two different paste thicknesses are required on the same board. For example, if fine pitch packages, larger tantalum capacitors, and other PLCCs and SOICs are to be mounted on the same board, each with a different paste thickness requirement, selective printing can be very desirable. This practice is becoming more common with the increasing usage of fine pitch devices, which require lower paste thickness.

Selective or multilevel printing is not feasible in screening but is easily accomplished in stencils. For making stencils with multilevel printing, as shown in Figure 9.20, first a larger area (larger than the land pattern area of the component to prevent solder skipping and damage to the squeegee) is etched to a desired thickness for the components that require lower paste thickness. For example, if the metal thickness for most of the board is 8

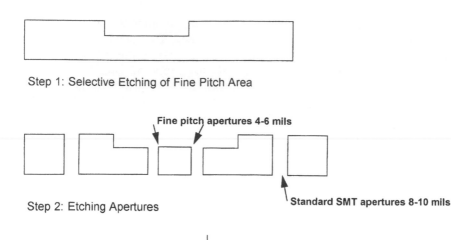

Step 1: Selective Etching of Fine Pitch Area

Step 2: Etching Apertures

Step 3: Printed Paste Deposit

Figure 9.20 A schematic of a multilevel stencil for dual-thickness printing.

mils, and only 6 mils for a fine pitch package, the metal area for the fine pitch device and additional 100 mils around the package will be first etched down to 6 mils. Then the entire stencil is processed in normal fashion. See Figure 9.21 for a photograph of a multilevel stencil.

With screens, off-contact printing is always used to ensure deposition of an adequate thickness of paste and to prevent smearing during the flood mode (screens use only flood/print mode). In stencil printing, both off-contact and on-contact printing are used. Smearing is not a problem because the flood mode is not used for stencils, only the print/print mode. Off-contact printing entails less risk of the substrate sticking to the screen. Since stencils are very easy to align with the board, hand printing (on-contact printing) can be used for quick prototyping jobs. Hand printing is difficult with screens because they are so difficult to align.

The metal mask and flexible metal mask stencils are etched by chemical milling from both sides using two positive images. However chemically etched stencils suffer from undercutting, which can lead to excess solder. The problem can be corrected by modifying the artwork. To avoid the problem of undercutting, laser cut stencils and stencils with additive processes are used.

Table 9.9 lists the pros and cons of screens and stencils. Stencils have more advantages for solder paste printing. They are used where

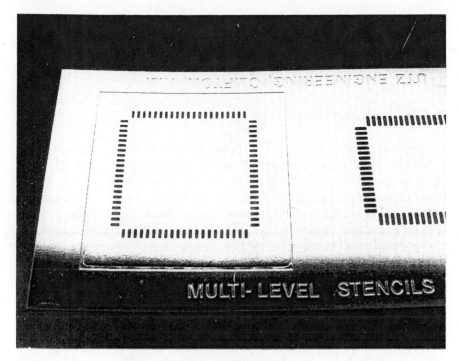

Figure 9.21 **Multilevel or selective etching in a metal mask stencil for depositing different solder paste thicknesses on the same board. (Photograph courtesy of Intel Corporation.)**

Table 9.9 **Advantages and disadvantages of screens and stencils**

STENCILS	SCREENS
ADVANTAGES	DISADVANTAGES
• Easy to set up	• Harder to set up
• On-contact and off-contact printing	• Off-contact printing only
• Can print by hand	• Cannot print by hand
• Wider usable viscosity range	• Narrow usable viscosity range
• More durable	• Less durable
• Do not plug easily	• Plug easily
• Easy to clean	• Hard to clean
• Allow multilevel printing	• Do not allow multilevel printing
DISADVANTAGES	ADVANTAGES
• Higher cost	• Lower cost

thicker deposition is required. A print of at least 8 mils thick is required for acceptable solder joints in most cases, especially given the general need to compensate for some board warpage and lead coplanarity and still produce acceptable solder joints.

9.3.5   Dispensing

Solder paste is dispensed by being squeezed through the needle of a syringe. Since paste is dispensed on one land at a time, generally, in a manual operation, it is a much slower process than screen printing. However, dispensing can also be done automatically or semi-automatically, and in such cases the speed can be increased substantially. Most syringes rely on pneumatic dispensing systems, and the problems most commonly encountered entail clogging of needles. Syringing requires a paste of lower viscosity than is used for printing.

Dispensing is not considered for speed of paste application but for its versatility. For example, the dispensing method can be used to print different shapes (squares, rectangles, circles, or dots) and different materials (paste, adhesive, or temporary solder mask), permitting the same tool to be used for several applications. Also, there is no vendor turnaround time involved, as is the case with screens and stencils.

Dispensing is appropriate when screening cannot be used—for example, in repair, since there are other components on the board. After the defective component has been removed, the solder paste is dispensed on the appropriate land pattern. The replacement component is then placed on the paste-covered land pattern and soldered.

Dispensing may also have special application for fine pitch packages that require lower paste thickness than other components, such as PLCCs and SOICs on the same board. However, as noted earlier, dispensing is not the only way to get different paste thicknesses on the same board. This can also be accomplished by selectively etching metal stencils (Figure 9.20) as already discussed. Screening and stenciling methods will continue to be the dominant methods for paste application. However, dispensing is needed for specialized applications for which screening is not viable. Chapter 8, Figure 8.12 shows a semi-automated dispenser that can also be used for dispensing solder paste.

9.4   PASTE PRINTING DEFECTS

Solder paste print defects can be classified in various categories as shown in Figure 9.22, namely smearing, skipping, ragged edges, and

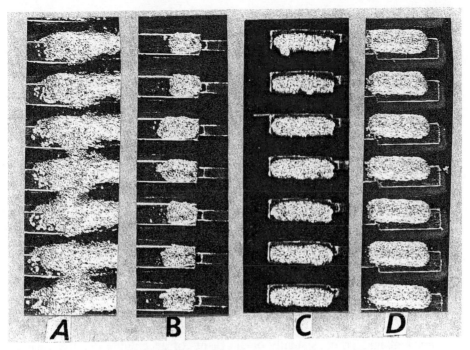

Figure 9.22   Common print defects during solder paste application: (a) smeared print, (b) skipped print, (c) print with ragged edges, and (d) misaligned print. (Photograph courtesy of Intel Corporation.)

misalignment. Smearing occurs when solder paste is printed on areas on which it is not required (e.g., in between surface mount lands). Smearing is unacceptable because it causes solder bridges and/or solder balls. Figure 9.22a shows a smeared print. Skipping (Figure 9.22b) occurs when there is insufficient deposition of the solder paste on the surface mount lands. Skipping is unacceptable because it will cause insufficient solder joints. A variation of skipping is scooping, caused either by too much squeegee pressure or using too soft a squeegee. The maximum variation in paste height should be about 20%. Figure 9.22c shows a print with ragged edges, a defect that will result in nonuniform solder joints. A misaligned solder paste print (Figure 9.22d) occurs when the land patterns and the openings of screens or stencils are not properly aligned before printing. Another common defect is slump, where the paste deposit loses its shape. For slump and misregistration, a maximum variation of 15% may be acceptable.

As one begins to print finer and finer pitches, the incidence of bridging increases. Printing at 25 and 20 mil pitches is relatively simpler, although one can expect to have more problems, relatively speaking, at 20 and 25 mil pitches than at 50 mil pitch. And even at 50 mil pitch bridging is not uncommon. To reduce the incidence of bridging one must control many parameters, not just one or two. In addition to the paste quality and rheology and equipment-dependent variables discussed earlier, there are many other variables that play a significant role.

The incidence of print defects can be minimized by regularly cleaning of the stencil, continuously monitoring the paste quality, and properly controlling the equipment- and non-equipment-related variables discussed next.

9.5 PASTE PRINTING VARIABLES

At the beginning of a typical production run of surface mount boards, various solder paste equipment and printing parameters are set to maximize the print quality (Tables 9.7 and 9.8).

Some of these parameters change over the course of a production run. For example, a stray vacuum leak will cause a drop in print quality. If vacuum is used for hold-down, the fixture plate should not allow any leakage. To consistently obtain prints of acceptable quality, various parameters must be maintained within a narrow range of values.

9.5.1 Solder Paste Viscosity

Earlier, we discussed viscosity and some variables that affect it, such as particle size, shear force, and temperature. We noted that the instruments commonly used for measuring viscosity are the Brookfield viscometer (Figure 9.5) and the spiral Malcom viscometer (Figure 9.6). In this section we discuss the impact of viscosity on the printability of paste. Solder paste viscosity is very critical for obtaining an acceptable print: smearing will occur if it is too low, and there will be skips if it is too high.

Depending on the method of application, solder paste should have the following viscosity, as measured at 25°C using a Brookfield viscometer (model RTV with a TF spindle at 5 RPM):

- Dispensing: 200,000 (centi poise) to 450,000 cP
- Screening: 450,000 (centi poise) to 800,000 cP
- Stenciling: 750,000 to 950,000 cP for 50 mil pitch components
- Stenciling: 900,000 to 1,200,000 cP for fine pitch (and mixed assembly even if there is only one fine pitch on the board)

We discussed the impact of temperature on viscosity in connection with Figure 9.4. Solder paste viscosity also increases with the number of boards printed. Generally, after only a 5°C change in temperature or three prints, the viscosity of the paste remaining on the stencil/screen may be degraded enough to affect print quality. Thus, to minimize viscosity changes induced by print quantity, fresh paste may have to be added to the stencil/screen after three prints. Not all pastes degrade so quickly, however, and this may be one of the factors to consider when selecting a paste.

To minimize temperature-induced viscosity changes, the paste should be printed in a temperature-controlled area. Some automatic printers provide temperature-controlled housing for the solder paste container. In any case, since the viscosity of the paste will increase as the rheological modifiers it contains evaporate, the paste should be removed from the stencil/screen at the end of each production run.

9.5.2 Print Thickness and Snap-Off

Print thickness determines the volume of solder in the joints. Too thick a print will result in excessive solder joints or even solder bridges, and too thin a print will result in insufficient solder joints. The thickness of the paste print is primarily determined by the thickness of the metal mask of the stencil (or the emulsion thickness and mesh number for a screen).

While it is true that stencil thickness primarily controls the paste thickness, other variables are also important. For example, the snap-off height (the distance between the bottom of the stencil and the top of the board, see Figure 9.18) and the condition of the printing equipment determine print height. The most commonly used snap-off height is 0.020–0.040 inch. Also, as discussed earlier, the reflow solder height is determined not by the paste height alone, but by the metal content of the paste as well (Table 9.2).

No matter what approach is needed, one must answer some basic questions. For example, why was the method (single thickness versus stepped stencil) chosen? What is the final paste thickness required? What is the snap-off distance? Are you using a metal or a rubber squeegee (see next section)? How is the stencil going to be made (etching versus laser versus additive process)? Have you looked into the implications of each approach? The basic idea is to get good fillets on fine pitch without getting insufficient fillets on the standard pitches.

Figure 9.23 provides a general guideline for paste thickness requirements for different pitches. However, one must keep in mind that boards

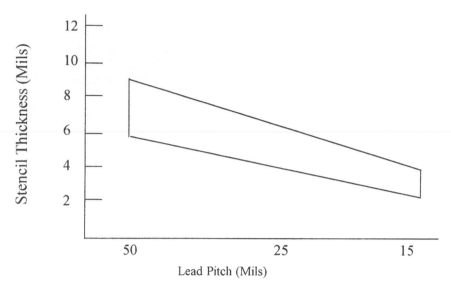

Figure 9.23 Recommended range of stencil thicknesses for various lead pitches.

will have a number of different pitches. In such a case the requirement is governed by the finest pitch on the board. Also there is not just one right thickness, as is indicated by the band width in Figure 9.23.

Generally, the print obtained from a stencil 8 to 10 mils thick will be thick enough to generate acceptable solder joints on 50 mil lead pitch packages using standard land dimensions.

For finer pitch (25 mil or less) packages, paste only about 6 mils thick is required. This requires the use of multilevel or selectively etched metal stencils as discussed earlier. (Refer to Figures 9.20 and 9.21.) The openings in the stencils for fine pitch packages can also be modified (staggered partial printing on fine pitch lands) to allow the same paste thickness for all packages.

Paste height requirements for ball grid arrays (BGAs) are roughly the same as for standard 50 mil pitch surface mount components. Refer to Section 9.6 for further discussion on printing thickness requirements for BGA, fine pitch, and through-hole components.

9.5.3 Squeegee Wear, Pressure, Hardness, Type, and Orientation

Squeegee wear, pressure, and hardness determine the quality of print and should be monitored carefully. For acceptable print quality, the squeegee edges should be sharp and straight. A low squeegee pressure results in skips and ragged edges. A high squeegee pressure or a soft squeegee will cause smeared prints and may even damage the squeegee and stencil or screen. Excessive pressure also tends to scoop solder paste from wide apertures, causing insufficient solder fillets. See Table 9.8 for recommended pressures.

Two types of squeegees are commonly used: rubber or polyurethane squeegees and metal squeegees. When using rubber squeegees, 70 to 90 durometer hardness squeegees are used. For screen 70–80 durometer hardness is recommended. For stencils, higher hardness (80–90 durometer) squeegees will give prints of acceptable quality.

Higher viscosity pastes (more than 750,000 centipoise) require harder squeegees (more than 80 durometers) especially for fine pitch. When using harder rubber squeegees, the pressure tends to be high (around 30 lb.). As mentioned earlier, this may scoop solder paste, especially from wider apertures. If that happens, the pressure should be lowered. *The general guideline is to start with a low pressure and increase it until a clean swipe on the stencil is achieved. There should be no paste on the stencil.*

When excessive pressure is applied, bleeding of paste underneath the stencil may cause bridging and will require frequent underside wiping. To prevent underside bleeding, it is important that the pad opening provide a gasketing effect while printing. This is dependent on the roughness of the stencil aperture walls. We discuss this issue in Section 9.5.7.

When using a rubber squeegee, it is important that the edges be kept sharp. If a radius develops on the edge, it is time to sharpen the squeegee. There are two common configurations of a squeegee blade: diamond for bidirectional printing and flat blade for unidirectional printing. See Figure 9.24.

Metal squeegees are also commonly used. Their popularity has grown with the usage of finer pitch components. They are made from stainless steel or brass in a flat blade configuration. They are used at a print angle of 30–45°. Some squeegees are coated with lubricating material. Since lower pressure is used, they do not scoop paste from apertures. And since they are metallic, they do not wear easily like the rubber squeegees do and hence they do not need to be sharpened. They do cost significantly more than the rubber squeegees. They can also cause wear on the stencil.

Metal squeegees are not suitable for screens. They are primarily used on stencils, but they are generally not recommended for stepped stencils. If metal squeegees are to be used on stepped stencils, it is important that there be sufficient clearance around the package (about 0.125 inch for each 0.002 inch step down).

One will notice a lesser amount of solder paste on pads that are

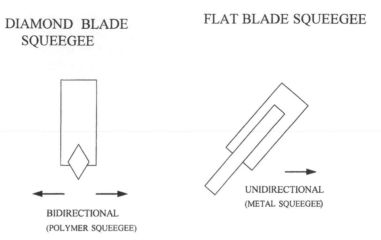

Figure 9.24 Bidirectional and unidirectional rubber squeegee blade configurations.

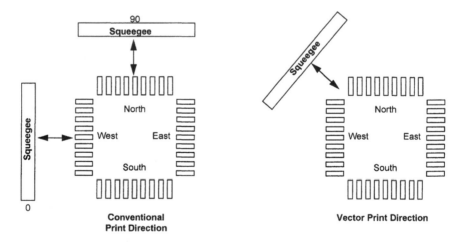

Figure 9.25 Vector printing—squeegee at an angle for uniform paste printing [17].

parallel to the squeegee than the pads that are perpendicular to the squeegee travel line. The squeegee is in contact longer with the pads in the longitudinal direction (N–S pads in Figure 9.25) than with the pads that are parallel to the squeegee (E–W pads in Figure 9.25). The N–S pads have a longer time for filling the aperture than the E–W pads. This may lead to clogging or insufficient paste on E–W pads and bridging on the N–S pads. To alleviate this problem, aligning the squeegee at 45° or "vector printing" is used at times [17]. This approach may provide a more uniform paste deposition on all pads. However, aligning the squeegee at an angle is not a very common practice since most printers do not have this capability.

9.5.4   Print Speed

Print speed is critical for obtaining good print quality. A fast print speed will cause planing of the squeegee, resulting in skips. The squeegee will not have enough time to fill the aperture, resulting in insufficient fill. The E–W pads in Figure 9.25 show this tendency. Though a slow speed is generally preferred for paste printing, too slow a speed will cause ragged edges or smearing in the print. See Table 9.8 for recommended print speeds.

9.5.5   Mesh/Stencil Tension

Screen mesh is used both in screens and stencils. If mesh is used with stencil, it is referred to as the flexible metal mask (Figure 9.13d and Figure 9.15). The mesh should be taut enough to maintain good registration between the land pattern on the substrate and the stencil/screen opening, adequately releasing the paste onto the lands by snapping back the stencil/screen and not smearing the print. However, the tension should not be high enough to elongate the screen/stencil mesh beyond its yield point.

When using an all metal stencil, providing adequate tension may be an issue. If this is the case, only on-contact printing may be possible. As discussed earlier, universal frames (Figure 9.16) have a sufficient number of screws to provide the needed stencil tension to allow off-contact printing. In on-contact printing, the stencil must release vertically without any lateral vibration to avoid any smudging. In off-contact printing, the stencil is in contact with the board only at one point (see Figure 9.18) and the possibility of smudging is reduced.

9.5.6   Board Warpage

Excessive board warpage will result in skips on areas where the top of the board does not touch the bottom of the stencil during printing. To minimize board warpage, the boards should be held flat, with vacuum pressure supplied from the underside of the base plate. Vacuum leaks should be minimized. Any stray vacuum leak will result in a poor quality

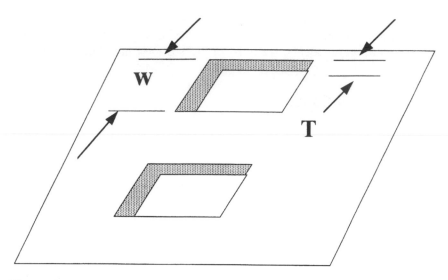

Figure 9.26 Stencil aperture width versus thickness for good print with minimum aspect ratio (W/T) of 1.5.

print. Tool plates with tooling pins are also necessary to hold boards in place for accurate paste printing.

9.5.7 Etched, Laser Cut, and Electroformed Stencils

One of the important variables in print quality is the accuracy and smoothness of the side walls of the stencil aperture. And one of the variables that control the accuracy of the aperture is the way data is transferred to the stencil manufacturer. There are three ways to transfer this data: by Mylar film, by glass master, or by Gerber file. The Mylar film is temperature and humidity dependent and can easily be damaged. The glass master can broken or scratched. Gerber file is the best and fastest way to transfer data (via modems).

It is important to maintain a proper aspect ratio between the width and thickness of the stencil. Figure 9.26 shows the recommended aspect ratio (aperture width divided by stencil thickness) of 1.5 for a stencil. This is important for preventing clogging of the stencil. Solder paste will tend to remain in the opening if the aspect ratio is less than 1.5 [18].

Both the smoothness and the accuracy of aperture walls are controlled by the process by which the aperture is made. There are three common processes for making stencils: chemical etching, laser cutting, and the additive process. Let us review them.

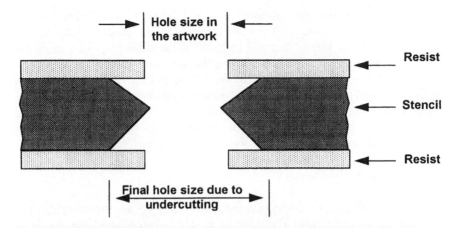

Figure 9.27 **Undercutting in chemically etched stencil producing walls that taper to an hourglass shape.**

9.5.7.1 *Chemically Etched Stencils*

The metal mask and flexible metal mask stencils are etched by chemical milling from both sides using two positive images. During this process, etching proceeds not only in the desired vertical direction but also in the lateral direction. This is called undercutting. Thus the openings are larger than desired, causing extra solder deposit. Since 50:50 etching proceeds from both sides, it results in almost a straight wall tapering to a slight hourglass shape in the center as shown in Figure 9.27.

The hourglass shape shown in Figure 9.27 is a problem due to undercutting, which in turn is related to the etch factor. A function of the thickness of the metal being etched, the etch factor is about 25% of the metal thickness. By providing artwork with reduced pad sizes at the photographic stage during chemical milling, the openings will be larger than shown in the artwork but will match the land patterns on the board. For example, the artwork for a 0.008 inch stencil must be reduced by 0.002 inch per side. This process is called compensation, since it compensates for the etch factor. Some vendors can correct for the etch factor at additional cost by modifying the artwork or the Gerber file. This is a more desirable option since the vendor is more familiar with his process and can make the necessary changes in the artwork or Gerber file to meet the final stencil thickness, and aperture shape and size requirements.

As shown in Figure 9.27, during the etching process, the walls may not be smooth or totally vertical. Electropolishing, a microetching process, is one method for achieving a smooth wall. Another way to achieve smoother side walls in the aperture is nickel plating. A polished or smooth surface is good for paste release but may cause the paste to skip across the stencil surface rather than roll in front of the squeegee. This problem can be avoided by selectively polishing the aperture walls without polishing the stencil surface. Nickel plating further helps improve the smoothness of the wall and provides some improvement in printing performance [19]. However, it does reduce aperture opening and requires artwork adjustment.

The problem of undercutting is avoided when making stencils with the laser and additive processes; these are discussed next.

9.5.7.2   *Laser Cut Stencils*

Like the chemical etching process, laser cutting is a subtractive process but without the undercutting problem. The stencil is produced directly from the Gerber data and hence the accuracy of the apertures is improved. The data can be adjusted in the Gerber file to change dimensions as necessary. Better process control also helps improve aperture accuracy. Another benefit of laser cut stencils is that the walls can be programmed to be tapered. Chemically etched stencils can also be tapered if they are etched only from one side, but the aperture size may be too large. A tapered aperture with opening slightly larger on the board side than on the squeegee side (1–2 mils to produce an angle of about 2°) is desired for easier paste release.

Laser cut is a process that is capable of producing aperture widths as small as 0.004 mil with an accuracy of 0.0005 inch, hence it is very suitable for ultra fine pitch component printing. Laser cut stencils also produce ragged edges since during the cutting process the vaporized metal is transformed into metal slag. This can cause paste clogging. Smoother walls can be produced by microetching the walls. Laser cut stencils cannot make stepped multilevel stencils without pre-chemical etching of the areas that need to be thinner. The laser cuts each aperture individually, so the cost of the stencil depends upon the number of apertures to be cut.

9.5.7.3   *Electroformed Stencils*

Chemical etching and laser cutting are subtractive processes for making stencils. The chemical etch process is the oldest and most widely used. The laser cut is a relative newcomer. The newest process for making stencils is the additive process. It is also referred to by other names such as electroformed or Galvano; the term electroformed is most common.

In this process, nickel is deposited on a copper mandrel to build the aperture. First a photosensitive dry film is laminated on the copper foil (about 0.25 inch thick). The film is polymerized by UV light through a photomask of the stencil pattern. After developing, a negative image is created on the mandrel where only the apertures on the stencil remain covered by the photoresist. The stencil is then grown by nickel plating around the photoresist. After the desired stencil thickness is achieved, the photoresist is removed from the apertures. The electroformed nickel foil is separated from the mandrel by flexing—a key process step. The mandrel can be reused. Now the foil is ready for framing as in other stencil making processes.

Electroforming is very similar to the process of making printed circuit boards except that it is done in a very clean room environment (class 1000) with operators in bunny suits. The bunny suits and the concept of making metal foils by nickel plating should give some idea of the precision and cost. Electroforming is a process for making stencils with very precise

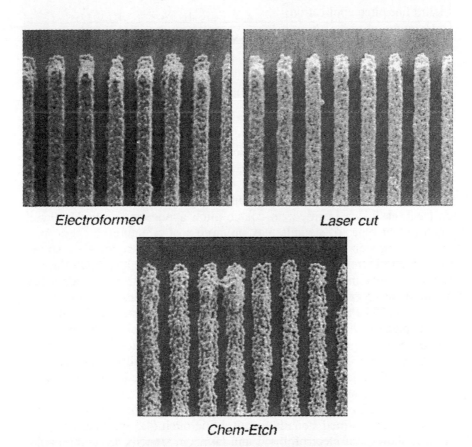

Electroformed   Laser cut

Chem-Etch

Figure 9.28   Ultra fine pitch paste deposited through 4.5 mil apertures made by electroformed, laser cut, and chemically etched stencils. (Courtesy of AMTX, Inc.)

Stencil Type / Key Features	Chemically etched Stencils	Laser cut Stencils	Electroformed Stencils
Aperture size control	△	○	◇
Aperture wall smoothness	○	△	◇
Aperture wall shape control	△	○	◇
Step stencil capability	◇	△	△
Foil strength	◇	◇	○
Gasketing effect	△	○	◇
Ultra fine pitch application	△	○	◇
Cost	Low	Medium	High

◇ = VERY GOOD   ○ = GOOD   △ = FAIR

Figure 9.29   **The key features of chemically etched, laser cut, and electroformed stencils**

tolerances (in microns). It should be no surprise that it is the most expensive process for making stencils. But in applications such as ultrafine pitch printing (below 20 mil pitch) or even in flip chip applications, where conductive adhesive patterns are printed directly on the silicon wafer, its cost can easily be justified. In coarse pitch applications (above 20 mil pitch), other types of stencils may be more cost effective.

Making step stencils by electroforming is not possible. Also, sometimes the foil strength of electroformed stencils, especially when the foil is under 0.004 inch thick, can be a concern [19]. As the process has matured, stronger electroformed stencils with a hardness of around 400 Knoop and 221 PSI tensile strength have been commonly produced. One of the biggest advantages of this process is that because of the close tolerances possible, electroformed stencils provide a good gasket effect, which minimizes seepage of paste under the stencil. This means that the frequency of underside stencil wiping is drastically reduced. This helps in reducing potential bridges.

Figure 9.28 compares the quality of print by chemically etched, laser cut, and electroformed stencils for ultra fine pitch (4.5 mil aperture size) [20]. It shows that electroformed and laser cut stencils are comparable, but the chemically etched stencil shows bridges. Figure 9.29 summarizes the key attributes of these three stencil making processes.

9.6 PRINTING FOR DIFFERENT TYPES OF COMPONENTS

The printing process is lead pitch dependent. As discussed earlier, the pitch determines the aperture size, and aperture size determines the powder size, rheology, and other printing variables. For simplicity, we will divide the printing requirements into three categories: standard SMT including ball grid array (BGA), fine and ultra fine pitch, and through hole. Let us review the printing requirements for these components.

9.6.1 Printing for Ball Grid Arrays

As far as selection of paste and stencil designs are concerned, printing for BGA is no different than printing for standard 50 mil pitch surface mount components. For BGA, Type 2 solder paste with a powder size of 45–75 microns is used. There are some differences in ceramic BGA and plastic BGA printing requirements, however.

For a ceramic ball grid array (CBGA), the volume of the solder is very critical since the balls do not melt during reflow. (Generally the reflow temperature is kept at a maximum of 230°C. CBGA balls made of 90% Pb and 10% Sn melt at 302°C.) For CBGA 8 mil thick paste, 7000 cubic mils of paste is nominal and 4800 cubic mils is the minimum required [21]. If the volume of paste falls below the minimum volume, the incidence of opens and insufficient solder increases.

For a plastic ball grid array (PBGA), the paste thickness is not as critical because the balls reflow completely. As a result of the natural collapse of solder balls during reflow, tight process controls are not required during PBGA printing. As a matter of fact, for PBGAs, even a flux only process works fine (assuming no warpage in the package or the board). So when using paste, about 8 mils works fine for most BGAs.

9.6.2 Printing for Fine Pitch and Ultra Fine Pitch

When printing for fine pitch, the powder requirement in the paste changes from 45 to 75 microns (Type 2 paste) to 20 to 45 microns (Type 3 paste). Type 4 fine particle powders (20–38 microns) may be required for ultra fine pitch (under 20 mil pitch) printing. This leads to the need for a higher paste viscosity. A paste viscosity of 1 to 1.3 million cps is common for fine pitch. When printing for ultra fine pitch, it also becomes important to control the smoothness and the dimensions of the apertures in the stencil. This means that the use of laser cut or electroformed stencils should be seriously considered.

When using fine pitch, it becomes necessary to consider dual-thickness printing (Figure 9.20). This requires added clearance around the fine pitch lands (100 mils minimum). Dual-step stencils call for 5–6 mils for fine pitch and 7–8 mils for other components. The same thickness must be used in laser cut or electroformed stencils since step stencils are not feasible with these processes.

Stepped stencils are excellent for providing the correct amount of paste at specific locations. In a board containing both fine pitch and 50 mil pitch components, we have obtained good results with 7 mil thick stencils for 50 and 25 pitch components and 5 mil stencils for 20 mil pitch components.

When only one thickness of paste is used for both fine pitch and standard SMT, there is some concern that there will be either too much paste on the fine pitch lands or too little paste on the standard surface mount lands. The use of stepped stencils alleviates this concern. However, when using a metal squeegee, stepped stencils may not be a good idea since it is difficult to apply the added pressure needed for stepped stencil.

Fine pitch also calls for higher squeegee hardness (>80 shore) or may necessitate the use of metal squeegees. The use of higher viscosity paste, step down stencils, and metal squeegees at slower print speeds (0.5 to 1.5 inch per second) are some of the process changes that need to be made for fine pitch. The finer the pitch, the slower the print speed. A vision system may also become necessary when printing for fine pitch. Also, frequent underside wiping of the stencil will become necessary (e.g., wiping after every four boards).

9.6.3 Printing for Through Hole in a Mixed Assembly

Paste printing is generally not used for through hole, but there are circumstances when it should be considered. For example, in a mixed assembly with double-sided SMT boards and very few through-hole components, it may be cost effective to print paste for through hole to avoid hand or wave soldering. When hand soldering is used for through hole, there is always the potential for internal damage. Refer to Chapter 14. However, through hole components must be able to withstand reflow soldering temperatures.

There are three ways to print for mixed assembly with both surface mount and through-hole components. The first approach is the same as discussed earlier using one stepped stencil as shown in Figure 9.20.

However, the main concern with this approach would be to provide sufficient clearance around the surface mount lands since the step is large (20 mil for through hole and 6–8 mils for SMT). Such a large step is hard to print, even after providing sufficient clearance.

One can also use two separate stencils for mixed assembly [19]. First the paste is printed on the surface mount lands in a standard way with 6–8 mil stencil. Then a second stencil (20 mil thick is used for printing only on the through-hole lands (Figure 9.30). The second stencil has etched relief (10 mil thick) on the contact side (not the squeegee side). It is essentially the opposite of the stepped stencil shown in Figure 9.20, which has etched relief for fine pitch on the squeegee side.

Also, the etched relief, as shown in Figure 9.30, does not have any aperture for paste printing on surface mount lands. Instead, the 10 mil deep etched relief is provided so that the previously printed paste on surface mount lands (under 8 mil thick) does not touch the bottom of the stencil when the paste is printed on the through-hole lands. In other words, the relief area is intended to prevent paste smearing of previously printed paste on the surface mount lands.

The benefit of two step printing is that optimum paste thickness is applied to both surface mount and through-hole lands. However, the use of two different stencils on the same board does increase the potential for paste smudging and smearing unless the alignment of the stencils is perfect each time. Also, when attempting to reflow through-hole components, one must make sure that they are not moisture sensitive. If they are, they need to be baked before reflow. In addition, one must ensure that they can withstand the reflow temperature. If they are temperature sensitive, they cannot be reflow soldered.

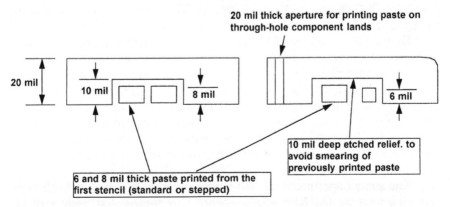

Figure 9.30 Schematic of a step stencil for printing on through-hole lands in a mixed assembly.

The third method for printing sufficient solder paste for reflow soldering of through-hole components is to use the same stencil thickness for both surface mount and through-hole components. The desired solder volume is achieved by printing paste with larger stencil apertures (square, rectangular, or round) than their target through-hole pads.

The volume of required solder paste is a function of the through hole diameter, lead diameter, and the thickness of the board (length of hole). In addition, as shown in Table 9.2, one needs to keep in mind that the volume of reflowed solder is about half the solder paste volume printed on the board.

Considering these factors, let us determine the volume of solder paste required. To achieve top and bottom solder fillets, the paste volume can be determined by the following formula:

$$RPV = \text{required paste volume} = 2 \times \Pi (D_h^2 - D_l^2) \times T/4$$

where 2 is the multiplication factor to compensate for 50% paste shrinkage
during reflow as per Table 9.2
D_h is the plated through hole diameter
D_l is the lead diameter
T is the board thickness (or the length of the hole)

It should be noted that the above formula does not take into account any paste that may be deposited into the hole during printing. So any paste deposited into the hole will provide added assurance of achieving the desired top and bottom side solder fillets.

The required stencil aperture area for a particular lead can be determined by dividing *RPV* by the thickness of the stencil. The exact aperture geometry (length and width or diameter) will have to be determined by the space available around the pad of that lead.

Belmonte and Zarrow [22] have found that there is an optimum size of plated through hole for a given lead size. For example, they have established that plated through-hole size for a round lead should be 12 mils (0.3 mm) larger than the maximum lead diameter and 10 mils (0.25 mm) larger than the maximum diagonal dimension of the lead. If the hole size is too large, it will require so much paste that it may be impractical to print it. This means that the plated through hole size needs to be established at the design stage if paste printing is to be used for through-hole components.

The stencil aperture sizes will be much larger than the through-hole pad area to meet the RPV requirement. This means that paste will be printed on the solder mask. Printing on solder mask may increase the potential for solder balls if the paste quality is not properly controlled.

In addition, due to oversized stencil apertures, the potential for bridging is increased because of reduced space between adjacent pads.

No matter which one of the three methods discussed in this section is used, as noted earlier but worth repeating, it should first be determine that the through-hole components can withstand the reflow temperature. In addition, it should also be determined if the parts are moisture sensitive. If they are, they must be baked before soldering for reasons discussed in great detail in Chapter 5, Section 5.7.

It should also be noted that if the board has wave solderable surface mount components on the secondary side, reflow soldering of through-hole components will not save any process steps. See Chapter 7, Section 7.4, Soldering Considerations, for applicability of different soldering approaches.

9.7 SUMMARY

The variables that control print quality are paste, screen printer, method of paste application, and printing process variables. Solder paste or cream is simply a suspension of fine solder particles in a flux vehicle. The composition of the particles can be tailored to produce a paste of the desired melting range. Additional metals can be added to change paste compositions for specialized applications. Particle size and shape, metal content, and flux type can be varied to produce pastes of varying viscosities.

Some of the major features to consider when selecting screen printers are the maximum board size handling capability, controls for accurate screen alignment and repeatability of print, and the board hold-down mechanism. Considerable progress has been made in the application methods, such as screening and stenciling for pastes on substrates. Each method of application has its pros and cons, but stenciling is more widely used for both low and high volume production.

There are three main ways of making stencils: chemical etching, laser cutting, and electroforming. Chemical etching is the oldest, most widely used, and least expensive option. Laser cutting and electroforming, especially the latter, find appeal in applications where precision is paramount. In addition to the type of stencil, the type of squeegee blade is very important. Metal blade squeegees are widely used because they require less maintenance than rubber squeegees.

In order to achieve good printing results, a combination of the right paste material (right rheology, i.e., viscosity, metal content, largest powder size, and lowest flux activity possible for the application), the right tools (printer, stencil, and squeegee blade), and the right process (good registration, clean sweep) are necessary.

Although the supplier is essentially responsible for providing the desired solder paste and screens or stencils and the squeegee blades, the user must control process and equipment variables to achieve good print quality. Even the best of pastes, equipment, and application methods are not sufficient by themselves to ensure acceptable results.

REFERENCES

1. J-STD-005. Requirements for soldering pastes. Available from IPC, Northbrook, IL.

2. Stein, A.M. "How to choose solder paste for surface mounting. Part 1." *Electronics*, June 1985, pp. 57–59.

3. Rooz-Kozel, B. Solder Paste, Surface Mount Technology. International Society for Hybrid Microelectronics Technical Monograph Series 6984-002, Reston, VA, 1984.

4. Peterson, G.G. "A practical guide for specifying printing equipment for electronic applications. SMT-III-21, *Proceedings of the SMART III Conference* (IPC/EIA), January 1987.

5. Jennie S. Hwang. "Solder/Screen Printing." *Surface Mount Technology*, March 1994, pp. 44–50.

6. J-STD-004. Requirements for soldering fluxes. Available from IPC, Northbrook, IL.

7. Seelig, Karl, "A new generation of no clean solder paste," *Proceedings of Surface Mount International*, August 30–Sept. 3, 1992, pp. 705–715.

8. O'Hara, Wanda B., and Lee, Ning-Chen. "How voids develop in BGA joints." *Surface Mount Technology*, January 1996, pp. 44–47.

9. Rosen, R.L. *Fundamental Properties of Polymeric Materials*. New York: Wiley, 1982, p. 202.

10. MacKay, C.A. "Solder creams and how to use them." *Electronic Packaging and Production*, February 1981, TR-1071.

11. Kevra, Joseph. "Solder paste rheology using spiral viscometer." *Proceedings of ISHM*, 1991, pp. 240–243.

12. Morris, J. "Solder paste tackiness measurement." *IPC Technical Review*, September–October 1987, pp. 18–24.

13. Mackay, C.A. "What you don't know about solder creams." *Circuits Manufacturing*, May 1987, pp. 43–52.

14. Heimsch, R.D. "Framework for the evaluation of precision screen printing equipment." *Hybrid Circuit Technology*, June 1986, pp. 19–24.

15. Atkinson, R.W. "An automated screen printer." *Circuits Manufacturing*, January 1983, pp. 26–29.

16. Borneman, J.D., and Rennaker, R.L. "Paste printing from a pro." *Circuits Manufacturing*, February 1987, p. 38.

17. Bennett, Rickey. "Controlling paste volume with vector printing." *Surface Mount Technology*, June 1995, pp 96.

18. Herbst, Mark D. "Metal mask stencils: The key components." *Surface Mount Technology*, October 1992, pp. 20–22.

19. Coleman, William E., "Photochemically etched stencils for ultra-fine pitch printing." *Surface Mount Technology*, June 1993, pp. 18–24.

20. Clouthier, Richard. "Appraising stencils for fine-pitch printing." *Surface Mount Technology*, March 1995, pp. 60–64.

21. Dody, Glen, and Burnette, Terry. "BGA assembly process and rework." *Proceedings of Surface Mount International*, August 29–31, 1995, pp. 361–366.

22. Belmonte, Joe and Zarrow, Phil. "Advanced surface mount manufacturing methods," *Circuits Assembly*, September 1996, pp. 36–41.

Chapter 10

Metallurgy of Soldering and Solderability

10.0 INTRODUCTION

The solder used in electronic assembly serves to provide electrical and mechanical connections. In through-hole mount assemblies we had to worry primarily only about obtaining sound electrical connections, because the plated through holes imparted sufficient mechanical strength. With the advent of SMT, the role of the surface mount solder joint has become very critical, because it must provide both mechanical and electrical connections. The solder joint strength is controlled by the land pattern design and a good metallurgical bond between component and board.

Surface mount land pattern design for providing adequate mechanical strength was covered in Chapter 6. In this chapter we concentrate on the metallurgical aspects of a reliable solder connection, as determined by the solder and the solderability of components and boards. A reliable solder connection must have a solderable surface to form a good metallurgical bond between the solder and the components being joined. An understanding of metallurgical bonding entails knowledge of phase diagrams and the concepts of leaching, surface finish, wetting, and oxidation of metallic surfaces.

Metallurgical phase diagrams are used to display the solubility limits of one metal into another and the melting temperatures for metals and their alloys, for a better understanding of intermetallic bonds. Phase diagrams can also be used for a better understanding of leaching or the dissolution phenomenon (of one metal into another). Finally, to produce solder joints in a cost-effective way, we need to know about the methods, requirements, and economics of solderability testing.

Tin-lead (Sn-Pb) solders are the most commonly used alloys for electronic soldering. However, due to environmental concerns about lead in solder, there is also considerable interest in various lead-free solders. So, while we will focus primarily on issues related to Sn-Pb solders, we will also briefly discuss lead-free solders.

The focus of this chapter is on the practical aspects of metallurgical issues related to the solderability of surface mount assemblies. The subject of soldering for surface mount assemblies is covered in Chapter 12. For an expanded coverage of basic metallurgical issues in electronics, refer to Wassink [1] and Manko [2].

10.1   PHASE DIAGRAMS

Phase diagrams are used in the electronics industry to determine the existence of intermetallic compounds and the melting points of metals and their alloys. These charts can indicate the solubility of one metal into another at different temperatures, as well as various phases of alloy compositions. Phase diagrams are graphic representations of the equilibrium interrelationships between elements and compounds. A multitude of phase diagrams have been established.

The most commonly used phase diagram in the electronics industry is the tin-lead phase diagram, because tin-lead is used for soldering, but tin-bismuth, indium tin, nickel tin, and silver-tin phase diagrams are also used. Some phase diagram systems are as follows:

1.   *Unary systems,* which contain only a single element or compound—for example, tin, copper, water. Unary systems show only the melting and boiling points of elements or compounds as a material transforms from the solid to the liquid phase.
2.   *Binary systems,* which contain two elements or compounds, are the most widely used. Examples of binary systems in the electronics industry include lead-tin, gold-tin, silver-tin, tin-bismuth, indium-tin, and nickel-tin.
3.   *Ternary systems,* which contain three elements or compounds, such as tin-lead-silver, are used in some solder pastes. These complex systems are more difficult to interpret.

The construction of a phase diagram, by plotting melting point versus compositions for the alloy under evaluation, is a laborious and time-consuming process. Fortunately, others have done the difficult job and we can enjoy the benefits as long as we can learn how to "read" a phase diagram. We begin by stating that as the composition varies, the microstructure and melting points vary. For most alloys, there is a range of melting points. This will become clearer if we examine Figure 10.1, a phase diagram for a binary eutectic system consisting of elements A and B. The Y axis represents temperature and the X axis represents

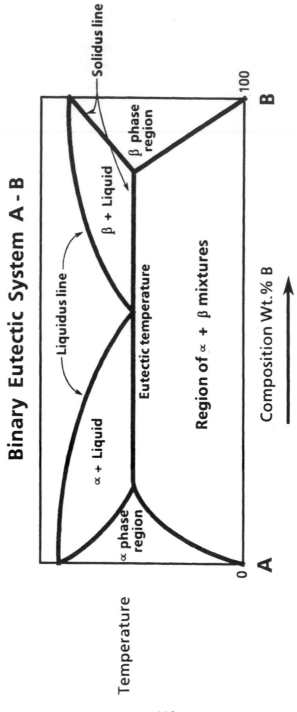

Figure 10.1 An example of a binary eutectic system of elements A and B.

composition. At point A, the composition is as follows: element A 100%, element B 0%. At point B, we have 100% element B and 0% element A.

The solid lines in a binary phase diagram represent an equilibrium between the two phases of the system, namely, the liquid phase and the solid phase. The line above which the alloy is entirely liquid is called the liquidus line, and the line below which the alloy is entirely solid is called the solidus line. The solid phase, lying below the solidus line, has two different phases: α and β. The α phase region shows the maximum solubility of element B in A at different temperatures, and, similarly, the maximum solubility of element A in element B at different temperatures is shown in the β region.

One should not confuse the individual elements with the different phases. For example, the α phase is not element A saturated with element B. It is very different from both elements A and B and has entirely different properties and microstructure. By the same token, the β phase also differs from each of its components. Between the two regions, however, is a mixture of the α and β phases.

The eutectic temperature is the lowest temperature at which melting is observed for the eutectic composition (α and β phases) of elements A and B. At temperatures above the eutectic melting point, the eutectic composition is liquid but any other composition is pasty. To the left of the eutectic composition (above the eutectic temperature), the pasty region contains the liquid and α phases; to the right, the pasty region contains the β and liquid phases.

At temperatures below the eutectic temperature, the alloy compositions (of elements A and B) to the left and right of the eutectic composition are called the hypoeutectic AB and hypereutectic BA compositions, respectively. For determining the percentages of different microstructural constituents (such as liquid, α, β, element A, element B) at different locations on the phase diagram, a rule known as the lever rule is used. These and other details of phase diagrams are not necessary for the points I want to make in this chapter. Readers interested in such details should refer to Wassink [1] and Manko [2].

If we say that element A is lead (Pb) and element B is tin (Sn), we have an actual Pb-Sn phase diagram, as shown in Figure 10.2, which can be used to obtain some useful information on solder alloys. The terminology and the principles are the same as those just discussed. Thus from Figure 10.2 we see that the α phase is the solid solution of tin in lead. The maximum solubility of tin in lead (or lead in tin) occurs at the eutectic temperature. The eutectic solder composition of 63% tin and 37% lead, which melts at 183°C (361°F), has a lower melting point than either lead (327°C) or tin (231°C), as shown in Figure 10.2.

The Sn-Pb phase diagram also shows the limits of variations in

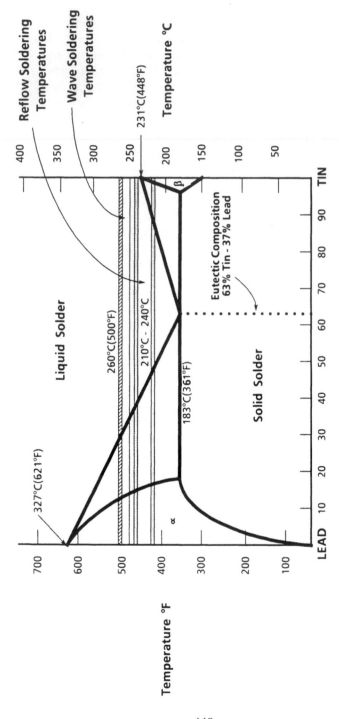

Figure 10.2   The tin-lead phase diagram.

448

compositions that can be easily soldered. The horizontal line superimposed as the reflow soldering temperature clearly shows that the solder composition should not vary too much from the eutectic composition. Otherwise the reflow soldering temperatures, normally around 220°C, would have to be considerably higher.

The rule of thumb is that the reflow and wave soldering temperatures should be about 30 to 50°C and 35 to 70°C, respectively, above the melting point of the alloy being used for soldering.

Figure 10.2 shows the reflow soldering temperature to be 210°–240°C. Wave soldering temperatures are even higher (generally 240°–260°C) because the solder not only should be molten but must have very low viscosity and suitable hydraulic flow characteristics to permit recirculation in the solder pot. Of course soldering can be accomplished at lower temperatures (above the melting point of the alloy) but, depending on the temperature, it may take longer. Thus a lower temperature may be technically acceptable but not practical in a manufacturing environment.

The noneutectic solder has a pasty range as defined by the temperature range between liquidus and solidus lines. The pasty range is also illustrated by two peak temperatures shown in Figure 10.3 by differential scanning calorimetry (discussed in Chapter 8). DSC is a quick way to determine the solidus and the liquidus points of a noneutectic solder. The first peak in Figure 10.3 (179°C) indicates the solidus line. According to the Sn-Pb phase diagram (Figure 10.2), however, this temperature should be 183°C for any Sn-Pb composition. What is the explanation for the difference?

It should be noted that phase diagrams represent equilibrium conditions and assume that sufficient time has been allowed for the alloy to melt completely. A DSC curve, on the other hand, shows only the first occurrence of melting; so at this point the entire sample may not have melted. The second peak in Figure 10.3 indicates the peak liquidus temperature. Between these two peaks, depending on its composition, the solder exists both in the liquid phase and in either the α or β phase.

Can we distinguish between α and β phases from a DSC curve? Not in general. In the case of Figure 10.3, however, it is possible to tell that the region between the two peaks is the β phase. The clue is provided by the maximum peak temperature (276°C) in Figure 10.3. If we refer back to the Sn-Pb phase diagram in Figure 10.2, we see that the maximum temperature for the β phase is only 231°C.

However, if the second peak temperature shown by Figure 10.3 were less than 231°C, the region between the two peaks could be either the α or the β phase and the liquid phase. Thus a DSC curve may not give the exact composition of an alloy, but it will provide an approximate range.

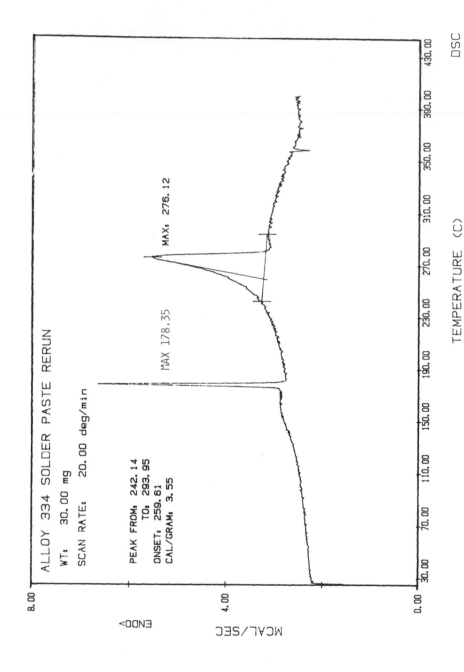

Figure 10.3 Melting points of noneutectic solder as determined by differential scanning calorimetry. Note two distinct melting point peaks.

The DSC curve for pure elements or eutectic compositions will show only one peak, since the melting point is fixed in these cases.

Although eutectic Sn-Pb solder is the most widely used composition, a tin-lead-silver ternary system consisting of 62% Sn, 36% Pb, and 2% Ag is also used. This is a eutectic composition of the ternary system with a melting point of 179°C (354°F). The equilibrium microstructure at room temperature is α, β, and silver-tin intermetallic (Ag_3Sn). The chemical formulas (e.g., Ag_3Sn) of compounds are not derived from the phase diagram but require other analytical tools. Refer to Wassink [1] for details. Also refer to Section 10.4 for a discussion on compositions, properties, and phase diagrams of lead-free solders.

10.2   METALLIZATION LEACHING IN PASSIVE SURFACE MOUNT COMPONENTS

Leaching is a kinetic metallurgical phenomenon in which one element dissolves into another at soldering temperatures. In surface mounting, the term "leaching" designates the dissolution of silver metallization or palladium in surface mount resistors and capacitors during soldering.

Why is leaching a problem? To provide electrical connections to the internal electrodes or terminations, surface mount ceramic chip capacitors and resistors require an adhesion layer of precious metal such as gold, silver, or silver palladium. (See Chapter 3, Figure 3.3.) A silver adhesion layer is more common. After a silver adhesion layer has been applied by a thick film process (printing of paste containing silver), the terminations are either dipped in solder or plated with tin-lead solder. During soldering, the silver is prone to dissolution in the tin. This causes the underlying ceramic surface to be exposed and results in a poor solder fillet or no fillet at all.

How is leaching avoided? By using a barrier layer under the terminations or by using silver in the solder paste. The barrier layer approach works for both wave and reflow soldering, but the addition of silver in solder paste is intended only for the reflow process. Let us discuss the barrier layer approach first. The rate of dissolution of various metals in solder varies greatly. For example, gold, silver, palladium, copper, nickel, and platinum dissolve at decreasing rates in solder, as indicated in Figure 10.4.

Bader [3] conducted an experiment by dissolving pure metal wires, 20 mils in diameter, in solder; different temperatures and times were used, and the rate of dissolution was measured. Figure 10.4, adapted from this study, shows relative rates of dissolution only. For actual rates of dissolu-

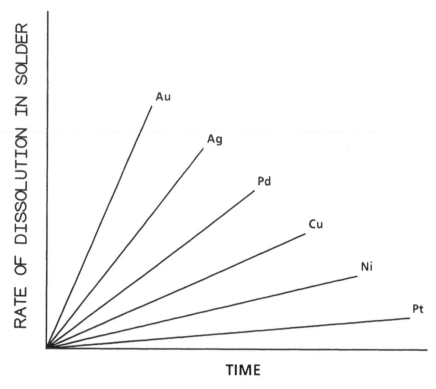

Figure 10.4   The rate of dissolution of various metals in solder [3].

tion for various metals in molten solder at various temperatures, refer to Table 10.1.

From Figure 10.4 we see that platinum and nickel have the lowest dissolution rates in solder and gold and silver have the highest rates. If platinum or nickel is used between silver and solder on surface mount resistors and capacitors, it will act as a barrier and prevent the dissolution of the silver metallization. Because of its high cost, platinum is not used as a barrier layer. Nickel is much cheaper and nearly as effective and therefore is commonly used.

When nickel is used to prevent leaching, the nickel-tin intermetallic compounds formed are Ni_3Sn_4 (liquid-solid reaction) and Ni_3Sn (solid state reaction). The formation of Ni_3Sn intermetallic compounds continues even when the solder has solidified. Once the solder joint has solidified, however, the reaction rate is much slower.

How thick should the nickel barrier layer be? This question can be answered by knowing the growth rate of nickel in tin as a function of time and temperature. This means that we need to take into account not

Table 10.1 Dissolution rate of various metals in solder at different temperatures [3]

METAL	TEMPERATURE (°F)	RADIAL DISSOLUTION RATE (microinches/ second)
Gold	390	35.0
	420	68.5
	450	117.9
	486	167.5
Silver	390	20.7
	450	43.6
	525	97.0
	600	190.7
Palladium	450	1.4
	525	6.2
	600	3.6
	700	14.0
	800	40.5
	900	103.1
Platinum	700	0.83
	800	5.0
	900	16.9
Copper	450	4.1
	525	7.0
	600	21.2
	700	61.5
	800	143.0
	900	248.0
Nickel	700	1.7
	800	4.4
	900	11.3

only the soldering time and temperatures but also the time and temperatures in service. As noted above, solid state reactions (diffusion) cause leaching to continue even after solidification, although at a much slower rate.

It is known that the rate of dissolution of silver at 525°F (near soldering temperatures) is more than 50 times the rate for nickel. It is difficult to precisely compare dissolution rates for silver and nickel, since

Table 10.1 does not show data at common temperatures for all metals. This is because the rate of nickel dissolution at temperatures under 700°F is so low that it is hard to measure. Even at 700°F, nickel dissolves at only 1.7 microinches/second, whereas the dissolution rate for silver at a lower temperature (600°F) is 190.7 microinches/second. The dissolution rates are higher at higher temperatures, where reaction rates increase.

Surface mount components spend 3–4 seconds in the wave and 30–90 seconds in reflow. Since the dissolution rate of nickel in solder is almost negligible at soldering temperatures and is 1.7 microinches/second at 700°C, a nickel barrier layer of about 25 to 30 microinches should be sufficient to prevent leaching of silver during wave and reflow soldering.

A nickel barrier of about 50 microinches is commonly used on all chip components to prevent leaching during soldering. This provides some margin for storage and for two soldering steps on double-sided assemblies. It should be noted that once an assembly has been soldered, the leaching phenomenon becomes less important.

If a component lacks a nickel barrier, a second approach to preventing leaching during reflow is to use a solder paste containing 2% silver, in a 36% lead-62% tin base. Leaching still occurs, but the sacrificial source for silver is the silver in the solder paste on the land, not the essential silver adhesion layer under the component termination. This approach can be used only for reflow soldering (vapor phase, convection, or infrared), not for wave soldering.

Leaching is a more serious problem in wave soldering than in reflow soldering because of the higher temperatures used in wave soldering. Can users add silver to the wave solder pot to prevent leaching? No. This would be wasteful if the same wave soldering machine were also used for through-hole assemblies, which are not affected by leaching, but even for dedicated wave soldering equipment for SMT, the price is too high.

The most appropriate and commonly used approach is to buy surface mount resistors and capacitors with about 50 microinches of nickel barrier from a component supplier. This allows flexibility in using the same components for either wave or reflow soldering. The nickel barrier approach has almost become an industry standard and should be named on the procurement specifications.

10.3   SOLDER ALLOYS AND THEIR PROPERTIES

Solder is most often thought of as an alloy of tin and lead. This is partly true. Tin and lead alloys are the most common, but many other compositions of solder are gaining acceptance, including some lead-free solders discussed in Section 10.4. Two important selection criteria for solder are melting point and strength; these properties for familiar alloys are shown in Table 10.2. Refer to Chapter 9 for commonly used solder compositions in a solder paste.

The properties of solder vary widely depending on composition and microstructure. Of all the properties, fatigue resistance is most important. Relative fatigue resistance values for commonly used solders are shown in Table 10.3 as developed by DeVore [4], who used eutectic solder as the control. DeVore also characterized the fatigue resistance of various solders based on their microstructure, as shown in Table 10.4 [4].

It should be noted from Table 10.3 that the fatigue resistance of 63/37 tin-lead solder is three times less than that of the rarely used 96/4 tin-

Table 10.2   Properties of electronic grade solders

ALLOY DESIGNATION		MELTING POINT OR RANGE (°C)	MECHANICAL STRENGTH (PSI)	
			IN TENSION	IN SHEAR
62 Sn/36 Pb/2 Ag	Sn 62	179		
63 Sn/37 Pb	Sn 63			
	ASTM 63A	183	7700	5400
	ASTM 63B			
60 Sn/40 Pb	Sn 60			
	ASTM 60A	183-189	7600	5600
	ASTM 60B			
96.5 Sn/3.5 Ag	Sn 96			
96 Sn/4 Ag	ASTM 96 TS	221	8900	4600
96 Sn/5 Ag		221-245	8000	
95 Sn/5 Sb	Sb 5			
	ASTM 95 TA	232-240	5300	5360
10 Sn/88 Pb/2 Ag	Sn 10	268-299		
10 Sn/90 Pb	ASTM 10B	268-302		
5 Sn/92.5 Pb/2.5 Ag		280		
5 Sn/93.5 Pb/1.5 Ag		296-301		
5 Sn/92.5 Pb/2.5 Ag		300		
5 Sn/95 Pb	Sn 5			
	ASTM 5A	301-314		
	ASTM 5B			
1 Sn/97.5 Pb/	Ag 1.5			
1.5 Ag	ASTM 1.5 S	309		
60 Sn/40 Pb		174-185	4150	
50 Sn/50 Pb		180-209	4670	2680

Table 10.3   Fatigue resistance of solders (life relative to Sn 63/Pb 37) [4]

ALLOY		FATIGUE LIFE (relative to Sn 63)
96/4	Sn/Ag	3.03
95/5	Sn/Sb	2.10
62/36/2	Sn/Pb/Ag	1.14
63/37	Sn/Pb	1.00
59/37/4	Sn/Pb/Sb[a]	0.66
40/60	Sn/Pb	0.61

[a]Approximate composition.

silver solder. Table 10.2 shows very little difference between the tensile strengths of these solders: 7700 and 8900 PSI, respectively. This shows that strength plays the least important role in the selection of solder, since the commonly used solders (63/37 and 60/40 tin-lead solders in Tables 10.3 and 10.4) have the poorest fatigue resistance.

For a given solder, the fatigue resistance of a solder joint depends on shear strain, frequency, and temperature, as shown in Figure 10.5. This graph, developed by Wild [5], shows that in eutectic solder, the fatigue resistance (mean cycles to failure) at a given shear strain is much lower at lower frequencies (cycles/day) at a given shear strain. The difference can be 2 orders of magnitude when the frequency is changed from 1 cycle per day to 300 cycles per day. Slower frequencies mean longer dwell times at test extremes (thermal or mechanical cycles), leading to complete stress relaxation, loss of ductility, and premature failure.

The most commonly used solder is eutectic solder with 63% tin and 37% lead; its melting point is 183°C. Since tin is more expensive than lead, the eutectic (63 Sn/37 Pb) solder costs slightly more than the 60 Sn/40 Pb solder. For wave soldering applications, the two forms are equally good. For surface mounting, eutectic solder is more common because of its slightly lower melting point. However, the noneutectic solder alloys with a pasty range are more effective than eutectic solder in minimizing solder wicking (the traveling of solder up the lead, leaving no solder on the pad and causing an open solder joint) in reflow soldering.

The noneutectic solders melt over a range of temperatures, and some of the solder paste remains on the pad by the time the lands have reached the melting point of the paste. Thus the incidence of wicking may be reduced considerably by using a noneutectic solder alloy. Wicking is process dependent, being more common in vapor phase than in infrared reflow. Wicking of solder joints can also be minimized by adding 2%

Table 10.4   Fatigue characteristics of solders [4]

PERFORM-ANCE	COMPOSITION (wt %)	MICROSTRUC-TURE
Poor	63 Sn, 37 Pb	2 phase, eutectic
	60 Sn, 40 Pb	2 phase, near eutectic
	62 Sn, 36 Pb, 2 Ag	2 phase, Ag hardened β
	65 Sn, 35 In	2 phase, variable composition
	42 Sn, 58 Bi	sition
	50 Sn, 50 In	2 phase, eutectic
	50 Sn, 50 Pb	2 phase, variable composition
		sition
		2 phase, eutectic plus proeutectic
Fair	15 Pb, 80 In, 5 Ag	—
	99 In, 1 Cu	2 phase, eutectic
	90 Sn, 10 Pb	2 phase, mostly proeutectic β
	99.25 Sn, 0.75 Cu	tectic β
		2 phase, eutectic
Good	99 Sn, 1 Sb	1 phase, solid solution
	50 Pb, 50 In	1 phase, solid solution
	100 Sn	1 phase
Excellent	48 Sn, 52 In	2 phase, eutectic
	96 Sn, 4 Ag	1 phase, particle hardened
	95 Sn, 5 Sb	ened
	80 Au, 20 Sn	1 phase, particle hardened
		ened
		2 phase eutectic AuSn + hexagonal close packed solid state

silver to produce a 62 Sn/36 Pb/2 Ag solder, but it is slightly more expensive. Selecting devices that meet coplanarity requirements and using infrared reflow soldering are other measures to reduce solder wicking. (See Chapter 12 for details on the types of infrared reflow and the mechanism of solder wicking.)

Solders can be divided into two broad categories: high temperature solders (having a melting point above 183°C) and low temperature solders (solder alloys melting below 183°C) [1]. The use of solders of different melting points is helpful in step soldering, in which one side of the board

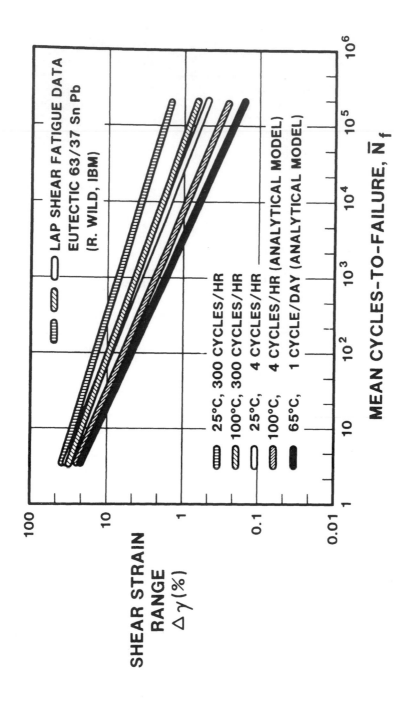

Figure 10.5 The fatigue behaviors of eutectic solder [5].

458

is soldered with high temperature solder first, then the secondary side is soldered with low temperature solder, thus preventing the melting of the high temperature solder.

Step soldering may be used if it is necessary to keep larger devices from falling off the board. Surface tension may not be sufficient to hold onto PLCCs larger than 84 pins [6]. Step soldering is rarely used, however. Instead, boards requiring very large devices should be designed such that these devices are on the side that is reflowed last.

In a discussion about solder composition, it is necessary to briefly consider the effects of solder contamination. The acceptable contamination levels (weight percent) in 63 Sn/37 Pb solder in QQ-S-571 are shown in Table 10.5. In eutectic solder the tin composition can vary from 59.5% to 63.5% and the remainder is lead (37%).

Contamination levels of aluminum above 0.004% (the specification allows 0.006%) and copper above 0.2% (specification allows 0.3%) cause gritty-looking solder joints. Entrapped dross in solder may also cause gritty solder joints. Noneutectic solder, because of its pasty melting characteristics, also contributes to a gritty solder joint appearance. These factors may act alone or together to cause grittiness.

A gritty appearance by itself is not a matter of concern and is to be expected with noneutectic solders. If it is due to dross or copper or aluminum, however, the solder joint properties may be degraded. Gritty solder joints are not common in reflow soldering. They are generally seen in wave soldering because the solder is prone to contamination as a result of repeated use of the same solder.

Gritty solder joints should not be confused with dull solder joints. Reflow solder joints are often dull, not bright and shiny like the wave

Table 10.5 Acceptable contamination levels in eutectic solder

METALS	MAXIMUM ALLOWED LEVEL
Aluminum	0.006
Antimony	0.5
Arsenic	0.03
Bismuth	0.25
Cadmium	0.005
Copper	0.30
Gold	0.2
Iron	0.02
Silver	0.1
Zinc	0.005
Nickel	0.01

soldered joints, because they cool at a slower rate than the wave soldered joints. A dull appearance is simply a reflow process characteristic and is not a matter for concern.

It is important to note that different inspectors will often fine a wide variation in defect rates on the same board. Wrong interpretations of solder joint quality based on appearance during visual inspection is one of the reasons for such wide variation. Extensive operator training in workmanship requirements (accept/reject criteria) and the use of automated inspection can minimize variation. See Chapter 14 for a detailed discussion on this subject.

10.4   LEAD-FREE SOLDER

As mentioned in the previous section, tin-lead solder is the most commonly used solder for electronic assembly. However, there are concerns about the use of lead due to its adverse effects on human health. Lead is linked to health hazards such as disorders of the nervous and reproductive systems and delayed neurological and physical development. Lead poisoning is particularly damaging to the neurological development of young children.

There are laws that control the use of lead. For example, lead usage in plumbing, gasoline, and paint is heavily regulated. And the use of lead in consumer paint has been banned since 1978 in the United States. Additional regulations to control the use of lead are under consideration in the United States and especially in Europe.

Table 10.6 shows the use of lead in various products [7]. Storage batteries account for most of the lead usage (80%). Electronic solder accounts for about 0.5% of all lead usage. So even if the use of lead in electronic solder were to be banned, it would not solve the lead poisoning problem. However, even the 0.5% lead used in electronics soldering amounts to significant lead use.

10.4.1   Lead Replacement Elements

The electronics industry is looking into lead-free solders that can replace the universally accepted and widely used tin-lead solder. Research and development efforts are focused on the study of potential alloys that provide physical, mechanical, thermal, and electrical properties similar to those of tin-lead eutectic solder. The metals that can replace lead and their relative costs are shown in Table 10.7.

In addition to cost, it is also important to understand the supply and

Table 10.6   Lead consumption by product

PRODUCT	CONSUMPTION (%)
Storage batteries	80.81
Other oxides (paint, glass and ceramic products, pigments and chemicals)	4.78
Ammunition	4.69
Sheet lead	1.79
Cable covering	1.40
Casting metals	1.13
Brass and bronze billets and ingots	0.72
Pipes, traps, other extruded products	0.72
Solder (excluding electronic solder)	0.70
Electronic solder	0.49
Miscellaneous	2.77

demand of elements that are being considered as replacements for lead. For example, as shown in Table 10.8, an alloy containing Bi may not be desirable from the availability standpoint. The current available supply of bismuth could be used up completely if this alloy were to be used extensively by the expanding electronics industry.

As is obvious from the relative costs of potential replacement metals for lead shown in Table 10.7, many of the lead-free solders will be much more expensive than the tin-lead solders they are attempting to replace. For example, indium (In) is one of the leading elements used to replace lead. But it is a semi-precious metal. It is almost as expensive as silver. We should note, however, that a high cost of solder is not as significant in determining the final product cost as it may first appear. Because of the small quantity needed, the cost of solder in an assembly is almost

Table 10.7   Alternative materials to replace lead and their relative cost [8]

REPLACEMENT ELEMENTS FOR LEAD	RELATIVE COST
Lead (Pb) for reference	1
Antimony	2.2
Bismuth	7.1
Copper	2.5
Indium	194
Silver	212
Tin	6.4
Zinc	1.3

Table 10.8 **Data from U.S. Bureau of Mines on world production and world capacity for different elements [9]. Note: Present world solder consumption = 60,000 tonnages or 6,600,000 liters**

ELEMENT	WORLD PRODUCTION (TONNAGES)	WORLD CAPACITY (TONNAGES)	"SPARE" CAPACITY (TONNAGES)
Silver (Ag)	13,500	15,000	1,500
Bismuth (Bi)	4,000	8,000	4,000
Copper (Cu)	8,000,000	10,200,000	2,200,000
Indium (In)	80–100	200	100
Antimony (Sb)	78,200	122,300	44,100
Tin (Sn)	160,000	281,000	81,000
Zinc (Zn)	6,900,000	7,600,000	700,000

insignificant in comparison to other cost factors such as components, bare board, and assembly. The properties of the selected alloys are very important however. They are discussed next.

10.4.2   Lead-Free Solders and Their Properties

Among thermal, mechanical, creep, fatigue, and other properties, melting point is one of the most important. Table 10.9 provides a select list of lead free solders that have been available for some time [10].

It should be noted that the compositions of the lead-free solders are still being optimized to achieve the desired properties. So the compositions of the solders shown in Table 10.9 may vary slightly from commercially available solders at different points in time. For example, Table 10.10 shows some of the commercially available solders from different suppliers under their trade names.

The lead-free alloys containing high indium (e.g., the first alloy in Table 10.10) have the potential incompatibility of indium with lead if it is present either on the board surface or on the component leads. In order to have a truly lead-free process, it may be necessary in some cases (e.g., if using alloys containing indium) to use a lead-free surface finish on the PCB and component soldering surfaces. Now the industry is focusing on developing alternate plating. Examples are Alpha Metals' AlphaLevel™ flash silver plating and Motorola's tin-bismuth plating for board and component lead finishes.

From Table 10.9, we can see that the lead-free solders have either much lower melting points or much higher melting points than tin-lead

Table 10.9   Examples of some lead-free solders and their properties [10]

LEAD-FREE SOLDER COMPOSITION	MELTING POINT RANGE	COMMENTS
48 Sn/52 In	118°C eutectic	Low melting point, expensive, low strength
42 Sn/58 Bi	138°C eutectic	Established, availability concern of Bi
91 Sn/9 Zn	199°C eutectic	High drossing, corrosion potential
93.5 Sn/3 Sb/2 Bi/1.5 Cu	218°C eutectic	High strength, excellent thermal fatigue
95.5 Sn/3.5 Ag/1 Zn	218°–221°C	High strength, good thermal fatigue
99.3 Sn/0.7 Cu	227°C	High strength and high melting point
95 Sn/5 Sb	232–240°C	Good shear strength and thermal fatigue
65 Sn/25 Ag/10 Sb	233°C	Motorola patent, high strength
97 Sn/2 Cu/0.8 Sb/0.2 Ag	226–228°C	High melting point
96.5 Sn/3.5 Ag	221°C eutectic	High strength and high melting point

eutectic solder. Table 10.10 shows mostly higher temperature lead-free solders.

Special flux is necessary when low temperature solders are used, because the standard flux may not be active at lower temperatures. Another problem associated with low temperature solders is the reduction in wetting properties caused by the lower fluidity at subeutectic temperatures.

For low temperature applications, solders containing indium (In) are gaining some acceptance. One indium alloy being used by some companies because it provides better rework/repair characteristics contains 52% In and 48% Sn. Since the alloy melts at 244°F (118°C), rework can be performed at lower temperatures many times without causing thermal damage. If the printed circuit board must be plated with gold as an antioxidant, indium solder can be used to prevent leaching of gold [11].

Another low melting point lead-free solder is 42 Sn/58 Bi. The phase diagram of Sn-Bi alloy, shown in Figure 10.6, shows the melting point to be 138°C. Bismuth is used in soldering alloys to achieve low soldering temperatures, but bismuth alloys generally exhibit poor wetting characteristics.

Table 10.10 Examples of lead-free solders from different suppliers (Courtesy of Dr. Raiyomand Aspandiar, Intel Corporation)

SOLDER NAME	SUPPLIER	COMPOSITION	MELTING POINT	COMMENTS
Indalloy™ 227	Arconium Specialty Alloy	77.2 Sn/20 In/ 2.8 Ag	187°C	Potential In-Pb incompatibility. Will require lead-free plating on PCB lands and component leads
Alloy H™	Alpha Metals	84.5 Sn/7.5 Bi/5 Cu/2 Ag	212°C	Liquidus temperature too high. Over 260°C wave temperature will be required
Tin-Zinc Indium	AT&T	81 Sn/9 Zn/10 In	178°C	Potential In-Pb incompatibility. Will require lead-free plating on PCB lands and component leads
Castin™	AIM Products	96.2 Sn/2.5 Ag 0.8 Cu/0.5 Sb	215°C	Liquidus temperature too high. Over 260°C wave temperature will be required
Tin-Silver-Copper	U.S. Dept. of Energy (DOE)	93.6 Sn/4.7 Ag/ 1.7 Cu	217°C	Liquidus temperature too high. Over 260°C wave temperature will be required

Many other alloys shown in Table 10.9 have much higher melting points than tin-lead eutectic which has a melting point of 183°C. The phase diagram of zinc-tin, a high temperature lead-free solder with melting point of 198°C, is show in Figure 10.7.

The high melting point solders will be incompatible with widely used board materials such as FR-4. In addition, higher temperatures necessary for rework can significantly increase the potential for board damage.

There are no drop-in replacement lead-free solders at this time, although some suppliers describe their solder as "near drop-in" [12]. Even

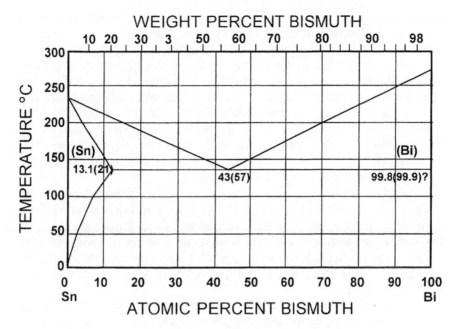

Figure 10.6 **Bismuth-tin phase diagram (note the melting point of Bi 57/SN 43 eutectic at 138°C—a low temperature lead-free solder).**

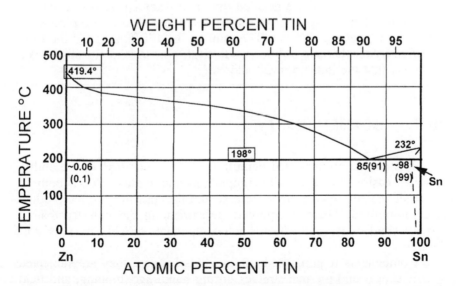

Figure 10.7 **Tin-zinc phase diagram (note the melting point of Sn 85/ Zn 15 at 198°C—a high temperature lead-free solder).**

these require a soldering iron temperature of 750°F (400°C) for rework. This may be too high a temperature in some applications and can cause potential thermal damage as discussed in detail in Chapter 14.

Also, one of the key problems in using the higher melting point solders shown in Tables 10.9 and 10.10 in wave soldering is that they increase the potential for capacitor cracking. The wave soldering temperature needs to be kept at about 230°–245°C, which is about 45°–65°C above the melting point of tin-lead solder. A lead-free solder with a melting point of 220°C will require 265°–280°C wave temperature. This increases the delta temperature between the preheat and the wave and hence the potential for capacitor cracking.

In general, almost all of the lead-free solders exhibit less wetting (spreading) than tin-lead eutectic and this causes an inferior solder fillet. To improve the wetting properties, special flux formulations are required. The fatigue properties of lead-free solders are also not as good, although in one study no degradation of solder joint integrity was observed after thermal cycling with high temperature 96.5 Sn/3.5 Ag (the last alloy in Table 10.9) [13].

Ideally the melting point of the selected solder should be around 180°C so that reflow temperatures of 210°–230°C, wave pot temperatures of 235°–245°C, and hand soldering temperatures of 345°–400°C (650°–700°F) can be used. Higher hand soldering temperatures can be used but only by very experienced operators in order to avoid thermal damage.

J-STD-006 provides a detailed list of tin-lead and lead-free solders [14]. None of the lead-free solders is considered a drop-in replacement for tin-lead eutectic, however. The industry is still looking for the right lead-free solder that can truly be a substitute for tin-lead eutectic. It is a challenge that the industry must address.

10.5   SOLDERABILITY

The commonly accepted definition of "solderability" is the ability to solder easily. This applies to components and boards alike. If both do not solder easily, rework costs increase and the reliability of the product is compromised. There is complete agreement in the industry that to provide good solder joints, boards and components should have good solderability.

Components require good solder joints because they are subjected to mechanical and thermal stresses during handling, shipping, and field service. This is especially critical in surface mount, since the solder joint provides both electrical and mechanical connections. The clinched leads and the plated through holes, which impart added mechanical strength in conventional assemblies, are absent in surface mount.

Surface mounting has made the solderability of components and boards even more critical by changing the soldering parameters (time and temperature of soldering) and by using less active and no-clean fluxes. In the rest of this chapter we discuss various aspects of such solderability issues as mechanisms of solderability, approaches to ensure solderability, and solderability test methods and requirements. There are three basic mechanisms of solderability: wetting, nonwetting, and dewetting. We shall deal with these in turn.

10.5.1   Wetting

Wetting of the surfaces to be soldered is important for the formation of a metallurgical bond. Just because a metallic surface is covered with solder does not mean that it is wetted. If there is an oxide layer between the solder and the metallic surface, the solder will not adhere to the surface and the joint will fail in service when subjected to mechanical stress.

In the soldering of electronic assemblies, the oxide is removed by the application of flux in the presence of heat, which not only activates the flux but also reduces the surface tension of the solder. The higher the temperature, the lower the surface tension of the solder and the better the wetting action. In the absence of good wetting action, either a nonwetted or a dewetted solder joint will result.

In electronic assemblies the wetting action takes place during soldering, in the interface between the copper and the tin. The reaction products of this wetting action are the copper-tin intermetallic compounds Cu_3Sn (near copper) and Cu_6Sn_5 (near solder) [15]. The reaction must take place rapidly to ensure the formation of a good solder joint.

Degrees of wetting also can be characterized by the angle of contact of the solder on the base metal. A smaller contact angle between two surfaces is an indication of better attraction, hence a stronger bond. On the other hand, a larger contact angle, as between water and a waxed surface, indicates very superficial attraction between the surfaces.

The concept of the contact angle is illustrated in Figure 10.8, where the circles surround solder joints and the vertical angle in each case can be assumed to be the lead. In the preferred condition, the solder contact angle is less than 90°; an angle of exactly 90° is acceptable, but a solder contact angle that exceeds 90° should be classified as an unacceptable, nonwetted solder joint.

Figure 10.9 schematically illustrates good solder wetting, with the characteristic concave fillet; the contact or wetting angle is less than 90°

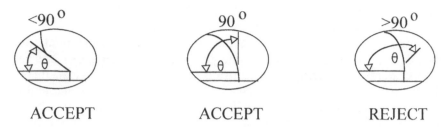

ACCEPT   ACCEPT   REJECT

Figure 10.8 Acceptable and unacceptable contact angles for solder joints.

with good feathering at the solder lead or termination interface. Also, in good wetting a uniform and smooth layer of solder stays on the surface. In Figure 10.10, a photomicrograph of a wetted solder joint, the surface is smooth and an intermetallic layer clearly indicates that a good metallurgical bond has been formed. There is generally no problem in identifying a wetted surface in concave fillets, but an excess of solder can make it impossible to tell whether a surface is well or poorly wetted.

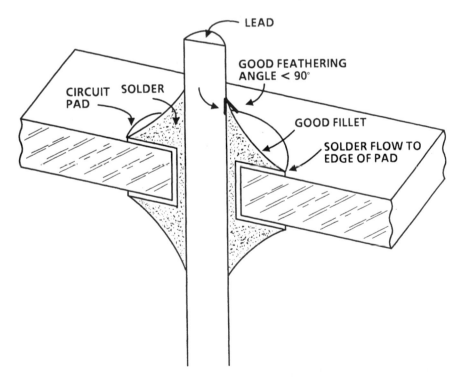

Figure 10.9 Schematic view of good wetting of a lead.

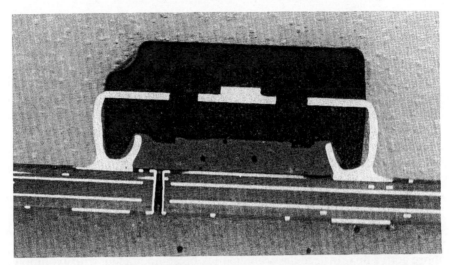

Figure 10.10   A photograph of a well wetted surface. Clearly visible are good inner and outer solder fillets along with the cross-section of the substrate and via hole (bottom of the photograph). The photograph also shows the silicon die over the lead frame paddle or die paddle (top center of photo). (Courtesy of Intel Corporation.)

10.5.2   Nonwetting

In nonwetting the solder simply does not adhere to the surface. This condition, the opposite of wetting, is caused by a physical barrier (intermetallic or oxide) between the solder and the base metal. The predominant cause is the total consumption of the layer of solder, which turns into an intermetallic compound of the base metal. In other words, if the solder coating is thin, the intermetallic layer will be exposed. Intermetallic compounds have poor wetting characteristics. The second mechanism of nonwetting is the formation of a thick oxidized layer that may be too thick for the flux to remove within the time allotted for soldering. Oxidation of the surface is generally not the primary cause, since oxidation is a very slow process. Nonwetting is a very easy condition to identify.

10.5.3   Dewetting

A dewetted surface is characterized by irregular mounds of solder caused by pulling back of solder, as if it had changed its mind about wetting. Dewetting is difficult to identify, since solder can be wetted at some locations while the base metal is exposed in other places. It may appear as a partial wetting condition in which some regions of the surface are wetted and others are nonwetted.

Given excess time at soldering temperatures, some base metal surfaces may, through excess intermetallic formation, exhibit dewetting. Interaction with the barrier metal (over the base metal) may also result in dewetting as the barrier metal is consumed. Dewetting can also result due to gas evolution during soldering. The source of gas can be either thermal breakdown or hydration of organic materials generated by the processes used for the plating of leads and boards [15]. The gas thus released leads to the formation of voids, which often remain dewetted, interfering with the integrity of the joint.

In severe cases, if enough organics are deposited during the plating process, the extent of gas evolution can be sufficient to passivate the surface, and a nonwetting condition will result. The problem is not corrected by increasing the time and temperature of soldering or by using a more active flux. Instead the problem may be compounded, with evolution of more gas, hence more dewetting and nonwetting. Pretinning such a lead may not prevent dewetting either. The problem can be corrected only

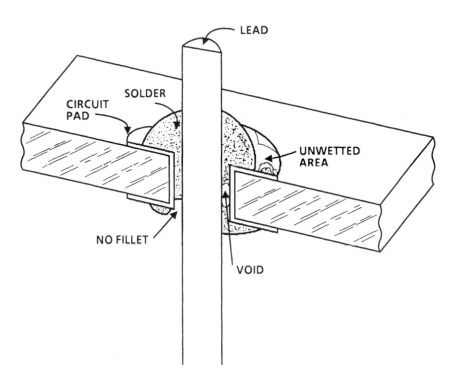

Figure 10.11   Schematic view of a dewetted solder joint.

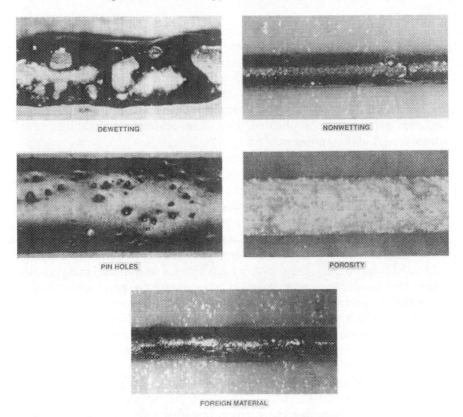

Figure 10.12   Types of solderability defects [16].

at the source, where parts are plated. An example of a dewetted condition is shown schematically in Figure 10.11.

Figure 10.12 shows several solderability defects, including a dewetted surface with gas voids [16]. Voids and pinholes in a joint can lead to stress concentration and eventual fatigue failures. Good substrate and component solderability not only provide good solder fillets, but also minimize voids and pinhole problems and reduce the level of overall touch-up and repair. Poor solderability is not the only cause of voids and pinholes; it can, however, compound these problems.

10.6   VARIOUS APPROACHES FOR ENSURING SOLDERABILITY

Good soldering results, free of dewetted or nonwetted joints, call for the use of substrates and components with good solderability. Achieving good solderability, however, is far from a simple technical matter; rather, it entails a complex mixture of various issues—technical (test method and requirements), quality and reliability (rework/repair, flux), financial (inventory, pretinning), standardization, and vendor/management (getting the users and suppliers to agree to common requirements). When parts are purchased from distributors, dealing with the original manufacturer becomes even more complicated. For these reasons, solderability is a controversial and complex subject.

The solderability problem has been with us since the advent of the electronics industry and it will remain at least for the foreseeable future. However, when assemblies are soldered using components of unknown or variable solderability, it is almost impossible to obtain good manufacturing yield and consistent quality. This leads to excessive touch-up, which not only costs money but may degrade product reliability.

Not surprisingly, there are many different approaches to solving solderability problems, each with its advantages and disadvantages. No matter how the problem is approached, parts must have good solderability at the time of soldering. This is becoming more critical as more and more companies switch to less active and no-clean fluxes. Now let us briefly discuss some of the approaches commonly being practiced in the industry.

The current approaches for solving solderability problems fall into six major categories: (a) using a very active flux, (b) doing in-house solderability testing and either pretinning the failed parts or returning the shipment to the vendor (i.e., forcing the vendor to supply solderable parts), (c) pretinning all incoming parts, (d) controlling solderability by source inspection at the vendor's facility to ensure receipt of solderable parts, (e) establishing a usable industry-wide, consistent solderability specification, and (f) requiring a surface finish on leads and substrates to ensure solderability. These approaches are interdependent and cannot be adopted in isolation.

In addition, another factor that affects solderability is the storage environment and the duration of storage. A part that is solderable when shipped by the vendor may lose its solderability after a certain period of time because of the slow growth of intermetallics at storage temperatures. Thus some users maintain a minimum inventory and do not store components for long periods. They also maintain a very controlled storage environment.

In this section we discuss the first four approaches; these offer only short-term fixes, but they are the most popular. Approach (e) is discussed in Section 10.7 and approach (f) is discussed in Section 10.8 and 10.9. Approaches (e) and (f) have the most direct effect on solderability and truly provide a long-term solution.

Theoretically speaking, almost any metallic surface can be soldered if we use flux that is active enough at the highest possible temperature and for the longest possible time. From a practical standpoint, we do not have the luxury of selecting these variables because higher soldering temperatures and longer times can increase the intermetallic thickness so much that the resulting joint will be unsolderable, brittle, or both. Also, as mentioned in connection with dewetting, if the underplating of the components and the board contains organics, a higher temperature and a longer time simply compound the problem of gas evolution [15].

A very active flux is inherently able to cause failures by corrosion, as discussed in Chapter 13. Thus if the components and boards cannot be soldered within the time, temperature, and flux requirements of the process being used, active fluxes are out of the question. Under schedule pressure, for example, many users either pretin parts or use poorly solderable parts and then touch up defective joints. One may not agree with this approach, but it is not uncommon. Other users rely on suppliers to provide solderable parts and do not do any in-house solderability testing or pretinning.

The companies that simply pretin all incoming parts consider this to be the cost-effective alternative. They tend to be users of parts of many different types, none in any significant quantity, and the solderability of components and boards varies from excellent to nonexistent. A fairly common practice among firms supplying the military sector is to pretin all incoming parts, store for a predetermined time, and then tin again, generally after 2 years of storage. This approach is limited in that it is difficult to retin an unsolderable part. For through-hole parts either the pretin or the touch-up option can be used, depending on the company's philosophy, as well as cost, volume, and schedule constraints.

For surface mounting, however, it is not practical to pretin components, especially small devices. For example, how can one pretin resistors and capacitors and package them back on tape and reel? For all practical purposes, controlling solderability at the vendor's facility is the only cost-effective solution for surface mounting.

Now there is a trend in the industry to require vendors not only to meet the solderability requirement but also to process parts with a specific finish to ensure solderability. This would seem to be a simple solution but it is not. The suppliers are not unwilling to perform the solderability tests or to provide the lead finishes requested. The problem lies in the considerable variations in board and lead finishes that exist in the industry.

10.7   SOLDERABILITY TEST METHODS AND REQUIREMENTS

Solderability is defined as the ability of a metal surface to be wetted by molten solder. The board and lead surfaces to be soldered must have good ability to be wetted by molten solder in the solder wave or by solder paste in the reflow oven. Surfaces with poor solderability cause soldering defects.

There are various solderability tests that are used by the industry. The two most commonly used tests are the dip and look test and the wetting balance test. The dip and look test is much more common than the wetting balance test. The third type of test, the globule test, is widely used in Europe. We will discuss these tests in some detail in the following sections.

J-STD-002 for components [16] and J-STD-003 for boards [17] are used as industry standards for evaluating solderability. They provide the details on test equipment, methods, and requirements for component and board solderability. These two standards have replaced various past standards. They are accepted by IPC, EIA, and the military.

First we will discuss general solderability requirements that apply for all the tests. We will follow this with the details of the dip and look test, the wetting balance test, and finally the globule test.

10.7.1   General Solderability Test Requirements

There are some requirements that apply to all solderability tests. Examples of these are steam aging, type of flux to be used, and solder composition requirements. Let us discuss them first before delving into the specifics of test methods.

10.7.1.1   Steam Aging Requirements

The purpose of steam aging the leads and boards before test is to simulate storage conditions before use. If a company is run with very little inventory, aging tests may distort the results by causing failures that might not have occurred in manufacturing. For this reason, one of the options is not to age leads and boards before testing.

For steam aging, distilled water in a noncorrosive container should be used, and the part should be suspended in a manner that prevents condensation on the leads. Components should be dried before the solderability test. Boards should be baked after steaming to ensure that moisture on the board surface does not influence the solderability results.

Table 10.11 provides the temperatures for steam aging. Prior to application of flux, it is required that all specimens to be tested be steam aged at a steam temperature 7°C below the local boiling point. This means that at sea level the steam aging should be about 93°C. At higher altitudes of about 5000–6000 feet the steam aging temperature needs to be only

Table 10.11 Steam temperature requirements at different altitudes [16, 17]

ALTITUDE	AVERAGE LOCAL BOILING POINT °C [°F]	STEAM TEMPERATURE °C [°F]
0–305 meters (m) [0–1000 ft]	100 [212]	93 +/–3 [200 +/–5]
305–610 m [1000–2000 ft]	99 [210]	92 +/–3 [198 +/–5]
610–914 m [2000–3000 ft]	98 [208]	91 +/–3 [196 +/–5]
914–1219 m [3000–4000 ft]	97 [207]	90 +/–3 [194 +/–5]
1219–1524 m [4000–5000 ft]	96 [205]	89 +/–3 [192 +/–5]
1524–1829 m [5000–6000 ft]	95 [203]	88 +/–3 [190 +/–5]

88°C. It has been found that aging at a temperature slightly lower than boiling point is more effective in oxidizing the surface to simulate long-term storage conditions.

The time of aging is also important. There are three categories of aging. In category 1, no steam aging is required if the boards or parts will be stored for a short time (less than 6 months). It should be noted however that parts may lose some solderability during this time period. In order to achieve conservative results, aging should be used.

In category 2, one hour steam aging is intended for nontin or non tin-lead surfaces such as OSP (organic solderability protection) surfaces, which do not do well when steam aged for a longer period of time.

In category 3, eight hour steam aging is used for tin-lead surfaces. In the past, steam aging requirements used to be much longer (16 to 24 hours). However, Wild [18] established that after 8 hours of aging, no appreciable deterioration in surface conditions takes place and major deterioration stops after 16 hours.

As the industry is moving more and more towards no-clean fluxes, it may be preferable not to use the category 1 option (no steam aging) unless the parts are not stored at all. A good tin-lead surface should have no trouble passing solderability after 8 hours of steam aging. This option (8 hour aging) should be exercised whenever possible since aging alone is rarely the cause of poor solderability.

10.7.1.2 Flux and Solder Requirements

The type of flux used for solderability testing is R (rosin) with solids content of 25% in 99% isopropyl alcohol. It is important to use the R flux to get conservative test results. The flux should be covered when not in use and should be discarded after 8 hours or should be maintained to a specific gravity of 0.843 and be discarded after one week of use.

Solder composition requirements are either Sn 60/Pb 40 or Sn 63/Pb 37 (eutectic). Solder composition may make a slight difference in results for marginally solderable components, but it is not a major issue.

10.7.2 Dip and Look Test

In the dip and look test, which also applies to the solderability testing of substrates, the component or substrate is dipped in a pot of molten solder and solderability is evaluated by determining the percentage of wetted area. The important test parameters in dip and look solderability tests are the aging requirement, solder temperature, dwell time, and type of flux (see Table 10.12).

This method is subjective, and it is generally difficult to estimate the nonwetted/dewetted area, especially if solderability is marginal. Estimating excessive or minimal dewetting is generally not a problem. Refer to Figure 10.13 as a guide for estimation of the percentage of defective areas in the field of view.

For the component or the board to be considered acceptable, no nonwetted or dewetted areas should be present. Obviously it is best to have 100% coverage, but 95% or greater is acceptable. Anomalies other than pin holes, dewetting, and nonwetting are not cause for rejection.

From Table 10.12 we see that some of the test parameters are different from the soldering parameters used in manufacturing. For example, although reflow soldering of surface-mounted components is performed at 220°C, the dip and look solderability tests are performed at 245°C. There is also inconsistency between the flux used for the test and the flux used in soldering processes. For example, a mildly activated RMA flux and even water-soluble organic acid flux in solder paste are commonly used for reflow and wave soldering.

So it is argued by some that to accurately assess solderability, the tests should be conducted using production conditions. Such an approach is not workable, however. Given the many reflow processes used by different companies, no supplier could possibly duplicate all the different manufacturing setups used by its customers. This absence of replicability could cause a real problem in the correlation of solderability results

Table 10.12   Dip and look solderability test parameters and acceptability requirements per J-STD-002 [16]

PARAMETERS	REQUIREMENTS
Aging	• No steam aging required, on • 1 hour +/−5 minutes steam aging, or • 8 hour +/−15 minutes aging
Baking/drying after aging	• Leads should be dried • Boards should be baked at 105+/−5°C (221 +/−9°F) for 1 hour
Flux type	• Nonactivated rosin (Type R) with 25% by weight water white gum rosin in 99% isopropyl alcohol with specific gravity of 0.843 +/−0.005 at 25°C
Solder composition	• Sn 60/Pb 40 or • Sn 63/Pb 37
Solder pot temperature	• 245°C +/−5°C (473+/−9°F)
Dwell time in solder	• 5 + 0, −0.5 seconds • Immersion and emersion rate of lead in the pot 25+/−6 mm (1.0+/−0.25 in)/second
Magnification for inspection	• 10X • 30X for fine pitch
Accept/reject criteria	• Minimum 95% surface coverage. See Figure 10.13 • Surface anomalies other than nonwetting and dewetting are acceptable

obtained by the users and suppliers. It would be hard, for example, to determine whether a solderability problem was caused by component, board, paste, or reflow equipment. Using a manufacturing process may be a good development tool for users but it could not feasibly be adopted as an industry standard.

The solderability test parameters are conservative and tend to accentuate differences between good and bad lots of boards and components within a reasonable set of test conditions. They serve to increase test sensitivity. The specific test parameters are shown in Table 10.12. Let us elaborate on some of the key points.

Only the areas under test should be coated (by dipping the part in

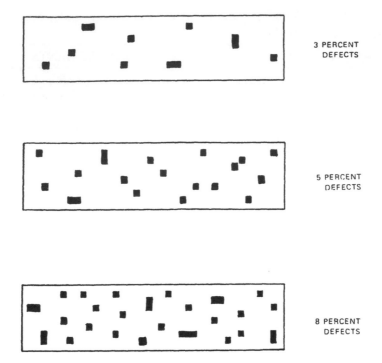

Figure 10.13 Guidelines for acceptable solderability in dip and look test [16].

flux for 5 to 10 seconds) at room temperature. The excess flux should be removed and dried for 5 to 20 seconds, then dipped into the solder pot without any dwell above the solder pot (no preheat). The solder temperature and dwell times are 245°C for 5 seconds.

The type of solder pot may affect the results significantly, however, and a static solder pot is recommended. The solderability results obtained with a static pot are more conservative than those obtained with an agitated solder pot.

After testing, the flux residue should be cleaned off using a suitable solvent such as isopropyl alcohol. The magnification to be used for inspection after testing is 10X under a microscope. However, for fine pitch the magnification requirement is much higher (30X).

Finally, what is the requirement for acceptable solderability? An argument can be made that since the user cannot ship a few percent defective assemblies, no dewetting or nonwetting should be allowed. All the components and boards should be perfect and should show 100%

wetted surface. Given the difficulty of evaluation, it is not realistic to require 100% coverage.

We are talking about a test to simulate good solderability, not an actual soldering process. After all, in a manufacturing environment, the components and boards are not aged. Also, some activators are used in the flux. In general it has been found that a part that shows less than 5% dewetted or nonwetted area during testing will provide an easily solderable surface.

10.7.3   Wetting Balance Test

In the second commonly used solderability test, the wetting balance method, the lead is dipped in a solder pot and the time required to reach the maximum wetting force is measured. The equipment has a means of recording force as a function of time on a data logger or computer. In this test the flux-covered steam-aged leads of surface mount or through-hole components are dipped in the solder pot (at 245°C as in the dip and look test) at the rate of 10 mm/second for a dwell time of 5 seconds. The flux used is rosin R as in the dip and look test. (For details on test equipment and procedures, refer to J-STD-002.)

Figure 10.14 shows a typical curve generated by the wetting balance test method. As indicated in the insct to Figure 10.14, when the specimen is first lowered into the solder bath, the upward or buoyancy force is recorded as the negative wetting force on the specimen. The specimen heats up in the bath and the flux becomes active, and then the wetting begins. The slope of the curve moving up, as the wetting continues, is very important. The time it takes to reach the maximum force or maximum wetting is defined as the wetting time.

The wetting balance method, by virtue of its interaction with the liquid solder, allows a quantitative measurement of the forces acting on the sample. The forces acting vertically on the specimen are measured as a function of time. The shorter the time to reach the maximum wetting force, the more solderable the component.

The wetting balance test method does not have established accept/reject criteria. It is intended for evaluation purposes only. The suggested criteria for evaluating solderability as shown in Figure 10.14 for leaded components are as follows [16]:

* The recorded signal trace should cross the buoyancy corrected zero at or before 1 second from the start of the test.

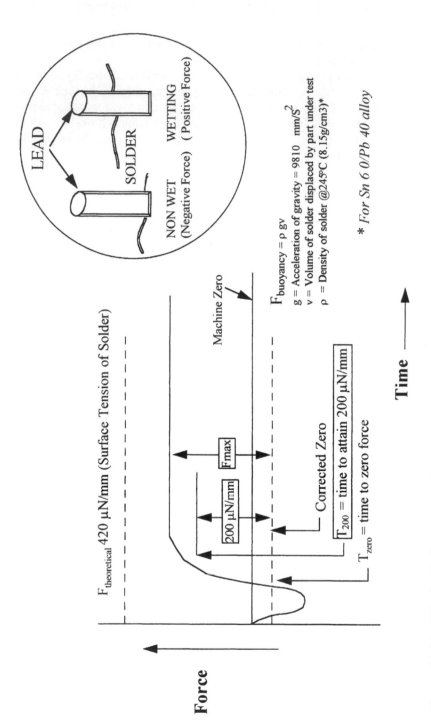

Figure 10.14 Wetting balance curve for leaded parts [16].

480

- The recorded signal trace should reach a positive value of 200 micronewtons/millimeter wetting force at or before 2.5 seconds from the start of the test.
- The recorded signal trace should still be above the positive value of 200 micronewtons/mm wetting force at 4.5 seconds from the start of the test.

The suggested criteria for solderability evaluation of leadless components are different. The coefficient of wetting (micronewtons/second) should be calculated for each termination or set of terminations tested in a single dipping operation. The calculation should be the equilibrium wetting force F_{eq} (in micronewtons) divided by the time T_{eq} (in seconds) it takes to achieve equilibrium wetting.

As shown in Figure 10.15, the suggested criteria for solderability evaluation of leadless components should show a coefficient of wetting greater than 150 micronewtons/second [16].

In this test for leadless parts, the flux-covered terminations of surface mount components are dipped into the solder pot at a rate of 5 mm/second. Contact in solder is continued until equilibrium wetting has been

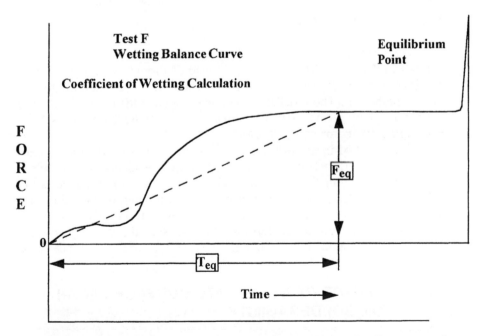

Figure 10.15   Coefficient of wetting calculation for leadless parts by wetting balance test [16].

achieved, plus approximately 5 seconds. At this time the contact with the molten solder is broken. For details on the test equipment and procedures refer to J-STD-002.

The basic difference between the wetting balance test and the dip and look test is that the latter is subjective, whereas the former is considered to be quantitative. Actually, however, the time requirement for definition of solderability in the wetting balance test is subjective at present even though the test method itself is quantitative. This is the reason why there are no accept/reject criteria established for this test.

Many companies use the wetting balance testing for process control, to determine whether or not to accept parts. Parts that fail the wetting balance test are rechecked with the dip and look method before being rejected.

Different countries have adopted different tests for solderability. For example, there has been something of a divergence in the approaches to solderability testing used in Europe and in the United States. Europeans have decided to concentrate on the wetting balance test because of its quantitative features, while American companies have continued to improve upon the dip test because it is simple and cheap.

10.7.4   Globule Test

The third type of test, the globule test, is widely used in Europe as defined in IEC-68-2-20. In this test, instead of dipping into a solder pot, a solder globule of prespecified size is paced on a pretinned aluminum block and heated. The test specimen (steam aged lead) is immersed into the molten globule, and the wetting time is measured by the timing needle. See Figure 10.16 for an illustration.

The speed of immersion into the globule should be 10 mm/second. The time elapsing between the moment that the lead bisects the solder and when the solder flows around and contacts the timing needle is the soldering time. Like the wetting balance test, this method also does not have any accept/reject criteria. It is used for evaluation only.

The globule test is generally used to measure the wetting time for through-hole components but it can also be used for surface mount devices.

10.8   EFFECT OF SUBSTRATE SURFACE FINISH ON SOLDERABILITY

In soldering surface mount or through-hole mount devices, the surface finishes of substrates and leads or terminations play a key role in determin-

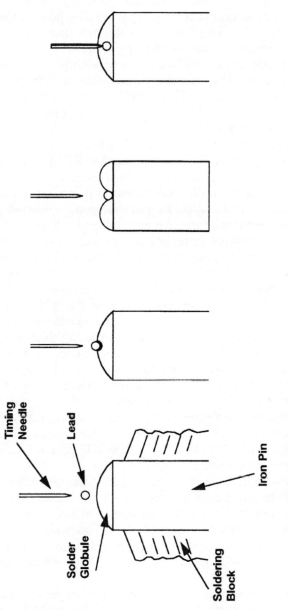

Timing Needle

Lead

Iron Pin

Solder Globule

Soldering Block

Figure 10.16   Globule test method [16].

ing the solderability of components and substrates. Generally, a high percentage of solderability problems are caused by boards. One problem area is found near the "knee" of plated through holes; weak knees are due to exposed intermetallics and their passivation.

For boards, the critical surface finish requirement is to ensure sufficient solder thickness. The general recommendation in the industry about maintaining at least 0.3 mil (300 microinches) is essential for long-term solderability, as found by Mather [19]. Solder coatings thicker than 0.3 mil have good long-term solderability and meet solderability requirements without difficulty even after aging for 24 hours.

On the other hand, solder coatings less than 0.3 mil thick dewet after as little as 4 hours of aging. Similar conclusions were reached by Mackay [20], who found that 5 μm (200 microinches) of tin coating and 7.5 μm (300 microinches) of 60 Sn/40 Pb provide good soldering even after storage in some artificial aggressive environments.

Given this, somewhat surprising results were found in an industry-wide round robin test program conducted by the IPC [21]. These boards retained solderability after 2 years of storage even when they had very little solder coating. Thus, solder coating thickness alone is not important but the quality of coating is very important. If the solder coating is not thick enough, the plating must be free of deleterious organic materials under the plated surfaces, and the plating or coating thickness should be uniform and continuous. Even when the coating quality is good, though, a coating thickness of less than 100 microinches should be cause for concern.

There are problems with both think and thick surface coatings, especially in surface mounting. Thinner surfaces (<100 microinches) are not desirable because they are almost entirely consumed during reflow, especially if the assemblies are reflow soldered twice (primary side and then secondary side). Since intermetallic compounds passivate easily if exposed, they have poor solderability.

Thicker coatings, especially with wide variations within the same land pattern of the same package, are not good because they compound the coplanarity problem for active devices. Thicker coatings are good for solderability, but anything exceeding 300 microinches is really not necessary. There is one exception, however. As discussed in Chapter 12, 500 microinches of solder is necessary for hot bar soldering of ultra fine pitch devices such as TAB.

HASL (hot air solder leveling in which the copper surface is dipped in hot solder and the excess blown away with a hot air knife) is the most commonly used process for the final board surface finish. Unfortunately, the main concern with this process is that it results in great variation in solder thickness. See also Chapter 4, Table 4.8 for the results of one study.

The alternative to HASL for achieving uniform solder thickness is the solder plating process. However, the plating process is not without its problems. For example, the tin composition of the solder is decreased as the current density is lowered. This can cause formation of a lead-rich solder, which is less solderable. Even with good control, however, weak "knees" are encountered in the plated through holes.

Other options for uniform surface finishes are OSP (organic solderability protection) and gold and palladium plating. Refer to Chapter 4, Sections 4.8 and 4.9 for details on surface finishes.

10.9   EFFECT OF COMPONENT LEAD OR TERMINATION FINISH ON SOLDERABILITY

Surface finishes are applied to component leads and terminations by plating or by hot dipping. The requirements placed on active devices for military applications favor hot dipping, but for commercial applications a plating process may be better. There are pros and cons to both approaches. For all applications an appropriate finish is necessary to achieve a good solder joint.

The hot dipping process provides sufficient solder thickness, but the coating is generally uneven, displaying a "dog bone" appearance with thicker terminations and a narrower body. Such a shape tends to cause problems during placement because it is difficult to accurately pick and place the "dog bone" parts.

Plating of terminations is preferred over dipping to achieve symmetrical shapes and uniform coating. However, if a plated surface is reflowed before shipment, most of the tin may be consumed in intermetallic formation, causing solderability problems. It should also be noted that reflowing exposes the edges of the base metal, which adds to solderability problems. Therefore plated terminations should be reflowed only at the time of soldering.

Let us briefly describe the process sequence in lead finishing for commercial and military applications and then discuss the pros and cons of the plating and hot dipping processes.

In commercial applications, the lead frames are molded in the plastic compound, and then the leads are plated. The packages may or may not be burned in before shipment. In burn-in, the components are kept operational at elevated temperatures (about 125°C) for certain duration to weed out infant mortality. For commercial applications, the burn-in time is relatively short (48–68 hours).

Plating produces uniform thickness, but the coating tends to be somewhat porous. Porosity can be reduced by controlling the process to produce

Table 10.13 Solderability test result comparison (failures in sample size of 22) of plated and hot dipped copper leads using dip and look tests [22]

STEAM AGE TIME	SOLDER PLATED FAILURES	SOLDER DIPPED FAILURES
9	0/22	3/22
24	0/22	6/22

a fine grain size. Also, the lead ends of plated leads, such as SOIC and gull wing fine pitch packages, have exposed copper because the tie bars holding the leads are cut after plating. The exposed copper may cause poor/dewetted outer fillets. They should be acceptable as long as a good inner fillet at the heels of the leads exists.

The solderability of plated leads has been found to be superior to that of hot dipped leads by Belani et al. [22]. Their dip and look and wetting balance results appear in Table 10.13 and 10.14, respectively. The tables also show the test conditions. Steam aging of 16 to 24 hours was used to simulate worse case storage conditions.

Table 10.13 shows that no solderability test failures were found in the plated leads, but the hot dipped leads showed 14% to 27% failures for zero and 24 hour aging, respectively. These dip and look results were confirmed by the wetting balance tests shown in Table 10.14. The wetting times for the hot dipped leads are considerably higher.

For military applications, the sequence of lead finish and burn-in is different. After die attach and glass sealing, the leads are tin plated. After that they are burned in for about 168 hours. Finally they are hot dipped. Sometimes plating is not used at all. Packages are hot dipped followed by burn-in.

Serious problems can result if leads are burned in after hot dipping, however. As shown in Figure 10.17, burn-in has a serious impact on lead solderability. The alloy 42 (58% iron, 42% nickel) leads, which were

Table 10.14 Solderability test result comparison of plated and hot dipped copper leads using wetting balance tests showing wetting time in seconds [22]

STEAM AGE TIME IN HOURS	WETTING TIME (SEC) FOR SOLDER PLATED LEADS	WETTING TIME (SEC) FOR SOLDER DIPPED LEADS
1	0.30	0.30
16	0.38	0.89
24	0.41	0.63

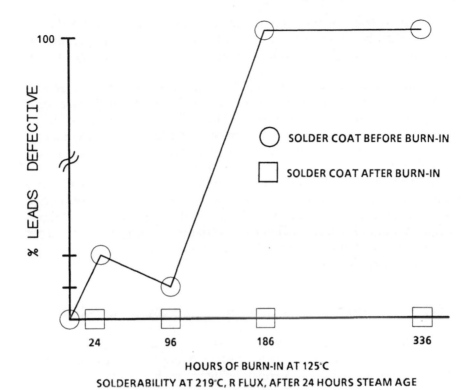

HOURS OF BURN-IN AT 125°C

SOLDERABILITY AT 219°C, R FLUX, AFTER 24 HOURS STEAM AGE

Figure 10.17 Impact of burn-in on solderability of alloy 42 leads before and after hot dip. (Courtesy of Intel Corporation.)

burned in after hot dip coating failed the solderability test miserably; but when tin-plated leads were burned in and then hot dipped in solder, as expected, no solderability defects were found.

There are two additional items that should be noted about hot dipped leads. In hot dipping, not only does the solder tend to pull away from the edges of the leads, but at the lower bends of the "J" in PLCCs, the solder thickness is minimal. So expect poor lead solderability at lead edges and lead bottoms in hot dipped leads. Also, hot dipping may compound the coplanarity problem.

The burn-in process uses burn-in sockets, which can also cause mechanical damage to the lead finish. Because such mechanical and metallurgical phenomena degrade the solderability of an otherwise good lead finish, it is better to perform the final lead finishing operation after burn-in.

In many cases, plating and hot dipping complement each other. In applications that call for a long burn, for example, both plating (before burn-in) and hot dipping (after burn-in) may be required to provide the needed solderability. This is true for both commercial and military applications.

How thick should the plated or hot dipped surface be? The requirements for leads and for boards should not differ. Thus whether the leads are plated or hot dipped, 300 microinches of solder is preferred. If the base material is pure copper (as opposed to alloy 42, for example) a thickness less than 100 microinches is insufficient because most of the solder is consumed as a copper-tin intermetallic compound.

In addition to solder thickness, the grain size influences solderability [23]. The difference in wetting times for fine and coarse grained microstructures becomes even more significant with the steam aging of leads before solderability testing. Grain size variation is seen essentially in the plating process with finer grain sizes resulting in denser finishes which are less susceptible to oxidation.

10.9.1   Effect of Ni-Pd Lead Finish on Solderability

Palladium plating on component leads is not as common as the tin-lead finish just discussed. However, with the widespread use of fine pitch, there is growing interest in alternative finishes such as palladium because solder dipping of fine pitch lead tends to cause bridging. Also, with increased interest in lead-free solders, alternative finishes are being considered by some companies. For example, in 1989, Texas Instruments developed a preplating process for lead frames in which a nickel plated lead frame is subsequently coated with a thin (3-4 microinch) Pd layer. The palladium coating process involves the following steps [24]:

- Nickel strike over copper base metal (copper lead frame), 0.1 μm (<5 μinches) thick
- Ni-Pd alloy strike, 0.1 μm (<5 μinches) thick
- Ni plate 1–1.5 μm (40–60 μinches) thick
- Pd plate 0.1 μm (minimum 3 μinches) thick.

The role of palladium is to protect the underlying nickel layer from oxidation during package assembly processes and subsequent storage. Being a noble metal, palladium is good for this function. During the soldering process, palladium dissolves in solder since it has a fast dissolution rate as discussed earlier (Section 10.2, Figure 10.4). Soldering takes place between the solder and the underlying nickel layer.

Even though palladium coated leads have been available since the early 1990s, they have not been universally liked and accepted. Lead solderability and post assembly inspections have been the main concerns. The solder joints formed with palladium coated leads have a very different appearance from the joints formed with solder coated leads. For example, there is not a smooth transition from lead toe or heel to the solder joint. Instead, there is a crack-like appearance between the lead and the solder. A cross-sectional analysis will generally show such a joint to be perfectly acceptable. However, inspectors are not used to seeing such solder joints, and you cannot be sure if this is a real case of nonwetting or simply a normal solder joint since metallographic analysis is not feasible on production boards. In most cases, this is more of a cosmetic issue and can be corrected by training.

There are some real concerns with palladium coated leads, however. The leads generally fail the dip and look test especially if they are steam aged. Also, one investigator found that palladium coated leads could not pass the J-STD-002 wetting balance test [25]. In order for the leads to pass the wetting balance test, RMA flux needed to be used. As discussed earlier, RMA flux cannot be used for solderability testing; only R (rosin) flux can be used. (See Table 10.12.) This is an important point since the industry is moving more and more towards less active no-clean fluxes as discussed in Chapter 13.

Part of the explanation for poor lead solderability is not palladium but the nickel underplating. As discussed above, palladium dissolves in solder while soldering takes place between the nickel and the solder. It should be noted that nickel is not a very solderable material. This is part of the reason why the solder joint appearance is not as smooth as one is used to seeing with solder coated leads. For this reason, it may be necessary to use higher temperatures during soldering. This may cause some problems with other solder joints (e.g., thicker Cu-Sn intermetallic) since the leads of the other components on a given board are likely to be coated with solder.

The nickel/palladium coated leads also show various levels of microcracking, with some cracks as deep as 10 μm (400 microinches), exposing the copper under the Ni-Pd surface finish [25]. Such microcracks are generally seen under the bottom side of the heel and most likely they occur during the component trim and form operation. Even a hint of a crack in a solder joint, especially around the heel fillet, is cause for concern. The heel fillet is the major fillet in gull wing type components and is very important for reliability.

Despite some of the real and cosmetic problems discussed in this section, palladium coated leads can be used and have been used by many companies. However, successful use of palladium coated leads requires changes in solderability test methods and requirements, a relatively active flux (along with its inherent problems), a higher soldering temperature, changes in inspection criteria, and most importantly, special training of the personnel involved. In other words, like lead-free solder, the use of palladium coated leads is not a drop-in process.

10.10   SUMMARY

The metallurgy of soldering is the key to understanding the mechanisms of intermetallic formation and solderability testing. A basic knowledge of phase diagrams of the metals used in soldering permits one to determine the solubility of one metal in another. The solubility, in turn, controls such important variables as alloy composition, melting point for soldering, and occurrence of leaching.

Tin-lead solder is the most commonly used solder for electronic assembly. However, the known health hazards of lead have caused the industry to look into solders that do not contain lead. None of the lead-free solders now available can be used as a drop-in replacement for tin-lead eutectic, however.

There is general agreement that for successful soldering results, components and boards must have good solderability. Now the industry has settled on common standards for solderability test methods and requirements.

The lead finish not only plays a key role in solderability but also has an impact on coplanarity in active devices and component placement in passive devices. To control the solderability of printed circuit boards or components, one must not only have adequate solder thickness but the thickness should also be uniform and continuous.

REFERENCES

1. Wassink, R. J. *Soldering in Electronics.* Ayr, Scotland: Electrochemical Publications, 1984.

2. Manko, H. H. *Solders and Soldering,* 2nd ed. New York: McGraw-Hill, 1984.

3. Bader, W. G. "Dissolution of Au, Ag, Pd, Pt, Cu, and Ni in molten tin-lead solder." *Welding Research Supplement,* December 1969, pp. 551–557.

4. DeVore, J. "Fatigue resistance of solder." *NEPCON Proceedings,* February 1982, pp. 409–414.

5. Wild, R. N. Some fatigue properties of solders and solder joints. IBM Technical Report 73Z000421, January 1973.

6. Hart, C. "Double-sided attach using reflow solder." *NEPCON Proceedings,* February 1985, pp. 46–53.

7. Lead and the electronics industry: A proactive approach. May 1995 report available from National Center for Manufacturing Sciences, 3025 Boardwalk, Ann Arbor, MI 48108, Tel. 313-995-0300.

8. Napp, Duane. "NCMS lead free electronic interconnect program." *Proceedings of Surface Mount International,* 1994, pp. 425–431.

9. Hwang, Jennie S. "Overview of lead free solders for electronics and microelectronics." *Proceedings of Surface Mount International,* 1994, pp. 403–421.

10. Socolowski, Norbert. "Lead free alloys and limitations for surface mount assembly." *Proceedings of Surface Mount International,* 1995, pp. 477–480.

11. Keeler, R. "Specialty solders outshine tin/lead in problem areas." *EP & P,* July 1987, pp. 45–47.

12. Seelig, Karl. "A study of lead free solder alloys." *Circuit Assembly,* October 1995, pp. 46–48.

13. Melton, Cindy. "How good are lead free solders." *SMT,* June 1995, pp. 32–36.

14. J-STD-006. Requirements for electronic grade solder alloys and fluxed and non-fluxed solid solders for electronic soldering applications. Available from IPC, Northbrook, IL.

15. DeVore, J. A. "Solderability." *Journal of Metals,* July 1984, pp. 51–53.

16. J-STD-002. Solderability tests for component leads, terminations, lugs, terminals and wires. Available from IPC, Northbrook, IL.

17. J-STD-003. Solderability tests for printed boards. Available from IPC, Northbrook, IL.

18. Wild, R. N. "Component lead solderability versus artificial steam aging, II." *Proceedings of the Naval Weapon Center Soldering Conference,* January 1987, NWC TP 6789 pp. 299–320.

19. Mather, J. C. A need for tighter requirements on circuit board solderability. IPC-TP-484, September, 1983, Available from IPC, Northbrook, IL.

20. Mackay, C. A. "Surface finishes and their solderability." *Brazing and Soldering Supplement, Metal Fabrication,* January/February 1979.

21. IPC-TR-462. Printed/board protective coating solderability evaluation over long term storage. Round Robin Test Program. October 1987. Available from IPC, Northbrook, IL.

22. Belani, J.; Sajja, V.; Mathew, R.; and Patel, A. "Plated lead finishing for reliable solder joints." *NEPCON Proceedings,* February 1986, pp. 467–470.

23. Geiger, A. L. "Solderability of capacitor lead wire." *Proceedings of the Naval Weapon Center Soldering Conference*, February 1986, NWC TP 6707, pp. 111–129.

24. Romm, Douglas W. Palladium lead finish user's manual. Available from Texas Instruments, 8360 LBJ Freeway, Center Complex, Dallas, TX 75265.

25. Finley, D. W., et al. "Assessment of nickel palladium finished components for surface mount assembly applications." *Proceedings of SMI,* August 29–31, 1995, pp. 941–953.

Chapter 11

Component Placement

11.0 INTRODUCTION

Surface mount components are placed on a printed circuit board after deposition of adhesive or solder paste. Generally solder paste and adhesive are deposited at a separate work station by a screen printer or dispenser. Some placement machines also dispense adhesives. Placement equipment is commonly referred to as "pick-and-place" equipment. Sometimes it is called "onsertion" equipment, to differentiate it from the insertion equipment used for through-hole components. The term "onsertion" did not become popular, however.

Several factors make it almost mandatory to use placement equipment for surface mounting. For example, many surface mount components, especially passive components, are not marked. Moreover, packages with finer and finer lead spacings or "pitches" and ball grid arrays, where the balls are hidden under the package, are being used. Finally, manual placement of surface mount components, which is neither reliable nor economical, should not be used for production.

The pick-and-place machine is the most important piece of manufacturing equipment for placing components reliably and accurately enough to meet throughput requirements in a cost-effective manner. Typically, surface mount pick-and-place equipment, including a full complement of feeders, constitutes about 50% of the total capital investment required for a medium volume surface mount manufacturing line.

Also, the throughput of a manufacturing line is primarily determined by the pick-and-place machine. The majority of manufacturing defects that require rework stem from placement problems (although other processes such as screen printing and reflow soldering along with product design play major roles). The type of feeder system used on the pick-and-place equipment also plays an important role in determining both throughput and the reliability of placement. Since no placement equipment is best for all applications, the effort required in selecting equipment with an appropriate feeder system should constitute a major part of the effort spent in selecting SMT capital equipment.

In this chapter, after briefly discussing manual placement, we concentrate on automated-placement equipment and its selection criteria. We will use some examples of currently available equipment to illustrate our selection criteria for SMT, fine pitch, TAB, chipscale packages, and flip chip. Using the selection criteria as a guide, the reader can approach the task of choosing placement equipment for individual applications, even though specific equipment models change almost every year. A detailed questionnaire is provided in Appendix B1 for this purpose.

11.1   MANUAL PLACEMENT OF PARTS

As mentioned earlier, manual placement of surface mount components is neither reliable nor economical, but it can be used for prototyping. Also, if the placement equipment requires considerable programming time or if the appropriate feeders are not on hand, manual placement may be used for quick turnaround prototypes. Some subcontractors have catered to quick turnaround prototype jobs for years without any pick-and-place equipment capability.

Since many passive components do not have any part markings, one of the main problems in manual placement is preventing part mix-up, and a procedure must be put in place to handle this problem. One common method is by using bins or containers marked with part numbers or values. If the parts so classified get mixed up, they must be positively identified or thrown away.

Another problem in manual placement is the increased potential for placing even marked components in the wrong orientation. This typically occurs when tantalum or other polarized passive devices such as SOICs or PLCCs are to be placed. The operator must know how to identify pin 1 of active devices (look for a dot over pin 1) and the polarity of tantalum capacitors (the positive terminal has either a beveled edge, a notch in the termination, or a welded stub). Most important of all, manual placement is operator dependent, and not everyone is well suited for the job. Only a person with a steady hand and good dexterity can consistently place parts accurately. Some operators can drop a fine pitch package exactly on location almost every time, but this is relatively rare.

Placement accuracy is more critical when placing components onto solder paste than onto adhesive because of potential paste smudging across the pads, which may cause bridging. It is not a good idea to rely on self-centering during reflow of a misplaced component on solder paste. If components do get misaligned, it is better to realign them and take a chance on solder paste smudging.

In component placement, there are two main functions: pickup and placement (hence the name "pick-and-place" for this type of equipment). In manual placement, the components parts are picked up either by tweezers or by a vacuum pipette. For passive components tweezers are adequate, but for multileaded active devices, a vacuum pipette is very helpful in dealing with component rotation. An example of a vacuum pipette used for placement of surface mount components is shown in Figure 11.1. When placing components manually, the following additional guidelines should be kept in mind:

1. Care must be taken to avoid mixing parts that look identical but may have different values. Parts that have been dropped and recovered should be positively identified or discarded.
2. Undue tension or compression on the components should be avoided.

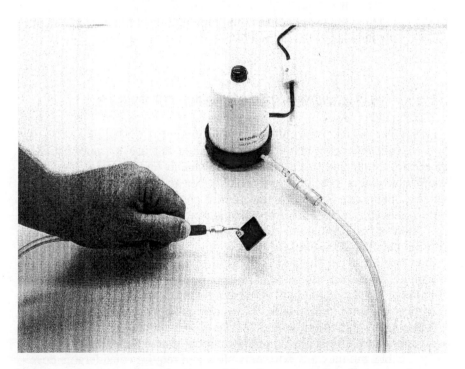

Figure 11.1   A vacuum pipette used for placement of surface mount components.

3. Tweezers or other tools that may damage the part should not be used to pick up components.
4. Parts should be gripped by their bodies, not by their leads or terminations.
5. Care should be exercised to keep tweezers free of adhesive or solder paste. Since some contact with adhesive or solder paste may occur during part placement, tweezer tips should be cleaned regularly with a solvent, such as isopropyl alcohol.
6. Incorrectly placed parts should be discarded or properly cleaned before reuse. During cleaning, damage to terminations or leads is a real concern.
7. Programmable devices such as programmable logic devices (PLDs) are handled manually for programming before placement. If this is not done properly, especially when the devices are pried out of the programming sockets, lead damage can cause significant lead coplanarity problems. This will result in solder opens during soldering. However, PLDs have become available from some semiconductor houses that are programmed after soldering by the tester. This approach not only saves on the inventory cost of various programmed part numbers but it also lowers manufacturing defects.

11.2 AUTOMATED PLACEMENT OF PARTS

Accuracy requirements almost mandate the use of autoplacement machines for placing surface mount components on the board. The placement machine is the most important piece of equipment required for surface mounting. It absorbs the highest capital investment, and it also determines the overall economy of manufacturing.

Placement equipment can be classified based on flexibility and throughput. The flexibility to place components of many different types comes at a price: the higher the flexibility, the lower the throughput. For example, robots are placement machines that provide the ultimate in flexibility. They can be used for placing surface mount components, for placing odd-form components, for solder mask or paste dispensing, for soldering, and for lead tinning. Their hardware cost is relatively low, but software and hardware development can be quite expensive. Robots are very flexible, but they are extremely slow and require considerable development for each type of application.

11.3   SELECTION CRITERIA FOR PLACEMENT EQUIPMENT

There are many different placement machines available, and new models are constantly being introduced in the marketplace. The cost can vary from a few thousand dollars for a benchtop model to almost a million dollars for a very high speed placer. The cheaper models are generally used for prototyping but serve in production in many companies. They have limited capacity and may not be very accurate. Production placement equipment can be either flexible, i.e., able to place surface mount components of all types, or dedicated, placing only components of certain types.

Evaluation and selection of a pick-and-place system is a very complex process. The selection of the appropriate autoplacement machine is dictated by many factors, such as the complexity of the device, the applicable packaging and equipment standards, the type and number of parts to be placed, and current and future needs for volume and flexibility. Thus, one must establish some guidelines for selection of a machine. A matrix detailing the desired features versus features of the available equipment will help reduce the number of choices for a given application. Here recommendations by existing users can be the most important selection criteria. Refer to Table 11.1 for a summary of selection criteria. Appendix B1 offers a detailed questionnaire that will help in the final selection of a pick-and-place machine. The questionnaire should be used to narrow the list, which can be further evaluated with some form of weighted points for features important for the given application. (For example, What kind of parts are to be handled? Will they come in bulk, in a magazine, or on tape? Can this machine accommodate future changes in tape sizes? How will the board be handled? Does the equipment come with an automated board handling system?)

Figure 11.2 shows an example of the use of weighted points for important features in the evaluation of pick-and-place equipment. Different weights are assigned depending on a feature's relative importance. The maximum total points should equal 100 to make comparisons easier. This method of relative ranking makes the evaluation less subjective, especially if the assigned points for each machine reflect the average of all points assigned by the selection team. After the ranking is finished, if the highest-rated equipment does not seem to be preferred, re-examine the points that have been assigned to each feature. It may be that too few points are assigned to a feature that is actually more important. When done in a team setting, this exercise can be very useful for clarifying your needs and selecting the equipment with the right features for your application.

Table 11.1 **Summary of selection criteria for pick-and-place equipment for surface mount components. (Refer to Appendix B1 for details)**

Company _____ Contact name _____
Address/Phone/FAX

Model number _____ Base price _____
Options prices _____ Feeder prices _____
Maximum board size/placement area _____
Maximum number of 8 mm feeder slots _____
Placement rate per hour: Standard SMT _____ Fine pitch _____
                          Chip components _____
Placement accuracy: X-Y _____ Θ (Rotation) _____
Programming: On line _____ Off line _____ Teach mode _____
Acceptable type of feeder input: 8 mm tape _____ 12 mm tape _____
                               over 12 mm tape _____
                               7 inch reels _____ 13 inch reels _____
                               Tube/stick _____ Bulk _____
Maximum JEDEC standard waffle pack _____
Need waffle pack handler? _____
Adhesive application: Yes __ No __
                               Adhesive dot dispensing per hour _____
Smallest chip components (0402/0603/0805/1206) _____
Largest component size _____
Acceptable component types:
MELF _____ SOT _____ PLCC _____ SOJ _____ SOIC _____ BGA _____
Others _____
Fine pitch: 25 mils _____ 0.5 mm (20 mils) _____
Ultra fine pitch: 0.4 mm _____ Below 0.4 mm _____
Special features: component on-line testing _____
Component missing verifier _____
Statistical process control _____
Management information system (components left/mispicks) _____
Software for feeder placement optimization _____
CAD down load features (acceptable data format): Gerber __ ASCII __
                               Pin 1 orientation _____
Unique features

Service/support/references:

SELECTION	CRITERION	MAX. POINTS	Machine 1	Machine 2	Machine 3	Machine 4	Machine 5
RISK ASSESSMENT	PARTS AVAILABILITY	5	5	5	4	5	5
	SERVICE/SUPPORT	10	9	8	4	10	10
	UPGRADE (Trade Up Machine)	5	3	5	5	2	4
	FEEDER (Forward Compatibility)	10	8	10	8	10	8
	REFERENCES	5	4	5	0	4	5
THROUGH-PUT	ACCURACY/ REPEATABILITY	5	3	5	5	2	5
	PLACEMENT RATE	5	5	5	3	4	4
	FEEDER CAPACITY	10	7	9	10	9	8
	SETUP TIME	5	4	5	4	4	4
	BOARD SIZE (max.)	5	4	4	5	3	4
SOFTWARE	EASE OF PROGRAMMING	10	6	8	9	7	10
	OFF LINE PROGRAMMING	5	5	5	5	5	5
	CPU IN THE SYSTEM	5	3	5	5	5	4
COST	PRICE	10	6	5	8	6	10
	WARRANTY	5	3	4	5	4	4
	TOTAL	100	75	88	80	80	90

Figure 11.2 Example of use of weighted points assigned to major selection criteria for of pick-and-place equipment.

499

11.3.1   Maximum Substrate Size Handling Capacity

Although many placement machines developed for the hybrid industry can handle only small substrates, for surface mounting it is not uncommon to use 12 inch × 18 inch or even 18 inch × 18 inch substrates. And, sometimes unique board sizes such as 1 inch × 46 inches are required. So when selecting placement equipment, the maximum size of the substrate or panel that the apparatus can handle is probably the place to start. This requirement alone may eliminate many machines.

Even when the substrate size is small, it is generally economical to use large panels containing multiple breakaway substrates. If fixtures are to be used for handling, their maximum size should be used instead of substrate size as the selection criterion for a pick-and-place machine.

Various types of fixtures are available to hold the printed circuit board in place while components are being placed. Figure 11.3 shows an example of a unique PCB holder with movable magnetic supports. Depending on the board, the magnetic supports can be moved quickly to desired locations within the fixture to accommodate different configurations and sizes. Plastic edge clamps (on magnetic supports) and tooling pins in the fixture frames are used to hold the PCB in place over the magnetic supports. To prevent board buckling in large boards, a few pin supports on the base of the fixture plate can also be used if necessary. (See Figure 11.3.) The fixture holding the board can be transported either on a conveyor belt or on rails and presented to the placement head.

11.3.2   Maximum Feeder Input or Slot Capacity

The maximum number of feeder input positions is another measure of placement equipment capacity. Since this number provides a measure of capacity to process an assembly with different part types (or different values of the same part type) in a single pass through the machine, it is a very important consideration.

To determine the slot or input capacity needed, one must first analyze the product requirements and ascertain the maximum number of part types that will be used. (By "part type" we mean not only parts of different mechanical outlines but also parts of different electrical values in a given outline.) Future expansion needs that can be anticipated to arise when almost all components become available in surface mount should be included in the study.

A standard measure of feeder input capacity is the 8 mm tape feeder. The more 8 mm slots a machine can accommodate, the higher its input capacity. However, only a limited number of parts come on an 8 mm

MAGNETIC
BOARD SUPPORTS

PIN SUPPORT

TOOLING PIN

Figure 11.3 A printed circuit board holder with movable supports to accommodate different sizes of boards. (Photo courtesy of Intel Corporation.)

tape. Larger parts supplied in a tube take up multiple 8 mm slots, and the same components on tape and reel may require two to three 8 mm slots. Components supplied in bulk or in waffle packs affect input slot requirements differently.

Another option for compensating for insufficient input capacity is to complete the placement of components in two or three passes or to sequence many parts of the same size but differing values in the same tube feeder. This error-prone approach is definitely not recommended for production, where its adverse effect on placement rate could not be tolerated.

While conducting the in-house study on slot requirements, side benefits may arise. Perhaps, for example, some part numbers can be totally eliminated (e.g., a 1% tolerance resistor if all you need is a 5% tolerance part), or some part numbers can be consolidated. Standardizing on one or two component sizes and values for chip components will further reduce part count. Using a minimum number of part types not only conserves input slots (and reduces the number of feeders that must be purchased) but also can reduce inventory costs and provide increased leverage with vendors due to higher volumes.

11.3.3   Types and Sizes of Components

Equipment selection is very dependent upon the types and sizes of components that need to be placed. For example, the equipment needed for 0805, 1206, and smaller leaded components are different than equipment that must place 0603, 0402, BGA, and fine pitch components. Also, the maximum size of components to be placed should be kept in mind. Every machine has its limit on the maximum size and weight of components it can place.

In many applications, depending upon the volume requirements, it may be wise to consider two machines in an integrated line to complement each other's capability. In other words, one versatile machine may not be as good a choice as two different machines each with complementary strengths.

11.3.4   Placement Rate and Flexibility

In selecting a placement machine or set of machines to meet production requirements, one must determine the product mix, the number and types of components per board, and the production volume. Knowing the current requirement is not sufficient. Future needs and manufacturing plans should also be factored in. Thus, the placement rate is very important.

Actual throughput will depend not only on the placement rate of the machine but also on the component mix and the types of feeders used. The placement rate, in turn, depends on the location of the feeders with the most widely used components. If the product mix changes significantly, it may be difficult to place feeders of the most widely used components to provide the shortest possible distance between the pick location and the board. Some machines come with software that optimizes placement head movement for the shortest overall travel time for the entire board.

Another factor that slows placement rate is component testing before placement, especially if testing is done on the fly. If vision is used to locate components, the placement rate is further reduced.

Placement rate is also affected by board size and by component and feeder types used. For example, a tube or waffle pack feeder that requires constant loading and unloading may cause interruption of operation of the machine. Larger boards with an increased number of component types on the machine will require greater travels, which in turn will slow the throughput of the machine. Most importantly, a machine that is often down or in need of repair, even though it runs very fast when working, will have an adverse impact on throughput.

It quickly becomes obvious that determining the actual placement rate is not easy. As a general guide, the rated placement rate quoted by the vendor should be derated by 40–50% to arrive at a conservative number. The experience of other users who have tried a similar product mix with the machine under consideration can also give an idea about the placement rate.

We see that placement rate cannot be considered in a vacuum. Dedicated machines with limited types of parts placement capability are going to be much faster than machines that can place components of all types and sizes, test them before placement, and ensure accurate placement using a vision system.

11.3.5 Placement Accuracy/Repeatability

As components with larger sizes and finer pitches come into common use, the need for accurate placement stands out even more. There are many ways to define accuracy, but one useful characterization is as follows: accuracy is the greatest tolerable deviation of the component lead from the center of its corresponding land after placement [1].

Many times suppliers specify the machine resolution, which is defined as the smallest discrete resolution the machine can discern. This does not mean that what the machine can discern can be repeated time after time, however. Also, machine resolution generally assumes the use of CAD data for programming. Thus if the machine is programmed by "teach mode" or by using a digitizer instead of the CAD-supplied coordinates, additional errors can be introduced during programming and "teaching" of the machine.

Repeatability instead of accuracy is a more useful guideline, then— i.e., the consistent ability of the placement head to place a part at the specified target within a specified limit. Accuracy, for all practical purposes, simply means the placement of components on the land pattern within the acceptable deviation or shift. Depending on the application, the maximum acceptable shift of leads or terminations from their pad generally varies from 25% (Class III) to 50% (Class I and II) of pad width [2]. A 50% shift is excessive and may degrade solder joint reliability; a 25% shift is the maximum acceptable, as long as the component does not shift any further during reflow soldering. However, in some designs, no lateral shift is acceptable if there is inadequate distance between pads of adjacent components or vias.

Some components such as BGAs self-align better than others due to solder surface tension. But regardless of the components used, the misplacement should not be any more than the equipment is rated for. After all, placement accuracy is one of the determining factors in a machine's performance and price.

For 50 mil center packages, the pad width generally used is 25 mils, and for fine pitch packages only 12 mils or less. Thus a 6 mil accuracy requirement should be good enough for 50 mil center packages, and 3 mil placement accuracy should be sufficient for fine pitch packages. This is the X-Y accuracy requirement. However, the X-Y requirement alone is not sufficient because it does not take into account rotational accuracy and the additional deviations caused by manufacturing processes such as handling (between placement and soldering) and reflow soldering.

Rotational accuracy is not the same as X-Y accuracy. The same degree of Θ or rotational deviation will produce a larger offset in some pads for bigger devices than it will for smaller devices. For example, a one-degree rotation error will displace a lead near the corner of an 84 pin, 50 mil center PLCC by more than 0.010 inch [1]. If we add to this the deviation caused by the artwork and PCB manufacturing processes, the problem becomes serious. The total component shift may exceed 50%, which is totally unacceptable.

The accuracy requirement will vary for different applications, but a 0.002 to 0.004 inch in X-Y repeatable accuracy from the target on a 14 inch by 18 inch substrate, irrespective of programming method, and 0.2 to 0.5 degree rotational accuracy should be sufficient for most production applications. Of course, the acceptable tolerance will be much tighter for finer pitch components (under 20 mil pitch) and for any components if the interpackage spacing is very tight. (See Chapter 7 for interpackage spacing guidelines.)

Placement accuracy is improved with the use of a vision system, as discussed in the next section.

11.3.6 Vision Capability

One way to offset inaccuracy in placement is to use a vision system to tell how far a component lead is from its corresponding land and to instruct the machine to correct for the discrepancy. To implement the instructions of a vision system, the hardware design of the placement machine must afford the needed resolution or repeatability.

The vision system is a good way to compensate for deviations in land pattern locations due to PCB manufacturing processes and poor tolerances of component manufacturing. It also can compensate for the relatively larger deviations from the reference location in larger boards.

When using a vision system, a set of alignment targets, or fiducials, should be designed on the board. Location of alignment targets on a board are shown in Figure 11.4, which shows the location of tooling holes along with the alignment target. The dimensions A, B, C, and D shown in Figure 11.4 will depend on the form factor for a given product. The basic intent of Figure 11.4 is to show the location of 3 alignment targets (also called global fiducials) on the board. The alignment target shapes and their sizes vary. See Figure 7.3 in Chapter 7. No matter what the shape, the alignment target must be free of solder mask and should be flat within one mil. Some equipment manufacturers require that the target be flat within 0.4 mil. If the hot air solder leveling (HASL) process is used, it becomes difficult to achieve flatness within 0.4 mil. One must check with the placement equipment manufacturer about the alignment target requirements. By referencing part placement to these targets instead of to tooling holes or board edges, one can avoid such tolerances as tooling hole/tooling pin fit, tooling pin size, tooling hole size, and tooling hole location with respect to the land patterns.

There are two types of vision systems: single-camera and two-camera systems. In a one-camera system, the land patterns are viewed and the placement coordinates adjusted appropriately. In a two-camera system, both the component and the land pattern are viewed and compensated for. For finer pitch packages with lead pitches of 25 mils or less, it is almost mandatory to use a two-camera system that can match the land pattern with the corresponding component leads.

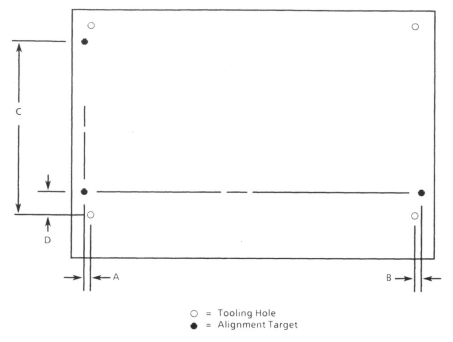

O = Tooling Hole
● = Alignment Target

Figure 11.4 Alignment target for the vision system in a pick-and-place machine.

The vision alignment uses either a binary or gray scale vision imaging to determine the offset between the component center and the placement nozzle center. Figure 11.5 compares binary and gray scale imaging systems [3]. The binary imaging system is the commonly used system. As shown in Figure 11.5, it uses back lighting to project the outline of the component to the camera. This system works fine for peripheral packages since the lead outline is projected as part of the package outline. The system adjusts for any offset between the nozzle center and component center and places the part accurately.

The binary system does not work well for ball grid arrays (BGAs). Since the balls are hidden underneath the package, it cannot correct for any variation in tolerances between the package edge and the location of the balls. It is of more concern in ceramic BGAs (CBGAs) because of their poor dimensional tolerance as a result of package shrinkage during their firing. So for BGAs the vision system must accurately align all the balls on the package with their corresponding pads on the board. To do this adequately without using the package edge as the reference point for alignment, a gray scale imaging system is required.

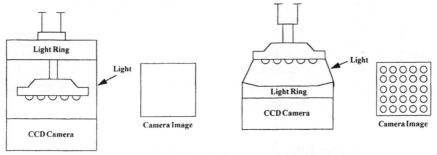

Figure 11.5   Comparison of binary and gray scale vision imaging systems for BGA placement [3].

As shown in Figure 11.5, the gray scale imaging system uses front lighting (as opposed to back lighting in the binary imaging system) which allows the camera to pick up any surface features along with a clear view of all the balls underneath the package. This allows alignment based on the locations of the balls instead of depending upon the location of the package edges. Dependence on package edges for ball location may be acceptable for plastic BGA (PBGAs) where the tolerance between the package edge and ball locations is reasonably good. But it is certainly not acceptable for CBGAs and not even for tape BGAs (TBGAs). (See Chapter 3 for a discussion of various types of BGAs.) As a side note, some misplacement in BGAs is not as critical as in fine pitch packages since due to the surface tension force of all the balls, they generally self-align even if misplaced almost by 50%. However, when one is trying to control all aspects of the process, it certainly does not help to use a system that is likely to misplace components.

The important key elements of vision systems are field of view, resolution, and the level of gray scale. The level of gray scale, generally up to 256 levels of gray scale imaging, is applicable only to the gray scale imaging system. The component lighting and the processing speed are also important for throughput. Most systems also come with lead coplanarity inspection as a part of the vision system.

For fine pitch, both the binary and gray scale systems will work, but for BGAs and any other area array packages such as CSP (chip scale packages) and flip chip, the gray scale vision system is required. No matter which system is used, the field of view and the resolution must accommodate the size and the pitch of the component. Most systems can accommodate 38 mm maximum size and 20 mil lead pitch.

In addition to a vision system, some design considerations on the board are also necessary for BGA and fine pitch. For example, for BGA and fine pitch parts, another set of local alignment targets or fiducials should also be used. See Figure 7.2 in Chapter 7. This is in addition to the global fiducials on the board. The local fiducials compensate differently in various regions of the same board and are critical for accurate placement of BGA and fine pitch components.

11.3.7   Adhesive Dispensing Capability

Nonconductive adhesive is used to hold components in place before wave soldering. The adhesive is dispensed, the component is placed, and the board is heated to cure the adhesive, thus ensuring that the components will withstand the rough action of the wave during wave soldering.

There are various ways to dispense adhesive as discussed in Chapter 8, but the most commonly used methods are stenciling and syringing. For syringing, an adhesive dispensing system using a syringe is generally an integral part of the pick-and-place equipment, although dedicated adhesive dispensers are also used. When the pick-and-place machine is used for adhesive dispensing, either dedicated adhesive dispensing heads with their own X-Y table can be used, or the component placement X-Y table can be shared for adhesive dispensing.

Many equipment suppliers offer both options—a dedicated X-Y table for adhesive dispensing or a common X-Y table for both adhesive dispensing and component placement. The latter option is cheaper but slower because the placement head will either be dispensing adhesive or placing components. Thus, selection should be based on production volume requirements. For very high volume applications, it may be preferable to use a stencil printer, which is less expensive, rather than tie up an expensive placement machine.

Adhesive dispensing parameters were described in detail in Chapter 8. It is simply worth noting here that having the desired adhesive dispensing capability may be one of the decisive factors in the selection of a particular placement system if the machine is intended for Type II or III SMT assembly. Figure 11.6 shows an adhesive dispenser as an integral part of the pick-and-place system.

11.3.8   Equipment Software Program

The software features of the pick-and-place machine determine the ease of programming for board placement. Different machines have differ-

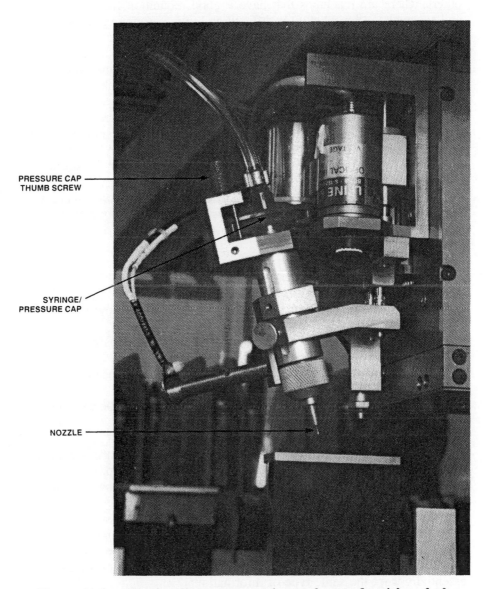

Figure 11.6   Adhesive dispenser as an integral part of a pick-and-place system. (Photograph courtesy of Intel Corporation.)

ent programming features. Some are easier and more user friendly than others. There are machines that have a good reputation for throughput and accuracy but for which it is difficult and time consuming to program and to edit existing programs. For example, some of these machines cannot easily accept CAD output. Cases where the data is entered manually, or where the "teach" mode of programming is used (in which you physically take the placement head to pick and place positions) are archaic, subject to potential mistakes, and time consuming.

In manufacturing there is always pressure to reduce the setup time. Not meeting the delivery schedule is a common problem in manufacturing, along with quality issues. The setup time is essentially driven by software and not by hardware. It is an important equipment selection feature that is often overlooked.

The general input requirements of any machine for component placement are similar. The key items that must be input either manually or through download from the CAD system are as follows:

- Board size and thickness
- Global and local fiducial type and size
- Whether the board is one up or is in a panel
- If in a panel, whether the machine should identify and not build boards with reject marks
- X, Y coordinates of component locations on the board
- Component orientation (pin 1 orientation)
- Component names, sizes, and thicknesses at pick locations
- Component orientation at pick positions
- Location and type of feeder (tape and reel, tube, waffle pack, etc.) holding a given component.

All this information essentially boils down to letting the machine know where and how many of a given component to pick up, from what type of feeder, and where and how to place them on the board. This is summarized in Figure 11.7.

Inputting all this information is fairly straightforward and is done by technicians with some training. One does not need programming experience and one does not have to be mechanically adept. The skill set is generally transferable from one machine to another since all machines essentially want the same type of information.

It should be noted that while the software primarily determines the setup time, no matter which software is used, the user can also standardize certain items to reduce setup time. For example, standardizing board parameters such as length, width, and thickness; and fiducial type, size, and locations, and using consistent design rules for component orientation

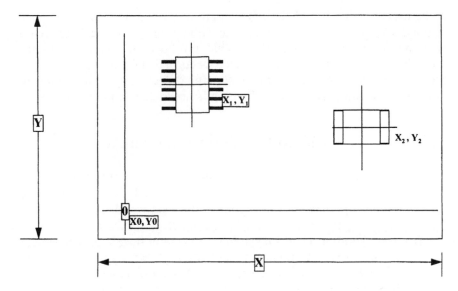

Figure 11.7 Examples of dimensions and coordinates that need to be programmed in a pick-and-place machine.

help in reducing setup time. By taking such steps and by developing internal software for the placement machine with the help of commercially available software *(Framework),* one company was able to reduce the setup time from 12 hours to 5 minutes [4].

11.3.9   Service, Support, and Training

The technical features discussed so far are important. But it is also important to keep in mind who you are doing business with and their service and support record. The placement industry is a highly competitive industry. There are many new suppliers coming on the market and many who no longer exist. One criterion to keep in mind is the installed base of the equipment you are considering. Will you be able to get the needed support for parts and service in time when the equipment is down? Being able to get the machine up and running quickly will be critical when one needs to meet a delivery schedule. If a company has an installed base of the type of equipment you are considering, you can get user references that should be weighed heavily in the selection process.

The time and quality of training on the equipment are very important. All companies provide some training. It is important that some level of training be provided at the supplier site before equipment shipment and at the user site after installation. The length of training should be adequate so that the operator is fairly comfortable with operating the machine. There are some companies that not only provide training on the particular equipment, but also provide training on other aspects of SMT process and design. Such training may be of real help in reducing defects, especially if you are new to this technology or if such training is not available internally.

11.3.10   Other Important Selection Criteria

Many other features of a pick-and-place machine are very important for most applications and thus should be considered. Their significance may be more important in some applications than in others.

1. The pickup mechanism can have a vacuum nozzle or mechanical jaws or both. Most machines pick up parts with a vacuum nozzle but use jaws to center components before placement.
2. In some placement equipment, both the placement head and the X-Y table (where the board is held) move during placement. In other cases the X-Y table remains stationary during placement and only the head moves. This is an important feature to keep in mind, because if the acceleration and deceleration of the X-Y table are not properly controlled, the tackiness of the adhesive or solder paste may not be sufficient to hold components in place. In general, the incidence of component misalignment is less in machines with stationary X-Y table design. This feature becomes even more critical in the placement of the larger fine pitch devices, in which case a moving table may cause component shift, hence repairs that otherwise would not have been necessary. It should be noted however that movement of the X-Y table by itself is not a disadvantage since there are some highly accurate placement machines on the market with moving X-Y tables. However, there are also machines where the moving X-Y table is a major contributor to part misplacement.
3. The pickup head should be able to sense that the part has been picked up. If this fails to happen, machine operation should halt. This is an important feature because components can be held too tightly or too loosely if tolerances of either the components or the tape cavity are poor.
4. The control system for the placement head should allow CAD downloading of component locations but should also allow for manual editing of adhesive placement locations and part pickup and placement locations. It should also provide management infor-

mation reports, such as the number of poor part picks and total components used. There are some very good machines widely used and known for their speed and accuracy but which do not allow downloading of CAD data. This adds significantly to equipment setup time.

5. The placement head should have programmable Z axis travel. The pickup head should be able to place the parts onto the adhesive or the solder paste within the preprogrammed Z axis displacement. The displacement force should be minimal, to avoid damage to the parts or the substrate.

6. Component test before placement is a nice feature. A test that can be performed in flight without much loss of placement speed is preferable. Testing in flight is generally accomplished on passive devices only. Active devices require a separate test nest, which can considerably slow the placement rate.

   Testing will prevent placement of wrong components if, as quite commonly happens, tapes are mounted at incorrect feed locations. To have a minimum effect on placement rate, the testing of passive devices may be programmed for only the few first devices to be placed. Thus the test "programming" effort for passives is trivial. Instruments are readily available to test resistance, capacitance, and inductance, but testing active devices requires a considerable test development effort. Also, the equipment cost for testing actives is at least an order of magnitude greater than that for testing passives.

7. Flexible placement systems should be able to accept such commonly used feeders as 7 and 13 inch reels, tubes, and bulk and waffle packs.

8. The advance mechanism for the reels should be adjustable in 4 mm increments because in a given feeder width, the feeder pitches vary in 4 mm increments. This allows the use of the same feeder for a given tape width but with varying tape pitches.

9. An important feature that may be very desirable is on-line programming capability for PLDs (programmable logic devices). Currently these devices are programmed off-line and then put in a tube for placement. Since PLDs generally have high fallout rates, the pick-and-place system must be able to test them before placement. The need for this feature may not exist if devices are programmed during final testing (after soldering).

10. Placement machines with intelligent feeders do not require programming of feeder locations on the machine. They can be placed anywhere and the machine will know their location. If such feeders come in banks of more than one (generally 16), even if one feeder

goes bad, the whole feeder bank becomes inoperative. So if one is considering intelligent or "smart feeders," as they are generally called, individual smart feeders are preferable to a bank of smart feeders. Even when using smart feeders the operator must still load the right reels of components for which the feeder is identified. The use of smart feeders is no protection against such mistakes.

11.4 SELECTION OF FEEDERS FOR PLACEMENT EQUIPMENT

Generally much attention is given to the selection of placement equipment and not enough to the selection of feeders. The importance of feeders, however, cannot be overemphasized. In the absence of reliable feeders that can feed components consistently and with minimum operator intervention, the placement rate suffers because the operator must constantly interrupt machine operation to correct the feeder problems.

The feeder system for the pick-and-place machine essentially determines reliability and throughput in placement. We use the term "feeder system" advisedly, because simple chip resistors, capacitors, and transistors do not use feeders of the same type required by the complex, multi-leaded devices such as PLCCs, SOICs, and fine pitch packages. In many cases there are overlaps, though, allowing the same type of feeder to be used for components of different types.

The determining factors in the selection of a shipping and handling medium should be desired quantity, availability, part identification, capability of the pick-and-place machine, cost of components, inventory issues, and potential for damage during shipping and handling. There are four types of component feeders: tape, tube, bulk, and waffle pack.

In general, tape and reel feeders are used for all surface mount components, but the bulk feeders are generally used only for very limited, simple components such as resistors and capacitors. The tube feeders are used for SOICs and PLCCs, but the waffle packs are limited to fine pitch devices, BGAs, and other moisture sensitive devices that may require baking before placement. There are exceptions of course. Each method has its inherent advantages and disadvantages. The details are discussed next.

11.4.1 Tape and Reel Feeders

Tape and reel is used for chip resistors, capacitors, MELFs, SOTs, SOICs, PLCCs, SOJs, and BGAs. It is by far the most popular and desirable feeding method, especially if large volumes of components are to be placed. As shown in Figure 11.8, the components are nestled in

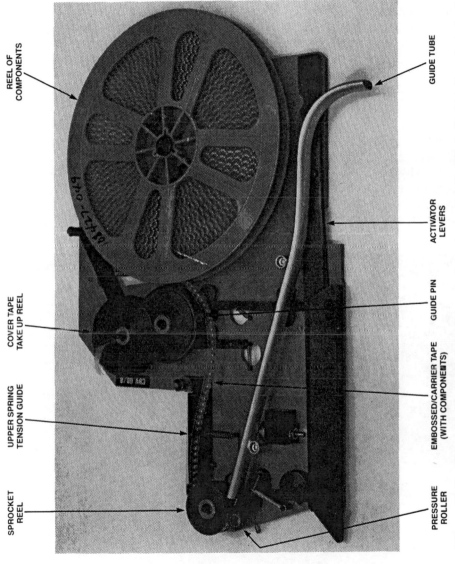

REEL OF COMPONENTS

GUIDE TUBE

COVER TAPE TAKE UP REEL

ACTIVATOR LEVERS

UPPER SPRING TENSION GUIDE

GUIDE PIN

EMBOSSED/CARRIER TAPE (WITH COMPONENTS)

SPROCKET REEL

PRESSURE ROLLER

Figure 11.8   Tape and reel feeder: the components are nestled in individual pockets in the tape and are covered with plastic. (Photograph courtesy of Intel Corporation.)

individual pockets in the tape and are covered with a plastic cover, which is peeled back at the time of placement. Tape widths vary from 8 mm to 56 mm for different components. Every tape width requires a different feeder. Figure 11.9 shows a row of tape feeders mounted on a pick-and-place system.

The reels protect the devices from damage during shipping and handling, save on loading and unloading time on the pick-and-place equipment, and prevent both part mix-up (by compensating for the lack of part markings on many surface mount components) and placement of components in the wrong orientation. Placement in the wrong orientation is seen even with tape and reel when some components are packaged incorrectly. This is not a common occurrence but it does happen. Placing components in the wrong orientation is more common when using tube feeders, which require frequent loading and unloading.

Why aren't the tape feeders used exclusively, then? First, not all components are available on tape and reel, and not all pick-and-place machines have tape and reel handling capability. Also, components such as PLDs that require programming before placement cannot be handled

Figure 11.9 A row of tape feeders mounted on a pick-and-place machine. (Photograph courtesy of Intel Corporation.)

on a tape and reel if they are preprogrammed or are programmable during final tests as mentioned earlier.

Another factor preventing the universal use of tape and reel becomes important especially if smaller quantities are desired, since tape and reel is best for large quantities. This factor is related to the standard that controls tape and reel construction, namely EIA-RS-481. Because the tolerances in this specification are not very tight, reel quality varies greatly from one supplier to another. Thus tapes often have poor tolerances, especially in the center-to-center distance of sprocket holes. This can cause variation in pick location, which may even prevent part pickup.

Also, there may be problems in cover tape removal. The EIA specification has poor tolerance in peel strength. Thus one of the most common problems in tape feeding is the wide variation in pull force required to peel off the cover tape to expose components for placement. Either all pick-and-place machines should be able to handle these differences or the peel strength tolerance should be tightened.

Still another problem involves the very broad reel size requirement. The EIA specification (EIA-RS-481) establishes the maximum reel size at 13 inches but gives no minimum. The de facto standards for passive and active components have become 7 and 13 inch reels, respectively. Since active components generally are available only in the 13 inch reel size, inventory cost may be a real problem for small users. The minimum quantities that come on a 13 inch reel are shown in Table 11.2. The user must be willing to purchase such a minimum quantity to use tape and reel at all. Table 11.2 also shows the quantity of components that could be handled if 7 and 10 inch reel sizes became available. Thus the desired

Table 11.2 Parts count in various sizes of reels for tape and reel feeders. (If active components become available on 7 or 10 inch reels, the number of components per reel may also be standardized)

TYPES OF PACKAGES	TOTAL COMPONENTS PER REEL		
	7 INCH[a]	10 INCH[a]	13 INCH
SOIC 8, 14, 16L	1057	2323	4158
SOIC 20, 24, 28L	341	851	1468
PLCC 18, 20L			
SOJ 20L			
PLCC 28	168	422	769
PLCC 44, 52	112	281	513
PLCC 68	52	140	259
PLCC 84	46	124	230

quantity generally is of prime importance in the selection of feeder equipment. If the desired quantity is large, tape and reel is the natural choice.

If small quantities are needed, selection becomes more difficult. Suppliers are generally not interested in offering smaller or partial reels because they would rather sell more on a larger reel than less on smaller or partial reels. Users who need small quantities end up buying in tubes. This may add another mil or two in lead coplanarity during shipping and handling.

It should also be noted that the IEC (International Electronics Council) allows reels of 7, 10, and 13 inches and other sizes. A change by EIA-RS-481 would promote compatibility with other standards, since components are purchased beyond the national boundaries. Even in the absence of standards for smaller reel sizes for active devices, the user has the option of packaging components in-house if he has such a capability. If not, independent companies and some distributors will package only the desired number of active components in tape and reel. This adds to the component cost, however.

The larger reel size is not an issue when handling larger quantities of surface mount components. However, there will always be some need for smaller reel sizes, even for larger users. For example, even a high volume product takes from 6 to 12 months for design and prototyping, and smaller quantities are needed during this phase. Every attempt should be made to use tape and reel whenever possible.

11.4.2   Bulk Feeders

Bulk feeders may be acceptable in prototyping and in very low volume applications, but they are very inconvenient for feeding components in automated assembly. Generally only passive devices are fed by this method, that is, from a vibrating bowl into a track. The track width limits the width of component that can be fed.

There are many problems associated with bulk vibrating feeders. For example, if the intensity of vibration is not properly controlled, the components sometimes jump out of the feeder. Since one or more components may not be placed when this happens, rework is likely to increase. In addition, termination solderability may be lost as a result of the constant rubbing of components against each other and against the bowl feeder. Bulk feeding should be avoided because of potential damage to terminations and leads, as well as misplacement and part mix-up problems.

11.4.3 Tube or Stick Feeders

Despite the advantages of tape and reel, tube feeders are widely used because of the large inventory cost of tape and reel for small volume users (and also because some components are not available in tape and reel). Also, many pick-and-place machines, especially the benchtop models, will not accommodate tape and reel feeders.

Tube feeders can be divided into three types: horizontal, stick-slope, and ski-slope. The horizontal tube feeders have multiple tracks that are machined to accommodate components of different widths. The number of tracks per feeder depends on the width of the component body to be used. Figures 11.10 and 11.11, respectively, show a horizontal tube feeder and a row of tube feeders mounted on a pick-and-place machine.

The stick-slope feeder is simply a modification of the horizontal feeder in which the plastic tube can be directly loaded into the individual tracks of the horizontal feeders. Since the components do not have to be taken out of their plastic shipping tubes, the stick-slope feeders have all the advantages of horizontal feeders without any of the disadvantages. Both the horizontal and the stick-slope feeders use some form of electromagnetic vibration to move components along to the pick position.

The ski-slope feeders are like the stick-slope feeders, but the components are moved along by spring action, not by vibration. Also, unlike the horizontal tube feeders and the stick-slope feeders, the ski-slope feeders do not use machined tracks for different components; rather, the tracks are an integral part of the equipment. The ski-slope feeders require tighter component tolerances and will jam up if the tolerances vary as a result of plastic flashes left on the component body.

The ski-slope feeders are more expensive and require additional space on the machine to accommodate their length. Greater length can be an advantage, however, because ski-slope feeders can take more components than either the horizontal or the stick-slope feeders, hence they do not need to be loaded and unloaded as frequently. Nevertheless, they are not as popular as the horizontal or the stick-slope feeders because of their higher cost.

The horizontal tube feeders are fairly reliable and relatively inexpensive; they can be used on almost all pick-and-place machines, both large and small. Since the tube feeders are short, however, parts must be constantly fed into the tracks. These feeders do not cause lead or termination damage because the components are not dumped in a bowl. However, the intensity of vibration must be properly controlled.

Both the horizontal and stick-slope feeders should use tracks that are machined to accommodate some variation in component dimensions.

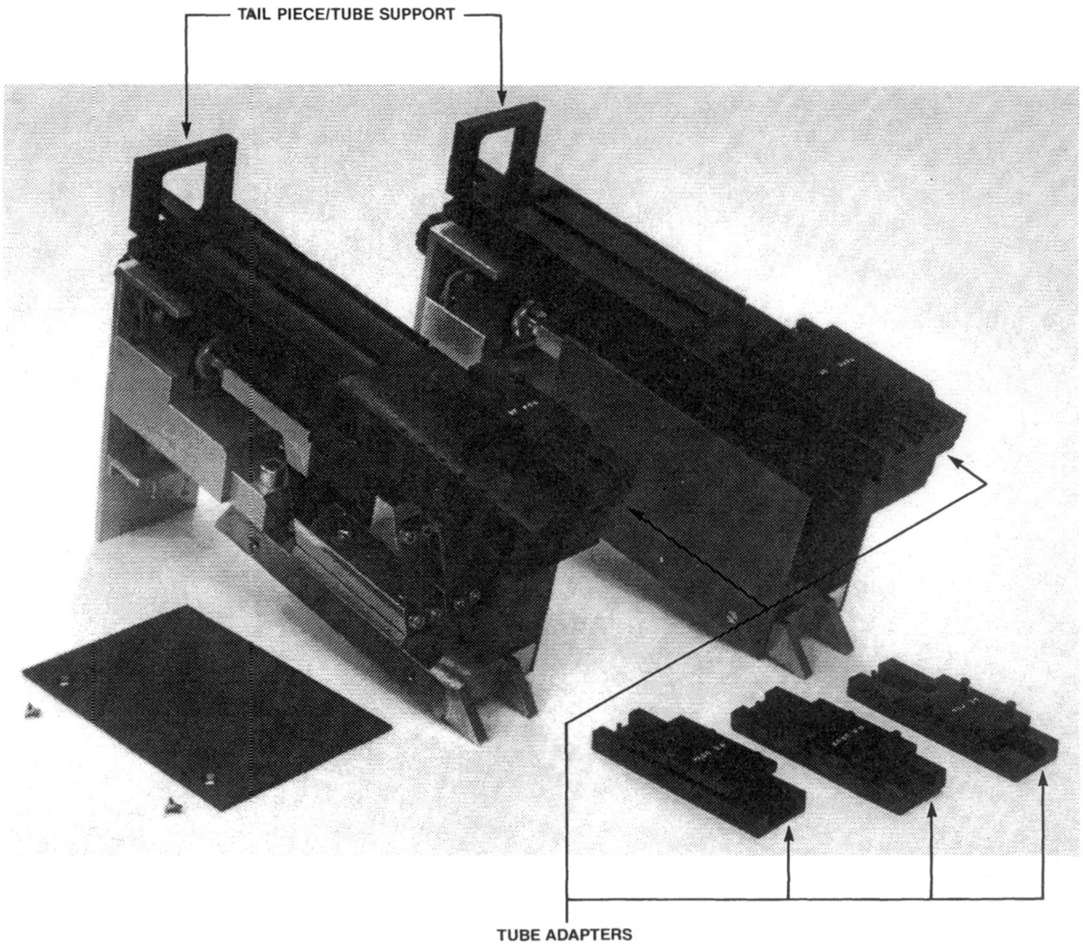

TAIL PIECE/TUBE SUPPORT

TUBE ADAPTERS

Figure 11.10   A short vibrating tube feeder. (Photograph courtesy of Intel Corporation.)

Figure 11.11   A row of short vibrating type feeders mounted on a pick-and-place machine. (Photograph courtesy of Intel Corporation.)

Since the tolerances in component dimensions are fairly loose, the tracks should be machined to accommodate the maximum and not the nominal component dimension. This will allow the use of components from different vendors that meet JEDEC standards. However, the track should not be wider than necessary, since excess track width may compound inaccuracies in placement by shifting the lateral pick positions.

One common problem experienced when using tube feeders is feeding components with the wrong orientation. It is very important to keep this possibility for error in mind, and operators should be trained to identify the location of pin 1. (Look for a dot near the beveled corner.)

11.4.4   Waffle Packs

Waffle packs, as the name implies, use flat machined plates with pockets or plastic carriers resembling waffles. Because waffle packs are large, they tend to consume a lot of input slot space. Also, the X-Y movement range of the pick-and-place equipment is generally limited. Since the pickup position is different for each individual component when using waffle packs, a very limited number of components can be picked, depending on the X-Y movement range of the placement head. In comparison, for a tube or for tape and reel, the pickup position is always the same for all components in a given tube or tape.

In addition, waffle packs have very small capacity and must be constantly replenished. Limited capacity may not be a serious problem, however, if only one or two large packages are used per board. To alleviate the capacity limitation imposed by waffle packs, waffle pack handlers can be used. They pick up parts from the waffle packs and bring them to a fixed location and thus maintain the same pickup position for pickup by the placement head, as in tubes or tape and reels. However, because the waffle pack handlers transport the part to an intermediate fixed location for pickup by the placement head, the overall placement rate is decreased (instead of one pickup, two pickup steps are involved for the same part and those steps can only be performed sequentially). The need for waffle pack handlers also increase equipment cost.

Despite the adverse effect of waffle packs on placement rate and equipment cost, there are instances when use of a waffle pack is the only viable option. For example, as discussed in Chapter 3, Japanese fine pitch packages generally do not have corner bumpers to protect the fragile leads (fine pitch devices made in the United States have corner bumpers to allow packaging in tubes). Waffle packs are the only option for handling these bumperless fine pitch devices, where part movement within the tube cannot be tolerated because of the danger of lead damage.

Waffle packs may also be the only option available for most BGAs, fine pitch devices, and larger PLCCs. Since these components are generally moisture sensitive, they may require baking. Waffle packs can withstand the baking temperatures required (125°C). Tubes can withstand temperatures of only 40°C. Components, if baked, require 24 hours at 125°C and eight days at 40°C.

11.5   AVAILABLE PLACEMENT EQUIPMENT

Commercially available placement machines can be classified into four major categories: (a) high throughput, (b) high flexibility, (c) high flexibility and high throughput essentially by combining two similar machines—typically one of them with a vision system, and (d) low cost and throughput but with high flexibility. I have simplified this issue considerably. One can make a case for many more categories depending upon various levels of features such as price, throughput, and flexibility.

No particular category of machine is the best for every application. Some companies require one machine from each category to meet various product needs. For example, one manufacturing line may need a flexible machine for product development and medium production, whereas another line may call for one machine with low or medium flexibility and one with high flexibility to provide a good balance between throughput and flexibility requirements.

The issues discussed in this chapter and the questionnaire in Appendix B1 should be used for a detailed evaluation of these or other machines before a final selection is made. In addition, it would be a good idea to refer to the latest equipment survey done by professional publications [5].

Since changes in the machine models are practically made to be timed with every major electronic show, we cannot discuss the latest model of any particular equipment. Instead we will use some examples of equipment currently available to illustrate the point about the general category of commercially equipment that should be considered for different applications. This general classification of equipment and their major attributes will be applicable to equipment from any number of suppliers.

11.5.1   Equipment with High Throughput

High throughput generally means less flexibility. In the past, a low flexibility machine used to be the one that placed only passive components such as resistors and capacitors. These machines were generally referred to as chip shooters. Despite a long-standing need for equipment that provides a good balance between flexibility and throughput, such machines did not exist until very recently.

Now very high speed placement machines are available that are still intended for passive devices but are modified to be able to place small active devices. They are still referred to as chip shooters but they have the flexibility to handle larger packages such as SOIC, PLCC, and SOJ. Their throughput can vary widely (10,000 to 60,000 components per hour) depending on the equipment make and model and component mix. The placement rate will be lower when larger components are placed.

An example of high placement equipment is shown in Figure 11.12. For maximum machine uptime, some machines have multifeeder banks. This allows feeders to be loaded on one bank while the equipment is still in operation. The machine shown in Figure 11.12 has two 80 feeder banks or four 40 feeder banks for added flexibility. As a reminder, the numbers 40 or 80 in the feeder bank refer to the maximum number of 8 mm reels that a particular bank can accommodate. So in effect this machine can accommodate 160 different 8 mm tapes. For maximum throughput, it is also important that the feeders be easily accessible and allow rapid reel changes.

The throughput is generally increased by multiple heads. Some ma-

Figure 11.12 An example of a pick-and-place equipment with high throughput (Photograph courtesy of Universal Instruments.)

chines have single heads with multiple nozzles, whereas others have multiple heads but only one nozzle per head. The machine shown in Figure 11.12 has multiple heads but only one nozzle per head. (See close-up in Figure 11.13.) It is important to remember that the nozzle size must be different sized components. For example a chip shooter placing 0402 and 0603 size parts needs to use a different nozzle than when placing 1206 or 0805 size parts or larger active devices. So the configuration of heads and the number of nozzles in each head determine the placement rate.

11.5.2 Equipment with High Flexibility

Equipment with high flexibility has lower throughput. Thus the equipment in this category is almost the opposite of the equipment described in the preceding section. The high flexibility machines are of the sequential type in which components are placed one at a time. The equipment can place almost all types of components, from a chip to a large PLCC,

Figure 11.13   Multiple heads commonly used in a high throughput pick-and-place machine. (Photograph courtesy of Universal Instruments.)

including fine pitch and BGAs. Their throughput depends upon the type of components being placed but it is relatively low.

The machines in this category are very accurate for placement; they are generally modular and can be used alone or in conjunction with equipment in other categories to meet differing manufacturing requirements. Some of the machines have unique features such as on-line testing of passive components and vision capability without much loss of placement speed for passive components.

Different machines have different levels of flexibility. For example, some equipment is limited only to different sizes and types of conventional surface mount components. Then there are machines that have a common platform for all types of components including flip chip and TAB. An example of such a machine is shown in Figure 11.14. It has a broad range of capability—from 0402 size chips to 304 pin fine pitch components as well as BGA. An example of BGA being placed by such a machine is shown in Figure 11.15. It can also place flip chip. Of course you need to purchase these different options.

Figure 11.14 An example of a pick-and-place equipment with high flexibility. (Photograph courtesy of Universal Instruments.)

Using the same platform, the machine shown in Figure 11.14 is so flexible that it can be fitted with different placement and soldering heads for TAB, as shown in Figure 11.16. For TAB bonding, there are also dedicated machines known as hot bar machines. (See Chapter 12, Figure 12.29.) As discussed in Chapter 12, TAB leads are very fragile and flexible and should be soldered immediately after placement. With application of force and heat, the thermode shown in Figure 11.16 solders the leads in

Figure 11.15   BGA being placed by a very flexible pick-and-place equipment. (Photograph courtesy of Universal Instruments.)

place. Sufficient space between packages must be allowed for the thermode. This process is generally the last step in the assembly of the board.

In buying a highly flexible machine, you must carefully consider whether you will ever use the options you are paying for. As is the case in buying gadgets, some options may appear appealing but may not be worth paying for because they are really not needed. There is no substitute for spending the effort to thoroughly evaluate your current and future needs before deciding on a particular machine. This is especially true when buying a placement machine since they generally account for about half of the total investment in the SMT line.

11.5.3   Equipment with High Flexibility and Throughput

As noted before, either a machine is fast but not flexible or flexible but not fast. In a manufacturing environment, many times one needs

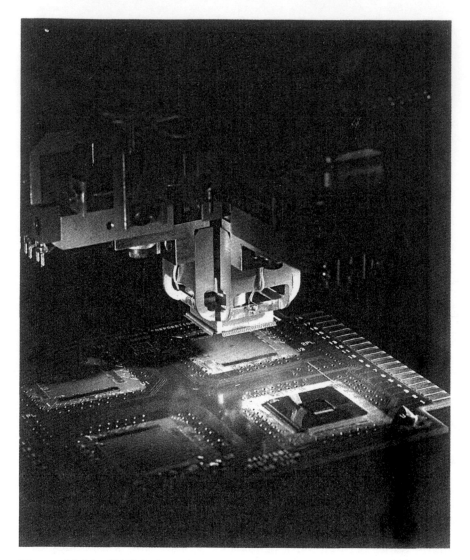

Figure 11.16 TAB being placed and soldered by a very flexible pick-and-place equipment. (Photograph courtesy of Universal Instruments.)

both speed and flexibility. To get both, two machines—one with high throughput and the other with great flexibility—are installed in the same line. For example, after the screen printer, a high throughput machine like the one shown in Figure 11.12 would be installed followed by a flexible machine like the one shown in Figure 11.14.

A more common approach to achieving both throughput and flexibility is to combine two complementary machines in a given line. Both machines are essentially similar except that one is fitted with a vision system. One machine is used for smaller components and the other machine is used for larger and fine pitch and BGA components. In this category there are numerous machines. As is the case with other categories, the cost of the combined machines can vary significantly among different manufacturers. The total throughput of the combined machines can vary widely (10,000 to 60,000 or more components per hour) depending on the equipment make and model and component mix.

An example of two machines acting as one machine is shown in Figure 11.17. One of the biggest advantages of combining two machines is that one can get sufficient feeder input slot to populate a board with over 100 different types of components. The placement rate will be lower when larger components are placed. Also, depending on the requirements and budget, only one machine can be purchased first and the second machine can be added later on as needed.

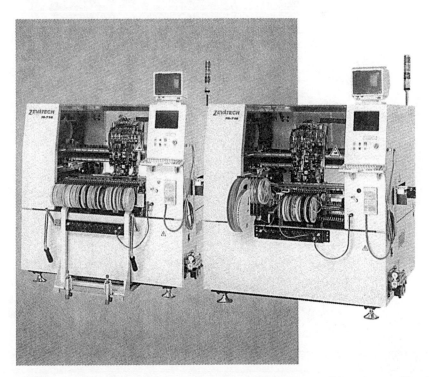

Figure 11.17 **An example of two pick-and-place machines in one line for both high flexibility and throughput. (Photograph courtesy of Zevatech.)**

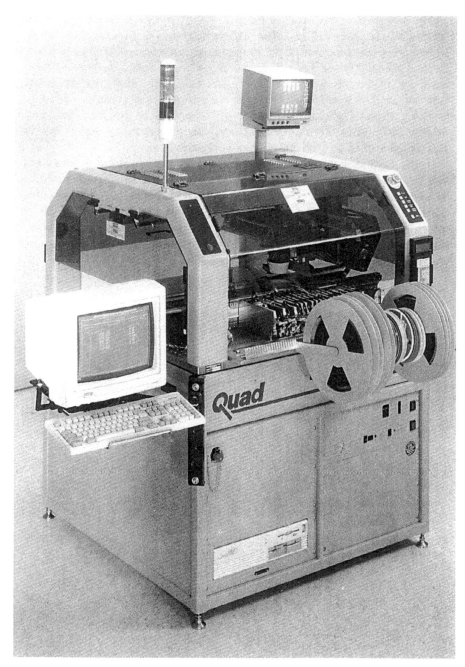

Figure 11.18   An example of a pick-and-place machine with low cost and throughput but high flexibility. (Photograph courtesy of Quad Systems.)

11.5.4 Equipment with Low Cost and Throughput but High Flexibility

So far we have been discussing expensive machines. There is need for low cost machines as well, for use in the laboratory, for quick turnaround prototype work, and also for low volume production. The cost of such machines varies widely but is considerably lower than the cost of the machines discussed in the preceding sections. Even though these machines are relatively inexpensive, if other equipment in the line are properly selected, the general rule of thumb about the placement machine costing about half the cost of the total line still applies.

Equipment in this category may offer an inexpensive way to get into the surface mount business, and when production volume picks up, it can be used for prototyping or as a short-term backup for the production machines. In many applications it can be used in production if volume requirements are low. Some companies use these machines in larger numbers as a substitute for an actual production machine; if one of them is down, the others can still get products out the door. Typically the machines in this category have throughput in the range of 1000–4000 components per hour.

An example of a placement machine in this general category is shown in Figure 11.18. It can accommodate over 100 feeders (8 mm) and large board sizes. For placing fine pitch components, it is important to purchase the vision option. As in other equipment categories, there are numerous suppliers in this category as well.

11.6 SUMMARY

The pick-and-place machine is one of the most important pieces of equipment for surface mounting. There are many types of placement equipment available on the market today with different placement rates and flexibility. Some machines can place all types of components, accurately but slowly; others are intended for speed or throughput but not flexibility, and some combine flexibility and throughput. Features such as off-line programming, teach mode, edit capability, and CAD/CAM compatibility may be very desirable. Special features such as vision capability, adhesive application, component testing, board and waffle pack handling, and reserve capability for further expansion may be of interest for many applications. Vision capability will be especially helpful for accurate placement of fine pitch and BGA packages. Reliability, accuracy of placement, and easy maintenance are important to all users.

The user must define current and future needs before embarking on the evaluation and selection of a placement system. A detailed questionnaire should be used to evaluate all candidates before making a final decision, since no single piece of equipment can meet all needs. In some companies, a combination of machines in various categories, as opposed to a single placement machine, may be needed to create a placement system to meet both flexibility and throughput needs.

No matter which category of equipment is finally chosen, it should be kept in mind that there are numerous domestic and off-shore suppliers and a wide variation in prices in any given category. In addition to prices and equipment features, service, support, and user references should play an important role in the final equipment selection.

REFERENCES

1. Amick, C. G. "Close does not count." *Circuits Manufacturing,* September 1986, pp. 35–43.

2. IPC-A-610. Acceptability of electronic assemblies. Available from IPC, Northbrook, IL.

3. J-STD-013. Implementation of ball grid array and other high density technology. Available from IPC, Northbrook, IL.

4. Harris, J. Duncan. "Automated dynamic placement generation for a multivendor SMT line." *Proceedings of Surface Mount International,* 1992, pp. 349–357.

5. Gentry, Teresa. "High volume placement equipment survey." *Circuits Assembly,* January 1996, pp. 36–49.

Chapter   12

Soldering of Surface Mounted Components

12.0   INTRODUCTION

Welding, brazing, and soldering processes for joining metals together differ basically with respect to the temperature at which the joining takes place. Welding is generally used for joining ferrous metals and is accomplished at temperatures of 815°–1093°C (1500°–2000°F). Brazing is used for joining nonferrous metals at relatively lower temperatures of 426°–538°C (800°–1000°F). Soldering, which is used mostly for electronic products, occurs at the lowest temperatures (210°–230°C or 410°–445°F) than those used for wave soldering (235°–260°C or 455°–500°F), but for a longer time.

In all cases, either two similar metals or two dissimilar metals or alloys are joined. The joining mechanism is governed by the formation of intermetallic compounds between the metals to be joined. A basic knowledge of metallurgical properties and of phase diagrams is necessary for an understanding of the principles involved in joining. The metallurgy of soldering and the concept of wetting are key to the understanding of soldering. These subjects are discussed in Chapter 10. In this chapter we focus on the processes and equipment used in soldering.

The earliest electronic products were hand soldered. In the 1950s hand soldering was replaced by dip, drag, and wave soldering processes for mass soldering of through-hole components. Then in the 1980s, reflow soldering came into widespread use for surface mount components, although it was used in the 1970s to some extent.

Wave soldering is the most commonly used process for soldering discrete components glued to the bottom side of a through-hole assembly. In the beginning most surface mount wave soldering equipment came from Japan. This is not surprising, since Japanese companies dominate the market for consumer products, which use surface mount discrete components extensively in their analog circuit designs. Now surface mount wave soldering equipment is also available from suppliers in the United States and in Europe.

The story of reflow soldering equipment is different however. The reflow soldering equipment market is essentially dominated by U.S. companies, where widely used reflow soldering equipment was developed. The digital circuit designs that are common in the telecommunications and computer industries feature active devices, which generally use reflow soldering, and this market is dominated by American companies.

In addition to wave and reflow soldering processes, hand soldering is also used for repair and prototype work. However, the hand soldering process requirements (temperature, time, and pressure on pad) are essentially the same as for through-hole assemblies, although special tools and tips are required for hand soldering of surface mount components. Some discussion on hand soldering is included in Chapter 14, but here we will concentrate on the details of commonly used production volume soldering processes for surface mount assemblies of various types.

Because of the widespread use of mixed assembly now and in the foreseeable future, both wave and reflow soldering processes will be used. The selection of a specific soldering process for surface mounting depends on the mix of components and the form of SMT assembly (Type I, II, or III) to be soldered. See Chapter 1 for definitions of assembly types. But throughput requirements, use of temperature-sensitive components, and the need for specialized rework, or soldering also must be considered in the selection of a soldering process. Each process has its pros and cons and none is suitable for all applications.

Wave soldering is well suited for mixed assembly. Actually no other process can compete with the cost effectiveness of the wave soldering process for through-hole and Type III SMT assemblies. However, for Type I SMT, wave soldering should not be considered. For Type I, reflow soldering remains the most commonly used option. For Type II SMT (mixed assembly), both reflow and wave soldering processes are used.

The basic differences between wave and reflow soldering lie in the source of heat and the solder. For example, in wave soldering, the solder wave serves the dual purpose of supplying heat and solder. The source for the supply of solder is unlimited because the wave pot holds plenty of solder. In reflow soldering, however, solder paste is applied first in a predetermined quantity, as discussed in Chapter 9, and during reflow, heat is applied to melt (i.e., reflow) the solder paste. Thus a more appropriate name for reflow soldering might be reflow heating. Various sources for the heat are used; namely, vapor phase, laser, hot bar, and infrared (IR).

There are three types of IR systems: radiation dominant systems, convection dominant systems, and systems combining radiation and convection. Many people tend to think that convection dominant systems are not IR, but it is important to remember that there is no such thing as a perfect IR or a perfect convection oven. All these processes are used for reflow soldering, but the convection dominant forced air systems are most widely used.

However, other processes will continue to play important roles. With the widespread use of ball grid arrays along with fine and ultra fine pitch and through-hole components on the same board, we have to deal with most of these soldering processes. Also, the use of reflow and wave soldering machines in inert environments such as nitrogen has become common. We will first discuss wave soldering and then move on to various reflow soldering processes including their pros and cons for different applications.

12.1   WAVE SOLDERING

Wave soldering is the main process used for soldering component terminations en masse in conventional through-hole mount assemblies. It is also the most widely used process for soldering surface mount discrete components (resistors, capacitors, diodes, etc.) glued to the bottom of Type II and Type III SMT assemblies.

Other surface mount components amenable to wave soldering are the small outline transistors (SOTs). Four-sided leaded packages, such as plastic leaded chip carriers (PLCCs) and plastic quad flat packs (PQFPs) are difficult to wave solder. Small outline integrated circuit (SOIC) packages can be wave soldered, but it is generally not desirable to wave solder PLCCs, PQFPs, or SOICs because of *potential* problems with reliability (active fluxes seeping inside the package through the lead frames) and manufacturability (shadowing and bridging). However, if the wave is set at 45° to the board travel direction, defects can be reduced considerably. Also, see Chapter 7 for design guidelines to avoid bridging in SOICs that are wave soldered.

Problems may be encountered in the wave soldering of surface mount components if appropriate action is not taken in design and manufacturing. Various wave soldering issues and some approaches for overcoming them are summarized in Table 12.1. Let us elaborate on the details of the potential problems and then suggest some solutions.

Table 12.1 Technical issues and their resolution in the wave soldering process

ISSUES	APPROACH
Accommodation of different board sizes and compatibility with in-house material handling system	• Evaluate trade-offs • Automated finger conveyor width adjustment • Standard board pallets • Robotic loader and unloader
Maintenance of specific gravity, activity, required flux wave, and foam height	• Use spray fluxer for low solids, low residue, no-clean flux to ensure uniform flux application • Evaluate automatic monitoring and control system • Flux density control • Dump aged or contaminated flux • Flux wave and foam height sensor and controller
Even preheat of boards	• Evaluate automatic monitor and feedback temperature control • Preheaters • Conveyor speed
Critical wave features for accommodating various types of board assembly features	• Nitrogen inerted wave • Wave height adjustment • Solder bath temperature control • Solder bath impurity control • Solder level and feed control • Investigate wave features best suited for various board assembly configurations • Smooth, rough, single, double wave • Dry, and oil intermixed wave

12.1.1   Design and Process Variables in Wave Soldering

In through-hole mount assemblies the lead provides an easy path for the wicking of solder through the plated through holes. As soon as the lead end touches the solder wave, the wetting force helps the solder climb up the lead. Inside the plated through hole the capillary action pulls up solder, and again on the top the wetting force of the pad spreads the solder onto the pad. Advancing solder pushes out flux vapors and air, thus filling the hole.

In surface mounting there are no holes or leads. Instead, flat or round components sit on flat pads, making soldering more difficult because of sharp corners formed by the component and the board surfaces. The classic solder wave with its laminar solder flow hits the underside of chip resistors and capacitors tangentially and may not always get to the corners formed by the rectangular components and the flat board surface.

Outgassing (gas bubbles released from wet flux causing solder skip or lack of fillet) and solder skips (lack of solder fillet due to misorientation of components) are the two main concerns during the wave soldering of resistors and capacitors. Outgassing (Figure 12.1), which is believed to be caused by insufficient drying of flux, can be corrected by raising the preheat temperature. Providing escape holes in surface mount pads can also alleviate outgassing problems by the same mechanism as in the through-hole mount soldering. Solder skips are due to the shadow effect of the part body on the trailing terminations (Figure 12.2). The shadowing effect also may be encountered if there are smaller components just behind larger components going over the wave. Shadowing of trailing component terminations will also occur when staggered components are not placed far enough apart.

Bridging (leads shorted with solder) and solder skips caused by shadowing in multileaded devices such as SOICs and SOTs are very common. To minimize solder skips, the board should be designed with the land patterns in the recommended orientation with respect to the board travel direction, as shown in Figure 12.3. Running the board over the wave in the recommended direction will prevent shadowing in chip components, SOICs, and SOTs. Failure to observe this convention can lead to solder skips.

If the layout of the components on the board makes it impossible to

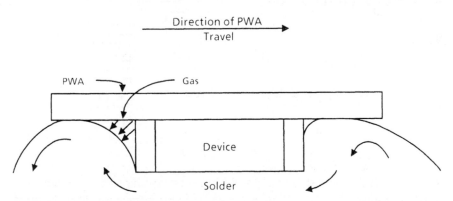

Figure 12.1   Outgassing in surface mount chip components during wave soldering.

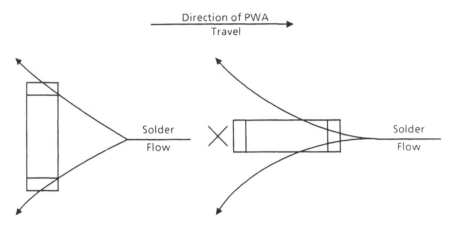

Figure 12.2   Solder skips ("×" in right figure) in surface mount components during wave soldering.

solder all terminations as recommended, the board should be oriented in such a way that the majority of terminations comply with the recommended orientation shown in Figure 12.3. Wrong orientation of components should not be allowed during the design phase. Even with the correct orientation of SOICs, bridging on trailing leads is fairly common. Adding extra dummy pads (also referred to as robber pads) on the trailing side takes

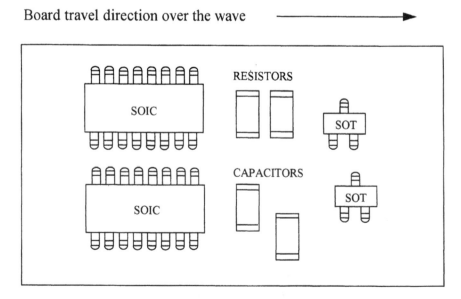

Figure 12.3   Recommended component orientation for wave soldering.

care of this problem. Bridging, if it then occurs, does not disrupt circuit operation because the bridged pad is not connected to a functional circuit. See Chapter 7, Figure 7.12 for recommended design and orientation.

Providing adequate interpackage spacing and orienting components in such a way that both terminations are soldered simultaneously solves most shadow effect problems. Refer to Chapter 7 for information on interpackage spacing and rules for staggering components to prevent shadowing of terminations during wave soldering.

In wave soldering, solder bridges occur beneath small chips such as 0805 (80 mils by 50 mils) resistors and capacitor chips. The problem is more serious under smaller components such as 0603 resistors and capacitors. Since these bridges are under the chips, they cannot be seen and they are generally caught during test. The cause of the problem is capillary action between the adhesive dots. This capillary action can be minimized by slightly increasing the gap between pads and by using solder mask between the pads or by adding a dummy pad which reduces the capillary cross-section area.

Only liquid photoimageable solder mask between the pads is recommended. Dry film solder mask should not be used in wave soldering because the acrylate material in the dry film solder mask absorbs chlorides from the flux. In such a case the board is likely to exceed the maximum allowed ionic contamination (10.06 µgram/in.² or 1.56 µgram/cm² of NaCl equivalent) and hence fail the cleanliness test. See Chapter 13, Section 13.8 for cleanliness test methods and requirements.

The other question that is asked is whether surface mount resistors and capacitors can withstand the 500°F (260°C) temperature when they are passed through the solder wave during wave soldering. I have found the maximum shift in tolerance of resistors and capacitors after wave soldering to be 0.2% [1]. This is a negligible amount, considering that the part tolerance of commonly used components is 5 to 20%. The components generally spend about 3 seconds in the wave, but they are designed to withstand soldering temperatures of 500°F for up to 10 seconds.

The impact of the solder wave on reliability may be different for active plastic packages such as SOICs, PQFPs, and PLCCs. In addition to potential shadowing and bridging problems, these packages may be adversely affected if flux seeps in through the lead frame during wave soldering. This can happen because the lead frame and the molding compound have different coefficients of thermal expansion. Seepage of flux inside the package is not a concern in reflow soldering because solder paste flux, which is used in the reflow process, does not touch the package body. Solder paste flux is generally confined to the package leads.

Passive components may undergo leaching of terminations during wave soldering; that is, there may be dissolution of the precious metal (silver or silver-palladium) adhesion layer on the ceramic body. If the precious metal adhesion layer is dissolved, the unsolderable ceramic body is exposed, preventing the formation of solder fillets or causing poor ones. Chapter 10 presents a detailed discussion of leaching and the rates of dissolution of various metals in molten solder.

Leaching can be prevented by a 50 microinch (minimum) nickel barrier underplating between the precious metal adhesion layer and the solder coating. Refer to Chapter 3, Figure 3.3. Nickel under plating should be required and should be part of the component procurement specification as discussed in detail in Chapter 3. However, a nickel barrier is not absolutely guaranteed to prevent leaching if the dwell time in the solder wave is too long. Dwell time in the solder wave thus should be kept to a minimum (3–4 seconds). Lower wave soldering temperatures (235°–245°C) will also reduce the leaching rate.

It is repeated many times in this book, and the point cannot be overemphasized, that to ensure process compatibility, *all components must be qualified for the processes to which they will be subjected.* This must be done before each component is designed into the product.

12.1.2   Process and Equipment Variables in Wave Soldering

A variety of equipment is available to minimize solder skips, outgassing, and bridging problems in the wave soldering of surface mounted components. Using an effective wave soldering machine is not by itself sufficient to achieve good yield, however. Many equipment- and non-equipment dependent variables play a role. Before we discuss the equipment itself, let's look at the key equipment variables.

The important variables in wave soldering, as suggested in Figure 12.4, are board handling, fluxing, solder profile (solder schedule), and solder wave geometry. These must be characterized and then controlled to achieve good soldering yield. In board handling, allowance must be made in the conveyor system for the thermal expansion of the board during preheating and soldering to prevent board warpage. A proper thermal profile and the use of pallets and stiffeners are also necessary to minimize board warpage.

In fluxing, flux density, activity (flux corrosiveness or flux potency for speed of soldering), and the ratio of flux foam to wave height must be closely monitored. A system must be in place to determine flux density, flux activity, and the timing for dumping the old flux. Machines that come with a flux density controller automatically add flux or thinner to maintain flux density within the specified limits; there are many such machines.

Flux activity monitoring must be done manually, either chemically

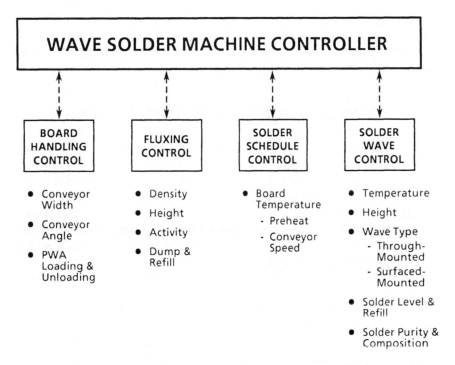

Figure 12.4   Common equipment variables in a wave soldering machine.

or by solderability tests on known good terminations or leads. When the flux gets old, the solderability of the known good lead will deteriorate and the flux should be discarded. The soldering equipment presently available lacks the capacity to monitor flux activity and give a signal for dumping inactive flux.

Monitoring of flux activity is not an issue if a foam fluxer is used, because the quantity of flux required is small. It may be more economical to dump the flux daily than to determine its activity. Similarly, monitoring flux activity in spray flux is not important because only fresh flux is sprayed. However, if a wave fluxer that contains many gallons of flux is used, monitoring flux activity may be desirable to control solder defects.

Solder wave geometry is the most important variable in wave soldering for minimizing solder defects. It must permit the effective soldering of both surface mount and through-hole mount components because both are present on the board in a mixed assembly. Wave geometry is important for preventing icicles (sharp needle-like protrusions of solder on lead ends) and bridges in through-hole mount components and for preventing outgassing and solder skips (shadowing) in surface-mounted components.

No matter which wave geometry is used, it must push the solder between tight spaces and into corners. There are various approaches to accomplish this, as we will discuss in Section 12.3. One of the most important variables in wave soldering is the thermal or solder profile, which we discuss next.

12.2   DEVELOPING A WAVE SOLDER PROFILE

The solder or thermal profile establishes the rate of heating of assemblies during the preheat and their immersion time in the solder wave. It is controlled by the temperature settings of various preheaters and the wave pot and the speed of the conveyor transporting the board.

The role of the solder profile in achieving good soldering yield cannot be overemphasized. A good solder profile helps not only to reduce solder defects but also to prevent cracking of surface mount capacitors, which are prone to thermal shock damage. To understand why this is true, let's examine a bad solder profile (Figure 12.5). Here the rate of heating is too rapid: the board has seen a maximum preheat of only 176°F (80°C) before hitting the first solder wave (first peak in Figure 12.5). To avoid capacitor cracking, the board should attain a top side preheat temperature of 220° to 240°F (104°–116°C) before it enters the solder wave.

Surface mount capacitors should be gradually heated, and the tempera-

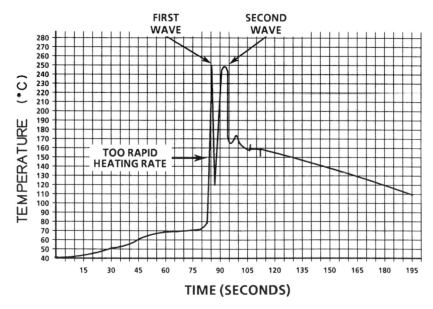

Figure 12.5   A bad example of solder profile for dual-wave soldering.

ture differential between the preheat zone and the soldering zone should be minimized. This not only prevents thermal shock in surface mount capacitors but also reduces solder skips caused by outgassing of undried flux.

Preheating a printed wiring assembly in two or three stages minimizes thermal shock damage and improves the service life of the assembly. Uniform preheating is achieved by developing a solder profile that specifies preheat settings and conveyor speed for each type of board.

Let us discuss the mechanics of developing a wave solder profile. First, select preheat settings at a given belt speed (e.g., 6 ft/min) and run the board without turning on the solder wave. Monitor the topside board temperature using a thermocouple, crayons which melt, or labels which change colors at different temperatures. If the target is not met (and this is likely the first time), adjust the preheat settings first. Remember that any adjustment of conveyor speed will also affect the board temperatures in the preheat zones. Thicker boards and multilayer boards with ground and power planes (a lot of copper surfaces) will require higher preheat settings and slower belt speeds.

Attaching thermocouples to boards and then to data loggers to monitor board temperature is an accurate but cumbersome and time-consuming process. Fortunately, many machines come with built-in thermocouples and software to monitor and plot the thermal profile. If the wave solder machine does not have profiling capability, portable hardware and software packages are commercially available to make thermal profile development an easy task. Examples of such commercial profile development packages are *MOLE* by ECD Inc. of Portland, Oregon and *KIC* by EDI, Inc. of San Diego, California.

When using thermocouples and data loggers, machine settings are changed and the board is run many times to achieve the desired profile. Most likely the first profile tried will not be the desired one; achieving the desired profile can take 1 to 2 hours. Hardware and software packages such as the *MOLE* and *KIC* profiles reduce the time needed for developing the profile. They also have prediction capability. Once the board is run with a given preheat and belt speed setting, the computer can predict the impact of different settings on the profile without running the board through the machine. This saves quite a bit of time. However, it is always wise to verify the final profile predicted by the software by actually running the board through the machine.

A slower ramp rate (heating rate) is needed in the preheat zone (2°–3°C/sec). As noted earlier, the maximum preheat temperature (top side) should be between 104°–116°C (220°–240°F). The wave pot temperature should be 235°C –245°C. And the delta between the wave and preheat should be 100°–120°C depending upon the type of capacitor being used. (Capacitors must be qualified for the delta temperature being used.)

The belt speed controls the immersion time in the wave. It should be such that the time in the wave does not exceed 3–4 seconds. This means that for a solder wave of about 3 inches, the belt speed should not be less than 5 feet per minute (3 inches × 60 seconds/60 inches = 3 seconds). Generally belt speed is kept at 6 feet per minute. For a belt speed of 6 feet or 72 inches per minute, the dwell time in a 3 inch wide wave will be 2.5 seconds. A longer immersion time in the wave creates the potential for leaching (dissolution of silver or silver-palladium under-plating) in the capacitors. A belt speed of less than 5 feet per minute (and hence longer immersion time in the wave) may be acceptable if the solder wave width is less than 3 inches or if the components are qualified by the supplier for the desired time and temperature in the wave (ideally the components and boards should be able to withstand 260°C for 10 seconds).

Once the desired preheat temperature is achieved, the wave should be turned on and the first board should be soldered. To assess the effectiveness of the new profile, check the quality of the solder joints (filled vias, top side fillets, bridges, icicles, etc.). Note that a random problem may be related to solderability but a consistent problem may be related to design or equipment. Solderable parts and a sufficient number of boards should be used to determine the trend. Only one parameter at a time should be changed.

The profile should document preheater settings (distance from conveyor, time for ramp, etc.), flux and fluxer parameters (wave/foam height, knife angle, pressure, thinner addition for density control, flux disposal, etc.), and solder wave parameters such as guideline for addition of fresh solder (e.g., 4 sticks when solder is 1 inch below wave pot edge), solder composition, and dross removal frequency. Once a profile has been set and determined to be the optimum one, it should be documented (along with solder defect results for future reference) and no changes should be allowed in the profile.

One profile cannot be used for boards of different thermal masses. This is a fairly common but wrong practice in the industry. It is important to remember that a unique profile is needed for each board or group of boards with the same thermal mass.

12.3   TYPES OF WAVE SOLDERING FOR SURFACE MOUNTING

Solder waves designed for the wave soldering of surface-mounted components are intended to solder both tightly packed and incorrectly oriented, difficult-to-solder trailing terminations. Does this mean that the components need not be oriented in the recommended direction as discussed previously? No. If components are incorrectly oriented, the terminations may be soldered by using specialized wave geometries, but oversized fillets will be formed on the trailing terminations. Such oversized fillets may stress and crack chip components.

The issue of wave geometry was introduced in Chapter 7 to emphasize the need to ensure correct orientation during the design cycle. We bring it up again here because even if the design rules discussed in Chapter 7 are followed, the process engineer should not count on dual-wave or any other wave geometry to solve a problem of uneven fillets. The board must be oriented correctly over the wave (Figure 12.3) to reduce solder skips and uneven solder fillets.

Many types of wave soldering equipment for surface mount components are available. The main differences among them lie in their wave geometries. The selected wave should ensure that the solder can reach into tight spaces to form the solder fillets.

Wave geometries commonly available on the market include unidirectional and bidirectional; single and double; turbulent or vibrating, smooth and dead zone; oil intermix, dry, and bubbled; and with or without a hot air knife. The wave geometries widely used in the industry can be classified as dual wave (with and without hot air knife) and vibrating wave. Let us discuss some of these wave geometries.

12.3.1   Dual-Wave Soldering

One of the solder waves on the market for reaching into the tight spaces between surface mount components is the dual-wave system. As the name indicates, there are two waves: a turbulent wave and a smooth or laminar wave, shown schematically in Figure 12.6. Figure 12.7 is a photograph of a dual wave while the wave is turned on.

In the dual-wave system, the turbulent section of the wave, which ensures adequate distribution of solder across the board, is the one that prevents solder skips. The solder is pumped through a narrow slit at somewhat high velocity so that the solder can penetrate between the tight spaces. The jet is pointed in the same direction as the board travel. The turbulent wave alone cannot solder components adequately. It leaves uneven and excess solder on joints; hence the need for the second wave.

The second laminar or smooth wave eliminates the icicles and bridges produced by the first or turbulent wave. The laminar wave is actually the same wave that has been used for a long time for conventional through-hole mount assemblies. The laminar wave can be either with or without oil intermix. When oil is used, a thin film of oil reduces oxide formation

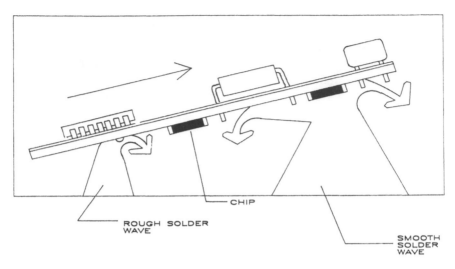

Figure 12.6   Schematic view of a dual-wave soldering system for SMT.

Figure 12.7   Actual view of a dual solder wave for SMT. (Photograph courtesy of Electrovert.)

and loss of tin (from the tin lead solder) as tin oxide, but an excessive amount can cause voids in solder joints. The two pots of the dual-wave system have independent electric motors so that the turbulent wave can be turned off when conventional assemblies are being soldered on the same machine.

In a common variation of the dual-wave system on the market today, the turbulent wave moves back and forth (oscillates perpendicular to the direction of board travel) and the solder jets come out of numerous tiny nozzles instead of one long and narrow slit as discussed above. The moving nozzles are more effective than the narrow slit in preventing solder skips because they create turbulence and inject solder in tight spaces to ensure solder fillets on all components.

It is not always necessary to buy an entire wave soldering machine to realize the benefits of the dual-wave system. Most conventional wave soldering machines can be retrofitted by just adding the dual-wave pots, at a fraction of the cost of a new machine. The rest of the equipment functions are essentially like those of conventional soldering equipment. Retrofitting may not be cost effective, though, if one needs the automated features available only in the newer machines.

12.3.2   Vibrating Wave Soldering

In the second method for minimizing outgassing and solder skips, nitrogen is injected into the solder wave. The nitrogen bubble wave was introduced by the Koki Company in Japan, but it did not become very popular in the United States. A variation of this concept is the oscillating wave with nozzles made by the Sensby Company of Japan. Another variation of the same concept is the ultrasonically vibrating wave known as the Omega Wave in the Electrovert system.

In the Omega Wave a transducer is introduced to generate low frequency (50 or 60 Hz) vibrations at a controlled amplitude as shown schematically in Figure 12.8 [2]. The vibration of the solder wave (Figure 12.9) breaks up the flux droplets to permit the solder to penetrate the tight spaces and corners between surface mount components and displace the gas bubbles that cause solder skips. As the board exits, the oscillations gradually decrease, and components travel through the conventional laminar wave to eliminate bridging and icicles.

The Omega Wave replaces the dual-wave (the combination of rough and smooth waves shown in Figure 12.7) system from Electrovert and is preferred by some for economic reasons because it uses less solder and is just as effective as the dual wave. In fact, one study has shown that

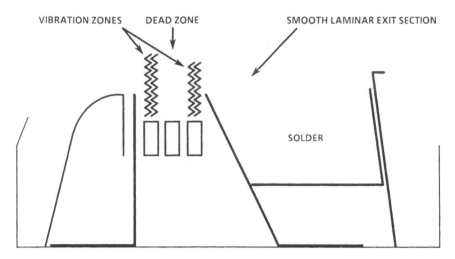

Figure 12.8   Schematic view of the Omega Wave vibrating single-wave soldering system for SMT.

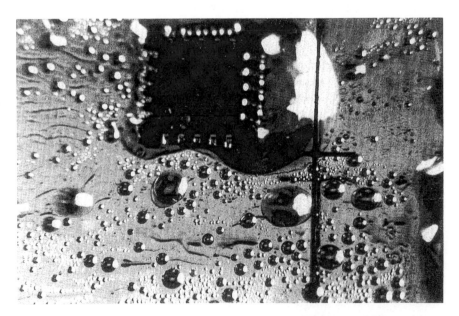

Figure 12.9   An actual vibrating single Omega Wave for SMT. (Photograph courtesy of Electrovert.)

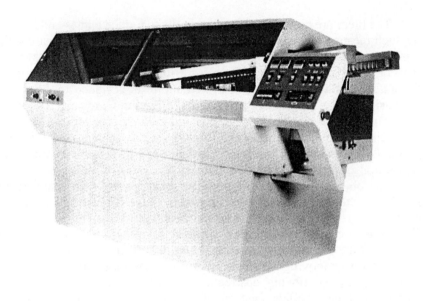

Figure 12.10   A wave soldering machine for surface mount and through-hole mount assemblies. (Photograph courtesy of Electrovert.)

the Omega Wave is more effective in filling holes and eliminating solder skips than the dual wave [3].

There will be more loss of tin in a pot with two waves because of the additional atmospheric exposure of solder. Thus, tin will have to be added more frequently to maintain the required percentage of tin in the solder pot. This may have a substantial impact on the initial and subsequent outlays of expense for solder.

However, in some applications a combination of the Omega Wave and the rough wave may be necessary in order to optimize the probability of eliminating all solder skips, icicles, and bridging. A wave soldering machine providing Omega Wave with rough wave option is shown in Figure 12.10.

12.3.3   Modified Wave Soldering

Although wave solder systems have long been used in the industry, they have been plagued by undesirable icicles and bridges. These defects are not acceptable. Icicles can pierce through the conformal coating and can also adversely affect printed board assembly performance, especially in high speed circuits. Bridges cause short circuits.

The literature chronicles many attempts to reduce icicles and bridges by such means as varying the amount of oil in the wave, preheating the boards, changing conveyor speed, and the use of the hot air knife system. But the wave geometry plays the most critical role. To prove this point the author conducted a series of systematic tests designed to determine the basic cause of icicles and bridges at the Boeing Company. The wave solder machines at various commercial and military production sites were producing icicles and bridges on 75–90% of the solder joints of some products. These defects had to be touched up, which not only increased cost but also impacted product reliability.

The objective of improving solder joint quality was achieved by modifying the conventional Hollis turbulent wave (Figure 12.11a) into a wave with a static or "dead" zone in the solder wave (Figure 12.11b). This was done by attaching a welded stainless steel dike (Figure 12.11c) to the far end of the wave where the assembly exits. The dike redirects waves moving upstream back downstream, so that they collide with other waves moving upstream. A static or dead zone is created where the waves collide (see Figure 12.11b). The static waves peel off much more slowly than the conventional waves, which fall rapidly toward the bottom of the solder pool. The modified wave virtually eliminates icicles and bridges.

The dike also improves top side fillets and prevents excess oil entrapment (oil entrapments cause voids in the solder joints). The modified wave creates a solder pool next to the dead zone. The pool postheats the bottom sides of the boards as they pass over it, preventing premature freezing of solder. This helps in formation of better top side fillets. Oil entrapment is reduced because the modified solder wave prevents the formation of icicles and bridges on the board, permitting the amount of oil required to be reduced substantially.

The dike is simple, easy to install, and costs under $100, yet the system is highly effective in eliminating icicles and bridges in addition to reducing oil entrapment and improving solder fillet quality. The dike was installed on all Boeing Hollis machines. For reducing the touch-up cost significantly, it was selected as one of the ten best inventions by the Boeing Company in 1982. So as mentioned earlier, the design of the wave geometry is most critical in reducing solder defects. This is true not only for conventional assemblies but also for surface mount assemblies and should be the primary consideration when selecting a new wave soldering system.

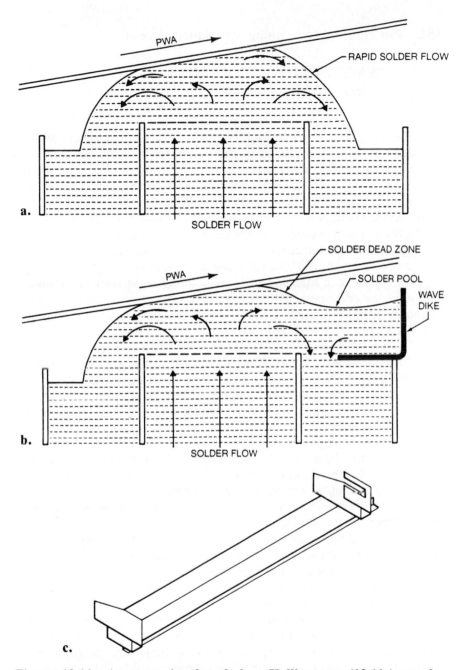

Figure 12.11 A conventional turbulent Hollis wave (12.11a) can be modified into a static wave with "dead" zone (12.11b) by a wave dike (12.11c). The dike, which virtually eliminates icicles and bridges, was developed by the author. It proves the importance of wave geometry in reducing solder defects.

12.4 WAVE SOLDERING IN AN INERT ENVIRONMENT

There are generally two ways to create an inert environment around the solder wave—by localized inerting just around the wave or by an inerted tunnel around the preheaters and the wave. Localized inerting creates an environment just around the wave where soldering takes place. As shown in Figure 12.12a, nitrogen is diffused from each side of the wave. As the board passes over the wave, the nitrogen is trapped between the PCB and the wave while the board is being soldered; so essentially the nitrogen is being injected over the wave from each side. Figure 12.12b shows a photograph of localized nitrogen inerting of the Hollis wave.

Another way of injecting nitrogen is to create an inert environment around all the preheaters and the wave by constructing a nitrogen tunnel as shown in Figure 12.13. This approach requires much more nitrogen than the localized inerting approach.

The use of nitrogen (N_2) has become very common since the industry has had to move away from chlorofluorocarbon (CFCs) solvents due to their deleterious effects on the environment (ozone depletion). When CFCs were available, one could use more active fluxes to remove oxides from the soldering surfaces. Now, no-clean and low activity fluxes are widely used and N_2 is used to compensate for the low activity of the flux. If flux activity is very low, oxidation of the solder surface will increase unless oxygen is kept away. Thus, the use of nitrogen helps improve solder yield. Nitrogen also makes the solder joint look bright and shiny; however, this is only a cosmetic improvement. Also, with the use of nitrogen, the amount of dross in the solder is significantly reduced and so is clogging of the nozzles.

Some have found an improvement in yield with nitrogen [4], but others have not realized measurable improvement [5]. There is a reason why yield improvement is not guaranteed. Solder yield is a function of many variables such as design, board and lead solderability, paste, printing, placement, and thermal profiles to name a few. The soldering environment is only one of many parameters and its influence is limited to its own effects. N_2 is not going to correct problems created by poor design, solderability, or rheology of paste, or by misprinting and misplacement. So the use of nitrogen in either wave or reflow soldering may not be beneficial in all applications.

Why use nitrogen instead of other inert gases? The answer is an economic one. As shown in Table 12.2, nitrogen is the least expensive widely available gas, although not the most inert gas. Table 12.2 lists properties of some commercially available gases [6]. It should be noted

(a)

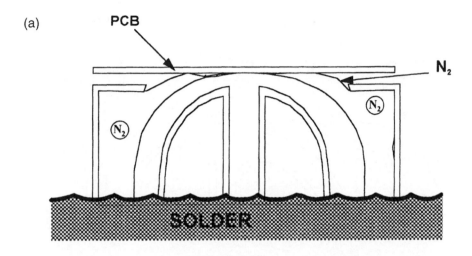

(b)

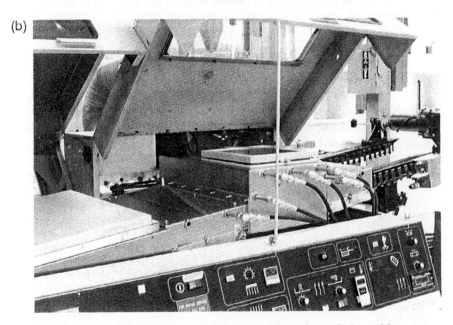

Figure 12.12 (a) **A schematic of localized inerting of the solder wave with nitrogen (CoN₂tour™ concept by Electrovert); (b) A photograph of localized inerting of the solder wave with nitrogen (courtesy Shore Concepts Inc.)**

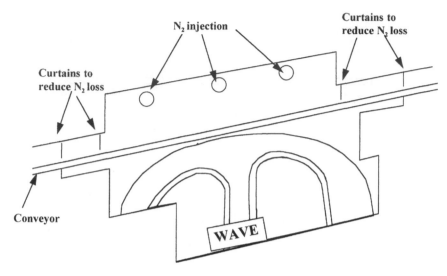

Figure 12.13   A schematic of a tunnel around the solder wave for nitrogen inerting.

that the relative cost shown in Table 12.2 for different gases may vary in different areas.

In addition to preventing the oxidation of circuit boards and components, N_2 also prevents oxidation of solder in the wave solder pot. Generally the nitrogen flow needs to be maintained at a level sufficient to maintain the oxygen level at 5 to 10 ppm (parts per million). This requires about 2000 cfm or 55 m^3/hr. Cfm is used to measure consumption of gas in cubic feet per minute. The cost will vary depending upon the purity level

Table 12.2   Properties of commercially available inert gases compared to air [6].

	AIR	HELIUM	NEON	NITRO-GEN	ARGON	CARBON DIOXIDE	KRYP-TON
Density (kg/m³)	1.21	0.17	0.842	1.17	1.67	1.85	3.51
Thermal conductivity (mW/cm°C)	255	1484	455	241	162	162	86
Cost relative to nitrogen	N/A	200	2500	1	6	2	1200

of the nitrogen, the source, and the location. The consumption level is also determined by the exhaust pressure, which should be controlled properly.

It takes about 10 minutes to achieve the 5–10 ppm oxygen level, after the nitrogen is turned on. Oxygen probes are used in various areas to determine oxygen level. Of course the key area of interest is the solder pot. The cost and accuracy of oxygen probes vary widely and hence these probes should be frequently calibrated.

Some machines have open entry and exit while others are based on a closed vacuum-lock concept. Machines with open entry and exit, as shown in Figure 12.13, allow the continuous flow of production boards necessary for high volume production. The curtains at the entry and exit points shown in Figure 12.13 are intended to retain the nitrogen. However, since they have to be raised (about 2 inches) above the board surface to allow for boards with components to flow freely, there is a considerable loss of inert gas. In the closed vacuum-lock system, the nitrogen consumption is reduced to as low as 500 to 600 cfm (15 m^3/hr), but so is production volume [7].

There are also shorter machines with a nitrogen tunnel that covers the preheat zone and the solder pot. Nitrogen consumption in such systems is about 1500 cfm (45 m^3/hr) [7].

12.5   SINGLE-STEP SOLDERING OF MIXED ASSEMBLIES

It is generally accepted that wave soldering and reflow soldering are the most effective means to solder Type III and Type I assemblies, respectively. However, mixed assemblies (Type II SMT) require both reflow and wave soldering processes. Generally the soldering of Type II SMT assemblies is accomplished in a two-step soldering process. The surface mount components are placed on solder paste on the top side and reflow soldered. Then the through-hole mount components are inserted and the entire assembly is wave soldered.

The driving force for a single-step soldering process is the cost savings realized by eliminating a soldering step and a cleaning step. The cost savings can disappear, however, if the surface mount components are sitting on solder paste shift when the through-hole mount components are inserted. In such a case increased touch-up may be required. Besides, soldering time is a small fraction of total assembly time; placement and testing times contribute most to the cost of assembly. Thus one should carefully evaluate the advantages and disadvantages of single-step soldering for mixed assembly before using it in production.

There are three approaches to soldering mixed assemblies in one step. The first two were attempted by wave soldering and vapor phase equipment manufacturers but never really took off. The third approach is used at times. Let us describe the first two just for historical record. They do make sense, at least theoretically; however, these two approaches are rarely used in actual practice.

In the first approach, a wave soldering machine is outfitted for reflow capability by the installation of infrared heating panels or hot air systems. Both wave and reflow soldering operations are then performed by the same machine. The surface mount components are placed on screened solder paste and then the through-hole mount components are inserted. The assembly goes through a wave soldering system for through-hole mount parts and immediately passes through a hot air reflow system for soldering of surface mount components. Such a system was once available but was not accepted by the marketplace.

Another approach to one-step soldering for mixed assemblies is to combine the wave pot and a vapor phase machine. In such a system, fluorocarbon liquid (FC-70 or FC-5311) heats the wave solder pot for soldering through-hole mount parts, and the surface mount components mounted on the top side are reflowed with the fluorocarbon vapor. This system was also stillborn.

The third approach, which is currently used by many companies, involves only a reflow system (vapor phase or convection, with convection being more common). In this approach solder paste is printed for both through-hole and surface mount components. Then parts are placed and reflow soldered. The key to this process is paste printing. The through-hole components need more solder paste (about 20 mil thick) than the surface mount components, which need 6–8 mil paste. This can be accomplished either by a one-step stencil or by two separate stencils—one for through hole and the other for surface mount. See Chapter 9, Section 9.6.3 and Figure 9.30 for details.

12.6   SINGLE-STEP SOLDERING OF DOUBLE-SIDED SMT ASSEMBLIES

Generally, the soldering of a two-sided full surface mount assembly is accomplished in two steps. In this case no adhesive is necessary because when the assembly is turned over after the first side has been reflowed, the surface mount components are held in place by the surface tension of the solder. In the two-step soldering process, it is recommended that lightweight components be reflowed during the first pass and the heavier components be reflowed in the second pass. Since the heavier components are on the top side, the potential for them falling off the board is eliminated. It also implies that the oven has a rails system (or a pallet may be used) to keep the bottom side components off the conveyor belt. It should be kept in mind that no matter how well the temperature is controlled, even the bottom side solder joints will reflow and hence should not be touching the belt.

To avoid the cost of two-step soldering some people have attempted to solder both sides in one single reflow step. This is accomplished by using adhesive on one side. In a single-step soldering process, the bottom side surface mount components are placed on the solder paste and also glued in place with adhesive, which is dispensed after screening of the paste. Adhesive is needed because the tackiness of the paste alone will not hold components in place when the assembly is turned over, and the solder cannot hold components in place by surface tension unless it has had a chance to reflow. The top side surface mount components are held in place on the solder paste; no adhesive is used.

When the assembly passes through the reflow system (vapor phase, convection, or IR), both sides can be soldered simultaneously. The process steps are as follows: screen paste, dispense adhesive, place components, and cure/bake the adhesive/paste; invert the board and repeat the process, except for dispensing the adhesive.

As in the single-step soldering of mixed assemblies, the soldering of double-sided surface mount assemblies saves money by skipping one soldering and one cleaning step. However, the time saved in one-step soldering may be spent in dispensing and curing glue to hold the bottom side components. In fact, depending on the number of components, even more time may be spent in dispensing adhesive than is saved on the soldering and cleaning steps.

There is another potential problem that may crop up in a single-step soldering process for a double-sided assembly. After the paste is printed on the first side, adhesive is dispensed and then the components are placed. The assembly is passed through an oven to cure the adhesive. The next step is to print the paste on the second side, but here is where the problem can arise. If the paste is misprinted, it must be washed off. This means that the components on the first side that were placed properly and glued with adhesive must also be taken off, and the process just described must be repeated. This can create quite a mess.

Thus the advantages of single-step soldering over double-step soldering are not quite obvious. It may make sense if the bottom side surface mount devices must be held with adhesive because of poor lead solderability. Components with poor solderability do not provide enough surface tension to prevent the bottom side components from falling off. However, instead of trying to use adhesive to solve a solderability problem, an attempt should be made to eliminate the solderability problem at the supplier.

12.7   VAPOR PHASE SOLDERING

Vapor phase soldering (VPS) has gone through changes in popularity. It was the process of choice in the early 1980s. But its use declined considerably for two reasons: problems with the VPS process itself and improvements in the IR processes. The problems with VPS are mostly in the area of higher defects as we shall see later. And the convection dominant IR systems provide efficient heating without the inherent problems of VPS.

As we discussed earlier, the use of nitrogen to accomplish soldering in an inert environment is increasing due to the widespread use of low solids and no-clean fluxes and solder pastes. Since VPS provides an inert soldering environment, the industry has been taking another look at this process. Also, since it is more difficult to provide uniform heating in larger ceramic and plastic BGAs even with convection dominant systems, VPS is being considered because of its very efficient and uniform heating characteristics. While it is not as common as the convection dominant systems, its use for reflow is far from over.

Vapor phase soldering, also known as condensation soldering, uses the latent heat of vaporization of a liquid to provide heat for soldering. This latent heat is released as the vapor of the inert liquid condenses on the component leads and PCB lands. In VPS the liquid produces a dense saturated vapor, which displaces air and moisture. The temperature of the saturated vapor zone is the same as the boiling point of the perfluorocarbon liquid. This fluid does not have any of the environmental concerns associated with CFCs. The peak soldering temperature is the boiling temperature of the inert liquid at atmospheric pressure.

VPS does not require control of the heat input to the solder joints or the board. It heats uniformly, and no part on the board, irrespective of its geometry, ever exceeds the fluid boiling temperature. VPS is suitable for soldering odd-shaped parts, flexible circuits, pins, and connectors, as well as for reflow of tin-lead electroplate on boards and surface mount package leads.

VPS is an easily automated process. Both in-line and batch-type equipment systems are used. Batch systems are designed for research and development or for low volume production. In a batch system, as shown schematically in Figure 12.14, the board is placed in a basket and lowered from the top into the primary vapor. In the past, a secondary vapor of CFC (Freon TE) was used to reduce the loss of primary fluid. Since CFCs have been banned, the secondary vapor is no longer used. After a

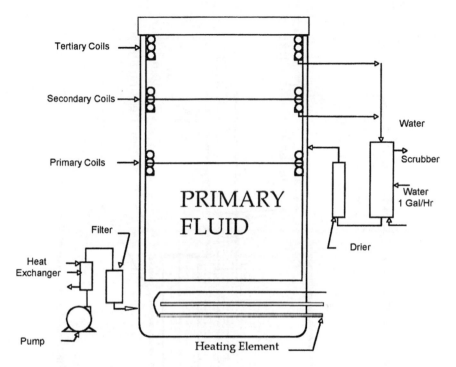

Figure 12.14   A schematic diagram of a single-phase batch vapor phase soldering system.

predefined dwell time in the primary vapor zone, the assembly is lifted above the primary fluid and kept there for about 30 to 60 seconds to drain off the primary fluid. The primary vapor is an inert perfluorocarbon, such as FC-70 or FC-5311 (from 3M Corporation). The boiling point for FC-70 and FC-5311 fluids is 419°F (215°C).

In-line systems are designed for mass production. In an in-line vapor phase system, the board travels on a conveyor from left to right; no secondary vapor is (or was) used. Generally in the older equipment, the conveyor belt is inclined before and after the reflow zone (i.e., the belt slopes down before entrance to the reflow zone and then slopes up when exiting the reflow zone). If the boards travel in the inclined position, however, the components, especially large ones, tend to slide. The new equipment allow boards to travel in a level position (Figure 12.15), thus eliminating this problem [8].

In both batch and in-line systems, water-cooled coils help reduce vapor loss. In addition, some equipment vendors have designed vapor recovery systems that have reduced fluid costs.

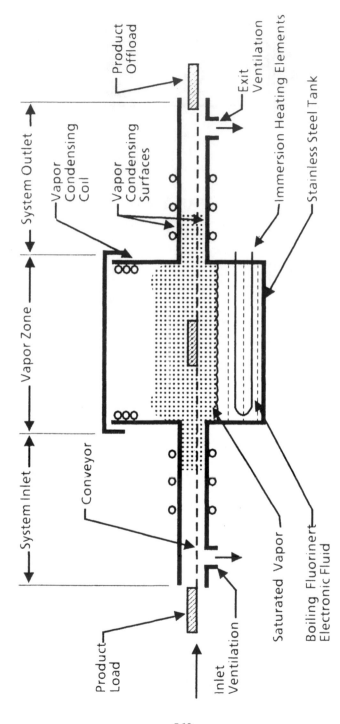

Figure 12.15 A schematic view of an in-line vapor phase soldering system for SMT.

Table 12.3 Technical issues and their resolutions in VPS process

ISSUES	POTENTIAL SOLUTION
Prevention of vapor loss	• Reduce effective throat of in-line system • Use water-cooled condensing coils • Control exhaust pressure
Prevention of part movement and other defects such as wicking	• Control solder paste rheology • Use dual-composition paste (with low and high melting points) • Increase preheat (without causing solder balling) • Integrate preheat systems • Minimize lead coplanarity • Use parts and boards with good solderability

Many of the VPS systems used now have infrared (IR) or convection preheat zones to provide the needed preheat before final reflow and to reduce the incidence of wicking (discussed in detail in Section 12.7.2). These preheat zones are either integrated into the VPS system, or a separate IR or convection oven is added in-line before the VPS machine. The technical issues in the VPS process and their resolutions are summarized in Table 12.3.

12.7.1 The Heat Transfer Mechanism in Vapor Phase Soldering

When the assembly to be soldered enters the primary vapor zone, the vapor condenses on its surfaces, releases its heat of vaporization, and forms a thin, continuous liquid film. The heat transfer rate to the surface of any part at any instant is controlled by the equation:

$$Q = hA \, (T_v - T_s) \tag{1}$$

where Q = heat transfer rate from vapor to part (Btu/hour)
h = heat transfer coefficient (Btu/hour $\cdot$ ft^2 $\cdot$ °F)
T_v = saturated vapor temperature (°F)
T_s = part surface temperature (°F)
A = product surface area (ft^2)

and h is determined by the thermal conductivity, viscosity, and density of the liquid condensate, and by whether the surface of a part is vertical, horizontal, or oblique [9].

It is apparent from equation (1) that only the maximum surface heating rate and the maximum surface temperature of any part on a surface mount assembly can be controlled during VPS.

The maximum surface heating rate of a part occurs when T_s is minimum, that is, when the part is initially immersed in the primary vapor zone. The initial T_s and hence the maximum heating rate of any part can be controlled by preheating the part before immersion in the primary vapor zone.

The maximum surface temperature of any part will depend on the dwell time of the assembly in the vapor zone. With increasing part dwell time in the primary vapor, T_s will asymptotically approach T_v, and the heating rate will approach zero. Given sufficient dwell time in the primary vapor, the surface temperature of all parts of a surface mount assembly will eventually reach a maximum of T_v.

The dwell time of the assembly in the primary vapor zone is fixed by the conveyor speed in in-line systems and by the elevator speed and dwell time in batch systems. As shown in Figure 12.16, if the dwell time

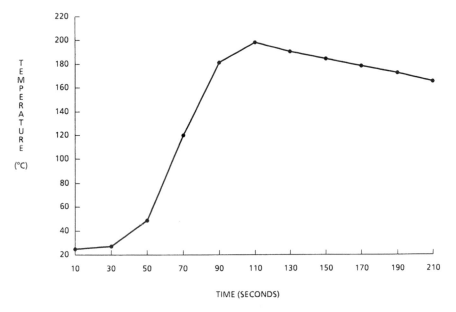

Figure 12.16 **The temperature-time soldering profile of a PLCC lead, illustrating the impact on maximum lead temperature of a short dwell time (10 seconds) in the primary zone of a batch VPS system.**

is too short, the surface of the lead does not reach the maximum possible temperature. Figure 12.17 shows that when the dwell time is sufficient, the lead temperature reaches the maximum possible (215°C). For successful soldering, it is recommended that all PCB land and component lead surfaces on a surface mount assembly reach a temperature 30° to 50°C above the melting point of the solder.

The total heat requirement from vapor condensation is determined by the thermal mass of the assembly. Assemblies with a larger thermal mass will require larger, or additional, heating coils. VPS systems with small capacities may be unable to supply enough vapor quickly enough to solder assemblies with large thermal masses because it takes some time for the liquid to boil.

Some VPS systems provide a constant supply of vapor by having large thermal heat sinks above the liquid level, inside the primary zone, in addition to heating coils below the liquid. These heat sinks act as thermal flywheels, keeping the vapor available in the quantity necessary to reflow larger masses of assemblies.

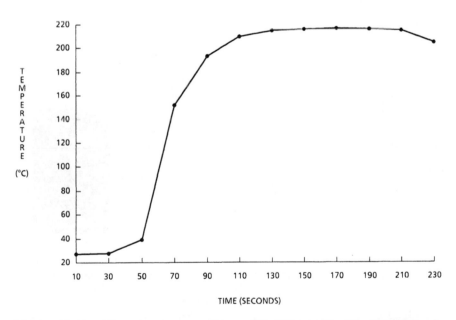

Figure 12.17   The temperature-time soldering profile of a PLCC lead, illustrating the impact on maximum lead temperature of a long dwell time (2 minutes) in the primary zone of a batch VPS system.

12.7.2 Solder Opens (Wicking)

One reason the VPS process lost its competitive edge was the excessive incidence of wicking where the solder wicks up the leads of J or gull wing devices, causing solder opens. The wicking phenomenon is caused by the lead and the land heating at different rates during VPS. The problem is exacerbated by noncoplanarity (the failure of the lead to touch the land). The lead and land surface temperatures are compared in Figures 12.18 and 12.19 for VPS and IR, respectively.

As shown in Figure 12.18, the surface of the lead reaches the melting point of the solder 16 seconds before the surface of the land during VPS. This causes the solder paste to melt, wet, and wick up the lead before the land becomes hot enough for the molten solder to wet it. By the time the land has reached the melting point of the solder paste, there is not much solder left on the land to form a good solder joint. The inset in Figure 12.18 displays the appearance of a solder open caused by wicking. Most of the solder is situated on the top part of the J lead.

On the other hand, during IR soldering, the lead and the land reach the melting point of solder (183°C) almost simultaneously, as shown in

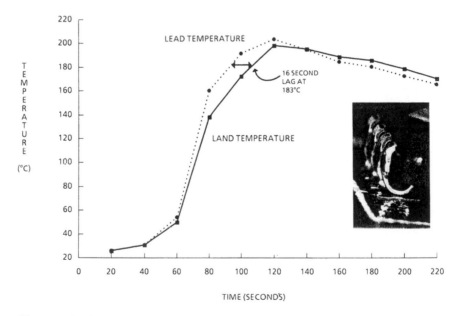

Figure 12.18   A comparison of the time-temperature soldering profiles of a PLCC lead and land without preheat during VPS in a batch system. The lead and land reach the solder melting point temperature (183°C) 16 seconds apart. Inset: a solder open caused by the wicking phenomenon.

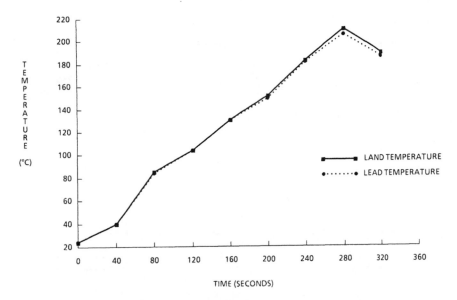

Figure 12.19 **A comparison of the time-temperature soldering profiles of a PLCC lead and land during IR. There is no lag between the lead and land temperatures.**

Figure 12.19. Hence, opens caused by solder wicking are almost nonexistent in a properly profiled IR process.

In equation (1) h is higher for a vertical surface than for a horizontal surface [9]. In any soldering process, typically the lead is vertical and the land is horizontal. Therefore, the lead heats faster than the land during the heating phase for VPS. If the land is vertical and the lead is horizontal (as when the board is placed vertically in the primary vapor zone), the heating rate differential between lead and land should diminish. This is confirmed by Figure 12.20, which shows that the period that elapses between the times the lead and the land reach the solder melting point decreases to 3 seconds when the board is vertical. However, keeping the board vertical during VPS is not a practical solution.

It is generally suggested that the practical way to reduce the incidence of solder wicking during VPS is to preheat the assembly [8, 10]. Preheating will reduce the problem but will not eliminate it—the leads heat faster than the land even after preheat. This is confirmed by Figure 12.21, which shows a difference of 10 seconds (as compared to 16 seconds in Figure 12.18) in the times needed for the lead and the land to reach the solder melting point after preheat.

Another approach, suggested by McLellan and Schroen [11], is to change the alloy composition of the solder paste. The basic concept is to

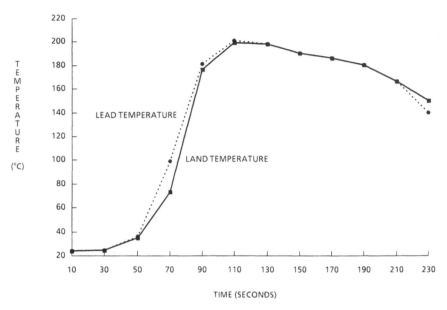

Figure 12.20   **A comparison of the time-temperature soldering profiles of a PLCC lead and land without preheat during VPS in a batch system. The board was lowered in the primary zone vertically; therefore the lead was horizontal and the land vertical.**

use a solder paste with particles made up of pure tin and 10/90 solder (10% tin and 90% lead) instead of particles of eutectic (63% tin and 37% lead) solder. However, the proportion of pure tin and 10/90 tin was chosen to make the net composition of the final solder joint eutectic. Using this special solder paste, McLellan and Schroen found the following dramatic improvement in solder opens: solder defects were reduced from 250–2000 to 0–130 opens per million joints. In addition, the incidence of voids was reduced when the paste containing 10/90 tin was used.

What is the reason for using two different compositions of solder powders that have higher melting points than the eutectic solder? Since the net final composition is still eutectic, the higher melting particles simply delay the melting time, but this is precisely the point. That is, the investigators wanted to delay the melting of paste until the pad had reached the melting point of the solder. Such an approach compensates for the lag time until the lands reach the melting point of solder. All the particles eventually mix by solid state diffusion. The amount of reflow is delayed by the rate of diffusion, which in turn is controlled by the sizes of the particles making up the paste.

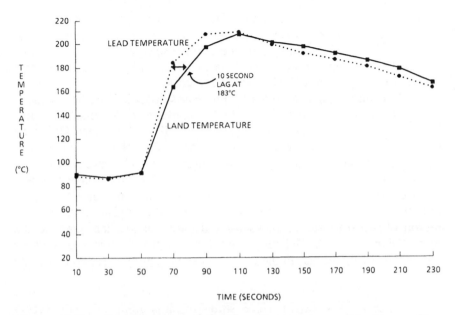

Figure 12.21 A comparison of the time-temperature soldering profiles of a PLCC lead and land with preheat during VPS in a batch system. The lead and land reach the solder melting point temperature (183°C) 10 seconds apart.

12.8 INFRARED REFLOW SOLDERING

Infrared reflow has changed considerably since its inception in the early 1980s. In the beginning, this process was not popular because it caused burning, even charring, or at least discoloration of boards. Now it has come to the point where it is the process of choice for reflow soldering. There are three types of IR processes as defined by SMEMA (Surface Mount Equipment Manufacturers Association) [10]. They are:

Class I: Radiant IR dominant systems
Class II: Convection/IR dominant systems
Class III: Convection dominant systems.

In any of these systems, there is always some source of radiation and some source of convection. There is no such thing as a perfect IR or perfect convection system. Among these three types of systems, the convection dominant systems are the systems of choice for new equipment purchases. This does not mean that if you have an IR system or a convection/IR system, you should scrap it and get a new convection machine. We should keep in mind that all of these systems are only heat sources. Any of them can be used to produce good results if one takes the time to develop the appropriate thermal profile. Refer to Appendix B3 for a detailed questionnaire of reflow oven selection.

These three major IR soldering processes differ in their heat sources and in their heating mechanisms. In the convection dominant systems, infrared radiation is the source of less than 20% of the heat. Quite the opposite is true for IR systems, where almost all the heating is achieved by infrared radiation in the short wavelength range. Convection accounts for less than 20% of the heating. For convection/IR, both convection and IR account for the heat source. In these systems, the direct infrared radiative heating from panels occurs in the middle infrared wavelength region (2.5–5 μm) and accounts for only 40% of the heat input to the parts being soldered. The remaining 60% of the heat is provided by convection of the hot air in the oven. These percentages are approximate of course.

Figure 12.22 shows a schematic diagram of a typical IR oven of any of the three classes. Figure 12.23 shows a photograph of a convection dominant system. Such systems have 3 to 20 independently controlled heating zones with built-in thermocouples for temperature control. These various zones are heated either by IR lamps or panels.

The assembly must be heated gradually in order to drive off the volatile constituents of the solder paste. The IR lamps or panels heat the board traveling on the conveyor from the top and/or bottom, as necessary. In the convection/IR or convection ovens, internal fans circulate hot air or an inert gas such as nitrogen. After an appropriate preheat time, the

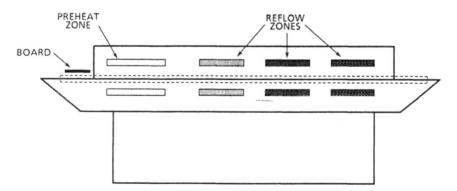

Figure 12.22   Schematic view of an in-line infrared reflow soldering system.

Figure 12.23 Photograph of a convection dominant system. (Courtesy of Research Inc.)

assembly is raised to the reflow temperature for soldering and then cooled. Cooling fans are generally used at the exit.

A minimum of three heating zones is required—a preheat zone to dry off the volatiles from the paste, a soak zone to bring up the temperature uniformly throughout the board to slightly below the reflow, and finally a reflow zone followed by rapid cooling.

More than three heating zones are not necessary from a technical standpoint, although a higher number of zones in an oven does make it easier to develop the desired thermal profile. For example, dedicating more than one heating zone to preheat, soak, and reflow zones makes it easier to achieve the ideal shape of the profile (discussed in Section 12.10) because each zone controls a smaller segment of the curve.

Contrary to popular belief, though, a higher number of zones does not necessarily increase throughput. For example, as discussed in Chapter 8, there is no significant difference in total adhesive cure time between 4-zone (Figure 8.17a) and 10-zone (Figures 8.17b and 8.17c) IR ovens. It takes about six minutes for each oven. Throughput in reflow ovens is generally not a major consideration anyway because reflow ovens, like screen printers, are rarely the bottleneck in production. Most pick-and-place machines cannot keep up with most screen printers and reflow ovens.

12.8.1 Heat Transfer Mechanism in IR Dominant Systems

Unlike conduction and convection, radiative heat transfer does not require physical contact. The two bodies do need to be in sight of each other and at different temperatures for radiation energy to transfer from one body to another. Heating of the earth by the sun is a good example of radiation—the bodies are at different temperatures, not in physical contact, and in sight of each other.

Gases such as air and nitrogen are transparent to infrared radiation and do not play any role in pure radiation heat transfer systems. The radiation heat transfer rate between two bodies at different temperatures is controlled by the following equation [12]:

$$Q = V \times E_s \times A_t \times K \ (T_s^4 - T_t^4) \tag{2}$$

where Q = infrared heat transfer rate (w/cm^2)
V = geometric view factor (0–1)
E_s = emissivity factor of the source (0–1, 1 for a black body [a perfect emitter])
A_t = absorptivity factor of the target (0–1, 1 for black body)
K = Stefan-Boltzmann constant (5.67 × 10^{-12} w/cm^2/K^4)
T_s = source temperature (K)
T_t = target temperature (K)

In SMT reflow, the geometric view factor is the amount of energy leaving the target and hitting the source. This number is generally very high (0.9 to 0.95). However, if a smaller part is adjacent to a larger part, it will be shadowed and the radiation energy will not reach it. This problem can be avoided by providing sufficient interpackage spacing, which is also needed for other reasons such as cleaning and rework.

The source emissivity factor E_s and the target absorptivity factor A_t are high in SMT applications. For example, solder paste is a good absorber of IR radiation, but a shiny surface of solder or gold is a poor absorber since it reflects radiation. For this reason one can expect that a shiny reflective surface will not get as hot as a dull black surface.

The wavelength of the radiation produced is determined by the emitter temperature as defined by Wien's displacement law given below:

$$W_p = K_w/T_e \tag{3}$$

where W_p = peak emission wavelength (μm)

K_w = Wien's constant (5215.6 μm°R), where °R is degrees Rankine

T_e = emitter temperature (°R)

From equation (3), shorter wavelengths will require higher temperature emitters. The IR dominant systems have lamp heaters, which produce shorter wavelength radiation because they operate at higher temperatures.

The wavelength range for infrared radiation is 0.78 to 1000 microns (μm). However, a wavelength range of 1–5 μm is used for the thermal processing of electronic assemblies; the 1–2.5 μm range is considered to be short or near wave IR, and the 2.5–5 μm range is considered medium wave IR.

Because the rate of heating for surface mount assemblies needs to be low (about 2°C/second), the radiant emitters need to be at lower temperatures so that the boards do not get fried. The lower temperatures will generate longer wavelengths which will be more susceptible to reflection and the shadow effect. It becomes quite tricky to find a good balance between the heating rate, wavelength, and shadow effect to accomplish uniform heating throughout the board. This is one of the reasons for the popularity of convection dominant systems, which do not have these problems. They are discussed next.

12.8.2   Heat Transfer Mechanism in Convection Dominant Systems

In a convection dominant system, some heat transfer occurs by the mechanism described in the previous section since there is a small amount (5 to 20%) of radiation energy. In addition some heat transfer also takes place by conduction where the heated board and components transfer their energy to adjacent areas. But the predominant mechanism is through convection as discussed below.

Convection heat transfer takes place when there is physical contact between a fluid (liquid or gas) and a solid body at different temperatures. The heat transfer occurs only on the layer of contact. A good example of this is a cool breeze blowing on a warm face. The following is a general equation for convection heating or cooling [12]:

$$Q = H \times \Delta T \tag{4}$$

where Q = heat absorbed per unit area (W/cm²)

H = convective film coefficient (W/cm²°C)

ΔT = temperature difference between fluid and object (°C)

Increasing the delta temperature between the fluid and the body and increasing the convective film coefficient (*H*) will increase heat transfer. The delta temperature is increased by raising the source temperature. *H* is a function of flow rate and angle of attack. *H* is similar to the wind chill factor. On a cold day it feels much colder when it is windy (high flow rate for heat transfer) than when it is not windy. In SMT soldering, *H* is raised by increasing the flow rate while keeping the delta temperature low. However, the flow rate cannot be so high that parts are displaced from the paste or liquid solder in the reflow zone.

It is not only the flow rate that controls *H* but also the angle of attack. A vertical flow rate hitting the board perpendicularly is more effective than horizontal flow. Horizontal air flow creates a stagnant flow which lowers heat transfer. This is the reason convection dominant systems use a turbulent vertical air flow, as shown schematically in Figure 12.24.

As shown in Figure 12.24, air or nitrogen enters in the various zones where it is heated by a series of heaters (nickel chromium wire for example). The gas is removed from the reflow zone by exhaust ducts. In all zones the gas is forced through a series of holes to impinge on the boards vertically. This turbulent flow improves convection heat transfer.

Each zone has a built-in thermocouple to monitor and adjust temperature as needed. The temperature of the board does not exceed the fluid

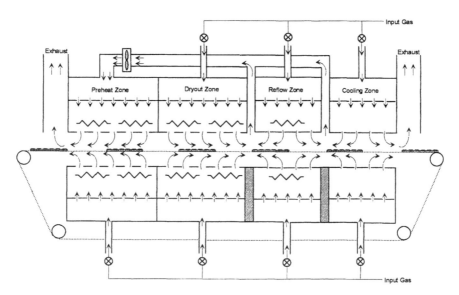

Figure 12.24   Air flow in a convection dominant system in different zones. (Courtesy of Research Inc.)

temperature. The delta temperature is kept low by keeping the source temperature low to ensure that the board does not overheat.

12.8.3 Heat Transfer Mechanisms in Convection/IR Systems

The heat transfer mechanism of convection/IR is essentially a combination of convection and IR radiation. The percentage of each component will differ based on air circulation and target temperature. The overall impact of radiation and convection is determined by the percentage of radiation and convection in the system. Before convection dominant systems became popular, a 60/40 ratio of convection to IR was widely used.

The panel heaters in most convection/IR systems run at lower temperatures, hence produce longer wavelength radiation. Many of the radiant dominant lamp systems have also adapted ceramic walls or reflectors to emulate what the panel systems do—emit mid to far IR wavelengths and present a larger surface area for heating.

The convection/IR system, by relying on both IR and convection energy, minimizes such once-common IR reflow soldering problems as shadow effects. Color sensitivity is also minimized, since the absorptivity of medium IR wavelength radiation is uniform across the color spectrum.

12.8.4 Pros and Cons of Various IR Systems

In convection dominant systems most of the energy is convection and is transferred by compressed air or nitrogen blowers with very little IR component for heat source. Some of the convection dominant systems are nothing but convection/IR systems with large blowers. Figure 12.24 shows the air flow direction in a convection dominant system in each zone.

One of the advantages of convection over IR is that in convection the temperature of the board will not exceed the air temperature. So there is never a danger of overheating the board. In most convection dominant systems, the temperature difference between the panel temperature settings and board temperature is much lower than in IR dominant systems.

Because of the nature of the heat source, each system has its pros and cons. In convection/IR systems, solder paste may not be completely cured because of a tough "skin" that forms as a result of circulating air. Formation of a skin can lead to "explosion" (solder splatter) during reflow [13]. In an IR dominant system, the short wave penetration into the board allows faster heating and complete curing of solder paste. This prevents solder splatter and minimizes solder balls [13]. These assertions may be more theoretical than real however since "explosions" are rarely seen unless the solder paste is heated rapidly in the preheat zone.

The IR dominant system has drawbacks too, namely, color sensitivity and the so-called shadow effect. Let us take color first. Short wavelength infrared radiation, prevalent in IR dominant systems, is not absorbed uniformly across the color spectrum. Hence an assembly consisting of different colored packages, such as ceramic and plastic component packages, will heat nonuniformly. Furthermore, since convection heating is virtually absent, regions on the board that are not directly exposed to the radiative heat sources will be cooler than those having a direct exposure. This, the shadow effect, results in nonuniform heating.

In convection dominant systems, heat transfer is primarily through convection with hot air or nitrogen being blown by high capacity blowers. Heating is rapid and uniform like the vapor phase process but without VPS's disadvantage of leads getting hotter than the land.

In any IR system, the product is heated by conduction, convection, and radiation. The convective portion is dominant in panel systems but plays a minor role in lamp systems. Variation in component sizes and colors can be overcome with IR dominant systems if a fine-tuned solder profile is developed. Thus, if the products in a company's line do not vary significantly, an IR dominant system may be entirely acceptable. However, for a company that makes a wide variety of products, an IR dominant system may require more work in solder profile development than a convection/IR or convection dominant system. By properly charac-

Table 12.4 Advantages and disadvantages of convection and IR dominant systems

CONVECTION DOMINANT SYSTEMS	IR DOMINANT SYSTEMS
ADVANTAGES	ADVANTAGES
• Uniform heating	• Efficient heating system (60–70%)
• Not sensitive to shading	
• Easier to develop thermal profile	• Easier to maintain N_2 purity level
DISADVANTAGES	DISADVANTAGES
• Inefficient heating (20–30%)	• Non uniform heating
• Higher and turbulent flow rate make it more difficult to maintain N_2 purity	• Potential for shading
	• Sensitive to color
	• More time consuming to develop thermal profile

terizing the process and by developing a detailed profile, any of the systems can give good results. Excellent results have been produced by IR dominant systems [15] as well as by convection/IR systems [16].

Uniform heating becomes more critical with packages such as BGAs, where any radiation energy will not reach to the inner rows of solder balls. In such applications, the convection dominant systems have a clear advantage. Since the industry is moving in this direction anyway and since there is no cost premium for such systems, most new machines purchased today are the convection dominant type. Table 12.4 summarizes the pros and cons of IR and convection dominant systems. The convection/IR systems will fall somewhere in between.

12.9 IR REFLOW SOLDERING IN NITROGEN

The use of nitrogen has become more prevalent due to the current widespread use of low solid and no-clean solder pastes. These solder pastes are not very active and hence are not very effective in removing oxides during reflow soldering. Nitrogen helps in preventing oxidation of solder powders and soldering surfaces of leads and boards during the reflow soldering process. With the increasing use of fine and ultra fine pitch packages, finer and finer solder particles are needed and oxidation of solder powders becomes a serious problem. (See Chapter 9, Section 9.1.3, on the subject of solder particle size and shape.)

The use of nitrogen in reflow is more common than the use of nitrogen in wave soldering. In wave soldering, there is only one major driving force for nitrogen—the increasing use of low activity fluxes (low solids and no-clean). But the reasons for using nitrogen in reflow are two-fold—the increasing use of low activity fluxes and of fine pitch packages.

The use of nitrogen in reflow can also result in other benefits—for example, the incidence of solder balling decreases and more consistent solder quality can be maintained. Also, the absence of oxygen prevents the flux from hardening at soldering temperatures. So if, for example, rosin fluxes are being used, the boards can be easily cleaned. However, different oven designs may produce different results. The ovens that can maintain optimum oven atmosphere provide the best results.

As shown in Table 12.5, the purity of nitrogen required for different purposes varies [17]. For example, board discoloration is significantly reduced when the oxygen level is below 10,000 ppm (parts per million). To prevent oxidation when using low residue fluxes, the oxygen level needs to be below 100 ppm. However, there is no additional measurable benefit below 30 ppm.

There is an extra cost for IR ovens with nitrogen capability. Instru-

Table 12.5 Required oxygen levels in reflow oven for different purposes [17].

DESCRIPTION	N$_2$ PURITY LEVEL
Significant reduction in board discoloration	10,000 ppm O$_2$
Ease of cleaning flux residues (RMA)	500 ppm O$_2$
Reduction in solder balling	200 ppm O$_2$
Reflow with low residue solder paste	<100 ppm O$_2$

mentation is required to monitor the flow of nitrogen in the oven and to monitor oxygen levels. Also, in order to reduce the cost of nitrogen gas consumption, some design changes are required to maintain an inert environment. This is where the oven design is crucial. The heating tunnel, which essentially means the perimeter of the oven, needs to be sealed and maintained at a positive pressure to prevent air from entering from the open ends of the machine.

It is more difficult to provide an inert environment in a convection dominant system than in an IR dominant system. In a convection dominant system, the heat transfer mechanism is essentially controlled by the flow rate. Since the temperature is relatively low, the flow rate is kept high. With a high flow rate, unless the system is closed loop, the cost of heating even just the air is very high. In order to make it cost effective to use nitrogen, the nitrogen must be recirculated. High velocity and turbulent flow are required to achieve good heat transfer and this increases the likelihood of oxygen leakage in the oven. In IR dominant systems, where the heat transfer is by radiation, there is no turbulent flow to worry about. This makes it easier to maintain a higher level of nitrogen purity and minimize nitrogen consumption.

12.10   REFLOW SOLDER PROFILE DEVELOPMENT

The solder profile, also known as the thermal profile, is one of the key variables in the manufacturing process that significantly impacts product yield. In this section we will discuss the details of developing solder profiles for IR and vapor phase soldering (VPS). The term IR profile refers to the profiles of all three types of IR systems—radiant dominant systems, convection dominant systems, and convection and IR dominant systems, although as mentioned before, the convection dominant systems are the most common.

No matter which system is used, if one does not take time to develop the appropriate thermal profile, changing to a newer machine is not going to solve the yield problem. The point was made earlier, but is worth repeating, that even after one has developed the appropriate thermal profile, if the source of the problem is poor design, poor solder paste, misregistration or misplacement of components, to name just a few items, a convection oven, or any oven or profile for that matter, is not going to eliminate the problems caused by the other variables.

Conveyor speed and panel temperatures are the only variables in thermal profile development. Experimental boards with representative thermal mass, size, and component density should be used for solder profile development. The solder profile is not only product specific, it is also flux dependent. Different pastes require different profiles for optimum performance. So it is important to consult the paste supplier before developing the solder profile.

For developing the profile, we need the loaded board for which the profile is being developed. At least four thermocouples (as narrow gauge as possible, 36 gauge wire is preferred) should be bonded with thermally conductive adhesive at desired locations across the board. Kapton tape or high temperature solder can also be used to secure the thermocouples to the board and leads.

We can start with a given belt speed (about 30 inches per minute or higher for desired throughput) and monitor the topside board temperature using thermocouples. Most new reflow ovens have built-in thermocouples and software packages to record the thermal profile. If not, commercial hardware and software packages, such as *MOLE*, by ECD Inc. of Portland, Oregon and *KIC* by EDI, Inc. of San Diego, California, are available that make thermal profile development an easy task.

When developing the profile, the top and bottom panel or lamp temperature settings in the oven should be kept roughly the same. We should note that panel temperatures for any setting will be different from the temperature seen on the board and the component leads.

The target shape of an IR solder profile is shown in Figure 12.25 using four zones: preheat zone, soak zone, reflow zone, and cooling zone. We will discuss each of these zones in a little more detail later on. It is important to control the shape of the curve in each zone. To achieve the target shape shown in Figure 12.25, each board or group of boards of similar thermal mass needs its own unique panel or lamp settings and conveyor speed.

In addition to meeting the shape of the profile, the oven settings must also show a narrow bandwidth (5°–10°C). The bandwidth is defined as the total temperature difference across the board in any given zone. The tighter the bandwidth, the more consistent is the yield. Figure 12.26 is an example of an actual profile with a very tight bandwidth (5°C) in a 4-zone convection dominant system.

If the shape of the profile does not resemble the shape of the curve

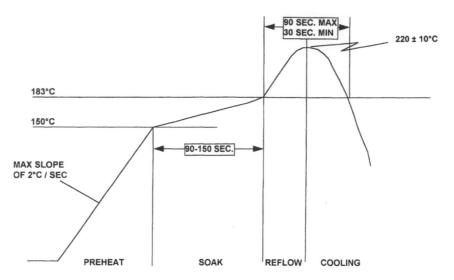

Figure 12.25 General solder profile for any type of IR system.

shown in Figure 12.25 or if the bandwidth is too wide (more than 10°C), adjust the panel settings. Generally two to four trial settings are required to meet the target shape and bandwidth. This could take 1–2 hours to complete. This is a significant improvement from the early days of only IR dominant systems with no built-in thermocouples (mid-1980s). Also, in those days, there were no commercially available profile development hardware and software packages with prediction features (after running the board at least one time through the oven), but these are common today. Developing profiles with data loggers and plotting the data used to be not only a cumbersome task but also time consuming (usually requiring a minimum of 4 hours for each board).

It should also be kept in mind that IR dominant systems require an added investment in engineering time to develop custom profiles because of potential problems with shadow effect and color sensitivity of the IR portion of the heating source. Because there is more uniformity of heating in a convection system, one can expect somewhat less time to be required to develop a solder profile. However, we should note that not all convection systems are created equal.

There is some misconception, mostly promoted by some suppliers, that if you buy their convection oven, there will be no need for developing a unique profile for each product. This is simply not true because each board has a different thermal mass and one may have different loading patterns (distance between boards as they are loaded in the oven). There is also a misconception that if you do need to change a profile, simply

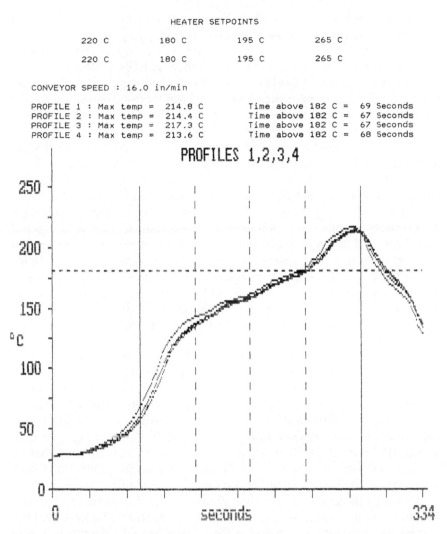

HEATER SETPOINTS

| 220 C | 180 C | 195 C | 265 C |
| 220 C | 180 C | 195 C | 265 C |

CONVEYOR SPEED : 16.0 in/min

```
PROFILE 1 : Max temp =  214.8 C     Time above 182 C =  69 Seconds
PROFILE 2 : Max temp =  214.4 C     Time above 182 C =  67 Seconds
PROFILE 3 : Max temp =  217.3 C     Time above 182 C =  67 Seconds
PROFILE 4 : Max temp =  213.6 C     Time above 182 C =  68 Seconds
```

Figure 12.26   An example of a solder profile in a convection dominant system. (Courtesy of Executone Information Systems.)

change the belt speed. Having to change only the belt speed is certainly easy, but it may not be the right approach. Changing the belt speed changes the temperature of the board in every IR zone.

In developing the thermal profile, it is important to rule out equipment dependent variables. For example, in IR ovens, especially the IR dominant systems, the board edges may heat up to 30°C higher than the center. It is easy to rule out the equipment related issues. Here is a simple guideline.

Select the most difficult board. A board with varying sizes and pitches such as BGA, fine pitch, and standard 50 mil pitch components along with resistors and capacitors may be a good choice. Mount thermocouples at different locations across the board. If there is BGA on the board, one of the thermocouples should be mounted near the inner rows of the solder balls. Also select a bare area. At least four thermocouples should be mounted across the board (center, edges). A good convection dominant oven will provide very small delta (3°–5°C) across the board.

Once the desired board temperature is achieved, run an actual production board with solder paste and components for reflow. After reflow, inspect the quality of the solder joints (fillet appearance and heights, wetting, nonwetting, solder balls, splatter, discoloration, etc.). Also look for differences, if any, in the solder joint appearance of smaller and larger components. A random problem only in a certain section of the board may be related to solderability, but a consistent problem in a given section may be related to the solder profile due to nonuniform heating (wide bandwidth). (Note too that consistent problems may also be related to paste quality and land pattern design.) Once the profile is found to give the desired results (assuming design and other material variables have been optimized), document the profile. After this point, no changes should be allowed in the profile.

Development of a soldering profile for VPS calls for the establishment of preheat temperature, elevator or conveyor speed, and dwell time in the primary zone. Like the IR systems, the vapor phase process also requires a custom solder profile for each product, but VPS profile development is much more straightforward because the fluid temperature is fixed. The only variable that requires adjustment in vapor phase is the conveyor speed. However, since IR is commonly used to preheat the board before it enters VPS in the reflow zone, the requirements for preheat and soak zones for IR as discussed earlier are equally applicable.

In both IR and VPS, the time above 150°C should be limited to 3 minutes maximum and the total time in the oven should be kept to about 5 minutes maximum. Glass epoxy FR-4 boards can be damaged if kept for an extended period above the glass transition temperature (125°C). This is generally not a problem. The total time is about 5 minutes even in smaller ovens with only three or four heating zones. Ovens with more heating zones can reduce this time slightly but not by much.

There is one more item that should be kept in mind. Even if the desired profile is achieved, it does not mean that one can always achieve the tight bandwidth established in the beginning. The oven needs to be regularly calibrated to ensure that it is functioning properly and that none of the heating elements or blowers is out of order.

Let us discuss in a little more detail the different zones of a thermal profile.

12.10.1 Preheat Zone

The heating rate in the preheat zone should be 2°–3°C/second and the temperature should be 100°–125°C. During preheat if the temperature ramp is too fast, the solder paste may explode and cause solder balls. Also, to avoid thermal shock to sensitive components such as ceramic chip resistors, the maximum heating rate should be controlled.

12.10.2 Soak Zone

The soak zone is intended to bring the entire board up to a uniform temperature. The ramp rate in this zone is very low, almost flat, as seen in Figure 12.25. The temperature is raised to almost the melting point of solder (183°C). The consequences of being outside the bandwidth in the soak zone are solder balls and solder splatter due to excessive oxidation of paste. The soak zone also acts as the flux activation zone for solder paste. In vapor phase there is no soak zone. The preheat and soak zones are part of the in-line IR oven used just before the VPS machine. If an oven has many heating zones, a significant number of zones should be dedicated to the soak zone.

12.10.3 Reflow Zone

After the preheat and soak zones, the board enters the reflow zone. In this zone if the temperature is above the bandwidth, boards may char or burn. If the temperature is too low, cold and grainy solder joints will result. The peak temperatures in this zone should be high enough for adequate flux action and good wetting. However, it should not be so high as to cause component or board damage or discoloration, or in worst case, charring of the board.

In some ovens, peak temperatures as high as 240°C are seen, but they should be avoided. The maximum temperature should not exceed 230°C. However, the peak temperature must be at least 20°C above the melting point of solder. Ideally the peak temperature should be between 210°C to 220°C for eutectic solder.

In VPS, the peak reflow temperature of the solder joint is determined by the boiling temperature of the primary fluid and cannot be controlled by the operator. The only controllable variable in VPS is the dwell time, which is controlled by the belt speed (in-line machines) or elevator speed (batch machines). The standard primary fluid used for surface mount assemblies has a boiling point of 215°C at sea level.

It is recommended that the solder at the joint be kept above its melting point for a minimum of 30 seconds and a maximum of 90 seconds. Extended duration above the solder melting point will damage temperature sensitive components. It also results in excessive intermetallic growth, which makes the solder joint brittle and reduces solder joint fatigue resistance.

12.10.4   Cooling Zone

The cooling rate of the solder joint after reflow is also important. The faster the cooling rate, the smaller the grain size of the solder, and hence the higher the fatigue resistance of the solder joint. So the cooling rate should be as fast as possible. However, from a practical standpoint, there is no control on the cooling rate other than making sure that the cooling fans at the end of the oven are operational. A cooling rate of 10°C/second can be easily achieved by most fans in most ovens. If the fans are nonoperational, the cooling rate will be much slower. This will increase grain size, causing relatively weaker solder joints.

12.11   COMMON REFLOW DEFECTS

The most common reflow solder defects are insufficient solder, excessive solder, opens, shorts, tombstoning, and part movement. The level of defects in general is higher in VPS than in the IR process. It is very difficult but not impossible to bring down the defect levels of vapor phase soldering to a rough equivalence with IR. For example, special solder paste (as discussed in Section 12.7.2) can be used for vapor phase to reduce the incidence of wicking common in VPS. Other ways to reduce the defect levels in VPS involve an adequate preheat profile and tighter lead coplanarity specifications. Tighter lead coplanarity and good thermal profile will help reduce the defect level in the IR process as well. Now let us examine some of the solder defects that are commonly seen in VPS and IR (convection, radiant, and convection/IR) processes.

12.11.1   Tombstoning and Part Movement

If the heating rates during reflow soldering are too high, the volatiles in the solder paste will rapidly evolve, possibly causing tombstoning (component standing on its end), drawbridging (component standing at an angle like a drawbridge), or lateral shift. We will refer to all these defects as part movements. The likelihood of such defects can be minimized through controlled ramp rates in the various heating zones.

Design of chip component land patterns also plays an important role in part movement. If the land width or the gap between the lands is too large, part movement can occur irrespective of the soldering process. For example, components with unequal termination widths are not uncommon. If such components happen to be used, they will cause unequal upward forces due to different levels of surface tension on the component terminations. This will cause part movement.

Poor chip component solderability can cause part movement as well, especially if the solderability of one end is inferior to that of the other or if a part is placed out of alignment. Such variables have the same effect as nonuniform termination width—unequal surface tension on each termination. Placement equipment also plays a role in part movement. For example, fast acceleration and deceleration of the X-Y table during placement can cause part movement before reflow, especially if the solder paste tackiness is not adequate.

12.11.2   Thermal Shock on Components

Rapid heating and cooling rates cause thermal shock because there is insufficient time for the center and the surface of the components to reach the same temperature. During application of heat, the surface temperature is higher than the internal body temperature. This discrepancy generates thermomechanical stress. This stress is compounded in materials that are poor conductors of heat. For example, chip resistors and capacitors, which are brittle and poor conductors of heat, are more susceptible to thermal shock. (See Chapter 14, Section 14.3.2: Component-Related Defects.)

The degree of thermal shock on components is higher in VPS than in IR. The IR soldering profile can be designed to heat the components at a rate of 2° to 6°C per second whereas only limited control can be exerted on the heating rate of components and boards in a vapor phase system. The maximum heating rate during VPS is typically between 15° and 50°C per second. Such high heating rates may not be acceptable for some components.

12.11.3 Solder Mask Discoloration

Some solder masks, especially dry film solder masks, tend to discolor in IR dominant systems. Solder mask discoloration is not an issue in convection dominant systems or in VPS. It is generally a cosmetic defect and can occur at temperatures as low as 160°C. In the case of one dry film solder mask, such discoloration was attributed to the oxidation of a polymeric dye in the solder mask.

Reduction of the oxygen partial pressure in the reflow zone by using an inert atmosphere (e.g., nitrogen) diminishes the potential for solder mask discoloration. New solder masks should be qualified before use to ensure their compatibility with the IR process and to confirm that any discoloration after soldering is only cosmetic and will not degrade board functionality.

12.12 LASER REFLOW SOLDERING

Laser soldering is a relative newcomer to soldering technology. It lends itself well to automation and is an ideal process for temperature-sensitive components since it heats only the solder joints and not the components. So heat-sensitive components that may be damaged in vapor phase or IR can be soldered easily by laser. In addition, moisture-sensitive components that are susceptible to cracking (popcorn effect) during vapor phase or IR can be soldered by laser without baking before reflow. Laser is also suitable for deleting circuit lines and adding components in a densely populated board without reflowing other joints. It is also used for through-hole components as a replacement for hand soldering (it is faster).

In vapor phase and IR, the glass epoxy substrate is heated above its glass transition temperature (125°C) for 3 to 5 minutes. This can warp or damage the board and form a thick layer of intermetallic. In comparison, laser soldering does not heat the substrate, and the reflow time is only about 300 milliseconds, which is not long enough to form a thicker layer of intermetallic.

Be that as it may, laser soldering is not a common process. It is a much slower process than other reflow processes such as VPS and IR. Also, laser is not suitable for ball grid array (BGA) or J leaded devices. It can be used for ultra fine pitch gull wing components. However, it is not as effective as the hot bar process even for ultra fine pitch devices that are prone to lead coplanarity. (Hot bar, as we will discuss in Section 12.13, pushes the leads down and hence lead coplanarity is not a concern.)

Generally two types of lasers are used for soldering: carbon dioxide

(CO_2) gas lasers and YAG-Nd (yttrium-aluminum-garnet/neo-dymium) lasers. The CO_2 laser is the more prevalent. Most laser soldering systems have controls for providing different amounts of heat input to different joints on the same board. In selecting a laser soldering machine, features such as wattage, type of laser (CO_2, YAG-Nd), and laser beam width control should be considered. In some new laser machines the laser beam will sweep back and forth across all the gull wing leads, on one side at a time, simulating a preheat and soak profile.

In every soldering process, there are certain issues of concern that need resolution. Being a relatively new process, laser soldering is no exception. The issues in laser soldering are summarized in Table 12.6.

Although laser soldering is capable of producing more reliable solder joints, it is slower than mass soldering methods such as wave and reflow soldering. As long as this is the case, cost considerations are likely to prevent its widespread use. However, the laser soldering function has been combined with the laser inspection function, hence according to the equipment manufacturer the technique has lost its cost disadvantage [18]. Despite this improvement, laser soldering is not a widely used process because laser inspection is rarely used. As the use of BGAs has increased, the use of x-ray laminography for inspection has become more common (see Chapter 14 for a discussion on x-ray laminography).

Laser soldering entails such process problems as potential damage to adjacent components (if the width of the laser beam is not properly adjusted) and solder ball formation, which can be even more prevalent in laser soldering than in VPS. Solder balls can be prevented by first fusing the solder paste or by buying boards with sufficient (0.3 mil) plated

Table 12.6 Technical issues and their resolution in laser soldering process

ISSUES	APPROACH
• Definition of a method for automating laser soldering	• Investigate method for supplying solder to the connection
• Prevention of solder balls	• Pretin mating surfaces before laser soldering
	• Fusion of solder paste on pad
• Prevention of laser damage to substrates	• Evaluate flux for damage control
	• Adjust beam width control mechanisms
	• Evaluate laser beam shielding techniques

or hot air leveled solder coating and then reflowing the solder joints with laser. This solution to a technical problem, however, makes the relatively expensive laser soldering process even more expensive.

Laser soldering has many advantages over other soldering methods, but its use will be confined to specialized applications until the process becomes economically justifiable. For laser soldering to move from the specialized soldering arena into general use, soldering and inspection operations must not only be combined in a single cost-effective system, as suggested above, to make the process relatively faster and hence cheaper, but automated inspection must be widely used in the industry. And for this to happen the inspection results must be more consistent than is currently the case. As discussed in Chapter 14 (Section 14.5: Solder Joint Inspection), laser inspection results are not always consistent.

A new system available from at least one manufacturer shown in Figure 12.27, however, has reportedly made some progress [18]. It uses an infrared detector to monitor soldering, and a YAG laser provides the power needed for soldering. As indicated in Figure 12.28, laser energy is targeted at the solder joint to melt it. The infrared detector monitors

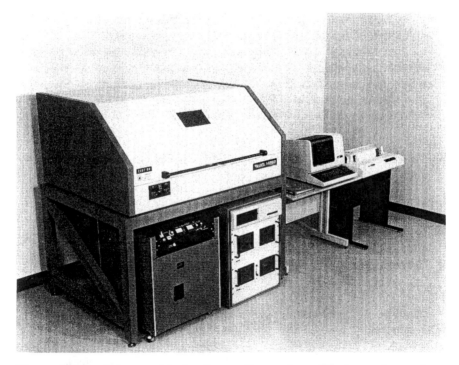

Figure 12.27   A laser soldering/inspection system. (Photograph courtesy of Vanzetti Systems.)

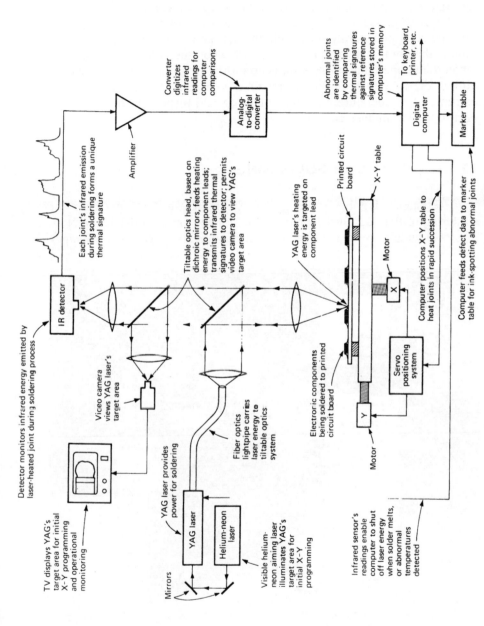

Figure 12.28 Process details for the laser soldering/inspection system shown in Figure 12.27. (Courtesy of Vanzetti Systems.)

587

the infrared energy that is emitted by each joint and forms its unique thermal signature. It them compares the thermal signature of each joint to a "known" good joint. The input for the known good joint is programmed into the computer during the developmental phase. The infrared detector directs the computer to shut the laser off when the solder joint melts or when the temperature of the joint is abnormal (high or low). This feature allows the system to mark for repair a joint that has experienced an extreme of temperature; thus the soldering and inspection functions are performed simultaneously.

12.13   HOT BAR SOLDERING

Resistance hot bar soldering has been in use for some time for soldering of flat pack surface mount components on 50 mil centers. Now it is also widely used, and for good reason, for ultra fine pitch packages

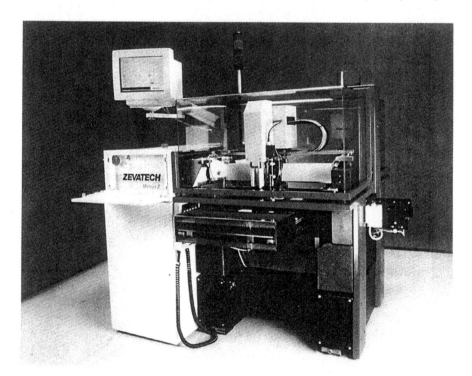

Figure 12.29   A dedicated TAB bonding machine. (Photograph courtesy of Zevatech, Inc.)

(under 0.5 mm or 20 mil pitch) and tape-automated bonding (TAB) or tape carrier packages (TCP). For ultra fine pitch packages ($\leq$ 20 mils), the conventional process of solder application by screen printing followed by IR becomes increasingly difficult.

The hot bar soldering process is based on resistance soldering of thermodes. It works in much the same way as a household electric stove burner. Electric current passing through the heating elements (thermodes) heats the thermode to soldering temperatures. The heated thermode is pressed against a fluxed solderable surface (gull wing lead on land). The temperature and the duration of thermode application at a predetermined pressure are such that the solder on the land melts to form a fillet.

Figure 12.29 shows a photograph of a dedicated TAB bonding machine. Also see Chapter 11, Figure 11.16 for a photograph of a very flexible pick-and-place machine outfitted with a TAB bonding head. The schematic of the TAB bonding head used for either of these machines is shown in Figure 12.30.

Pai and Thomas have demonstrated through their research that the resistance hot bar soldering process is superior to both VPS and IR for soldering fine pitch packages [19]. They have also used this process for rework/repair of fine pitch packages. The driving force for using the hot bar process is the susceptibility of fine pitch leads to lead bending and

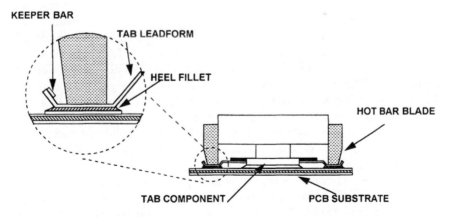

Note: TAB (tape automated bonding) is also referred to as TCP (tape carrier package) or COT (chip on tape).

Figure 12.30 Schematic of hot bar reflow thermode on TAB lead.

damage during shipping and handling. Since pressure is applied during the hot bar soldering process, it can tolerate lead coplanarity better than any other soldering process.

Since no solder paste is applied in this process, sufficient solder coating must exist on the pad. Solder coating is generally applied by solder plating, although other methods can also be used. For example, in some applications, solder paste is applied first and is fused in vapor phase or IR. The application of sufficient thickness of solder on the fine pitch lands is one of the key process steps. For ultra fine pitch, the desired thickness is 700+/−150 microinches. Less than 500 microinches or half mil (12.5 μm) of solder may cause problems (reliability related to Au-Sn intermetallic) especially if the TAB lead is plated with gold. Why so much solder on the lands? For acceptable solder joints, heel fillet formation should extend 8 mils (0.2 mm) up the lead. Toe fillets should be visible on all leads. Such acceptance criteria require that sufficient solder be on the land.

Hot bar is a far more expensive process than IR or VPS because it is much slower. Also the equipment itself is generally 7 to 10 times more expensive than a typical convection dominant IR oven. Why? The hot bar machine not only serves as the reflow machine, it also performs the function of a very accurate placement machine with vision and can be used as a rework tool as well. Considering the fact that a hot bar machine is a complete ultra fine pitch assembly line, the equipment cost may not be considered excessive.

The process sequence for ultra fine pitch using hot bar is as follows. Reflow standard surface mount and fine pitch components followed by wave or hand soldering of through-hole components on the board. Components such as ultra fine pitch packages and TAB are generally received in coin stack tubes which are first loaded into assembly equipment. The leads are excised and formed into gull wing shape. In the case of TAB, die attach adhesive is dispensed to the PCB if it is needed to provide a thermal and electrical path to the board. Finally the leads are soldered simultaneously to the PCB by the heated thermodes, one side at a time or two sides at a time or, in some cases, all four sides at the same time.

Figure 12.31 shows a typical reflow profile for the hot bar process for TAB. Table 12.7 provides specific guidelines for pressure and temperature. As noted in Table 12.7, when soldering fine pitch devices, thermode bonding pressure is much higher (about 30 lb. maximum) than when soldering TAB leads (3 lb. maximum). Also, it should be noted that in order to develop a repeatable profile, regularly scheduled thermode cleaning and grinding is necessary.

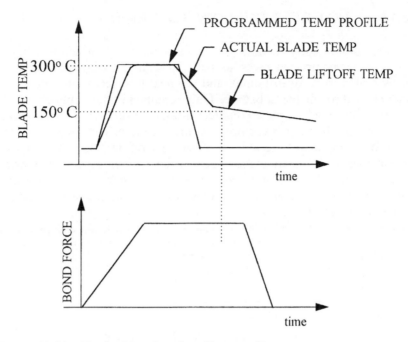

Figure 12.31   Typical hot bar bonding profile.

When using hot bar, there are some design guidelines to keep in mind. There should be no vias closer than 25 mils from the edge of the land in order to prevent solder damage into vias. Also, since hot bar is the last assembly process, enough clearance around ultra fine pitch packages such as TAB should be provided for the thermode. Clearance on the bottom is also needed for support. Also refer to Chapter 6 for land pattern design guidelines for TAB.

Table 12.7   Recommended thermal profile for hot bar for TAB soldering

Blade temperature	275°C
Temperature variation across blade	5°C
Blade-to-blade temperature variation	5°C
Blade force	3 lb. max. for TAB and 30 lb. max. for fine pitch
Dwell time	5 seconds

12.14 HOT BELT REFLOW SOLDERING

One of the early soldering processes for the reflow soldering of surface mount components was hot belt reflow. In this process, a conveyor belt is heated in an oven tunnel, and the heat transfer occurs mostly by conduction through the substrate. Only single-sided assemblies can be soldered, since the heat must conduct through the surface of the substrate.

In the hot belt conveyor process, the substrates can get excessively hot before components are able to reflow. Therefore it is generally used for components on ceramic substrates, which do not experience thermal damage. A typical temperature profile for a board in the hot belt conveyor process is shown in Figure 12.32. As is obvious from the profile, the rate of heating is very rapid.

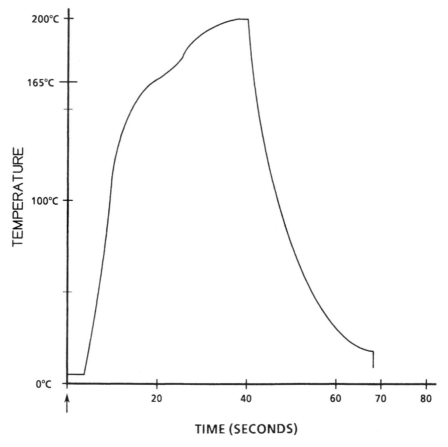

Figure 12.32 A typical soldering profile for a hot belt reflow soldering.

12.15 SELECTING THE APPROPRIATE SOLDERING PROCESS AND EQUIPMENT

As we have discussed in this chapter, all soldering processes have some technical concerns but there are ways to address those concerns. Table 12.8 shows cost-effective soldering options for different applications. As is clear from the table, no single soldering process can meet all needs. Instead, multiple soldering processes will be required for a mixed assembly. For example, through-hole and surface mount components need both wave and reflow soldering processes. If ultra fine pitch components such as TAB are also used on the same board, a third process such as hot bar or laser may also be required.

The selection of wave soldering for surface mounting need not entail much difficulty if a conventional soldering system is already in place, since it may be economical to retrofit the existing machine with a new wave solder pot, as discussed earlier. In such a case the equipment generally is purchased from the present vendor because a wave pot from a different vendor might not be retrofittable. However, depending upon the application and budgetary constraints retrofitting may or may not be the desirable option.

When it comes to reflow soldering, the use of convection dominant systems has become most common. So when purchasing a new oven, the convection dominant system is generally the oven of choice, especially

Table 12.8 Cost-effective commonly used soldering options for different applications

	WAVE	VAPOR PHASE	CONV IR	LASER	HOT BAR
Throughput	X	X	X		
Surface-mounted components	X	X	X	X	X
Most common SMT soldering process			X		
Most common through-hole soldering process	X				
Odd-shaped assembly		X		X	
Moisture/temp. sensitive (through hole)				X	
Moisture/temp. sensitive (SMT)					X
Most common ultra fine pitch/ TAB (<20mil pitch) soldering process					X

since there is no cost premium for a convection dominant system over an IR dominant system. However, this does not mean that one should go out and replace an existing IR or convection/IR dominant systems with a convection dominant system. If one is able to achieve the desired profile, the existing oven may be just fine. As long as one can achieve the desired profile, the heat source is irrelevant. However, in some cases, depending on the flux and other controls on the factory floor, it may be desirable to consider purchasing a new machine with nitrogen inerting option.

Even though the vapor phase process is essentially out of vogue, there are cases where it may make sense to keep the existing VPS system. For example vapor phase may be better for ceramic BGAs. However, it is rare for only BGAs to be exclusively used on a board; there are other components on the board as well. As we discussed in Section 12.7, if one is experiencing a higher incidence of overall solder defects in VPS, it may be cost effective to replace the VPS system with a convection dominant system.

Looking again at Table 12.8, it becomes clear that all soldering processes can be used for generic surface mount components. However, for ultra fine pitch (<20 mil lead pitch) and TAB surface mount devices, resistance hot bar soldering is the predominant process, although hot gas and laser soldering can also be used. On the other hand, wave soldering remains the most cost-effective soldering process for through-hole components.

Once a decision has been made on a particular process, the next step is to decide on the supplier of the equipment. There can be great variation from one machine to another. In addition to the usual issues of cost, some of the main areas of evaluation in selecting these machines are as follows: defect rates, level of automation features desired, compatibility of a given machine with existing or future equipment, service and support, and input from current users.

Finally, the equipment should be selected on the basis of a specification that establishes performance requirements and sets the acceptance criteria. This is very important because it not only helps you objectively determine your equipment requirement but also sets an equipment specification which can be used to evaluate the machine at the vendor's facilities before taking delivery. This is true not only for wave soldering machines but for any surface mount equipment as well. Developing the needed equipment in an involved effort but is well worth the time. It ensures that the equipment finally selected will not only meet the current requirements but will also take into account the strategic view of future requirements as well.

12.16 SUMMARY

There are many soldering processes available for surface mount devices but none is perfect for all applications. All processes have technical concerns and there are ways to resolve those concerns. Because it offers higher yield and lower operating cost, convection dominant IR soldering has evolved as the preferred process for reflow soldering.

Vapor phase will not disappear, however. It will continue to be used in niche applications. For some specialized jobs, other reflow soldering processes such as laser and hot bar resistance soldering are also used. These processes are intended not to replace either vapor phase or IR but to complement them. The process ultimately used should be selected on the basis of the specific requirements of the intended application, the solder defect results, and the overall cost.

Since the industry will remain in a mixed assembly mode, the use of through-hole components will continue for the foreseeable future. There is no process more cost effective than wave soldering for through-hole components. The use of reflow with solder paste will also be preferred for some applications.

Because of the widespread use of low solids or no-clean flux, the use of nitrogen has become common for both wave and reflow processes. However, one should not expect nitrogen to be a panacea for solder defects. It will help only to the extent of its influence on soldering yield. Nitrogen is not going to solve issues related to other parameters such as design, flux activity, solder paste rheology, print quality, and solder profile, just to name a few.

The selection of a process depends on the mix of components to be soldered. Various soldering processes will complement each other instead of replacing them. Even hand soldering is not going to disappear entirely.

REFERENCES

1. Prasad, Ray, Introduction to Surface Mount Technology, Intel application note AP-403, January 1987, p. 14. Available from Intel Corporation, Santa Clara, CA.

2. Biggs, C. "An overview of single wave soldering for SMT applications." *Surface Mount Technology*, June 1986, pp. 8–10.

3. Avramescu, S., and Down, W. H. "Omega wave: A new approach to SMD soldering." *NEPCON Proceedings*, February 1986, pp. 640–650.

4. Hymes, Les, and Pitsch, Tim. Controlled atmosphere soldering: A case study of cost/benefit. *Proceedings of Surface Mount International*, August 30–September 1992, pp. 835–838.

5. Fremd, Eric T. Quantitative quality comparison of wave soldering in inert nitrogen versus air, IPC Technical Review, January–February 1993, pp. 29–36. Available from IPC, Northbrook, IL

6. Stratton; Adams, S. M.; and Saxena, N. "Minimizing inert gas consumption." *Circuits Assembly*, October 1995, pp. 58–64.

7. Elliott, Donald A. "Q & A: Nitrogen wave soldering." *Circuit Assembly*, October 1991, pp. 55–59.

8. Pignato, J. "New phase in vapor phase." *Circuits Manufacturing*, September 1987, pp. 71–76.

9. Wenger, G. M., and Mahajan, R. L. "Condensation soldering technology, Part I: Condensation soldering fluids and heat transfer." *Insulation/Circuits*, September 1979.

10. Hutchins, C., and King, S. "The surface mount solder reflow process." *Printed Circuit Assembly*, April 1987, pp. 20–24.

11. McLellan, R. N., and Schroen, W. H. "TI solves wicking problem." *Circuits Manufacturing*, September 1987, pp. 78–85.

12. Cox, Norman R. Reflow technology handbook, 20 pages. Available from Research, Inc. Minneapolis, MN.

13. Flattery, D. K. "Minimizing solder defect rates through the control of reflow parameters, Part 2." *Surface Mount Technology*, December, 1986, pp. 7–12.

14. Dow, S. "An examination of convection/infrared SM reflow soldering, Part I." *Surface Mount Technology*, February 1987, pp. 26–28.

15. King, S. R. "SMT module improvement with infrared reflow." *Proceedings of NEPCON West*. February 1986, pp. 25–27.

16. Prasad, R., and Aspandiar, R. F. "VPS and IR soldering for SMAs." *Printed Circuit Assembly*, February, 1988, pp. 29–37.

17. Lea, Colin. "The significance of oxygen concentration in the atmosphere." *Proceedings of Surface Mount International*, August 25–29, 1991, pp. 159–177.

18. Vanzetti, Ricardo, and Traub, Alan C., Inspecting while soldering—The laser soldering option, *Proceedings of Surface Mount*

Technology Association Conference, August 1988, pp. 149–154. Available from SMTA, Edina, MN.

19. Pai, D. K., and Thomas, R. J., Surface mounting fine pitch gull wing VLSI carriers, *Proceedings of SMART III Conference*, Technical Paper SMT III-25. Available from IPC, Northbrook, IL.

Chapter 13
Flux and Cleaning

13.0 INTRODUCTION

The selection of flux and cleaning processes plays a critical role in the manufacturing yield and product reliability of electronic assemblies. The subjects of flux and cleaning are interrelated; one cannot be discussed without the other. The function of flux is to remove oxides and other nonmetallic impurities from the soldering surfaces and prepare a clean surface for joining. After soldering, cleaning may be required to remove flux residues.

The types of flux residues or contaminants that require cleaning are determined primarily by the type of flux used, but halides, oxides, and various other contaminants are introduced during storage and handling as well. The use of aggressive fluxes makes soldering easier even if components and boards are slightly oxidized and contaminated. However, most flux residues must be removed by cleaning. The cleaning process to be used is selected on the basis of type of flux, types of contaminants, and type of assembly—that is, mixed assembly (Type II and III SMT) containing both through-hole and surface mount components or full surface mount (Type I SMT). For example, mixed assemblies may need one cleaning process after reflow soldering and another one after wave soldering.

The cleaning process selected may use solvents or deionized (DI) water or a combination of the two. In the past, the commonly used solvents were CFCs (chlorofluorocarbons) such as Freon. However, as a result of an agreement of the international community, commonly referred to as the Montreal Protocol of 1987, the use of CFCs was banned by the end of 1995. CFCs have been determined to cause ozone layer depletion—an environmental hazard for humans and other species. Since the passage of the Montreal Protocol, various alternative solvents have been developed to replace CFCs. The industry has had no choice but to use either an alternative solvent or water-soluble fluxes and pastes for cleaning or to move on to a "no-clean" process by using low residue or no-clean fluxes and pastes.

Current technology using no-clean or low residue fluxes is eliminating the need for cleaning. However, the use of no-clean flux requires a clean work environment and a culture change that not only affects the users but their suppliers as well. In addition, the use of no-clean fluxes may require a controlled soldering atmosphere to provide a process window compatible with their lower activity.

In this chapter we explore the reasons for cleaning and discuss fluxes, cleaning processes, and equipment for both through-hole and surface mount assemblies. We focus on no-clean pastes and fluxes and strategies for their implementation because the use of no-clean flux is not a drop-in process. In addition, we will outline test methods and requirements for cleanliness to help us determine how clean is clean. We must note that except for cosmetic visual requirements, all fluxes (including no-clean) must meet the same cleanliness requirements.

13.1 CONCERNS IN SURFACE MOUNT CLEANING

The main concern in cleaning either through-hole or surface mount assemblies is adequate removal of flux to prevent corrosion problems in the field. For surface mount assemblies the job is difficult because flux may be trapped in the tight spaces between surface mount components and the substrate. Also, as a result of higher temperatures and relatively longer times during reflow soldering, the fluxes, especially rosin fluxes, become more tenacious, hence are more difficult to clean.

Figure 13.1 shows the area underneath the components along the X axis. It also shows an imaginary line below which components are more difficult to clean. Components with finer and finer pitches have lower stand-off heights and a "picket fence" is created around the package which makes cleaning difficult. In addition, the EIAJ fine pitch packages, as shown in Figure 13.1, have essentially no stand-off height. They sit practically flush with the board. Fortunately, though, ball grid arrays (BGAs) are increasingly replacing very high pin count fine pitch devices. BGAs are easier to clean because of their higher stand-off. The plastic BGAs (PBGAs) have a stand-off height of 18–22 mils and ceramic BGAs (CBGAs) have a stand-off height of about 35 mils. The 30 mil PBGA eutectic solder balls collapse after reflow to 18-22 mil stand off. The high temperature 35 mil CBGA balls do not melt during reflow. The micro BGAs have a low stand-off height of only 3.5 mils. It will be difficult to remove flux from underneath micro BGAs.

Whereas typical leaded components, like dual in-line packages

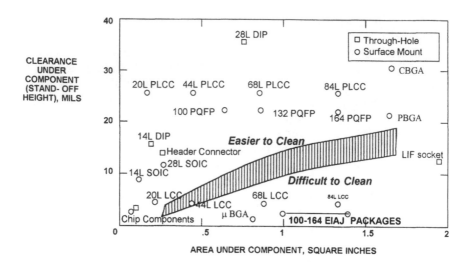

Figure 13.1   Area and stand-off height of different components.

(DIPs), are spaced 0.030 and 0.050 inch off the substrates, surface mount components are spaced much closer. Small passive components such as ceramic resistors and capacitors and large active components such as leadless ceramic chip carriers (LCCCs) may be spaced only 0.001 to 0.005 inch off the substrate. Other active surface mount components, such as small outline integrated circuits (SOICs) and plastic leaded chip carriers (PLCCs), are usually spaced from 0.01 to 0.025 inch off the board. Not all through-hole mount devices have higher stand-off heights than surface mount components. For example, as shown in Figure 13.1, 14-lead DIPs have lower stand-off heights than many surface mount components.

When chip resistors and capacitors are attached with adhesive, the adhesive itself will fill most of the space between the chips and the board, leaving little room for flux entrapment between the adhesive perimeter and the pad metallization. However, rapid curing of the adhesive may result in voids in the adhesive and will absorb flux that is almost impossible to remove. Chapter 8 (Section 8.6.1.2: Adhesive Cure Profile and Flux Entrapment) tells how to prevent voids in adhesive during cure. When no adhesive is used, as with LCCCs, SOICs, and PLCCs, flux and other residues are expected to end up under these components. The problem is especially acute if the gap is tight and the components are large: components the size of LCCCs not only have small stand-off heights, they also have a large area underneath, which increases the potential for flux entrapment and makes it difficult for solvents to penetrate beneath them to remove the flux.

The difficulty or ease of cleaning is graphically illustrated in Figure

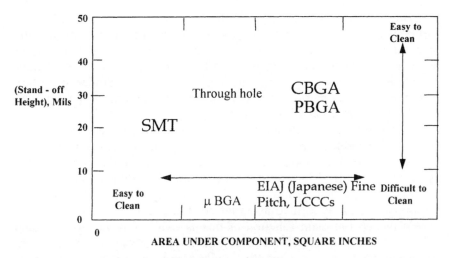

Figure 13.2 Relative difficulty of cleaning different types of components.

13.2. Clearly components on the right-hand side (larger area, smaller stand-off heights) are more difficult to clean than components on the left-hand side (smaller area, smaller stand-off heights). Thus area and stand-off heights should be considered together to ascertain the potential difficulty in cleaning. This is confirmed by Musselman and Yarbrough, who state that cleanability is proportional to the stand-off height (h^4) and inversely proportional to package area [1].

13.2   THE FUNCTION OF FLUX

All metals, except pure gold and platinum, oxidize in air at room temperature. The rate of oxidation can increase with an increase in humidity and temperature. Different metals have different affinities for oxygen, hence they differ in their susceptibility to oxidation, which is a form of corrosion. Some metal oxides, such as those of silver and palladium, revert to pure metals at high temperatures. Oxide surfaces can be either compact or loose. The compact oxides protect the underlying surfaces from further oxidation, as chromium oxide (Cr_2O_3) protects stainless steel surfaces. Quite the reverse is true for pure iron, which forms a loose oxide that allows diffusion of oxygen until all the metal has been consumed.

The basic mechanism of joining two metallic surfaces is controlled by the solubility of one metal into the other. As discussed in Chapter 10, the degree of solubility of one metal into the other and the temperatures at which they dissolve and form an intermetallic layer can be determined by inspecting the phase diagram of the two metals to be joined. For the formation of an intermetallic bond, two fresh metallic surfaces must come in contact. Since metal oxides are barriers to the formation of such a bond, they must be removed, by the use of flux, regardless of whether the joining process is welding, brazing, or soldering. Soldering differs from the other joining processes only in the type of oxides to be removed and the temperature at which joining takes place.

In soldering, the function of flux is to chemically react with oxides and quickly produce a fresh, tarnish-free surface at soldering temperatures so that intermetallic bonding can take place. When similar surfaces such as solder-coated leads and solder are to be bonded, they simply dissolve into each other. When dissimilar metals such as copper and tin are soldered, a tin-copper alloy is formed as an intermetallic layer (Cu_3Sn_6).

Although the main function of flux is oxide removal, it must also be able to counter the effects of pollutants and grease collected on the surface during storage and handling and to clean the surface for soldering. Tackiness of flux (in solder paste) is essential for preventing part movement during handling between placement and reflow soldering operations.

13.3   CONSIDERATIONS IN FLUX SELECTION

The activity in a soldering flux is generally provided by halides (chlorides, bromides) present in the flux, although there are some halide-free fluxes containing amino acids or other organic acids. The higher the activity or the halide content in a flux, the more effective it will be in performing its function or removing oxides from the soldering surfaces. Why not use only the active fluxes? The more active a flux, the higher the probability that undesirable corrosive by-products of the flux reaction will be left behind on the board, potentially causing reliability problems in the field. A more aggressive flux may perform its function well by reducing the amount of touch-up, but if the area is not properly cleaned after soldering, reliability may suffer.

Thus there are two major and contradictory considerations in selecting a flux for electronics products: it should be inactive at room temperatures before and after soldering, but active at soldering temperatures, to promote easy removal of tarnish and oxides. Actually, it should be active slightly below the soldering temperature, to ensure that a fresh surface is ready for soldering at the soldering temperature. If the flux residue is inactive after soldering, it will not jeopardize reliability if left on the board. No-clean flux comes close to meeting this requirement but it is relatively inactive even at soldering temperatures. The selected flux should provide a good balance between activity and cleanability.

As discussed in Chapter 10, if there is sufficient solder coating on the soldering surface to prevent exposure of the intermetallic surface, and if the components are properly stored, solderability can be preserved for a long time. If the components are not solderable, they can be pretinned to restore their solderability. It is generally not practical or cost effective to preclean and pretin boards and components before soldering. Hence if the solderability of surfaces is questionable, a more aggressive flux is needed to counter the effects of these conditions. It is tempting to use the most active flux to achieve the highest possible yield in manufacturing, but one must choose carefully according to the application. For example, in many military applications pretinning is performed at increased cost to restore solderability. An active flux is generally not used because of reliability concerns. The other option is to use low residue or no-clean fluxes which are halide free. They use weak organic acids to boost flux activity. They do require boards and components with good solderability, however. This may require pretinning and the use of a just-in-time inventory system to ensure good solderability. In addition, an inert environment such as nitrogen to prevent oxidation of soldering surfaces during the soldering process may be necessary.

13.4   FLUX CLASSIFICATION

Fluxes are classified based on their activity and their constituents, which determine activity. Flux activity, in turn, is an indication of its effectiveness in removing surface contaminants. Keeping flux activity as a measure, the fluxes are generally classified as inorganic acid, organic acid (OA), rosin, and resin (no-clean). Table 13.1 shows flux classifications based on their activity [2].

According to Table 13.1, another way to describe a flux may be by the letters A through Y. For example, the full description of flux A will be rosin or RO-H0 to indicate it is a halide-free rosin flux. Other halide-free fluxes in this category will be RO-M0 and RO-H0, but they are considered relatively more active than RO-L0. The halide content alone is not an indication of the activity level since other constituents may be substituted for halides. The classification of Table 13.1 describes both the activity of the flux and the activity of the flux residue as follows:

L = Low or no flux/flux residue activity
M = Moderate flux/flux residue activity
H = High flux/flux residue activity

Table 13.1 Flux classification based on material composition and halide content [2]

FLUX TYPE SYMBOL	FLUX MATERIAL OF COMPOSITION	SYMBOL	FLUX ACTIVITY LEVEL (% HALIDE)	FLUX TYPE
A	Rosin	RO	Low (0%)	L0
B			Low (<0.5%)	L1
C			Moderate (0%)	M0
D			Moderate (0.5–2.0%)	M1
E			High (0%)	H0
F			High (>2.0%)	H1
G	Resin	RE	Low (0%)	L0
H			Low (<0.5%)	L1
I			Moderate (0%)	M0
J			Moderate (0.5–2.0%)	M1
K			High (0%)	H0
L			High (>2.0%)	H1
M	Organic	OA[a]	Low (0%)	L0
N			Low (<0.5%)	L1
P			Moderate (0%)	M0
Q			Moderate (0.5–2.0%)	M1
R			High (0%)	H0
S			High (>2.0%)	H1
T	Inorganic	IN	Low (0%)	L0
U			Low (<0.5%)	L1
V			Moderate (0%)	M0
W			Moderate (0.5–2.0%)	M1
X			High (0%)	H0
Y			High (>2.0%)	H1

[a]Organic acid flux is also referred to as OR.

In each category there are three levels of flux activity or corrosiveness: low, medium, and high. And in each of these subcategories there are further classifications indicated by the numbers 0 and 1. Number 0 indicates halide-free flux. Number 1 indicates less than 0.5% halide in the low activity flux category, 0.5 to 2% halide in the moderate activity flux category, and more than 2% halide in the high activity flux category.

There are various tests such as copper mirror, halide content, corrosion, wetting balance, and spread, just to name a few, that these fluxes must pass to be classified in a particular category. Refer to J-STD-004 for details [2]. Briefly, the fluxes to be classified as L and M (low and moderate activity) must pass more tests than fluxes with H (high) prefixes. Similarly, the fluxes with 0 suffixes in the L, M, or H category must pass more tests than fluxes with 1 suffixes in the same category.

In general, these fluxes can be classified as highly corrosive (inorganic acid fluxes), corrosive (OA), mildly corrosive (rosin based), and noncorrosive (no-clean or low residue fluxes). However, as is clear from Table 13.1, there are different levels of corrosiveness in any flux category.

The highly corrosive inorganic acid fluxes of any category are seldom used in the electronics industry, and the mildly corrosive fluxes are generally used only in commercial electronics. The RO (rosin) and RE (resin) fluxes in the mildly corrosive category have activity comparable to that of the OA fluxes and are designed for solvent cleaning, whereas the OA fluxes are meant for aqueous cleaning. The RE fluxes in the low residue and no-clean category will have very low activity, however. (We should note that organic acid fluxes are sometimes referred to as OR fluxes. The OR designation has not really caught on, though, since the term OA has been in use much longer.)

The rosin (RO) fluxes are also referred to as rosin (R), rosin mildly activated (RMA), and rosin activated (RA). The rosin fluxes can be cleaned by either aqueous or solvent methods. RA fluxes are rarely used in solder paste. For reflow, in addition to rosin, both OA and no-clean pastes are also used. However, for wave soldering, RMA, RA, OA, and no-clean fluxes are used. No matter which category of flux is used, it must provide a good balance between the level of activity needed for the job and board cleanliness requirements. In the sections that follow we will discuss inorganic, organic, rosin, and no-clean fluxes in more detail.

13.4.1   Inorganic Fluxes

The inorganic fluxes are highly corrosive, being comprised of inorganic acids and salts such as hydrochloric acid, hydrofluoric acid, stannous chloride, sodium or potassium fluoride, and zinc chloride. These fluxes are capable of removing oxide films of ferrous and nonferrous metals such as stainless steel, Kovar, and nickel irons, which cannot be soldered with weaker fluxes.

The inorganic fluxes are generally used for nonelectronics applications such as the brazing of copper pipes. They are, however, sometimes used for lead tinning applications in the electronics industry. Inorganic fluxes should not even be considered for electronics assemblies (conventional or surface mount) because of potential reliability problems. Their major disadvantage is that they leave behind chemically active residues than can cause corrosion and hence serious field failures.

13.4.2   Organic Acid Fluxes

The organic acid fluxes are stronger than rosin fluxes but weaker than inorganic fluxes. They provide a good balance between flux activity and cleanability, especially if their solids content is low (1–5%). These fluxes contain polar ions, which are easily removed by a polar solvent such as water. Because of their solubility in water, the OA fluxes are environmentally desirable, although the use of no-clean flux may be environmentally more desirable. As the name implies, the OA fluxes contain organic acids commonly found in food and dairy products, such as citric acid, lactic acid, and oleic acid. Since fluxes of this type are not covered by government specifications, their chemical content is controlled by the suppliers. OA fluxes are available with or without the use of halides as activators.

The OA fluxes have generally been shunned even for conventional assemblies because of the term "acid" flux. As indicated previously, however, the fact is that even the so-called noncorrosive rosin fluxes contain halides that will cause corrosion if not properly removed.

The use of OA fluxes may be justified for mixed assemblies (Type II and III) for both military and commercial applications. It is incorrectly believed that a change from OA to rosin-based fluxes (RA and RMA) is mandatory when wave soldering Type II and III SMA boards.

Contrary to popular belief, OA fluxes have been successfully used on military programs. Examples include electronic assemblies for the Air-Launched Cruise Missiles and Airborne Early Warning Control Systems manufactured by the Boeing Aerospace Company. Many other leading companies in the commercial, industrial, and telecommunications sectors are using OA fluxes to wave solder surface mount chip components glued to the bottom of the board. OA fluxes have been found to meet both military and commercial requirements for cleanliness. See Section 13.8.3.2 for results.

OA flux materials have served successfully as flux coatings for the solder doughnut used in the reflow soldering of leaded through-hole components. Even after going through the reflow soldering operation, they are easy to clean with water. Now water-soluble solder pastes are widely used. In the past they were not as tacky as rosin-based fluxes, but the tackiness problem was solved a long time ago. (Tackiness in solder paste is necessary to prevent part movement during placement.) Because of environmental concerns associated with the cleaning of rosin-based pastes with CFCs, water-soluble pastes have become even more prevalent for applications that require cleaning or in applications where one has yield problems with low residue or no-clean pastes and fluxes.

13.4.3 Rosin Fluxes

Rosin or colophony is a natural product that is extracted from the stumps or bark of pine trees. The composition of rosin varies from batch to batch, but a general formula is $C_{19}H_{29}COOH$. It consists mainly of abietic acid (70–85% depending on the source) with 10–15% pimaric acids. Rosins contain several percent of unsaponifiable hydrocarbons; thus for removal of rosin flux, saponifiers (a form of alkaline chemical to make the water soapy) must be added.

The rosin flux is primarily composed of natural resin extracted from the oleoresin of pine trees and refined. Rosin fluxes are inactive at room temperatures but become active when heated to soldering temperatures. They are naturally acidic (165–170 mg KOH/g equivalent). They are soluble in a variety of solvents but not water. This is the reason for using solvents, semi-aqueous solvents, or water with saponifiers to remove them.

The melting point of rosin is 172° to 175°C (342°–347°F), or just below the melting point of solder (183°C). A desirable flux should melt and become active slightly below the soldering temperature. A flux is not effective if it decomposes at soldering temperatures, however. This means, as shown in Table 13.2, that synthetic fluxes can be used at higher temperatures than rosin fluxes, since the former decompose at higher temperatures [3]. In general, rosin fluxes are weak, and to improve their activity (fluxing action), as shown in Table 13.1, the use of halide activators is required.

The general formula for oxide removal by rosin is the following:

$$RCO_2H + MX = RCO_2M + HX$$

Table 13.2 Temperature at which flux decomposition becomes significant [2]

SUBSTANCE	TEMPERATURE (°C)
Rosin	273
RMA Flux	275
FSW 32 flux[a]	258
Synthetic resin	375
RMA-type synthetic resin	373

[a]German specification

where RCO_2H is rosin in the flux ($C_{19}H_{29}COOH$ mentioned earlier)
     M = Sn, Pb, or Cu
     X = oxide, hydroxide, or carbonate

Note: See Section 13.5.3 for chemical equations applicable to oxide removal with halide.

As mentioned earlier, the rosin fluxes are also referred to as rosin (R), rosin mildly activated (RMA), and rosin activated (RA). The various categories of rosin fluxes differ basically in the concentration of the activators (halide, organic acids, amino acids, etc.). R and RMA types are generally noncorrosive, hence safe. R and RMA fluxes are not even cleaned in some applications even though they are not classified as no-clean. However, without cleaning, the reliability of assemblies may be compromised because the sticky rosin can attract dust and harmful contaminants in the field during service.

13.4.4   Low Residue or No-Clean Fluxes and Solder Pastes

Many companies, especially in Europe, have considered the rosin (R and RMA) fluxes as no-clean and have not generally cleaned them. When not cleaned, they were sticky but did not pose any reliability concerns, especially if they were halide free (R fluxes). To avoid bed-of-nails test problems, sometimes they were cleaned by brush just on the bottom of the board. However, in the United States, rosin fluxes have generally required cleaning to meet cleanliness requirements (visual, ionic, and surface insulation resistance). Thus, Europeans have been in the vanguard of the no-clean revolution.

Since the ban on CFCs, no-clean fluxes have become more common worldwide than fluxes that require cleaning. In the United States, major companies such as Intel, Northern Telecom, AT&T, and many others have adopted no-clean fluxes and solder pastes. AT&T was the developer of a precision spray fluxing concept to minimize flux residue and was one of the first major companies to use no-clean flux and solder paste. However, the use of no-clean fluxes is rare for military and high reliability applications.

As discussed earlier, rosin (colophony) is used in rosin fluxes as the solid. The abietic acid in rosin along with halides, if used, serve as the activators. In the no-clean fluxes, resins such as pentaerythritol tetrabenzoate are used. The chemical structure of pentaerythritol tetrabenzoate is shown in Figure 13.3. It has low ionics and leaves behind very low residues since it decomposes during soldering. The principle of chemical reaction with these resins will follow the essentially similar chemical reactions discussed earlier under rosin fluxes. These resins generally do not react with metal oxides below 100°C but react rapidly at temperatures above 100°C. They volatilize and decompose quickly. Like rosins, they have low solubility in water.

The no-clean fluxes are formulated with various levels (1–35%) of resins which are generally referred to as solids. The high solids content no-clean fluxes will leave excessive residues on the board which may be unacceptable cosmetically even if they pass electrical tests. However, they provide good solderability and do not require the use of inert environments for reflow. At the other extreme, there are no-clean fluxes that will not leave any visible residue. Such no-clean fluxes are also referred to as low solids fluxes (LSFs) or low residue fluxes (LRFs). Now the trend is moving toward low residue fluxes with 5%, 2%, and even 1% solids content. These very low solids fluxes have low residues and low corrosion potential.

The no-clean solder pastes are not as low in solids content as the liquid fluxes since the paste must contain some gelling agents to prevent settling of solder powders in the flux and solvents. Even though these thickening or gelling agents are in very small quantity (about 0.5%), they will leave more residue on the board than is achievable with a liquid flux.

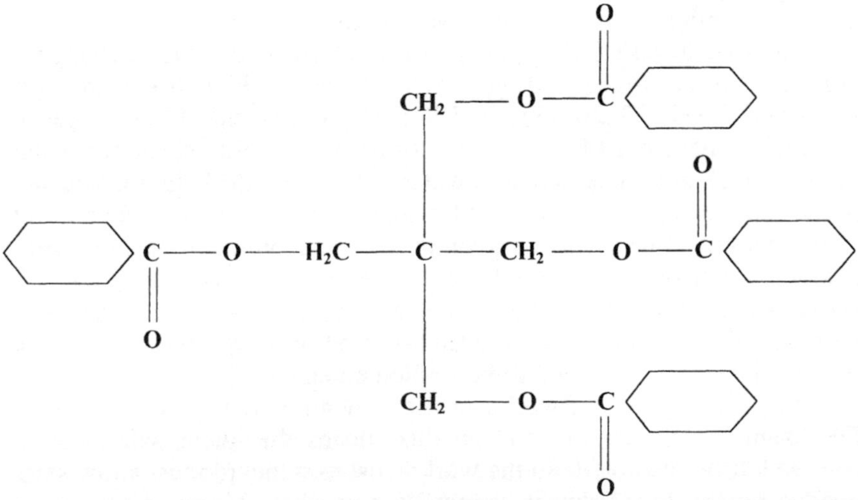

Figure 13.3 The chemical structure of pentaerythritol tetrabenzoate resin used in no-clean fluxes.

No-clean fluxes are generally halide free. They get their potency or speed of soldering action from nonhalide additives such as low molecular weight dicarboxylic acids or amines. The residues of these additives decompose during soldering and have very low potential for corrosion.

The driving force for using no-clean fluxes and solder pastes is that they save not only on cleaning costs but also on capital expenditure and floor space. In addition to these savings, environmental considerations are of major concern at most companies. CFCs have been banned after years of use and, similarly, the semi-aqueous and aqueous methods of cleaning which are acceptable today may pose problems with disposal later on. In a closed loop system, one may not be discharging effluents through the drain, but the filtered heavy metals and solder balls do need to be disposed of eventually.

13.4.4.1   *Concerns About No-Clean Flux*

No-clean fluxes are not without their problems. For example, no-clean fluxes do not require board cleaning, but fixtures such as wave solder pallets and stencils do require cleaning and the disposal of effluents. However, they do not pose anywhere near the same environmental concern as do fluxes and pastes that require cleaning.

It is very important to understand that the use of no-clean flux or paste is not a simple drop-in process. The no-clean pastes and fluxes require a very clean assembly environment and good board and component solderability. This means that suppliers also need to maintain a clean process in order to ensure solderable boards and components.

In order to take full advantage of no-clean fluxes, first a change in company culture is required. For example, the work environment needs to be very clean. The use of hand lotions and makeup that may get on the circuit board must be controlled. In order to use no-clean fluxes and pastes, which are not as active as water-soluble or RMA fluxes, steps are required to ensure that boards and components have good solderability. This means that one needs to implement either incoming inspections using solderability tests or source audits at the vendor sites to ensure solderable boards and components. Some change in inventory management may also be required. For example a FIFO (first-in first-out) system or just in time inventory system may need to be implemented.

Some changes in the workmanship standard may be required as well. For example, the use of no-clean flux means that there will be some residue left on the board. So the workmanship standard must allow some residue on the board, but it cannot be corrosive. It must pass solvent extract and surface insulation resistance tests. A good no-clean flux should not have any trouble passing these tests. The selected flux or paste must provide a balance between good solderability and an acceptable level of ionics left on the board.

One problem generally encountered when switching to no-clean flux for wave soldering is that the incidence of solder balls or webbing may go up. This happens even when nitrogen is used. The solder ball and webbing problem may be related to the undercuring of the solder mask or to some constituents in the solder mask. So compatibility of the selected flux and the solder mask must be established first. In addition, the flux must be compatible with the board surface finish, namely OSP (organic solderability protection).

In wave soldering, flux can be applied with either a foam fluxer or a spray fluxer. The foam fluxer uses air pressure through a porous stone to develop a stable head of flux foam. This means that moisture in the air will mix with the flux and, if not dried adequately, will cause spattering. Since a large area of flux is exposed, constant evaporation changes the flux density, which must be closely monitored and adjusted. Also, the air knife should be adjusted to ensure that excess flux is not being applied. For these reasons, most companies prefer spray fluxers, which apply a controlled amount of unadulterated flux uniformly. In general, spray fluxers apply less flux and hence the boards look cleaner.

No matter which application method is used, more problems will be encountered using no-clean flux in wave soldering than in reflow soldering where solder paste is used. For example, it is generally more difficult to prevent the formation of solder balls in wave soldering than in reflow soldering. If solder ball problems are encountered in reflow, it is easier to prevent their recurrence by conducting solder ball tests and correcting the thermal profiles as discussed in Chapters 9 and 12.

The no-clean process may not be compatible with conformal coating. It will also make bed-of-nails testing a little more difficult. For example, if sharp probes are not used, one may not get good contact. Also the frequency of cleaning the fixture may have to be doubled.

As mentioned earlier, because of the lower activity of no-clean fluxes and pastes, it may be necessary to use nitrogen to improve yield. For example, in wave soldering, the dross formation will be drastically reduced and hence the frequency of cleaning the machine will also be reduced. However, one should not expect to solve board and component level solderability problems by using nitrogen. It will only minimize further oxidation during the soldering process; it will not remove existing oxides from the soldering surfaces. (See Chapter 12 for further discussion.)

In summary, the no-clean process requires many changes in manufacturing procedures regardless of which type of flux is used. No-clean flux does not give us an option of being careless in handling boards or of using components and boards with less than perfect solderability. However, if adequate preparations are made, the switch to no-clean flux can be a very rewarding experience.

13.5 CONTAMINANTS AND THEIR EFFECTS

After soldering, various residues are left behind on the board surface. The overriding reason for removing them is to prevent potential electrical failures due to electromigration. The impact of different types of contaminants on electronic assemblies is summarized in Table 13.3. The residues to be removed can be broken down into three main categories: particulate, nonpolar, and polar.

Table 13.3   Impact of contaminants on printed circuit boards

CONTAMINANT	EFFECT
Polar or ionic	• Dielectric breakdown • Electrical leakage • Component circuit corrosion
Nonpolar or nonionic (rosin)	• May attract ionic contaminants through dirt • Poor conformal coating • Mealing • Poor electrical contact during bed-of-nails testing • Poor electrical contact with surface mount or edge connectors

13.5.1   Particulate Contaminants

Dust, lint, and solder balls are examples of particulate materials that must be dealt with in surface mounting. These residues are best removed by mechanical action such as the use of spray or nozzle pressure or ultrasonic cleaning.

Solder balls are a type of solder defect, not a flux residue. They are included as "contaminants" because they can be removed by cleaning. Solder balls are fairly common in surface mounting and can cause electrical shorts if equipment vibration results in the collection of many small balls at one location. (See chapter 9 for a discussion of the causes of solder balls.)

13.5.2 Nonpolar Contaminants

As summarized in Table 13.4, nonpolar and polar contaminants require nonpolar and polar solvents, respectively. Some fluxes produce only polar contaminants, some only nonpolar contaminants, and others produce both types. For example, rosin fluxes contain both polar and nonpolar contaminants; hence their complete removal calls for a solvent that is both polar and nonpolar. The following are typical nonpolar (or nonionic) insoluble residues:

- Rosin residues
- Synthetic resin
- Organic compounds from low residue/no-clean flux formulations
- Plasticizers from core flux
- Reaction products
- Oil used in wave soldering
- Grease or oil from pick-and-place and insertion machines
- Hand lotions or makeup from operators during handling
- Release agents on components
- Insoluble inorganic compounds (oxidation products)
- Rheological additives to solder pastes.

Nonpolar contaminants are removed by nonpolar solvents such as organic solvents, semi-aqueous chemicals (e.g., various Petroferm solvents), and saponifiers which dissolve them. If not removed, nonpolar contaminants will attract dust and other foreign particles in the field. The dust itself may bear deleterious polar contaminants that could cause corrosion or electromigration under the influence of temperature and humidity. Nonpolar flux residues left on the surface are an indication of poor process control in cleaning. Moreover, they may hide such defects as board damage or solder joint opens.

Table 13.4 General classification of contaminants and their solvents

CONTAMINANTS	SOLVENTS
• Polar or ionic (activators such as halides)	• Polar solvents (alcohols, water)
• Nonpolar or nonionic (rosin, resin)	• Nonpolar solvents (semi-aqueous such as Petroferm EC 7R, Prozone, or non-ozone-depleting solvents such as 3M HFE 7100 or DuPont 43-10 Vertrel)

Nonpolar flux residues are insulating by nature. They can make testing of assemblies difficult by preventing good electrical contact between the test probe and the test nodes during bed-of-nails testing. Flux on gold edge fingers or selectively gold-plated pads will also interfere with good electrical connections on surface mount connectors.

Any nonpolar flux residue left on the board will also cause poor adhesion of the conformal coating. This is visible when the board is subjected to thermal cycling and under high humidity as a mildew-like appearance called mealing or vesication. To avoid mealing, the board surface must be very clean before the conformal coating is applied.

13.5.3   Polar Contaminants

Polar residues or contaminants are defined as residues that form ions when dissolved in water. For example when a typical polar residue, "salt" or NaCl (sodium chloride), from a fingerprint on a circuit board dissolves in water, the sodium chloride molecule dissociates in water into positive sodium ions and negative chloride ions:

$$NaCl = Na^+ + Cl^-$$

Examples of polar contaminants, in addition to salts, are activators, such as halides and acids. In their ionized form, they will increase the conductivity of water. The more ions that are formed, the higher will be the electrical conductivity. They are the main causes of concern in electromigration, which can degrade the insulation resistance between conductors or, at worst, cause complete corrosion of circuits.

Electromigration produces dendritic growth between conductors and causes shorting. Three things must exist for dendritic growth: a voltage difference between the leads or terminations, the presence of moisture, and ionic material. The ionic material combines with moisture to form a conductive path when voltage is applied between the affected terminations. Here is an easy experiment to determine whether this phenomenon has occurred: apply a low voltage between two circuit lines having some flux residues; if there is any contaminant present, the dendritic growth causing an electrical short can be seen.

As we will discuss later, electromigration is the basis for one of the key cleanliness measurement methods, namely surface insulation resistance (SIR). The higher the SIR value (i.e., no electromigration), the cleaner the assembly.

In extreme cases, failure to remove corrosive flux residues may lead to corroded circuit lines [4]. There are various mechanisms of electrical failures. For example, the lead (Pb) in solder oxidizes and combines with the hydrochloric acid (HCl) in the flux to form lead chloride ($PbCl_2$), as explained by equations (1) and (2) below. This is consistent with the function of flux as noted earlier—that is, to remove oxides, in preparation of the fresh surface needed for a good intermetallic bond. However, if the lead chloride is not removed during cleaning, it reacts with the moisture (H_2O) and carbon dioxide (CO_2) in the air to form lead carbonate ($PbCO_3$), and hydrochloric acid is regenerated, in accordance with equation (3) below. The regenerated hydrochloric acid keeps the chemical reaction going until all the lead in the solder has been consumed. This weakens the solder joint and will eventually cause electrical failure. The chemical reaction proceeds as follows [5]:

$$Pb + 1/2 \; O_2 = PbO \tag{1}$$

$$PbO + 2HCL = PbCl_2 + H_2O \tag{2}$$

$$PbCl_2 + H_2O + CO_2 = PbCO_3 + 2HCl \tag{3}$$

The hydrochloric acid also attacks copper, where present, to form copper chloride [5]:

$$CuO + 2HCl = CuCl_2 + H_2O$$

Note: This equation is equally applicable to Pb, Sn, and Cu surfaces.

Dendritic growth and corrosion are the more serious effects of ionic contamination left on the board. A more benign and more common result of leaving some ionics on the board is reduced surface resistivity. This will happen in the presence of moisture. It may or may not affect the performance of the board, depending on the circuit design.

In any of these three possible scenarios of ionics on the board, the performance of the board may potentially be affected. It is not possible to eliminate ionics completely. We will discuss the level of cleanliness required in Section 13.8: Cleanliness Test Methods and Requirements.

13.6   MAJOR CONSIDERATIONS IN THE SELECTION OF CLEANING MATERIALS

There has been a great change in the selection of solvents due to the ban of chlorofluorocarbons (CFCs) at the end of 1995. CFCs were the most common solvents before it was established that they were the cause of ozone layer depletion. Now various types of solvents have been developed to replace CFCs. We will discuss them in Section 13.7. In this section our focus is on general considerations in the selection of cleaning materials.

The selection of cleaning materials involves consideration of the economics of usage, compatibility with substrates and components, performance, soil capacity, surface tension, streaking characteristics, toxicity, and environmental factors. Among all these considerations, the environmental factor has been the most important. We will discuss it first before delving into other selection issues.

13.6.1 Environmental Considerations

Chlorinated fluorocarbons (CFCs) are nonflammable, inert chemicals of low toxicity. Their typical uses, before their ban, were as follows:

- Air-conditioning and refrigeration systems   50%
- Flexible foams for furniture, bedding, foam insulation, and packaging   34%
- Cleaning of electrical and electronics components   12%
- Sterilization of medical supplies and instruments   4%

The specific CFCs and Halons that harm the environment are shown below:

NAME	FORMULA	COMMON USE
Freon 11 (trichlorofluoromethane)	$CFCl_3$	refrigerants
Freon 12 (dichlorodifluoromethane)	CF_2Cl_2	refrigerants
Freon 112 (tetrachlorodifluroethane)	$C_2F_2Cl_4$	electronics industry cleaning (mixed with alcohol)
Freon 113 (trichlorotrifluoroethane)	$C_2F_3Cl_3$	electronics industry cleaning
Freon 114 (dichlorotetrafluoroethane)	$C_2F_4Cl_2$	rarely used-aerosol propellants
Freon 115 (chloropentafluoroethane)	C_2F_5Cl	rarely used-aerosol propellants
Halon-1211	CF_2BrCl	fire extinguishers
Halon-1301	CF_3Br	fire extinguishers
Halon-2401	$C_2F_4Br_2$	fire extinguishers

It is important to note that only production of CFC is banned. Use of already produced CFCs is not banned. Even the production of CFCs is not banned in all countries. For example, third world countries are allowed to produce CFCs until the year 2010. In return for their approval of the ban on CFCs, the third world countries needed time to develop alternative chemicals indigenously instead of importing them from the West. Since they did not create the mess, they argued, they did not want to pay for the added cost of alternative chemicals, especially for refrigerants.

This loophole is being exploited by some unethical importers in the West. Since CFC 12 is cheaper than the alternative refrigerants, CFC 12 is being smuggled into the West. It is being used by unsuspecting consumers who have no idea whether their air conditioners and refrigerators are being recharged by ozone-depleting chemicals like CFC 12 or by non-ozone-depleting alternative chemicals. Let us briefly review the mechanism of ozone depletion.

In the formation of the ozone layer, first oxygen molecules (O_2) are split into two oxygen atoms (O) when they absorb the sun's ultraviolet radiation. These oxygen atoms then combine with oxygen molecules to form the ozone (O_3) layer. Ozone is a pungent, slightly blue gas—a close cousin to molecular oxygen (O_2). About 90% of the earth's ozone is located above the earth's surface in a frigid region of the atmosphere known as the stratosphere. This ozone layer acts as a shield against ultraviolet radiation.

CFCs have been proven to deplete the ozone layer. When they reach the stratosphere, ultraviolet radiation breaks chlorine atoms from the CFCs. A freed chlorine atom reacts with an ozone molecule (O_3) to form chlorine monoxide (CO) and molecular oxygen (O_2). This is the basic mechanism of ozone layer depletion. However, the chlorine atom is then split from chlorine monoxide to repeat the process. Oxygen atoms (O) formed by the splitting of O_3 by ultraviolet radiation as mentioned earlier, combine with chlorine monoxide to form O_2 and a free chlorine atom. Now the same chlorine atom is free again to repeat the formation of chlorine monoxide to repeat the cycle. So the freed chlorine atom repeatedly reacts with the ozone layer, causing its depletion. It is believed that a single chlorine atom causes depletion of 100 ozone molecules. Since some CFCs have a life time of 120 years, one can only imagine the extent of environmental damage caused by CFCs.

The ozone layer protects living things by absorbing solar ultraviolet light. Depletion of the ozone layer increases the incidence of skin cancer, and cataracts, suppresses the human immune system, and harms aquatic systems and biological organisms. Ozone depletion also compounds the greenhouse effect which heats up the climate.

CFC 113, the most commonly used CFC, was valued because of its stability and compatibility with materials used in the electronics industry. However, the very stability of this compound causes depletion of the ozone layer. Other chlorofluorocarbon compounds also deplete the ozone layer, but they are used in refrigeration systems, not in the solvent cleaning of electronics assemblies. Halons, used in fire extinguishers, also deplete the ozone layer.

On September 16, 1987, twenty-four countries, including the United States and members of the European Economic Community, signed the Montreal Protocol to control the use of CFC compounds that deplete the ozone layer. On March 14, 1988, the U.S. Senate ratified this treaty, which covers CFC 11, CFC 12, CFC 112, CFC 113, CFC 114, CFC 115, and Halon-1211, Halon-1301, and Halon-2401. Then the EPA (Environmental Protection Agency) issued guidelines for the control of these five CFCs and three halon compounds that deplete the ozone layer. State and local governments issued additional (even stricter) guidelines to control the use of these compounds.

Figure 13.4 summarizes the provisions of the Montreal Protocol to reduce consumption of CFCs. The initial plan, as shown by the dotted line in Figure 13.4, was to reduce CFC consumption in three stages to minimize the impact of the agreement on the industry. The first step involved freezing the consumption of CFCs to the level used in 1986. The second step was to reduce the consumption by 20% by mid-1993, and the third step was to reduce consumption by 50% by mid-1998. For the second and third stages, the CFC consumption in 1986 was to be used as the reference point.

The plan changed drastically once it was discovered that the problem was more serious than originally thought. As shown by the solid line in

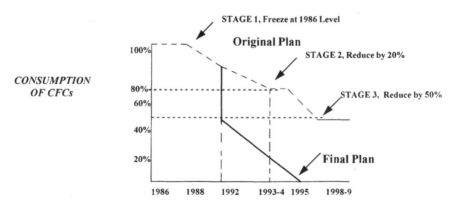

Figure 13.4 The original plan (dotted line) and the final plan (solid line) for elimination of CFCs for cleaning of electronic assemblies.

Figure 13.4, instead of reducing the consumption of CFCs to 50% of the 1986 level by 1995, the use of CFCs was banned at the end of 1995. Industry developed various alternative solvents. Some companies quit the solvent business. New companies sprang up, providing environmentally friendly alternative solvents.

Today one has various options, including drop-in replacements for CFCs that can use the same degreasers. There are also semi-aqueous and aqueous cleaning materials. These options solve the ozone depletion problem but they do add some newer problems such as disposal of solvents. For example, at 65°C, lead will dissolve 0.9 mg/L in 5 hours in water. Adding alkaline saponifiers (pH 12–13) increases this by five times. The semi-aqueous solvents will also dissolve lead. Some semi-aqueous solvents meet the EPA requirement of 0.43 mg/L and some do not, even exceeding this limit by tenfold [6]. When choosing cleaning materials, one must consider potential restrictions on disposal or recycling of solvents and water used in the cleaning machines.

Where do we go from here? We should keep in mind that even if a given material or process is acceptable today, it may not be acceptable tomorrow due to changes in environmental regulations. So in selecting a particular cleaning solvent, one should look into the future as well. This is one reason for the appeal of the no-clean process. However, no-clean flux does still require some cleaning. For example, tools such as the wave solder pallets and stencils must be cleaned and the effluents will have to be disposed of. However, this is minimal compared to using fluxes that require cleaning.

13.6.2   Other Considerations in Selecting Cleaning Materials

1. *Economics of usage.* After environmental considerations, cost is one of the most important factors, especially if more than one solvent meets performance and other technical requirements. For example, aqueous cleaning may appear to be more economical than solvent cleaning, but it may cost 20% more than solvent cleaning if all costs are taken into account. Costs of solvent cleaning also vary widely. One must consider not only the solvent cost but the total cost including the cost of equipment, energy, rework, inert environment if used, test fixtures, etc. For example, the no-clean process may appear to be the least expensive, but if the manufacturing yield, the cost of using an inert environment such as nitrogen, and the cost of converting to a clean environment

are considered, not every company will come to the same conclusion. Each application should be evaluated separately.

2.   ***In-house technical expertise.*** Some companies do not have experienced materials and process engineers to deal with the various problems that will arise with new or no-clean solvents. For example, one should expect to deal with yield problems when using no-clean flux. The problem can be more serious in wave soldering than in reflow soldering. Switching to no-clean flux increases the likelihood of solder balls on the board for both reflow and wave soldering, but the solution is more challenging in wave soldering. Similarly, when dealing with different cleaning materials, a good understanding of the chemistries involved is very important. One must have access to technical expertise before making a drastic change.

3.   ***Compatibility with equipment and components.*** A key consideration, especially with the newer semi-aqueous solvents, is compatibility of the solvent and the equipment. For example, if the tubing in the equipment is plastic, a solvent that requires stainless steel tubing would be incompatible. The semi-aqueous solvents will generally not be compatible with cleaning equipment intended for CFCs because they require a decanter to separate solvent from water. There is generally no issue in using old aqueous equipment with new aqueous solvents. When committing to a new solvent, especially a semi-aqueous solvent, one should get a list of compatible equipment from the solvent supplier.

4.   ***Performance.*** If a solvent does not perform its intended function well, all other factors, including cost, are meaningless. For surface mounting, performance is especially critical because of the tighter gap between components and substrate. Hence the user may want to emphasize cleanliness tests as a measure of performance to determine the efficacy of the cleaning solution before committing it to production.

5.   ***Surface tension.*** The lower the surface tension of the cleaning solution, the more easily it can penetrate the tight spaces between the surface mount components and the substrate. For reference, the surface tension of water is 72 dynes/cm. When a saponifier is added to the water, the surface tension is reduced to 30–35 dynes/cm. Most solvents have a surface tension of about 20 dynes/cm. Low surface tension alone is not enough, however. The cleaning solvents should not only penetrate the tight spaces between component and substrate, but should also dissolve the residue at these locations. These dissolved residues, of course, must be removed, but physical factors such as low surface tension, which

draws the solvent into tight spaces, also oppose its removal from these areas. Hence, high pressure sprays are necessary to remove solvents and the dissolved residues from under the components.

6. *Streaking properties.* Streaking is any visible dirt that remains on the board after cleaning. The amount of streaking is related to the boiling point of the solvent. Solvents with lower boiling points cause more streaking than those with higher boiling points because they tend to evaporate faster, leaving behind some of the dirt.

7. *Odor.* The odor of the solvent is a very important consideration. Some semi-aqueous solvents have quite a strong scent and may not be conducive to a good working environment. The solvents made from orange peel may be of special concern. Petroleum-based solvents do not have any issue with odor.

8. *Miscellaneous solvent properties.* When considering solvents one needs to take into account other properties such as their boiling point, flash point, and fire hazard rating. For example, a lower flash point poses a fire hazard. The Petroferm solvents, commonly used as semi-aqueous solvents, had serious problems with low flash point in their early development. Even today this is a problem with some solvents. The lower flash point solvents must be pressurized in a nitrogen environment to reduce their potential as a fire hazard. Reference 7 provides the properties of various commercially available semi-aqueous solvents.

13.7   CLEANING PROCESSES AND EQUIPMENT

The selection of a cleaning process depends upon the type of flux being used. As shown in Figure 13.5, rosin and resin fluxes can be cleaned by various types of solvents such as organic solvents or aqueous and semi-aqueous solvents. When cleaned by aqueous solvents, additives are needed. (See Section 13.7.3.) If no-clean fluxes are to be cleaned (sometimes they are), they can also be cleaned with these solvents, although, at times, special formulations may be required. Water-soluble fluxes can be cleaned with water with and without additives.

Table 13.5 provides a list of commonly used cleaning materials. For the latest list of qualified solvents and their suppliers, the user should contact industry organizations such as IPC and EMPF [8]. Let us briefly review these solvents and their associated cleaning equipment.

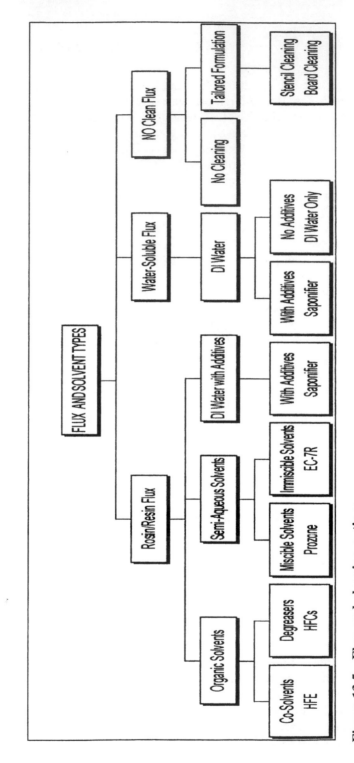

Figure 13.5 Flux and cleaning options.

Table 13.5 Examples of CFC alternative cleaning materials. Contact IPC and EMPF for the latest approved solvent and vendor contact lists [8]

Organic solvents	• Petroferm solvating agent 24 and cosolvents and rinse agents such as
	  • 3M HFE-7100 hydrofluoroether
	  • Du Pont HFC 43-10 mee hydrofluorocarbon

Semi-aqueous solvents	• British Petroleum Prozone • Dr. O.K. Wack Zestron FA (Germany) • Envirosolv KNI-2000 • Exxon Actrel ED 11 and Actrel ES • ISP Micropure CDF • Kyzen Ionox MC • Petroferm Axarel 32 and 36 • Petroferm BIOACT EC-7R, EC-Ultra
Aqueous solvents	• Alpha 2110 saponifier • Church and Dwight Armakleen E-2001 • Lonco RADS (Reactive Aqueous Defluxing System)

13.7.1   Organic Solvents (CFC Alternatives) and Cleaning Equipment

In the organic process, two solvents are used. The first is called the solvating agent and the second is the rinsing agent. In the past, rinsing agents were perfluorocarbons, such as 3M 6000, which were not ozone depleters but still required justification for their use because of their impact on global warming. Now newer rinsing agents have become available that do not have global warming potential and are also not ozone depleters. The industry is calling them true drop-in solvents as replacements for CFCs [9–12].

One of these new more environmentally friendly hydrofluorocarbon (HFC) solvents is 2,3 dihydrodecafluoropentane compound, also known as HFC 43-10 mee. It is being marketed by DuPont under the commercial name Vertrel SMT. It is a hydrofluorocarbon or HFC solvent. The hydrogen atoms in this compound help minimize its global warming potential [9]. Since it does not contain either chlorine or bromine, it does not pose any ozone depletion danger. Another example of a drop-in alternative to CFC 113 is the HFE or hydrofluoroether (C_4F9OCH_3) from 3M [10]. The commercial name for this solvent is HFE-7100. These solvents are used in the process called cosolvent cleaning, where organics such as rosin and other nonpolar solvents are dissolved by a solvating agent such as Petroferm agent 24 and the ionic contaminants are removed by rinsing agents such as HFEs and HFCs.

Since these solvents have thermal and physical properties very similar to those of CFC 113, they come closest to being "drop-in," where no modification of existing CFC degreasing equipment is necessary to use them [11, 12]. However, these replacement organic solvents are very expensive compared with what was charged for CFC 113 in the past or any semi-aqueous solvents in use today. Their use may be justified only in specialized applications, such as where the parts have blind holes or capillaries to trap water which cannot be removed. (If water cannot be removed, it may cause corrosion inside the part.) Their use may also be justified if ionic contamination requirements cannot be met by any other method or if the parts cannot withstand the heat of the drying cycle used in aqueous cleaning. In any event, such solvents are not widely used. These solvents also have some level of toxicity, so precautions must be taken in their use. The use of such solvents will be in niche applications where other methods of cleaning are not adequate.

13.7.1.1   Batch Equipment for Organic Solvents

CFCs were widely used in vapor degreasers (batch process) and in-line systems. Vapor degreasers essentially disappeared from use after the ban of CFCs but now they are coming back. In fact, some companies are using them without any modification.

A schematic of a vapor degreaser is shown in Figure 13.6. The boiling sump in the degreaser contains a mixture of the solvating agent and the rinsing agent. In order to use such a degreaser, the rinsing agent must be 100% soluble in the solvating agent. The rinse sump contains only the rinse agents such as HFEs or HFCs. Boards are passed from the boiling sump to the rinse sump. The boiling sump dissolves the rosin flux and other nonpolar contaminants and the rinse sump removes the ionics (polar contaminants).

Since these organic solvents are much more expensive than were CFCs, it may be cost effective to modify the equipment to minimize solvent loss by using additional cooling coils, a longer free board, and cover plates.

Batch cleaners or the vapor degreasers just discussed are used in the laboratory or for small volume production, as well as to clean reworked assemblies and stencils. For high volume production, in-line cleaners are used. Since in-line systems are not operator dependent like batch systems, they make it possible to achieve consistent and repeatable results. However, it should be noted that in-line solvent cleaners are not common.

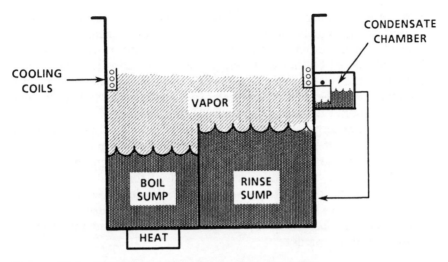

Figure 13.6   Schematic view of a batch solvent cleaner.

Organic solvents are used mostly in low volume applications where batch degreasers appear to be adequate to meet the need. In-line cleaners would be used only if justified by the volume requirements.

13.7.2   Semi-Aqueous Solvents and Cleaning Equipment

Semi-aqueous solvents are commonly used to remove rosin flux residues. Some of these solvents have been found to be more effective than the banned CFCs as measured by ionic content on the board surface and surface insulation resistance after cleaning [13].

There are various commercially available semi-aqueous solvents. Some of these solvents are immiscible in water and have lower specific gravity than water. They will tend to float over the water surface in the cleaning machines. Examples of immiscible semi-aqueous solvents are various Petroferm semi-aqueous solvents sold under such commercial names as EC-7R and Axarel 36. On the other hand, some solvents are 100% soluble in water. They have a specific gravity very close to that of water and cannot be separated. Examples of Multicore Prozone developed by British Petroleum and Kyzen Ionox MC.

The semi-aqueous solvents that are immiscible in water are used in low concentrations (3% to 5%) in water. The solvent dissolves the rosin flux and the water is used to carry it away. The solvent does not mix in water but it forms a white emulsion when mixed. (An emulsion is a mixture of two immiscible, not soluble, liquids kept in suspension with each other.) The emulsion can be thought of as the cleaning agent with flux residue in it surrounded by water. Since the flux residue dissolves only in the solvent but not in water, it does not redeposit on the circuit board during the rinsing cycle. The solvent tends to float to the top since it has a lower specific gravity than water so both the water and the solvent can be reused after decanting (separating) in a decanter.

The semi-aqueous solvents that are 100% soluble in water are used in concentrations of 5% to 85% but most commonly in a concentration of about 35%. These solvents can be used for both wash and rinse cycles although deionized (DI) water can also be used for rinsing. When the solvent is used for both wash and rinse cycles, the design of the cleaning equipment may have to be different than for the partially miscible solvents discussed above. The 100% soluble solvents can be used for cleaning stencils as well as boards.

13.7.2.1 Semi-Aqueous Cleaning Equipment

In semi-aqueous cleaning, both batch and in-line cleaning machines are used. They are more commonly used than the solvent cleaners but less commonly used than the aqueous cleaners discussed in Section 13.7.3. Let us discuss the batch semi-aqueous cleaners first.

Batch cleaners are used for low volume production and in-line cleaners are used for high volume production. A schematic diagram of a semi-aqueous batch cleaning process is shown in Figure 13.7. Different

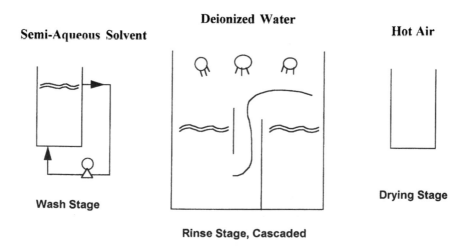

Figure 13.7   Schematic of a semi-aqueous batch cleaning system.

Figure 13.8   A photograph of a three stage semi-aqueous batch cleaning system. (Courtesy of ECD, Inc.)

equipment have different configurations but all use multiple tanks. Sometimes multiple tanks are installed in one machine and sometimes they are separate. When the drying stage is separate from the wash and rinse stages, throughput is greatly increased since different batches of boards can be in the wash and dry cycles simultaneously. Whatever their configuration, these cleaners all have a wash stage, a rinse stage, and a drying stage.

In the wash stage the assembly is lowered into the solvent or in some cases into alcohol at an elevated temperature [14]. Purging with nitrogen may be necessary if the temperature of the equipment cannot be maintained at least 10°C below the flash point of the solvent. Then the assembly is moved to the rinse tank. As shown in Figure 13.8, some systems have two rinse tanks where the deionized water from the second tank flows into the first tank. In the drying stage, the water is removed with hot air. The cleanliness results will vary depending upon the method of agitation and dwell time in the wash and the rinse tanks.

There may be further variation of such equipment. For example, as shown in Figure 13.9, equipment can emulsify the solvent for more efficient cleaning [15]. In this process, the board is washed in hot, highly agitated deionized water mixed with 3–5% semi-aqueous solvent followed

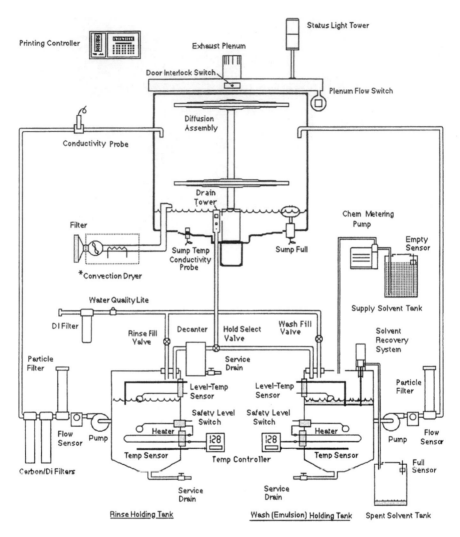

Figure 13.9 Schematic of a semi-aqueous batch cleaning system known as Emulsnator. (Courtesy of ECD, Inc.)

by a hot water rinse. The wash cycle goes through a particle filter to remove flux, solder balls, and other contaminants. The rinse cycle removes the solvent in a decanter. Then the organics and ionics in the water are removed by carbon particles and DI filters. The solvent is manually drained off the top of the water by a valve from the decanter.

Some batch and in-line systems come equipped with ultrasonic vibration capability to remove contaminants from under tight spaces. They are used not only in aqueous cleaning but also in organic solvent and semi-aqueous cleaning equipment. Although ultrasonic cleaning is controversial due to potential damage to the internal wire bonds in components, it is being used very successfully by many companies in the United States and in Japan. One reason is that now components are much more robust than they were in the past. Also, now much higher frequencies, in the range of 40 kHz rather than 25 kHz, are used. The higher frequencies are gentler to the parts.

The most prevalent method of damage was due to resonance initiated by the ultrasonic. This problem may still occur in components such as quartz clock oscillators. However, now a sweeping frequency is commonly used which reduces the risk of resonant damage by not staying at a single frequency long enough to set up the damaging resonant pattern. So now even the military allows the use of ultrasonics in certain applications.

In-line cleaning systems have multiple cleaning zones such as pre-wash and wash, followed by multiple zones of rinsing and hot air or infrared drying at the end. A schematic of in-line equipment is shown in Figure 13.10. In an in-line system the board is dipped in solvent, as opposed to a batch system where the board is lowered into the solution of water and about 5% solvent.

As shown in Figure 13.10, the solvent laden board enters the wash section, which is also called the emulsion rinse chamber. This chamber is isolated from the second and final rinse chambers and is connected to a decanter (if the solvent is immiscible in water, e.g., Petroferm EC-7R). The decanter separates the solvent and the water and sends them to the waste and emulsion rinse chambers, respectively. After the final rinse, the board enters the dryer.

In a closed loop system, the water from the second rinse chamber is recycled through carbon filters (to remove heavy metals and ionics) and an ion exchange to the final rinse chamber. Some machines use

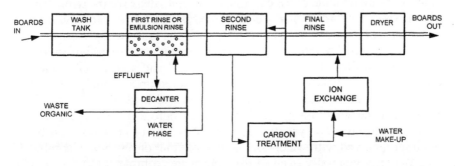

Figure 13.10   Schematic of an in-line semi-aqueous cleaning system. (Courtesy of Petroferm, Inc.)

proprietary membrane systems to purify water for reuse in the rinse sections.

If the solvent is miscible in water, there is no decanter to separate flux from water. As more and more flux dissolves in the solvent, its specific gravity increases. The equipment can be programmed to discard the solvent at a desired specific gravity by diverting it to a separate chamber for disposal. This applies to both batch and in-line systems.

13.7.3   Aqueous Cleaning Processes and Equipment

Aqueous cleaning of electronic assemblies has been used since the 1950s. Water works well for cleaning electronic assemblies, as we will show in Section 13.8.3.2. Aqueous cleaning has become even more common since CFCs have been banned. However, aqueous cleaning, like other cleaning processes, may also pose air and water pollution challenges. Lead dissolves in the wash and rinse water so users must be aware of the available means of recycling lead in an environmentally acceptable manner.

In general, water cleaning is cheaper than other methods of cleaning. However, this may not be true in all localities because of high energy demands to heat the wash and rinse water and to dry the parts. A closed loop system tends to reduce energy cost by using preheated water. The closed loop system is also very desirable from an environmental standpoint since no drain is used. However, the resin beds used for deionization and removal of heavy metals do have to be disposed off safely.

Water is a better solvent for polar (ionizable) materials than organic solvents but is a poor solvent for nonpolar materials such as rosin, grease, and oils. When used to clean rosin pastes and fluxes, a saponifier is added to the water. Saponification means "making soap." As mentioned earlier, saponifiers are alkaline chemicals that are used to dissolve the rosin. These amine-based agents chemically change the rosin and other carboxylic acids into a form of neutral soap and lower the surface tension of water.

The chemical reaction between the rosin and the saponifiers involves conversion of the ester groups in rosin to acid and acid to salt which is soluble in water. This saponification reaction requires a high pH (12–13) which causes dissolution of some of the tin and lead in water [6]. This means that the water either must be treated before disposal or must be used in a closed loop system. The rosin that cannot be saponified must be either removed with water-soluble glycol ether or physically scrubbed.

Water is generally used to dissolve water-soluble (OA) fluxes and pastes. Because of their high activity, these fluxes and pastes are very effective in providing good soldering results. However, high activity also implies corrosiveness of the flux and flux residues which must be completely removed.

Table 13.6 summarizes the effectiveness of aqueous cleaning in the removal of particulate, nonpolar, and polar contaminants. When water-soluble fluxes are used, contaminants that are not soluble in water are produced as a result of thermal decomposition of the flux. Neutralizers containing amines or ammonia and surfactants (i.e., surface-active agents) are added to provide better cleanliness levels. Surfactants are used to lower the surface tension of water. A detergent contains surfactants and other agents needed for a specific application. Soap is a familiar surfactant. When using these additives, increased wetting due to reduction in surface tension of water results in better cleaning. However, it does put an increased burden on waste management.

13.7.3.1 Deionization of Water for Cleaning

Water contains various compounds of magnesium and calcium that make it "hard"; hardness is measured in parts per million of calcium carbonate. Softeners are added to allow soap to be effective and to prevent the formation of scales inside cleaning systems that could block the nozzles. In addition, halides, sulfides, and other contaminants are present in water. Hence, to be an effective solvent, the water itself must be pure.

For purification of water, two processes—deionization and reverse osmosis (RO)—are generally used. In deionization, water is passed over resin beds to remove ionic contaminants through an ion exchange system. In this process, the dissolved minerals (ions) are removed from water by cation and anion resins in an exchange process. The cation resin removes all the positively charged ions (calcium, sodium, etc.) and replaces them with hydrogen ions (H^+). The anion resin removes all the negatively charged ions and replaces them with hydroxyls (OH^-). When H^+ and OH^- combine, they form purified water.

The cation resins and anion resins may be used in one tank or in their own separate tanks. Using separate tanks is more common. The tanks are replaced when the resistance of the DI water falls below a certain resistance level (about 2 megohms). The exact resistance level chosen depends on how the DI water is to be used. The purer the water, the higher its resistance.

Deionized water is corrosive to most metal surfaces. However, it should be used when the cleanliness level it provides is justified. It is important that only deionized water of very high resistance be used. Generally the resistance of the incoming deionized water in the equipment is over 10 megohms. It is discarded (recycled) when the resistance falls below 2 megohms. In RO systems, the water is passed through tubes of

Table 13.6 Relative effectiveness of aqueous detergents

CONTAMINANTS/ RESIDUES	CONTAMINANT REMOVAL		
	EFFEC-TIVE	MODERATELY EFFECTIVE	INEFFEC-TIVE
Polar			
Fingerprint salts	X		
Rosin activators	X		
Activator residues	X		
Cutting oils		X	
Temporary solder masks/solder stops	X		
Soldering salts	X		
Residual plating	X		
Residual etching salts	X		
Neutralizers	X		
Ethanolamine	X		
Nonpolar			
Resin fixative waves	X		
Waxes		X	
Soldering oils	X		
Cutting oils	X		
Fingerprint oils	X		
Flux resin, rosin	X		
Markings		X	
Hand cream		X	
Silicones			X
Tape residues			X
Temporary solder masks/solder stops	X		
Organic solvent films	X		
Surfactants	X		
Particulates			
Resin and fiberglass debris			X
Metal and plastic machining debris			X
Dust	X		
Handling soils		X	
Lint		X	
Insulation dust		X	
Hair/skin		X	

special materials at high pressures to separate contaminants by osmotic activity (diffusion of fluids through porous partitions).

From the disposal standpoint, there are two different ways to use the DI water in a closed loop system as shown schematically in Figure 13.11. In addition to carbon filters (Figure 13.11a), chelating resins are also used (Figure 13.11b).

Chelation is a process whereby an insoluble product is dissolved with the aid of a complexant. EDTA (ethylenediaminetetraacetic acid) and its sodium salt are typical chelating agents [16]. Chelation is used to dissolve heavy metal salts which are otherwise insoluble in water. The resultant complex or chelate is water soluble.

There are different implications of using or not using chelating agents. The user must consider the disposal options before choosing chelating. For example, when chelating agents are not used (Figure 13.11a), the

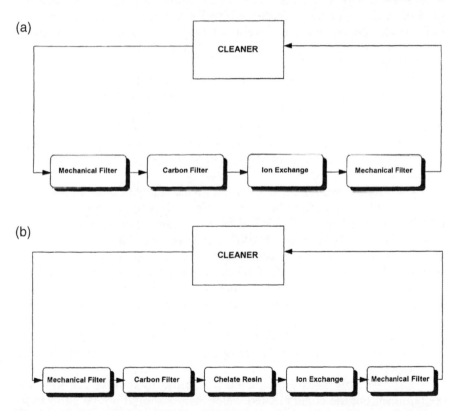

Figure 13.11   Schematic of a closed loop design (a) without chelating agent; (b) with chelating agent [16].

spent resin is not considered hazardous and can be handled by a nonregulated carrier.

On the other hand, when chelating agents are used (Figure 13.11b), the spent resin is considered hazardous solid waste and must be handled by a licensed waste hauler who will dispose of it in a special landfill. The benefit of using chelating agents is to prevent heavy metal ions from getting into the waste water system.

13.7.3.2   Aqueous Cleaning Equipment

Batch and in-line systems are the most common types of aqueous cleaning systems in use. A batch aqueous system is very much like a household dishwasher: the assemblies are loaded vertically like dishes, detergents or saponifiers are added, and cleaning and drying are accomplished in different timer-controlled cycles. Aqueous batch systems are not operator dependent like solvent batch systems; instead, the cleaning cycles are automatically controlled. Batch systems are good for low volume production.

In-line systems are very much like batch systems except that the board passes through different modules rather than different cleaning cycles. In-line systems are used for high volume production. The different modules of an in-line system are as follows: incline or horizontal loader, prerinse, recirculated wash, final rinse, air knives, drying stage, and assembly unloader. A schematic of the different modules is shown in Figure 13.12.

The prerinse module, which is used for the removal of gross contaminants, is kept separate from the recirculating stages to prevent contamination. The recirculating wash needs detergents (saponifiers) for rosin fluxes. Water heated to 140° to 160°F (60°–72°C) is used. Generally additives are not required for OA fluxes. There may be additional recirculating stages if detergents are used, if faster conveyor speeds are required, or if higher levels of cleanliness are necessary. It is very critical to have high pressure streams of water at different angles in the recirculating stage for effective removal of contaminants. Pressures from 10 to 30 PSI are used. Pressures in the high end of the range (30 PSI) are preferred if available. In addition to high pressure, the volume of water flow is also important.

The assembly goes through the final rinse to remove not the flux contaminant but the contaminated water, which should not be allowed to dry on the assembly. The clean water is blown away with air knives (more than one air knife is generally used). Finally, the clean assembly is dried and unloaded. The final air knife and the drying stages can be combined

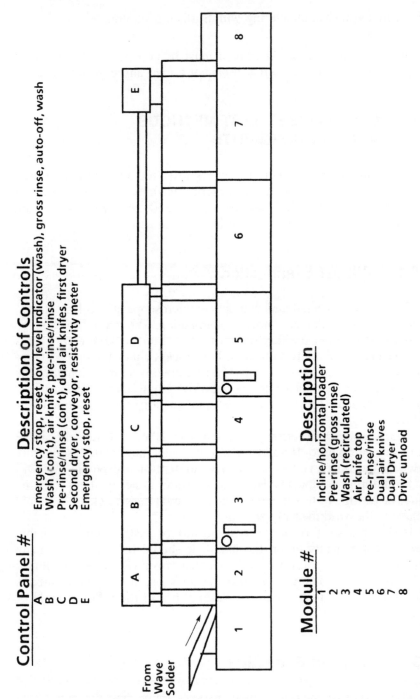

Control Panel

Description of Controls

A   Emergency stop, reset, low level indicator (wash), gross rinse, auto-off, wash
B   Wash (con't), air knife, pre-rinse/rinse
C   Pre-rinse/rinse (con't), dual air knifes, first dryer
D   Second dryer, conveyor, resistivity meter
E   Emergency stop, reset

Module

Description

1   Incline/horizontal loader
2   Pre-rinse (gross rinse)
3   Wash (recirculated)
4   Air knife top
5   Pre-rinse/rinse
6   Dual air knives
7   Dual Dryer
8   Drive unload

Figure 13.12   Schematic view of an aqueous cleaner.

635

to reduce the energy cost. The board must be completely dry before it emerges from the cleaner so that it cannot attract any contaminants.

13.8   CLEANLINESS TEST METHODS AND REQUIREMENTS

There are three different cleanliness test methods commonly used in the electronics industry. As we describe visual examination, solvent extraction, and surface insulation resistance (SIR), we will also discuss the validity of each approach for surface mount assemblies.

13.8.1   Visual Examination

Boards are inspected under an optical microscope at 2 to 10X magnification for flux residue and other contamination. The main disadvantage of this method is that the flux residues trapped under large components cannot be inspected microscopically. For process characterization, therefore, the components must be removed to allow visual examination. The visual method is qualitative and is valid for gross contamination levels only. It is difficult to visually detect and quantify minute amounts of residue.

The acceptance criterion for the visual method will depend on the type of flux being used. If no-clean or low residue flux is used, some flux residue on the board is acceptable. The amount of acceptable flux will vary for different companies. How much residue is acceptable should be decided at the time of flux qualification, and any necessary changes should be made to the workmanship standard.

If OA or rosin fluxers are used, no flux or paste should be visible under the components or on the board. It should be noted that many companies, especially in Europe, do not clean rosin (R and RMA) fluxes and hence accept visible rosin flux residues on the board. If flux is found, appropriate steps should be taken either in the reflow profile or cleaning process.

13.8.2   Solvent Extraction

The solvent extraction method involves immersing the board in a test solution and then measuring its ionic conductivity in terms of micrograms of NaCl equivalent per square unit of board area. For this method to be effective, the test solution (isopropyl alcohol and deionized water) must remove the contamination from under every component.

Without sufficient agitation of the solution, it is questionable whether all the flux residues, or indeed any, are being removed from under the components that reside close to the board surface. Recently, equipment has become available that allows agitation of solvents during the test. Commercial names for such equipment include Omegameter and Ionograph. This method is commonly used to monitor the cleanliness of conventional assemblies.

For process characterization, the component should be removed after cleaning and subjected to the solvent extraction test. The test should be conducted on a sampling of actual products, as well. If the Omegameter is used, the NaCl equivalent per square inch of board area should be less than 18 µg. Some companies have a lower requirement ($10-14$ µg/inch^2). J-STD-001 requires ionic contamination to be less than 10.06 µg per square inch (1.56 µg/cm^2) [17]. This applies to all fluxes including no-clean fluxes.

13.8.3 Surface Insulation Resistance (SIR)

The primary advantage of the SIR measurement method over other cleanliness measurement tests for surface mount assemblies is that it is direct and quantitative. Unlike the Omegameter method, which averages the contamination present over the entire board surface area, SIR can detect the presence of flux contamination of any kind that affects electromigration or conductivity even in localized regions of an assembly. The major disadvantage of SIR measurements is the need to design additional circuitry on the surface layers of the printed board to conduct the measurements effectively. (See Section 13.9: Designing for Cleaning.)

SIR measurements give very useful results when used on boards with an aggressive flux. If the boards are not properly cleaned, they will fail the SIR requirement. The J-STD-001 requirement for SIR values is 100 megohms after soldering or cleaning [17]. This applies to all fluxes including no-clean fluxes.

SIR tests can also flag adhesive cure problems. For example, if adhesive is cured too rapidly, the voids generated in the adhesive may entrap flux, which will cause SIR failures.

The SIR test method is widely used for determining the insulation or moisture resistance of laminates and other printed board materials and the compatibility of fluxes with printed board materials, as well as for cleanliness testing of printed board assemblies. Temperature and humidity test requirements are set in IPC-TM-650 for SIR measurements [18].

The trace pattern on which the SIR is measured for the above-

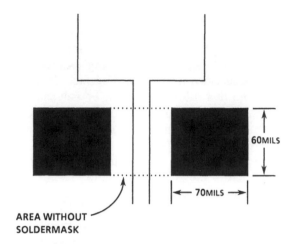

Figure 13.13   Y pattern for SIR measurements under chip components.

mentioned applications is standardized to either a typical Y pattern or a typical "comb" pattern. Since different component sizes make it difficult to standardize on one particular SIR pattern on the board, to obtain a fair indication of the contamination under the components as much of this area as possible needs to be filled with the SIR pattern. (See Figures 13.13 through 13.15).

A Y pattern (Figure 13.13) can be used under components of small area, such as chip capacitors and resistors. Comb patterns (Figure 13.14)

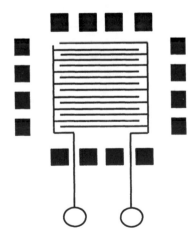

Figure 13.14   Comb pattern for SIR measurements under PLCC/ SOIC components.

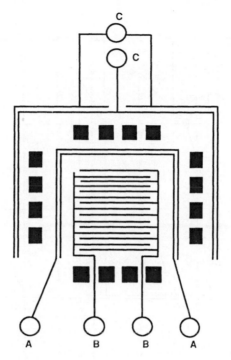

Figure 13.15 Another comb pattern for SIR measurements under PLCC/SOIC components.

are used for larger components, such as PLCCs and SOICs. The comb pattern shown in Figure 13.14 does not surround the pads, where flux may be present in excessive quantity. For this reason, another comb pattern (Figure 13.15) may be used. The pattern shown in Figure 13.15 surrounds all the pads where the paste and the flux are present.

13.8.3.1 *SIR Measurement Test Conditions*

The SIR values are dependent on test conditions. For example, they go down as the relative humidity (RH) is increased, as shown in Figure 13.16. Also, as shown in Figure 13.16, the controls have higher SIR values than the chip sites. Therefore, tolerances should be established for SIR test conditions. The test conditions that will have an important impact on SIR values are relative humidity and temperature in the test chamber, time in the chamber, purity of the deionized water used in the chamber, and the test voltage.

The IPC standards measurement conditions are listed in Table 13.7.

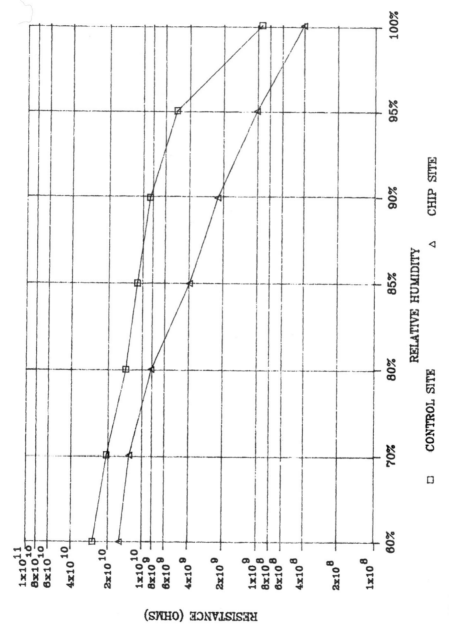

Figure 13.16 Effect of relative humidity on surface insulation resistance at 65°C on control sites and on chip sites for a Type III assembly soldered with OA flux.

Table 13.7   SIR measurement conditions for flux qualification for Class I, II, and III assemblies [2, 18]

TEST PARAMETERS	VALUES
Temperature	85 +/– 2°C
Relative humidity	85% +/– 2% RH
Polarization voltage (voltage bias)	45–50 volts DC
Test voltage	–100 volts DC
Duration	168 hours (seven days)
Test Coupon	IPC B24 test board

These test conditions are applicable for process qualification only. Some form of process control for cleanliness is necessary, especially when aggressive fluxes are used. The test conditions shown in Table 13.7 could not be used in a manufacturing environment on a production board. For example, a product could not be left in a humidity chamber for 7 days if the SIR test were to be used for daily process monitoring. The following test conditions are recommended for monitoring SIR values in a manufacturing environment on real products (or dummy test samples for process qualification):

1. RH = 86 ± 5%.
2. Temperature = 85 ± 2°C for Class III (high reliability products).
3. Time in chamber = 1 hour minimum.
4. Only pure water (deionized) with resistance greater than 4 megohms should be used in the humidity chamber. Otherwise, there may be a SIR failure due to the dirty water used for the test, not due to a bad board. Note: Resistance of water for cleaning should be greater than 2 megohms. The reason for specifying higher resistance water for test is to ensure accurate cleanliness measurements.
5. Polarization voltage (application of voltage while the board is in the chamber, generally referred to as bias) of 45–50 volts should be used.
6. The polarity must be reversed and 100 volts DC should be applied when taking the readings.
7. SIR values should be measured at both control and chip sites. A failure at the control site indicates either a serious cleaning problem or malfunctioning of the test equipment.

The equipment for measuring SIR values consists of high resistance meters, generally referred to as megohm meters, and humidity chambers.

Table 13.8 SIR requirements for *bare board* cleanliness requirements [19]. Note that J-STD-001 requires 100 megohm SIR values and 10.06 mg/in.^{2} of NaCl for all classes of assemblies.

CLASS	SIR VALUES
I	Maintain electrical function
II	100 megohm (100×10^6 ohm)
III	500 megohm (500×10^6 ohm)

To minimize operator-dependent variables, one can select equipment that combines the humidity chamber and the resistance meters in one system. If such a system is not used, an operator must monitor the test chambers to ensure that relative humidity and temperature requirements are consistently met (see next section).

How clean is clean? As mentioned earlier, J-STD-001 requires 100 megohms of SIR values for all classes of boards. However, Table 13.8 shows different cleanliness requirements for Class I, II, and III bare boards [19].

The SIR values measured on different SIR patterns with different trace geometries can be compared by normalizing the measured SIR values. Normalization is usually achieved by reporting the SIR as ohms per square. SIR in ohms per square is equivalent to surface resistivity or sheet resistance. The measured resistance in ohms is converted to resistance per square by multiplying it by the total length of one trace pattern set and dividing it by the average spacing between the two sets of traces. For example, if the trace is 0.060 inch long and the gap between the traces is 0.012 inch, the ohms per square is given by ohms (measured) times (0.060/0.012). In other words there are five squares in a rectangle of 60 × 12 mils. Thus for this case, ohms per square is 5 times the measured resistance. So if the measured value is 100 megohms, it becomes 500 megohms/square.

To take another example, if the length of the traces were 70 mils and the gap between them was 10 mils, the number of squares in a rectangle of 70 × 10 would be (70/10) = 7. In such a case the multiplication factor for the measured resistance would be 7 (or 700 megohms/square).

13.8.3.2   Application of the SIR Test

It is wrongly believed by some in the electronics industry that OA fluxes, hence water cleaning, should not be used for the wave soldering of mixed surface mount assemblies because these assemblies cannot be adequately cleaned. It is incorrectly argued that because of its high surface tension, water cannot penetrate the tight spaces between the surface mount components and the board. On the contrary, however, thousands of mixed surface mount assemblies have been shipped to customers for commercial applications with no reported problems to date [20].

For conventional assemblies, OA flux with aqueous cleaning with deionized water has been successfully used for more than a decade by leading U.S. companies such as AT&T, Boeing (yes, for military applications), Intel, IBM, Motorola, and Xerox, to name just a few. The current trend, especially in the commercial sector, is to move towards aqueous cleaning and no-clean fluxes.

As shown in Figures 13.17 and 13.18, OA fluxes can give cleanliness results similar to those of RMA rosin fluxes for mixed surface mount assemblies [4]. Table 13.9 gives the test conditions for the results shown in Figure 13.17 and 13.18.

The results shown in Figure 13.17 and 13.18 are derived from a Type II SMT assembly that had surface mount active components and through-hole components on the top side and surface mount passive components on the bottom side. The top side components were reflow (vapor phase and infrared) soldered using RMA flux in the paste. After reflow soldering of the top side surface mount components, wave soldering was used to solder the through-hole components on the top side and the chip components glued to the bottom side. Both RMA and OA fluxes during wave soldering were used to compare the cleanliness results.

Figure 13.17 and 13.18, respectively, show the SIR test results under chip components using the Y SIR test pattern and the results under PLCCs and SOICs using a comb pattern similar to the one shown in Figure 13.13 and Figure 13.14. The results in Figures 13.17 and 13.18 are the average of 28 SIR patterns (12 SIR comb patterns under the PLCCs, 12 comb patterns under SOICs, and 4 Y patterns under 1206 chip resistors). The SIR patterns were not covered with solder mask. When taking SIR measurements, no polarization voltage was applied.

The SIR results discussed so far are generally valid for water-soluble fluxes only. For example, Bredfeldt [21] found that rosin fluxes do not show SIR failures. It is believed that rosin tends to repel water and that electromigration does not take place when the assembly is subjected to high humidity conditions. Thus the SIR test may not be suitable for rosin fluxes. Nevertheless various industry and government task forces are currently using the SIR test methods for evaluating the effectiveness of various solvents and new fluxes. Note that it is possible for a rosin flux to fail the SIR test if the halide (ionics) content is high. It should also be noted that test boards that may pass the SIR requirements for rosin flux may fail the visual and the solvent extraction tests. For this reason, all three (visual, solvent extraction, and SIR) cleanliness tests should be used.

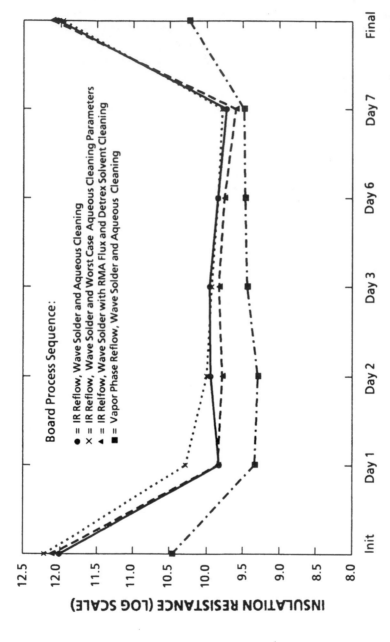

Figure 13.17 SIR results under chip resistors of a mixed Type II SMT assembly using rosin flux for reflow and rosin and OA flux for wave soldering [4].

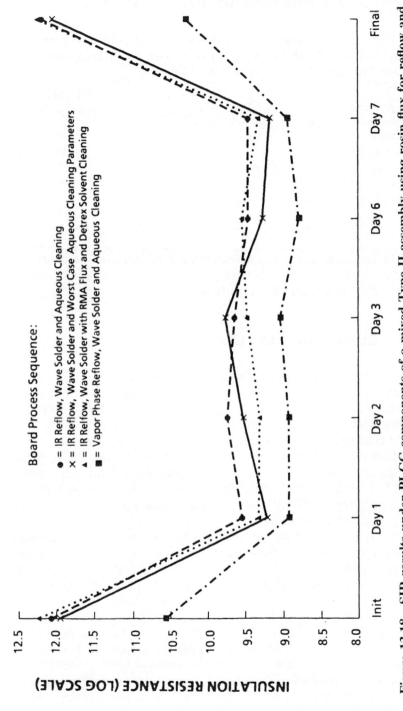

Board Process Sequence:

● = IR Reflow, Wave Solder and Aqueous Cleaning
✕ = IR Reflow, Wave Solder and Worst Case Aqueous Cleaning Parameters
▲ = IR Relfow, Wave Solder with RMA Flux and Detrex Solvent Cleaning
■ = Vapor Phase Reflow, Wave Solder and Aqueous Cleaning

INSULATION RESISTANCE (LOG SCALE)

12.5 — 12.0 — 11.5 — 11.0 — 10.5 — 10.0 — 9.5 — 9.0 — 8.5 — 8.0

Init — Day 1 — Day 2 — Day 3 — Day 6 — Day 7 — Final

Figure 13.18 SIR results under PLCC components of a mixed Type II assembly using rosin flux for reflow and rosin and OA flux for wave soldering [4].

Table 13.9 SIR measurement conditions for test results shown in Figures 13.17 and 13.18 for SMT Type II assemblies compared with conditions for SMT Type III assemblies [20]

PARAMETERS	TYPE III ASSEMBLY	TYPE II ASSEMBLY
Temperature, °C	70	85
Relative humidity, % RH	85	90
Test voltage, volts	10/100	100
Polarization voltage, volts	0/40	0
Temperature conditions	Static	Cyclic
Duration, days	10	7

After SIR measurements, components should be removed and visually inspected for entrapped flux. Naturally such a test can be conducted on process control or dummy test boards only.

13.9   DESIGNING FOR CLEANING

As the use of finer and finer pitch dense assemblies is increasing, all aspects of SMT processes including cleaning are affected. The dense assemblies and leads act as a picket fence to block the flow of solvents and contaminants. In addition, many of these fine pitch devices, especially the EIAJ (Japanese) packages sit almost flush with the board. Once flux gets under these packages (through the vias underneath them during wave soldering), no cleaning can remove the contaminants. Fortunately, many of these fine pitch devices, especially the large ones, are being replaced by ball grid arrays (BGAs) which have adequate stand-off for ease of cleaning. Refer to Figure 13.1.

There is not much that can be done about the trends in component package usage. However, if there are choices, it is best to use components that have adequate stand-offs. If not, one has to find ways to ensure cleanliness. For example when the low stand-off fine pitch packages must be used, one could minimize the number of vias underneath the package. But this is not always feasible. So after reflow soldering, the vias should be masked with a material that can dissolve in the solvent to be used or a temporary solder mask should be used to prevent seepage of flux through the vias.

One of the elements in designing for cleaning is proper component orientation on the board when the assembly enters the cleaning equipment. Designing for cleaning also should take into account shadowing of adjacent components. The guidelines for component orientation and interpackage spacing recommended in Chapter 7 apply for cleaning as well as soldering.

The use of solder mask is an important consideration. The selected solder mask should be compatible with the cleaning process. For example some dry film solder masks absorb halides from the flux and hence the board fails the cleanliness test. Another element of designing for cleaning entails the use of solder mask between surface mount lands and over traces (conductors). Using solder mask between lands of packages with 50 mil centers is not a problem. Now solder mask can be applied even between the lands of 20 mil pitch components. However, it is not common to provide solder mask between the lands of ultra fine pitch (under 20 mil pitch) packages. Hence, with only bare substrates and spacing of less than 0.010 inch, electromigration is a real possibility when ultra fine pitch packages are used. Extreme care is necessary in cleaning sites that do not have the protection of solder mask. Refer also to Chapter 7 (Solder Mask Considerations).

This brings us to another important element of designing for cleaning, namely, daily process monitoring. Some form of process control for cleanliness is necessary, especially when using aggressive fluxes. For example, based on some sampling plan, a given number of boards (either dummy test boards or actual products) should be checked for cleanliness.

It is easy to check for cleanliness by visual and solvent extraction methods. However, the use of SIR patterns must also be considered to ensure reliable products. Two SIR patterns, as shown in Figure 13.13 should be used. The traces used for SIR testing must not be covered with solder mask because the mask will prevent the flux from reaching the traces. A test pattern covered with solder mask will always pass.

At least two SIR test patterns must be used on each board. One pattern serves as the control and should have no component mounted on it. The second test pattern should be mounted with only a passive device, such as a 1206 resistor. If the SIR readings are unacceptably low (<100 megohms) on the control test pattern, either the test is invalid or the cleanliness process is really out of control.

The SIR test patterns should not be connected to any circuit. Incorporation of these test patterns will require inclusion of one additional non-functional passive component on the schematic and other necessary documentation used in the company. A passive 1206 component is generally selected because it is cheap and represents the worst-case flux entrapment problems because of its low stand-off height.

The use of a chip capacitor at the SIR site can also serve another function. Ceramic chip capacitors are susceptible to cracking in wave soldering. So instead of using a resistor, a chip capacitor could be used. This allows, in addition to testing the resistance between the SIR patterns, the checking of the capacitance values to determine if the capacitors have cracked. Cracking of capacitors indicates either a process out of control (too high delta between preheat and wave, too much pressure during placement or handling) or a poor capacitor. See Chapter 14 for further discussion.

SIR test patterns are not really necessary for reflow soldering if a milder flux is used in the solder paste. If activated rosin flux or organic acid fluxes are used in wave soldering, cleanliness test patterns should be provided on the secondary side of the board.

Insulation resistance as measured with two parallel traces at least 0.005 inch apart under the device should meet the requirements of Table 13.8 (>100 megohms or 500 megohms/square after cleaning and after a 7 day temperature-humidity test) as measured under test conditions shown in Table 13.7.

13.10   SUMMARY

We have discussed the function of flux, the principal types of flux, and the motivation for cleaning, as well as cleaning processes and equipment, no-clean flux, and cleanliness test methods and requirements. These are the basic areas for a theoretical and practical understanding of the subject.

When no-clean fluxes are used, boards do not require cleaning of course, but it may be advisable to clean the stencils for good printing. Both semi-aqueous and aqueous solvents work well for stencil cleaning. One should expect better results when stencils are cleaned.

The use of no-clean fluxes is increasing due to the environmental concerns of using fluxes that require cleaning and the disposal of used solvents containing lead. However, no clean is not a drop-in process. To be effective, it requires a change in company culture and will impact not only the internal operation but the operations of the suppliers as well.

No-clean flux is not as active as other types of flux and hence the soldering results may be less than desired unless adequate steps are taken. The use of relatively active flux helps improve the soldering yield, but the flux residues must be cleaned and the used solvents must be properly disposed of.

The selection of a particular cleaning process depends on the types of contaminant present, and these are determined by the type of flux and by the soldering and handling methods. Generally, for reflow soldered surface mount assemblies, rosin, organic acid, and no-clean fluxes are used.

For wave soldering of either mixed or through-hole mount assemblies, no-clean, rosin, and water-soluble fluxes are used. The water-soluble fluxes require water for cleaning, whereas the rosin fluxes are generally cleaned with semi-aqueous and organic solvents. Selection of a cleaning solvent should follow a careful evaluation of technical, economic, and environmental considerations.

In general, cleaning surface mount assemblies is more difficult because the tight gap between the board and components may entrap flux which may be difficult to remove during cleaning. However, if proper care is taken in selecting the cleaning processes and equipment, and if the soldering and cleaning processes are properly controlled, cleaning surface mount assemblies should not be an issue even when aggressive fluxes are used. It does need to be emphasized, however, that good process control is essential when using aggressive water-soluble fluxes.

It must be noted, however, that there is no such thing as the best flux, the best cleaning method, or the best method for determining cleanliness. These variables depend on the application. Thus, using the guidelines discussed in this chapter, the user must establish requirements for flux, cleaning, and cleanliness testing based on empirical data for a particular application. This means that the cleanliness tests (SIR, solvent extraction, and visual) should be performed on randomly selected cleaned assemblies as a check on the process. There is no substitute for good process control because if a bad board passes the cleanliness test, the failed assembly lot cannot be recalled, recleaned, or retested.

REFERENCES

1. Musselman, R. P., and Yarbrough, T. W. "The fluid dynamics of cleaning under surface mounted PWAs and hybrids." *Proceedings of NEPCON West*, February 1986, pp. 207–220.

2. J-STD-004. Requirements for soldering fluxes. Available from IPC, Northbrook, IL.

3. Rubin, W. "An inside look at flux formulations." *EP&P*, May 1987, pp. 70–72.

4. Aspandiar, R.; Prasad, P. D.; and Kreider, C. "The impact of an organic acid wave solder flux on the cleanliness of surface mount assemblies." *NEPCON Proceedings*, February 1987, pp. 277–289.

5. Romenesko, B. M. Cleaning, Surface Mount Technology. International Society for Hybrid Microelectronics Technical Monograph Series 6984-002, 1984, Reston, VA, pp. 239–243.

6. Stach, S., and Jhanji, A. Solubility of lead in new CFC replacement solvents, *Surface Mount International Proceedings*, 1994, pp. 553–555.

7. IPC-SA-61. Semi-aqueous cleaning handbook. Available from IPC, Northbrook, IL.

8. EMPF (Electronics Manufacturing Productivity Facility), 714 N. Senate Ave., Indianapolis, IN, 46202-3112.

9. Ramsey, R., and Merchant, A. "Considerations for the selection of equipment for employment with HFC-43-10mee." *Proceedings of NEPCON West*, 1996, pp. 359–369.

10. Lynch, M., and Infanti, T. "Beta site testing and qualification of HFE cleaning agents in high volume applications." *Proceedings of NEPCON West*, 1996, pp. 370–378.

11. Renee, M.; Hanson, B.; and Croes, J. "Performance testing of ODS free cleaner for precision cleaning and defluxing." *Proceedings of NEPCON West*, 1996, pp. 379–388.

12. Warren, K., and Owens, J. "The use of new fluorinated solvents in cleaning, coating, and manufacturing operations." *Proceedings of NEPCON West*, 1996, pp. 359–369.

13. Y., Erin. "A comparison of optimized process options for low, moderate and high volume semi-aqueous cleaning." *Nepcon West Proceedings*, 1993, pp. 281–292.

14. T., David. "Evaluation of batch cleaning systems for use with cyclic alcohol materials." *Surface Mount International Proceedings*, 1994, pp. 542–547.

15. E., Marilyn. "Semi-aqueous emulsion cleaning: A user's perspective." *Surface Mount International Proceedings*, 1994, pp. 548–552.

16. IPC-AC-62A. Aqueous post solder cleaning handbook. Available from IPC, Northbrook, IL.

17. J-STD-001. Requirements for soldered electrical and electronic assemblies, page 18. Available from IPC, Northbrook, IL.

18. IPC-TM-650. Test Method 2.6.3.3. Available from IPC, Northbrook, IL.

19. IPC-RB-276. Qualification and performance specification for rigid printed boards. Available from IPC, Northbrook, IL.

20. Aspandiar, R. F.; Piyarali, A.; and Prasad, P. "Is OA OK?" *Circuits Manufacturing*, April 1986, pp. 29–36.

21. Bredfeldt, K. "How well can we qualify cleanliness for surface mount assemblies?" *NEPCON Proceedings*, February 1987, p. 165.

Chapter   14

Quality Control, Inspection, Repair, and Testing

14.0   INTRODUCTION

Quality control is the key to effective competition in the international market. The dominance of the consumer electronics, steel, and automotive industries by Japan is largely attributed to the manufacture by Japanese firms of products of consistent quality. An actual difference in cost effectiveness will result when manufacturable designs are produced in tightly controlled processes.

How should one accomplish this goal? First, statistical quality control (SQC) should be used to minimize the defects in assemblies. We shall introduce the concept of SQC and then discuss the establishment of quality requirements, which generally are laid down in the areas of workmanship and material and process specifications.

Either visual or automated inspection is used to ascertain the quality of products in accordance with established requirements. When those requirements are not met, repair is needed to ensure compliance. Defects that slip by during incoming and assembly inspections are generally found by electrical tests.

The issues of quality control, inspection, test, and repair are interrelated and we must treat them as a system. If we are to manufacture reliable products at the lowest possible cost, we cannot view these areas in isolation. The resources spent in pursuit of goals in one dimension directly affect the others.

We not only need specific tools such as SQC, inspection, repair, and test to control product quality but a quality system should be in place to ensure that products are built to meet the required quality levels on a consistent basis. ISO 9000 has become one of the widely used quality systems that many companies have adopted. At the end of this chapter, we discuss ISO 9000 and the key issues in its implementation.

14.1   STATISTICAL QUALITY CONTROL

It is almost universally agreed that statistical quality control (SQC), also known as statistical process control (SPC), is the best way to collect data, analyze data, and take corrective action. SQC began as a conscious discipline in the 1920s with the in-house applications at Bell Labs invented by Walter Shewhart. During World War II the U.S. government adopted SQC wholeheartedly, and private applications proliferated. Thereafter, general disillusionment with government regulations led many American companies to discard the SQC system as a generator of unnecessary paperwork.

In Japan, however, strict regulations and the associated paperwork were seen as the path to overcoming a reputation for poor quality. SQC was not only adopted, but became a source of national pride. Now more than one-third of major Japanese companies practice SQC. The remarkable post-war recovery of Japanese industry and its subsequent dominance over worldwide markets are largely attributed to SQC.

Among other things, the Japanese followed the advice of an American named W. Edwards Deming. Dr. Deming taught Japanese managers that SQC can be a powerful tool that allows the user to measure process/product variation and then to correct a given process. The major benefit derived from SQC comes from identifying, measuring, and reducing process complexity.

Dr. Deming has proven that with the use of SQC, a slack manufacturing operation can be transformed into a smooth and efficient one, while reducing defects and increasing productivity. Japanese and American companies that have used SQC consistently have found that they can do the following [1, 2]:

1. Reduce defects by one or two orders of magnitude.
2. Shorten manufacturing times by a factor of 10.
3. Cut inventories by a factor of 2.
4. Cut floor space by a factor of 2.
5. Cut labor required on a specific job by a factor of 2.

What is SQC? According to Peter Gluckman, a student of Dr. Deming [3], SQC consists of three major elements:

- Process analysis to understand the system.
- Inductive reasoning to measure the system.
- Leadership to change the system.

Let us look at the definition of SQC in another way. The three elements of SQC are: Statistics, Quality, and Control. "Statistics" can be defined as a body of concepts, methods, and tools used for making decisions. The second element of SQC is quality. What is quality? It is conformance to requirements or fitness for use. To measure the system (the second element of Gluckman's definition), the requirements must be established first, as discussed later in this chapter. The third element of SQC is control, which Gluckman calls leadership to change the system. It involves the administration and management of systems, processes, operations, people, and machines.

To practice all three elements, Gluckman advises thinking of manufacturing operations as a series of processes rather than as a collection of unique events. He believes that reducing process complexity is faster and less costly than increasing process efficiency.

Dr. Deming believes that only 15% of an organization's problems are related to employees; the rest are built into the process and only management can address them. A simple process may involve a series of separate steps. By using SQC to gain an understanding of the various process steps, we can take action to eliminate unnecessary or unproductive ones. SQC can expose the roles played by various process steps in several

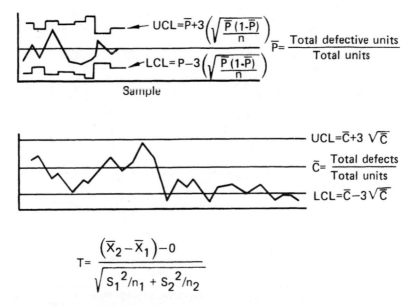

$$UCL = \bar{P} + 3\left(\sqrt{\frac{\bar{P}(1-\bar{P})}{n}}\right)$$

$$\bar{P} = \frac{\text{Total defective units}}{\text{Total units}}$$

$$LCL = P - 3\left(\sqrt{\frac{\bar{P}(1-\bar{P})}{n}}\right)$$

Sample

$$UCL = \bar{C} + 3\sqrt{\bar{C}}$$

$$\bar{C} = \frac{\text{Total defects}}{\text{Total units}}$$

$$LCL = \bar{C} - 3\sqrt{\bar{C}}$$

$$T = \frac{(\bar{X}_2 - \bar{X}_1) - 0}{\sqrt{s_1^2/n_1 + s_2^2/n_2}}$$

Figure 14.1 Examples of common SQC analysis tools: P charts (top) showing fraction defective plot (n = sample size) and C charts (bottom) showing defects per unit plot.

ways. Here we discuss the use of control charts, omitting such other SQC tools as Pareto analysis and cause-and-effect diagrams.

The control chart, a graphic record of data, is used to monitor the natural precision of a process by measuring its process average and the amount of variation from that average. Upper and lower control limits are defined based on the average. Examples of common SQC analysis tools, including control charts, are shown in Figure 14.1

The two commonly used SQC tools are the P chart (fraction defective plot) and the C chart (defects per unit). The formulas for determining upper control limits (UCL) and lower control limits (LCL) for P and C charts are shown in Figure 14.1. The details on constructing a control chart can be found in American Society for Testing and Materials (ASTM) standard STP-15D [4]. The effectiveness of SQC is best described by providing some actual examples, as discussed next.

14.2   APPLICATION OF SQC: A CASE HISTORY

SQC can be used in various applications, such as administration and management of systems, processes, operations, people, and machines. One practical application in electronics was used by the author to troubleshoot soldering defects. We will take this case history to illustrate the use of SQC to bring a soldering process under control. The example happened

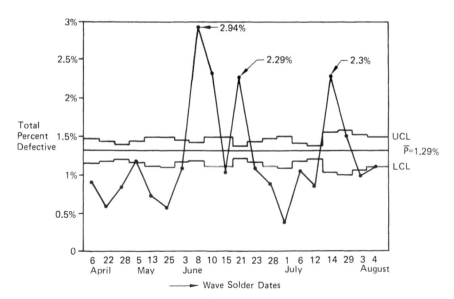

Figure 14.2   Control charts for Program A: April to August, 1983 [5].

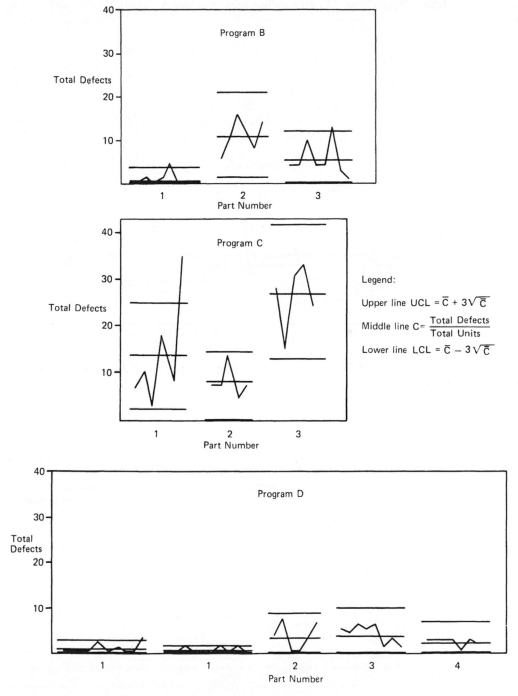

Figure 14.3   Control charts with process under control [5].

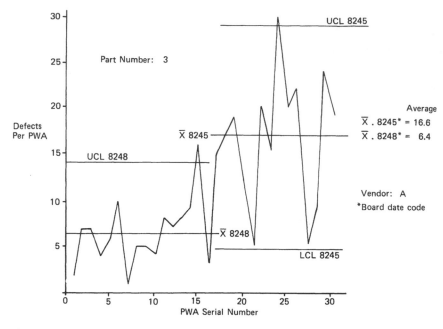

Figure 14.4 **Defect versus board serial number for PWAs with 0.3 inch lead extension [5].**

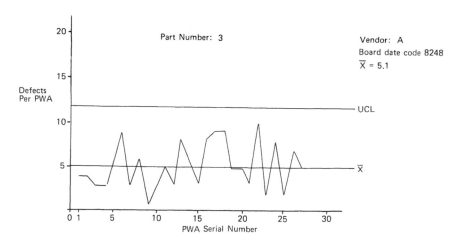

Figure 14.5 **Defect versus serial number for PWAs with 0.1 inch lead extension [5].**

to involve a through-hole board, but it can be instructive with respect to SMT applications as well. In this study, in addition to determining the various causes of wave solder defects, the human variables in inspection were identified.

First we regrouped major solder defects in various categories and identified all possible machine-controllable and non-machine-controllable process variables. We used SQC to determine the relationships, to the extent possible, between wave solder defects and wave solder variables. We collected the data most likely to be related to wave solder defects and sifted it to see which factors were correlated with fluctuations in quality.

One program, designated Program A here, had defects outside the control limits (Figure 14.2); in other programs, defects were within the control limits (Figure 14.3). Hence, we focused on Program A and three differences appeared: (a) the boards were designed with a metal core that was not present with other programs, (b) the component leads were not pretinned before soldering, as was the case on the other programs, and (c) the bare boards were purchased from outside suppliers, whereas the other boards were made in-house.

Further analysis of defects showed a predominance of voids in the Program A boards. Plated through-hole quality was found not to be the cause of voids. When the preheaters were adjusted to raise the top side board temperature from 200°F to 230°F, however, voids dropped by 50%. Program A boards had much longer leads protruding from the bottom side. Experiments were run with both long and short leads. While the defects for longer leads were outside the control limits (Figure 14.4), the defect level dropped within control limits for boards with shorter leads (Figure 14.5).

We also found that the lot that was out of control limits had poor board solderability and the lot within control limits had good board solderability. The details of this study, conducted at the Boeing Company, have been published elsewhere [5].

14.2.1   Implementing Statistical Process Control

How can SQC be implemented on the shop floor? The help of a statistician may be necessary in setting up control charts, although the people who use them daily do not have to be statisticians. (One does not need to be a mechanic to drive a car.) Because many people are intimidated by statistics, Dr. Warren Evans, who advocates statistics-free SQC, has found that developing SQC classes for a vertically integrated team consisting of managers, engineers, and operators responsible for the quality of product is the most effective method, based on five principles [6]:

1. Perfect training materials cannot compensate for improper management support.
2. The simplest SQC tools (control charts, Pareto analysis, cause-and-effect diagrams) identify 80% of the process problems, but problem-solving methods fix the problems.
3. You don't have to be a statistician to use statistics.
4. SQC requires teamwork, so people should be trained in teams. Teams need to be empowered by management to accomplish change.
5. Training should be practical and relevant and should reinforce the fact that calculating data is a minor part of the process improvement.

With these guidelines, Dr. Evans found that one team was able to reduce defects so substantially that its work station was reduced from 12 people to 2 people in 15 weeks, saving the company hundreds of thousands of dollars in rework cost [6]. Similar improvements in process are reported elsewhere [7]. This is possible when the team is motivated and is fully supported and, most importantly, assured by top management that no jobs will be lost. The real improvement starts with people because automating an out-of-control process only produces defects faster. This is especially true in SMT because automation is natural for the technology.

The road to total SQC implementation in American companies is not an easy one. The cost of data collection and reporting is generally considered to not be cost effective. This misconception can be eliminated by the appearance of tangible benefits like reducing defect rates and such intangible benefits as improved quality and better customer relations. It is better to change the process so that there is less spilled milk than to spill the milk and then try to save it. Prevention is better than cure.

In the rest of this chapter we discuss inspection, testing, and repair, three process steps that add no value to the product. If we can get the process under control, inspection, repair, and testing will be largely unnecessary. It is not realistic to assume that these functions can be entirely eliminated. However, it is realistic to assume that fancy and expensive inspection, repair, and test systems will not be necessary if the potential benefits of SQC can be tapped.

14.3   DEFECTS RELATED TO MATERIALS AND PROCESS

A good quality product calls for the establishment of design standards, materials and process standards, and solder joint quality standards. The design issues were discussed in Chapters 5, 6, and 7, the materials issues (adhesive, solder paste and solder metallurgy) were detailed in Chapters 8, 9 and 10, and the process issues were discussed in Chapters 11 (placement), 12 (soldering), and 13 (flux and cleaning).

Solder joint quality standards set the goal for quality, defining what is acceptable and what is not. The design and materials and process requirements provide the road map for achieving that goal. Some of the materials and process requirements apply to incoming materials supplied by the vendors, and others apply to in-house manufacturing.

No matter which category defects fall into, their cause and their impact on reliability and cost should be investigated. Even minor defects may be an indication of a serious problem and should be corrected, no matter what the application. According to Ralph Woodgate, it is no more expensive to produce a perfect joint than it is to produce a bad joint [8].

In through-hole mount components, the plated through hole provides electrical connection and added mechanical strength. Surface mount components, on the other hand, rely on solder joints on the surface of the board for their mechanical integrity. Thus, it is of paramount importance that every solder joint be of good quality.

14.3.1   Substrate-Related Defects

Generally speaking, the substrate is more a custom product than the components. Since, moreover, unlike the supplier of components, the substrate supplier does not enjoy the benefit of fine-tuning his process over very large volumes, users should expect more quality problems with the substrates than with components.

For substrates the major areas of concern are uniformity and thickness of solder on pads, trace width and spaces, finished via hole sizes, selective gold plating (if used), and the type and quality of solder mask to be used. Surface mounting generally forces one to use finer trace widths and smaller vias, especially for densely packed assemblies that contain primarily logic devices.

In substrates, warp and twist are of special concern for surface mounting. Warpage of 1%, which is generally acceptable for conventional assemblies, compounds the coplanarity problem for surface mount assemblies. Excessive warpage makes Z axis adjustment difficult for the pick-and-place machine and can cause solder skips in wave soldering.

If the board is used with dry film solder mask, solder skips can be expected because of the cratering effect of the thicker solder mask film around the passive devices glued to the bottom side. Dry film solder mask is also susceptible to absorption of halides or activators in wave soldering and may adversely affect the cleanliness of the board. This is especially critical if water-soluble organic acid (OA) flux is used.

Since it is difficult to apply solder mask between the lands of ultra fine pitch packages (under 20 mil pitch), these areas may be free of solder mask. If OA flux is used for the wave soldering of mixed assemblies containing fine pitch devices, absorption of OA flux in the laminate material and reduced spacing between adjacent lands may result in corrosion between the lands.

When hot air leveling is used for the application of solder to the board, solder thickness can vary from 50 to 1400 microinches. The thinner coatings can cause solderability problems and the thicker coatings can compound problems of lead coplanarity.

Thus, to minimize substrate-related defects, warpage in boards should be less than 0.5%, dry film solder mask should not be used if the assembly is going to be wave soldered, and, whenever possible, plating instead of hot air leveling should be used for substrate finish. In addition, when using fine pitch packages on boards, it may be beneficial to use alternative finishes such as OSP (organic solderability protection). (See Chapter 4, Section 4.9.) The requirements for warpage, twist, solder mask, and hot air leveling should be specified in close coordination with the supplier to ensure that they can be met.

14.3.2 Component-Related Defects

Both active and passive devices pose unique problems in surface mounting. For passive components the critical requirements are a nickel barrier underplating, solderability, and component markings. For active devices, one of the critical issues in surface mount packages is moisture induced package cracking. The plastic packages may need to be baked before reflow soldering to prevent this problem. (See Chapter 5, Section 5.7 for details.) Coplanarity of leads is also an issue for active devices. Nonplanar leads beyond 4 mils can cause solder opens or solder wicking. The lead coplanarity problem, hence the solder wicking problem, is compounded by improper component handling. Tapes and reels are generally preferred over other forms of shipping containers to preserve factory-shipped lead coplanarity.

Poor board or lead solderability can also cause open solder joints. Solder-dipped terminations or leads provide good solderability but may exacerbate the coplanarity problem. Either dipped or plated solder coatings are acceptable as long as they meet coplanarity and solderability requirements. Refer to Chapter 3 and Chapter 10 for additional details on component procurement and component lead finish, respectively.

One commonly experienced component-related defect for passive components is the cracking of ceramic multilayer capacitors. The cracks are generally so small that they are difficult to see, but they can grow in service and cause failures. (See Chapter 3.)

There are various causes of cracking, but component quality plays an important role. Some dielectric materials (e.g., Z5U) are more suscepti-ble to cracking than others (e.g., X7R). Some pick-and-place equipment exerts excessive pressure on components and may crack them. Susceptibil-ity to cracking is also determined by the thickness of the dielectric (thicker capacitors are more susceptible).

Another cause of cracking in ceramic capacitors is thermal shock during soldering. Different materials differ in their resistance to thermal shock. Bobrowsky [9] and others have used thermal shock parameters for engineering materials as defined by the following formula:

$$\text{thermal shock parameter} = \frac{K\sigma}{\alpha E}$$

where K = thermal conductivity of material
α = coefficient of thermal expansion (CTE)
σ = tensile strength
E = effective modulus of elasticity

Materials with a high thermal shock parameter can withstand thermal shock without cracking.

A multilayer ceramic chip capacitor consists of metallic electrodes, nickel-silver solder terminations, and ceramic dielectric (see Figure 3.3). The higher thermal conductivity of the electrodes and terminations causes these components to heat up more rapidly than the ceramic body. In addition, the stresses generated by CTE mismatch can crack the component because the tensile strength of ceramic is low and its modulus is high. That is, ceramic has a very low thermal shock parameter value. However, good quality ceramic materials with higher thermal shock parameters (higher thermal conductivity, lower elastic modulus, and higher tensile strength) can reduce susceptibility to cracking.

Elliott [10] states that "most manufacturers of chip capacitors use the highest quality materials available and these components easily survive the three seconds in the solder wave since they are designed to withstand soldering temperatures of 260°C (500°F) for up to 10 seconds." However, Elliott continues, "some manufacturers of chip capacitors either do not have the same process control in the fabrication of their chips, or they are using lower cost and inferior materials which cannot withstand the thermal shock in the wave. As a result, some capacitor manufacturers specify 2°C per second ramp rate during preheat and a 100°C (180°F, not 212°F) maximum difference between the end of preheat and wave temperature." Various IPC specifications also require that chip components must withstand 260°C (ambient to 260°C solder dip) for 10 seconds.

There is no problem in meeting vendors' recommendations for ramp rate and 100°C differential for infrared reflow. However, it may not be possible to follow these recommendations for every soldering process. For example, for vapor phase the reflow temperature cannot be controlled because the fluorocarbon fluid boils at a fixed temperature (215°C), and if the preheat temperature is much higher than 100°C, solder balling may reach unacceptable levels. This means that the difference between the preheat and the reflow temperatures in vapor phase stays about 115°C, and the user can do very little to reduce this differential without risking excessive solder ball problems.

This temperature difference may be even higher (120°C–125°C) in vapor phase since the board cools down somewhat after preheat, which may not be an integral part of the vapor phase system as is the case in wave soldering. If a ceramic cracking problem is found in vapor phase, poor quality of the components is the likely cause. In other words, the vendor needs to improve resistance to thermal shock by selecting materials with a high thermal shock parameter.

In wave soldering, it is difficult but not impossible to maintain the 100°C differential between preheat and wave temperatures. For example, in wave soldering the solder pot temperature is generally held around 500°F (260°C). Lowering the solder pot temperature below 475°F (246°C) for eutectic solder or raising the *top side* preheat temperature above 250°F (bottom side temperature around 260°F, or 126°C) will reduce thermal shock but may cause other problems—for example, a lower solder pot temperature may increase solder defects. Also, a preheat temperature above 126°C may exceed the glass transition temperature of the laminate material, causing board warpage.

In any event, since the component body and substrate do not reach the wave pot temperature due to short exposure (generally about 3 seconds), the 100°C differential requirement can be met on most assemblies for 250°F maximum top side preheat and 475°F wave pot settings.

A temperature differential of about 120°C between preheat and wave pot settings should be acceptable for wave soldering if component quality is good. This temperature difference is similar to what is experienced in vapor phase soldering. Thermal shock generally is not an issue for IR or convection reflow soldering. Refer to Chapter 12 for the soldering profiles of various soldering processes.

The cracking problem can be prevented by a concerted effort by component vendors, placement equipment suppliers, and users. When the land pattern design is correct (excessive solder fillet may also cause cracking), component quality is properly controlled by the vendor, pick-and-place equipment does not exert undue pressure on the components, and a correct soldering profile is used, cracking should not be a problem. One should follow the vendor's recommendation to keep the differential between preheat and reflow as low as possible but maintaining a 100°C differential is not necessary. Many major companies do not maintain a 100°C differential (they range from 120–140°C) and they do not have cracking problems. For this to happen, most importantly, only components that can meet the requirements of the manufacturing process should be used. For example, capacitors must be certified by the supplier to withstand higher delta temperatures (above 120°C). The X7R dielectric should be used when feasible instead of Z5U because the former, which contains more electrodes, heats more evenly. Also, capacitors larger than 1206 should not be wave soldered. Refer to Chapter 3 for details on dielectric materials for ceramic capacitors.

14.3.3   Adhesive-Related Defects

In surface mount assemblies that use components glued to the bottom side for wave soldering, adhesive-related defects are predominant. The defects stem from such adhesive properties as stringing characteristics, viscosity, and dispensing characteristics. Related to dispensing characteristics is the dispensing of improper amounts of adhesive: too much or too little.

Too little adhesive causes loss of components in the wave. The components may even fall off if the solder mask is changed to a formulation that is not compatible with the adhesive being used. Whenever a new solder mask is introduced on the board, its compatibility with the adhesive should be checked.

The most common adhesive-related defect is dispensing of too much adhesive. The excess adhesive can get on the lands and prevent formation of solder fillets. The circuit and part terminations should be free of adhesive. Ease of application without stringing reduces adhesive-related solder defects. Thus, it is important to be aware that any variation in dispensing characteristics or use of an adhesive with an expired shelf life can cause the adhesive to string, increasing solder defects significantly.

One of the most critical but less commonly known adhesive-related defects is board corrosion. If adhesive is cured too rapidly, voids may be formed during the curing cycle. Open voids in adhesive can trap flux that is almost impossible to remove during cleaning. If flux is not completely removed, however, exposed conductors on the assembly may corrode and cause failure in service. An appropriate adhesive cure profile is needed to prevent such a failure. See Chapter 8 (Section 8.6.1.2: Adhesive Cure Profile and Flux Entrapment). Also refer to Chapter 8 for the adhesive properties required to prevent other defects caused by this material.

14.3.4   Defects Related to Solder Paste

Solder paste requirements are critical and will vary depending on the manufacturing process [7]. Solder balls are the most common of the defects due to solder paste. They are more prevalent in vapor phase soldering than in infrared or convection soldering. Solder balls are generally caused by excessive oxide content in the paste, but misprinting and rapid heating of paste before reflow are also responsible for this phenomenon. Because the quality of solder paste changes very quickly, it must be tested daily to guard against balling. Solder balls are of special concern when no-clean flux is being used because there is no cleaning process to remove them.

Other important properties of the paste are its viscosity and slump characteristics, which affect screening and dispensing. For example, if the viscosity is too high, inconsistent or incomplete printing may result. If solder paste deposits appear to be more rounded than square, the paste may have poor slumping properties caused by low viscosity or low metal content. These properties are even more critical when printing paste for fine pitch devices. If the screen printing properties are not properly controlled, there will be excess solder bridging during soldering. Refer to Chapter 9 for the tests to be performed either by the vendor or the user to qualify a solder paste.

14.3.5   Process-Related Defects

The defects that are attributable to soldering processes include solder opens, solder wicking (solder open), tombstoning, bridges, misalignment, part movement, and solder balls. There are many causes for these defects—for example, poor solderability, lead coplanarity, poor paste printing, component misplacement, and poor soldering profile. There may be one cause or there may be several causes. Solder wicking, as discussed in Chapter 12, is a good example: it can be caused by lead coplanarity, by inadequate paste thickness, or by improper soldering profile.

Wicking is expected if lead coplanarity exceeds 0.004 inch, but vapor phase soldering may compound the problem, and the same component may provide an acceptable solder joint in an infrared or convection reflow process. Is this a component-related problem or a soldering problem? Or did insufficient paste thickness cause the problem? The correct answer in many cases involves a combination of all these.

Part movement, tombstoning, and misalignment are related with respect to their causes. All these defects are seen after soldering, but the root cause could be at the upstream processes such as solder paste printing or component placement. For example, if the paste was not tacky, a part might have moved during placement. Or the pick-and-place machine might not have been placing components correctly. Or the heating profile might have been too rapid.

Bridging is a commonly seen problem, especially in fine pitch and closely spaced chip components such as 0603 and 0402 resistors and capacitors. Reducing the stencil thickness and/or reducing the stencil aperture size and using solder mask between adjacent pads will minimize bridging.

Without belaboring the point, it is obvious that conditions generally thought to indicate a soldering problem may indeed represent either a design problem, a component problem, or a materials problem. Refer to all the relevant chapters cited at the beginning of Section 14.3 for details before deciding on the cause of a given defect and implementing corrective action.

14.3.6 Design-Related Defects

The causes of electronic assembly defects fall into three major categories: (a) defects related to incoming materials such as paste, adhesive, flux, components, and board; (b) defects related to equipment and processes; and (c) design-related defects. In the previous sections we discussed materials- and process-related defects; now we will briefly review some design-related defects.

There are work-arounds to solve some design-related problems in manufacturing. However, to achieve a permanent and cost-effective solution, it is important to design the product correctly in the first place so that manufacturing can focus on solving materials- and process-related problems.

Many of the problems discussed in the previous section may be design related. For example, poor land pattern design, incorrect component orientation, and inadequate interpackage spacing can cause bridging between the leads of a component or bridging between adjacent components. Lateral shifts, insufficient solder, tombstoning, and drawbridging can be the consequences of poor design, and poor design can make it difficult to clean, inspect, and rework an assembly. For details, see the discussion of correct design principles in Chapters 6 and 7.

A specific example of a design-related problem is bridging between the terminations of chip resistors and capacitors glued to the board for wave soldering. This problem is more prevalent in smaller chip resistors and capacitors but it has even been seen in 1206 size components.

Small resistors and capacitors (0805 and smaller) are especially susceptible to bridging underneath the component. If there is an inadequate gap between the lands, solder is drawn in by capillary action between the adhesive dot and the end termination during wave soldering. This results in bridging between the terminations underneath the component.

The occurrence of such problems has proven that 25–30 mils of gap between the lands of 0805 components may be acceptable for reflow soldering but is inadequate for wave soldering. The problem can be minimized by ensuring that the gap between the lands of 0805 chip resistors and capacitors is increased to about 40 mils. However, one can easily imagine that it is difficult to allow 35–40 mil gaps between the lands of 0603 and 0402 resistors and capacitors because of their overall small size. So it may be advisable to stay away from wave soldering such components. In addition, the use of solder mask between the lands of chip components helps minimize bridging underneath the terminations.

Some commonly seen defects, their acceptance criteria, and their potential causes are discussed next.

14.4   SOLDER JOINT QUALITY REQUIREMENTS

Workmanship standards set the goal for quality or what is acceptable and what is not. Woodgate suggests that the only acceptable joint is the perfect joint. Anything else requires corrective action [8]. What is a perfect joint? A perfect joint is a joint showing complete wetting of the lead and the pad. This requirement applies to surface mount and through-hole, and to commercial as well as military applications: reliability is important in any application. If a joint is not perfect, do not worry about repair but find the cause of the problem and fix it.

A simple solder joint standard that allows only perfect solder joints does not exist and will remain far from reality, at least in the near future. Thus we have to deal with more specific solder joint criteria. However, one does not need to develop one's own unique workmanship requirements. It is very cost effective to use industry standards.

For assembled boards, IPC-A-610 and J-STD-001 are the commonly used workmanship standards by the industry, including the military

[11,12]. IPC-A-600 is commonly used for bare boards [13]. *These standards are updated often so be sure the latest revision is being used.*

Even when using industry standards, some requirements can be modified or added if the supplier and the user agree. This may be more cost effective than developing (and maintaining) your own standard. In the event of conflicting requirements, the following order of precedence should apply:

1. Procurement as agreed upon between customer and vendor.
2. Master drawing or master assembly drawing reflecting the customer's detailed requirements.
3. When invoked by the customer or per contractual agreement, J-STD-001 and/or IPC-A-610 can be adopted or used.
4. Other documents may be used to the extent specified by the customer.

The requirements discussed in this section are based on industry standards IPC-A-610 and J-STD-001. According to these standards, the solder joint acceptability requirement depends on its use environment. The classes of use are defined as follows:

*   Class I: Consumer (radio, TV, games)
*   Class II: Commercial (computer, business)
*   Class III: High reliability, critical mission

It should be noted that both IPC-A-610 and J-STD-001 have similar requirements. J-STD-001 establishes the minimum acceptability requirements for printed board assemblies soldering. IPC-A-610, when addressing soldering, is a companion and complementary document providing pictorial interpretations of the requirements of J-STD-001. IPC-A-610 has additional criteria defining handling, mechanical requirements, and other workmanship requirements. It is useful to have both these documents easily available on the shop floor for quick and easy reference.

Before delving into some of the key accept/reject criteria established in industry standards, let us discuss some general classifications of defects. Solder defects are generally classified as major, minor, and cosmetic. Major defects such as bridges and dewetted or nonwetted joints affect circuit performance and must be corrected to permit the circuit to operate as intended. Dewetted or nonwetted joints may pass the electrical tests but the joints are sure to fail in the field.

Minor defects, such as discoloration of solder mask, white spots or measles, solder balls, component shift on pads, and insufficient or excessive solder when joints are fully wetted, may not cause failure in circuit performance. However, minor defects for one application may be major defects for another. This is where philosophical differences generally arise. For some, the so-called minor defects are an indication of the process gone haywire and are in fact considered major. For others, they are not serious defects, and accept/reject criteria may be dependent upon customer requirements.

Defects in the cosmetic category generally entail some overlap with defects in the minor class. Cosmetic defects include rough and grainy joints, exposed weave textures on the substrate, and exposed (i.e., not covered with solder) lead ends. Some examples of solder defect acceptance

Table 14.1   Solder joint requirements for rectangular chip components [11, 12]

DIMENSION	REQUIREMENT
Side overhang (dimension A)	• Target condition: No side overhang • Acceptable condition: 1/2P or 1/2W or 1.5 mm, whichever is less (1/4 for Class III)
End overhang (dimension B)	• No overhang permitted
End Joint Width (dimension C)	• Target condition: Same as P or W • Acceptable condition: 1/2P or 1/2W, whichever is less (3/4 for Class III)
Side joint width (dimension D)	• Target condition: Equal to termination length T • Acceptable condition: Not required as long as some termination is on pad (dim J)
Maximum fillet height (dimension E)	• Not specified as long as solder does not get on component body
Minimum fillet height (dimension F)	• 1/4H (termination height) or 0.5 mm, whichever is less (roughly 15 to 30 mil high)
Minimum solder thickness (dimension G)	• Wetted joint (for Class III 0.2mm or less if good cleaning can be demonstrated)
Minimum end overlap (dimension J)	• Dimension J applies only to rectangular and cylindrical components. There is no specific requirement as long as the component is long enough to bridge the gap between the leads (reasonable)

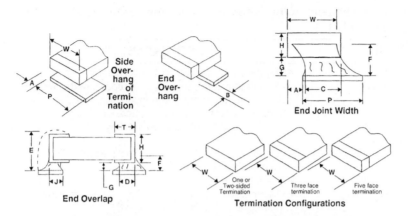

W = Width of Termination Area • T = Length of Termination • H = Height of Termination • P = Width of Land

Figure 14.6   Joint illustrations for rectangular chip components [11].

criteria are discussed below. These criteria can easily be met if the design and process guidelines discussed in the previous chapters are followed and the processes are in control.

Now let us discuss specific requirements for different types of components. For the sake of simplicity, the components are classified as rectangular, cylindrical, gull wing, J lead, butt lead, and leadless ceramic chip carrier (LCCC). We discuss the solder joint quality requirements for these components next.

14.4.1   Solder Joint Requirements for Rectangular Components

The acceptance criteria for rectangular components such as resistors and capacitors are shown in Table 14.1. The applicable illustrations are shown in Figure 14.6. Some rectangular components have solderable terminations only on the bottom (i.e., not on the sides). Such devices do not have fillet height requirements.

14.4.2   Solder Joint Requirements for Cylindrical Components

The acceptance criteria for cylindrical components generally referred to as MELF (metal electrode leadless face) are shown in Table 14.2. The applicable dimensions are shown in Figure 14.7.

Table 14.2 Solder joint requirements for cylindrical (MELF) components [11, 12]

DIMENSION	REQUIREMENTS
Side overhang (dimension A)	• Target condition: No side overhang • Acceptable condition: 25% W (compare this to 50% for rectangular chip components. Reason: Bonding area decreases rapidly for cylindrical comp)
End overhang (dimension B)	• End overhang not allowed for any class
End joint width (dimension C)	• Target condition: Greater than or equal to W • Acceptable condition: 1/2W
Side joint width (dimension D)	• Target condition: Equal to termination length T • Acceptable condition: D=50%T or termination plating length (75% for class III)
Maximum fillet height (dimension E)	• Not specified as long as solder does not get on component body
Minimum fillet height (dimension F)	• Properly wetted fillet (G+ 1/4W for class III)
Minimum solder thickness (dimension G)	• Wetted joint
Minimum end overlap (dimension J)	• For dimension J, there is no specific requirement as long as the component is long enough to bridge the gap between the lands (same as in rectangular components)

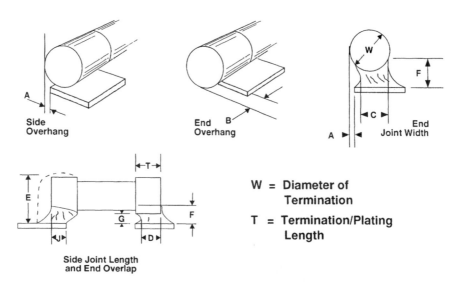

Side Overhang   End Overhang   End Joint Width

Side Joint Length and End Overlap

W = Diameter of Termination

T = Termination/Plating Length

Figure 14.7 Joint illustrations for cylindrical (MELF) components [11].

Table 14.3   Solder joint requirements for gull wing components [11, 12]

DIMENSION	REQUIREMENTS
Side overhang (Dimension A)	• Target condition: No side overhang • Acceptable condition: 1/2W or 0.5 mm, whichever is less (1/4W for Class III)
Toe overhang (dimension B)	• Acceptable for any class if minimum design conductor spacing is not violated
End joint width (dimension C)	• Target condition: C=W (lead width) • Acceptable condition: C=1/2W or 0.5 mm, whichever is less (3/4W for Class III)
Side joint length (dimension D, top of toe fillet to top of heel fillet)	• Target condition: Equal to lead width: W plus evidence of heel fillet • Acceptable condition: Equal to W
Minimum fillet height (dimension F)	• G (at toe) + 1/2 T (G+T for Class III)
Minimum solder thickness (Dimension G)	• Properly wetted fillet (no dimension)

14.4.3   Solder Joint Requirements for Gull Wing Components

The acceptance criteria for gull wing devices such as SOIC and fine pitch are shown in Table 14.3. The pitches of these devices can be 50 mil, 25 mils, and below (fine and ultra fine pitch). The applicable dimensions arc shown in Figure 14.8.

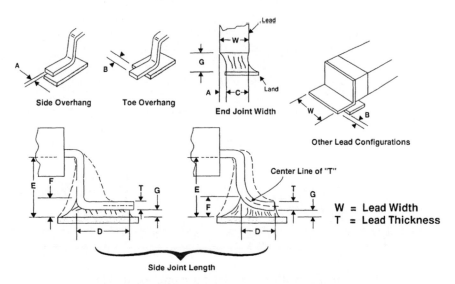

Figure 14.8   Joint illustrations for gull wing components [11].

Table 14.4 **Solder joint requirements for J-lead components [11, 12]**

DIMENSION	REQUIREMENTS
Side overhang (dimension A)	• Target condition: No side overhang • Acceptable condition: 1/2W (1/4W for Class III). W is the lead width
Toe overhang (dimension B)	• Not specified for any class • Toe sticks up, is underneath the package and looks like the heel fillet
End joint width (dimension C)	• Target condition: C=W • Acceptable condition: C=1/2W (3/4W for Class III). C is measured at thinnest point
Side joint length (dimension D, measured at the top of toe and heel fillet)	• Target condition: > 150% lead width • Acceptable condition: Equal to 1.5 W
Maximum fillet height (dimension E)	• Solder fillet not touch package body
Minimum fillet height (dimension F)	• 1/2 T (G+T for Class III)
Minimum solder thickness (dimension G)	• Properly wetted fillet

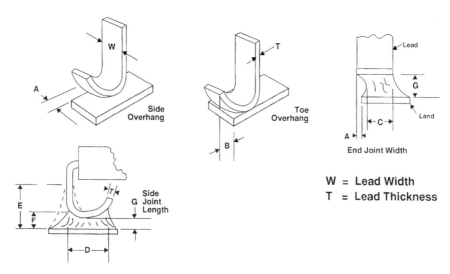

Figure 14.9 **Joint illustrations for J-lead components [11].**

14.4.4   Solder Joint Requirements for J-Lead Components

The acceptance criteria for J-lead components commonly referred to as the PLCC (plastic leaded chip carrier) are shown in Table 14.4. The applicable dimensions are shown in Figure 14.9. These types of components have lead counts from 18–84. When the lead count exceeds 84, gull wing fine pitch components are used.

14.4.5   Solder Joint Requirements for Butt Lead Components

The acceptance criteria for butt lead components are shown in Table 14.5. The applicable dimensions are shown in Figure 14.10. This type of component is generally a through-hole component which has been converted into surface mount by trimming the leads. Butt leads are used to save a process step such as hand or wave soldering. We should note that no side overhang is allowed in butt joints. One needs all the fillet possible to ensure a strong solder joint. Butt joints are not allowed for class III applications.

Table 14.5   Solder joint requirements for butt lead components [11, 12]

DIMENSION	REQUIREMENTS
Side overhang (dimension A)	• Target condition: No side overhang • Acceptable condition: No side overhang allowed (1/4 W for Class I)
Toe overhang (dimension B)	• No overhang allowed as in chip components
End joint width (dimension C)	• Target condition: C=W • Acceptable condition: C=3/4W
Side joint length (dimension D)	• Target condition: = pad length • Acceptable condition: Not specified
Maximum fillet height (dimension E)	• Solder fillet not touching package body
Minimum fillet height (Dimension F)	• 0.5 mm
Minimum solder thickness (dimension G)	• Properly wetted fillet

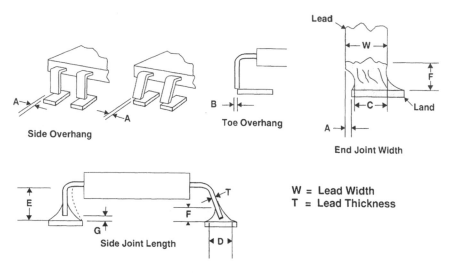

Figure 14.10 Joint illustrations for butt lead components [11].

14.4.6 Solder Joint Requirements for LCCCs

Leadless ceramic chip carriers (LCCCs) are primarily used in military applications where cooling is done through the board. It should be noted that because of CTE (coefficient of thermal expansion) mismatch between the board and the package, the solder joints are susceptible to cracking regardless of the quality of solder joint. This problem can be prevented

Table 14.6 Solder joint requirements for LCCC components [11, 12]

DIMENSION	REQUIREMENTS
Side overhang (dimension A)	• Target condition: No side overhang • Acceptable condition: 50% side overhang (25%W for Class III)
End overhang (dimension B)	• No overhang allowed for any class
End joint width (dimension C)	• Target condition: C=Castellation width W • Acceptable condition: C=1/2W (3/4 W for Class III)
Side joint length (dimension D)	• Acceptable condition: 50% minimum fillet height (F)
Maximum fillet height (dimension E)	• Solder fillet not specified for any class
Minimum fillet height (dimension F)	• Castellation height (H)
Minimum solder thickness (dimension G)	• Properly wetted fillet (for Class III 0.2 mm min for good cleaning plus wetted fillet)

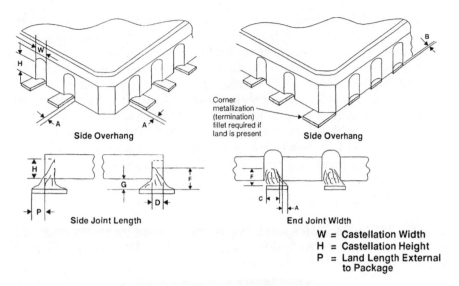

Side Overhang

Corner
metallization
(termination)
fillet required if
land is present

Side Overhang

Side Joint Length

End Joint Width

W = Castellation Width
H = Castellation Height
P = Land Length External
to Package

Figure 14.11   Joint illustrations for LCCCs [11].

by using compatible substrates with compatible CTEs. Refer to Chapters 4 and 5 for details.

The acceptance criteria for LCCCs are shown in Table 14.6. The applicable dimensions are shown in Figure 14.11.

14.4.7   Generic Solder Joint Requirements

In the previous sections we discussed dimensional requirements for solder joints. It should be noted, however, that even though the standards specify specific dimensions for solder fillets, the intent is not to physically measure every dimension and reject and then rework boards because a given dimension is slightly out of specification. The real intent is to look for a trend and take corrective action to bring the process under control so that the target condition is met. For example, if solder joints are consistently either excessive or insufficient, appropriate action in board and stencil design should be taken to correct the problem.

When in doubt, the normal practice is to touch up the joint. This is not recommended. Instead, if the joint shows evidence of good wetting, it should be accepted. The conditions leading to rejection should be solder dewetting, bridges and icicles, solder balls, solder wicking, component misalignment, and tombstoning.

Defects should be continually monitored and appropriate corrective action taken in three major areas: design, in-house processes, and vendor material quality (as exemplified by good design for manufacturability practice, board and component solderability, lead coplanarity, and adhesive and solder paste quality). As discussed earlier, design, materials, and process requirements must be established to prevent these defects.

In general, solder joints should show evidence of solder wetting all along the component-to-land interface. The termination/leads are completely wetted with solder and exhibit a concave fillet indicative of good wetting. Figure 14.12 shows photographs of preferred (top) and insufficient (bottom) solder joints in a fine pitch gull wing leaded device.

Now let us review some conditions that deviate from the preferred requirements:

1. *Minimum solder joint.* We discussed some of the dimensional requirements earlier for minimum solder fillet. As long as the

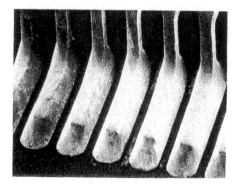

Figure 14.12   Photographs of preferred (top) and insufficient (bottom) solder joints in a fine pitch gull wing leaded device. (Photograph courtesy of Control Data Corporation.)

metal surfaces of the component terminations or leads are wetted, solder joints are acceptable.

2. ***Insufficient solder joint.*** The solder fillet is not evident or a gap is observed at the component termination/lead and land interface. Insufficient solder is the opposite of bridging; it is also a less severe case of wicking. In wave soldering, insufficient fillet is generally caused by wrong component orientation over the wave. Insufficient solder joints are not acceptable, especially if they are caused by insufficient wetting. (See Figure 14.12.)

3. ***Excess solder joint.*** On chip resistors and capacitors, solder bulges past the edge of the land and exhibits poor wetting at the land edge of the top of the part. On SOICs and PLCCs, solder bridges between the part body and the land, or eliminates the stress relief bend.

    In the wave soldering of chip components excess solder on trailing terminations is common if the components are not properly oriented during wave soldering.

    Excess solder is generally seen when the same height of solder paste is deposited on the lands of standard, fine pitch, and small components such as 0603 and 0402 on the same board. If the paste thickness is just right for the standard component, it may be too much for the smaller and finer pitch components. In such cases, either a dual-thickness stencil should be used, or the stencil apertures should be modified to deposit different volumes of paste on different lands.

4. ***Component misalignment.*** As we discussed earlier, 25–50% misalignment is acceptable. Given the accuracy of pick-and-place machines available today, this requirement is too loose. However, 50% misalignment for fine pitch is reasonable since it becomes very difficult to determine if a 6–8 mil wide fine pitch lead is off by 25% or 50%.

    The misalignment requirement should be taken into account when specifying test pads because misaligned components reduce the gap required for test probes.

    The best way to prevent misalignment is to use an accurate placement machine. Rapid heating of assemblies during reflow soldering can also cause misalignment.

5. ***Solder bridges.*** Solder bridging, another form of excess solder, is a major defect and must be corrected. A common cause of bridging of adjacent conductors is insufficient spacing between terminations, or lands or conductors.

    In reflow soldering, bridging may be caused by excessive paste thickness or excessive metal content. Alternatively, the paste may be slumping or the paste viscosity may be low. Misalignment of components or misprinting of paste can also cause bridging.

In wave soldering, bridging may be design related (e.g., the conductors are too close together) or process related (e.g., the conveyor speed is too slow, the wave geometry is improper, or there is an inadequate amount of oil in the wave or insufficient flux). Flux specific gravity and preheat temperature also influence bridging. Solder mask is generally used to achieve smaller gaps between conductors without causing bridges.

6. *Icicles.* Icicles are generally seen on the lead ends of through-hole mount devices. Their cause is generally the same as for bridges. The most effective way to eliminate icicles is to use the appropriate wave geometry.

7. *Tombstoning or drawbridging.* Tombstoning or drawbridging is a severe form of component misalignment where the components stand on their ends. Uneven lead termination width, bad land pattern design, or improper solder paste application are some of the causes. Variations of this defect include components that turn upside down or on their sides.

8. *Solder wicking.* In solder wicking, the paste travels up the lead during reflow and very little, if any, is left on the pad to form a good fillet. In extreme cases, solder wicking can cause a solder open, which becomes a major defect. Solder wicking is commonly seen in vapor phase soldering.

9. *Adhesive contamination.* Adhesive contamination is one of the major defects in wave soldering of devices glued to the bottom of the board. Adhesive that has poor dispensing characteristics (stringing/viscosity) may end up on solder pads, causing a lack of solder fillets. The same defect may result from poor dispensing accuracy in the placement/dispensing system, movement of components on the adhesive after placement, or simply an excess quantity of dispensed adhesive that ends up on the lands. See Figure 14.13.

10. *Voids in solder joint.* There is lot of concern about voids in solder joints. If solder joints are cross-sectioned, it is very common to find voids both in surface and through-hole mount joints. In general, voids in solder joints are unacceptable if the bottom of the void is not visible or if the x-ray inspection shows more than 10% of the area directly under a part is composed of voids. (X-ray inspection is not required in manufacturing.)

Only solder pastes that are not prone to void formation should be selected. In wave soldering, voids could be an indication of serious problems in the plated through holes, such as cracked

ADHESIVE ON PAD (NOT ACCEPTABLE)

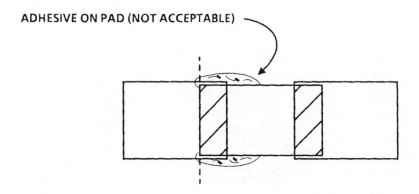

VOIDS (NOT ACCEPTABLE)

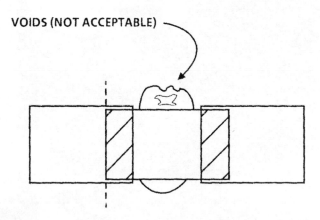

Figure 14.13 Adhesive contamination (not acceptable). Note adhesive on pad (top) and adhesive with voids (bottom).

barrels or moisture in the substrate. Commonly, boards are baked before soldering to prevent moisture-related voids. Voids may also indicate a dewetting problem. Voids in a through-hole joint are shown in Section 14.5.

11. *Other criteria.* Accept/reject criteria for other areas of assemblies are as follows:
   - Splicing wire or lead conductors with solder is not acceptable.
   - Parts, wire insulation, or printed circuit board resins that have been charred, melted, or burned should not be accepted.
   - Heater bar, probe, or hold-down tool marks should not be cause for rejection.
   - Lifted lands should not be accepted.

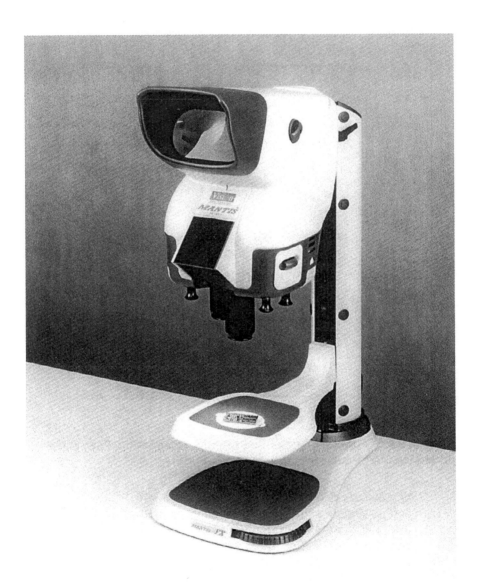

Figure 14.14   Microscope for visual inspection. (Photograph courtesy of Mantis Corp.)

14.5   SOLDER JOINT INSPECTION

Solder joint inspection is an after-the-fact step. A more effective practice is to take preventive action: that is, to implement process control to ensure that the problem does not occur. Does this mean that inspection is not necessary? Far from it. Inspection will have to continue to complete the loop on defect collection, to monitor the process and to implement corrective action so that the problem does not recur. There are two methods of inspection: visual and automated. We discuss them in the following sections.

14.5.1   Visual Inspection

The most common and widely used method of inspection is visual inspection at 2 to 10X magnification using a magnifier or a microscope (Figure 14.14). J-STD-001 requires inspection at 2–4X for all devices with lead pitches greater than 20 mils. 10X magnification is required for fine pitch devices with lead pitches 20 mils and below. Higher magnification should be used for reference only.

The main problem with visual inspection is that it is operator dependent and hence subjective. For example, in the SQC case history study discussed earlier, the same operator performed all wave soldering operations but at least ten inspectors inspected the assemblies. When it appeared that these inspectors would report different quality levels even if looking at the same assembly, we analyzed the data for three months (Figure 14.15) and found

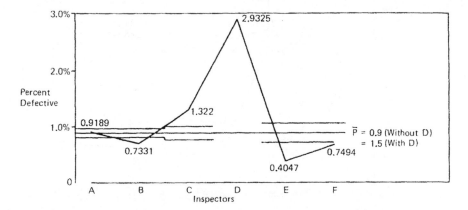

Figure 14.15   Defect rate versus inspectors for Program A, June to August, 1983 [5].

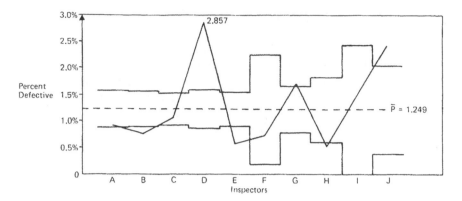

Figure 14.16 Defect rate versus inspectors for Program A, April to October, 1983 [5].

that inspector D reported three times more defects than the average of all inspectors[5]. When this analysis was expanded to defect data for nine months, a wide variation in reported defects still existed (Figure 14.16).

Further analysis of defect data was conducted by charting defects by each inspector for different part numbers of the same program. The results of only four inspectors are plotted in Figure 14.17 for clarity. Again inspector D consistently reported more defects than the others. Inspector E consistently reported fewer defects than the others, and the reports of inspectors C and G showed wide fluctuations.

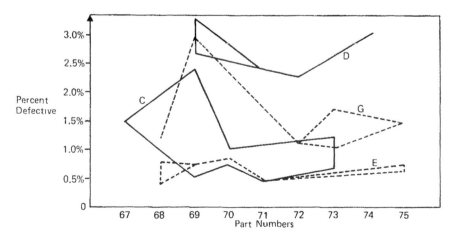

Figure 14.17 Defect rate versus part number for different inspectors for Program A, April to October, 1983 [5].

A normal response to this situation would be to reduce the human factor and switch to one of many automated inspection systems, discussed next.

14.5.2 Automated Inspection

Whether inspection is visual or performed by an automated system, should the goal be to point out the need for corrective action or to achieve process control? It is often suggested that inspection is for process control (i.e., preventive action), especially when an automated inspection system is used. Yet inspection is performed in a manner that leads to corrective action, so that the defect does not compromise product integrity.

Sometimes it is wrongly believed that automated inspection systems can be used for process control by changing the appropriate variables to correct defects on a real-time basis. This may be wishful thinking at present, since many of the changes necessary to prevent the problems require human intervention.

There is no system that can really pinpoint a given defect, hence identify its cause. Analysis of a defect requires human intervention and engineering judgment. For example, the auto inspection systems to be discussed in this section largely depend upon the mass and density of solder (insufficient, excess, or none). Defects having more than one cause, however, could never be placed into only one of these three categories to meet the needs of an automated inspection system. For example, insufficient solder can be caused by voids, by insufficient fillet, or by insufficient solder paste. The corrective actions for these defects are different, calling for human judgment and intervention.

In any event, the automated inspection systems do serve a useful purpose of at least finding the problem in an objective basis. It is up to the process engineer to determine the cause.

Despite the availability of various types of inspection systems, many companies use automated test equipment (ATE) not only for testing but also as an alternate inspection system. The limitation of the ATE equipment for inspection is that they can be used primarily only to find opens and shorts, as far manufacturing defects are concerned. So there is some push by inspection system manufacturers to replace the ATE equipment with their inspection systems. In reality, the functions of each of these equipment are different; they complement each other.

In any event, the problem in inspection is not the inherent "subjectivity" of visual or automated inspection, but in a lack of definition and understanding of what constitutes an acceptable solder connection [14].

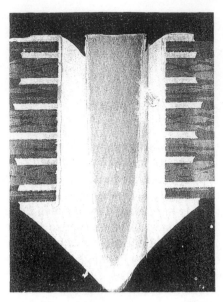

Figure 14.18 An example of unacceptable solder joint by x-ray and visual inspection but considered good by laser inspection.

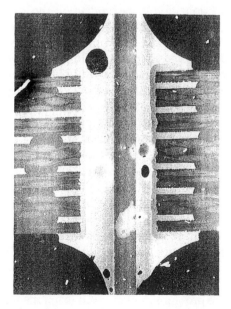

Figure 14.19 An example of a visually acceptable solder joint considered poor by laser and x-ray inspection.

This problem is compounded because the accept/reject criteria are usually not based on product performance.

Data collection can be done with either visual or automated inspection. The changeover to an automated inspection system is primarily a financial decision but the technical issues do need to be resolved first.

As the industry has been moving towards finer pitches and ball grid arrays, visual inspection of solder joints is either very difficult or impossible. Also, as just discussed, visual inspection even when feasible is very much operator dependent. So the industry has been moving towards automated inspection systems. Examples of automated inspection systems are laser inspection (e.g., Vanzetti system), transmission x-ray (e.g., IRT and Nicolet systems), and the x-ray laminography system for Four Pi-Hewlett-Packard. Let us review these systems with an emphasis on x-ray laminography, since it has become popular with the emergence of BGA technology.

14.5.2.1   Automated Laser Inspection

The laser inspection system patented by Vanzetti uses a laser to heat the solder joint. An IR detector monitors the peak solder joint temperature. Connections that become hotter than a known good joint are considered unacceptable. The Vanzetti system is very effective for soldering, but one evaluation of the Vanzetti laser inspection system showed inconsistent results compared with visual and x-ray methods. For example, the solder joint in Figure 14.18 shows no solder fillet by visual inspection and x-ray inspection revealed internal voids. This solder joint is unacceptable by the foregoing criteria, but the laser system passed it as a good solder joint.

The joint in Figure 14.19, on the other hand, was found to be visually acceptable, but x-ray and laser systems rejected it because of excessive voids. Voids are a fairly common phenomenon, especially if oil is intermixed in the solder wave. Voids are also commonly seen in surface mount assemblies.

Also, since the results of a laser system are tied to the volume of solder, a perfectly wetted joint with minimum solder may show up as hotter and hence unacceptable. Thus, the automated laser system may be "subjective" as well.

14.5.2.2   Transmission X-Ray and Scanned Beam Laminography Inspection

The transmission x-ray systems are more commonly used, but they have their share of problems as well. The images generated by transmission x-ray systems are due to the combined attenuation of the x-ray beam by every feature along its path. This type of system is good for detecting insufficient solder and voids in joints (as in detecting cavities in teeth) but it cannot detect a nonwetted joint.

Also, when inspecting the high temperature and high density noncollapsible solder balls (90 Pb/10 Sn) used in ceramic BGAs, this system has difficulty detecting problems in the lower density eutectic solder material (37 Pb/63 Sn). The high-lead content balls obscure the eutectic solder fillets. Transmission x-ray systems also cannot reliably assess the quality of joints on double-sided boards since intervening board and component material can significantly diminish the joint images.

To overcome the problem of transmission microscopy, Four Pi-Hewlett-Packard makes what is commonly referred to as the scanned beam laminography (SBL) system. It is also referred to as the x-ray laminography system. A photograph of this system is shown in Figure 14.20. SBL is a technique incorporating a microfocusing x-ray system to automatically generate and analyze cross-sectional images of solder joints.

SBL creates horizontal cross-sectional images by scanning an x-ray beam about a vertical axis in sync with a rotating x-ray detector screen

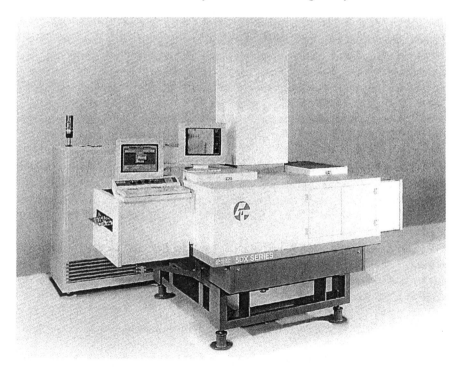

Figure 14.20 Photograph of the scanned beam laminography (SBL) inspection system. (Courtesy of Hewlett-Packard.)

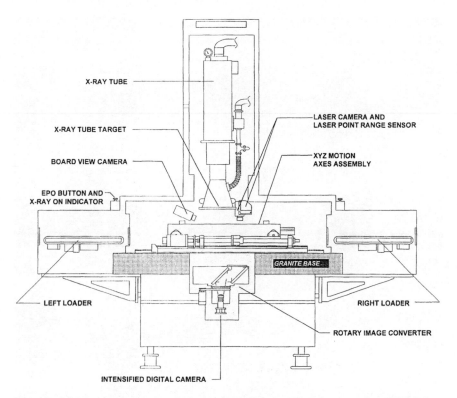

Figure 14.21 Schematic of the scanned beam laminography (SBL) inspection system. (Courtesy of Hewlett-Packard.)

[15]. By taking a cross-sectional image at different heights (pad, lead/ball and component), it can measure the amount and location of solder joint. Solder is imaged as darker pixels while less dense material (insufficient solder or voids) is lighter. The thickness of an unknown object is compared to the thickness of a set of known test samples.

As schematically shown in Figure 14.21, the off-axis image formed on the detector is integrated through one or more rotations to generate a cross-sectional image of the horizontal plane where the x-ray beam intersects the vertical axis. In addition, SBL defocuses planes above and below a given plane to blur features outside the focal plane. This is very critical in isolating the plane of interest in solder joints from other material above and below it. For example, a laminographic image (cross-section taken at pad level) detects insufficient solder (middle joint, Figure 14.22) and vias disappear because they are not in the cross-section.

On the other hand, transmission x-ray images of the same board (Figure 14.23a) cannot detect the insufficient joint because the BGA ball

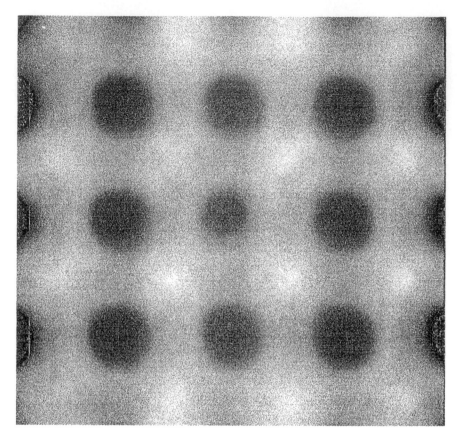

Figure 14.22   Laminographic image cross-section taken at pad level.

hides it. This will be especially true if high temperature balls are used. Figure 14.23a also shows all the vias on the board.

However, transmission x-ray images can and do provide indications of insufficient solder if the solder balls reflow. For example, as shown in Figure 14.23b, since the vias were not covered with solder mask, solder drained off into vias causing insufficient solder. As is clear from the picture, the inner row pads connected to vias are about one third the size of normal pads (outer row pads not connected to vias), indicating insufficient solder. A cross-sectional analysis showed the smaller joints to be an hourglass shape.

In most cases, in order to establish the actual size and shape of the hidden solder joint, transmission microscopy will need to be complemented with cross-sectional analysis—a destructive test. In x-ray laminography, moving the board in small vertical increments through the focal

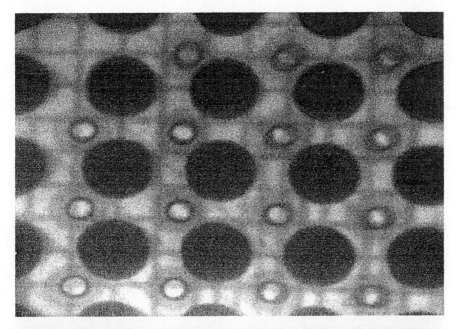

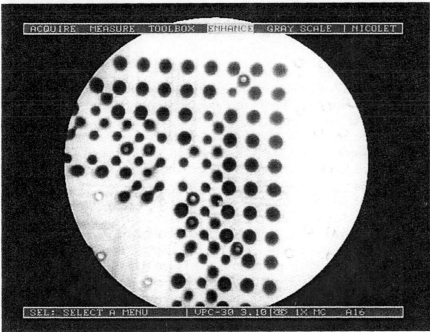

Figure 14.23 (a)   Transmission x-ray image of board in Figure 14.22; (b) Transmission x-ray of another board with insufficient solder in pads connected to vias not covered with solder mask (second row top right and inner rows). (Photograph courtesy of Intel Corporation.)

plane, allows different cross-sections of the same solder joint to be examined nondestructively. Large vertical movements of the board through the focal plane are used to examine solder joints on both sides of a double-sided board.

Using such a system, the user can establish tolerance bands for certain types of defects, such as insufficient solder, and can identify marginal joints that should be further examined. The drawback is that it takes considerable time to develop the necessary accept/reject criteria. In addition, if such a system were to be used for 100% in-line inspection, it could become a bottleneck in high-volume manufacturing.

However, many users have found the X-ray laminography system to be a good process control system when used to inspect boards on a sample basis or to inspect only a few of the critical components such as ultra fine pitch and BGAs on every board. For example, one study found the use of X-ray laminography to be a valuable tool in BGA process development and process control [16]. When used in this manner, it can be a great tool for process development, characterization, and monitoring to ensure that the processes such as paste printing, placement, and reflow are under control.

14.6   REPAIR EQUIPMENT AND PROCESSES

Despite all the effort put into design, process, and material control, there will always be some assemblies that fail to meet the requirements and must be repaired. This is true in every company in every country. Even with the conventional through-hole assemblies which have been around since the 1950s, some level of repair is performed by every company. Surface mounting is no exception. There is no reason to think that the need for rework/repair will ever completely go away.

Fortunately, contrary to popular belief, the repair and rework of surface-mounted components is easier and less damaging than it is for through-hole mount. There are many reasons for this SMT advantage. For example, repair of through-hole assemblies requires considerable operator skill, especially for removing multileaded through-hole devices, and the process itself is prone to cause board damage [17]. The major causes of thermal damage in repair/rework of through-hole assemblies are pressure on the pad, and uncontrolled temperature and time of desoldering and soldering. For example, as shown in Figure 14.24, depending on the size of the hand soldering iron tip, the tip temperature can vary widely from the set temperature. And a soldering iron tip temperature above

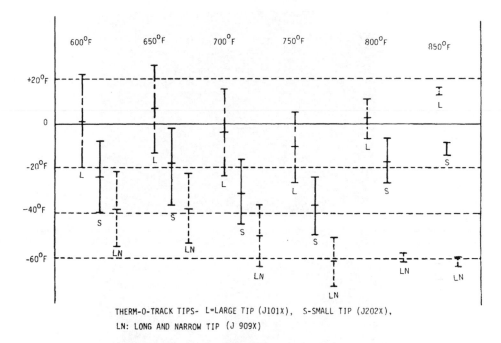

THERM-O-TRACK TIPS- L=LARGE TIP (J101X), S-SMALL TIP (J202X),
LN: LONG AND NARROW TIP (J 909X)

Figure 14.24   Variation in soldering iron tip temperatures for different sizes of tips at dial settings from 600–850°F [17].

700°F has a much higher probability of thermal damage than a tip temperature below 700°F [17].

Hand soldering, especially soldering temperatures above 700°F, can cause charring of the board surface and delamination around the pad as shown in Figure 14.25 (top figure). The bottom figure shows the cross-section of the board and clearly shows internal delamination. Pressure on the pad is the most harmful. In surface mounting, however, most repair systems use hot air, which totally eliminates this pressure.

Also, in through-hole assemblies, most of the damage is caused during removal of components. When the leads are removed from the plated through holes, they exert mechanical and thermal stress on the internal connections and can cause internal failure, as shown in Figure 14.26. The top of Figure 14.26, taken at 40X, shows the cross-section of a through-hole joint. The bottom figure, taken at 100X, clearly shows a break between the pad and circuit line. In surface mount assemblies, however, there are no plated through holes, hence no damage due to removal of leads from the board.

In addition to being less damaging, repair of surface mount devices is

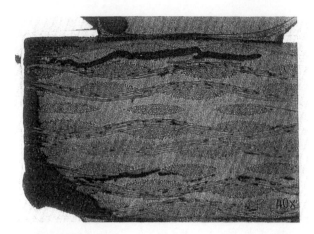

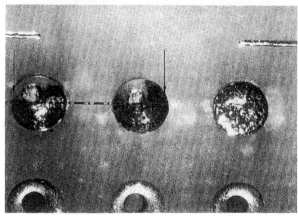

Figure 14.25 Thermal damage on board due to temperature and pressure of soldering iron. The bottom photograph shows the cross-sectional view of the top photograph. Note delamination in the bottom photograph. Due to absence of pressure, such damage is rare during rework of surface mount assemblies with hot air systems [17].

easier than repair of through-hole assemblies because operator-dependent variables are reduced in the hot air systems commonly used in SMT. Once the repair variables have been defined, it is easier for an operator to consistently make repairs with acceptable quality.

There are some problems associated with the repair of surface mount assemblies, nevertheless. For example, the interpackage spacings on surface mount boards are much smaller than those encountered on conventional boards. This means that when removing or replacing surface mount

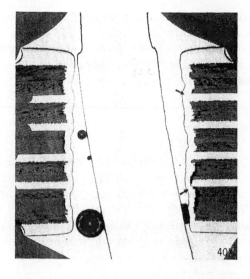

**A
Internal Pad-Tubelet
Separation**

**B
Magnified (100X) view of A**

Figure 14.26   Internal pad-tubelet separation: at 40X magnification (top) and 100X magnification (bottom) [17].

components with hot air, the correct size of hot air nozzle must be used to ensure that adjacent devices are not adversely affected.

Other repair methods such as conduction and infrared are also used for SMT. However, with these methods, the potential thermal damage to either the component or the board or both is significantly increased.

14.6.1   Repair Requirements

The repair equipment and processes for surface mounting fall into two main categories: hot air devices and conductive tips. The latter are not as common, but they are inexpensive, whereas the cost of hot air systems can be significant. Infrared rework systems are also used. We will discuss some specific rework equipment in Section 14.6.2. No matter which method is used, the following requirements should be observed to minimize thermal damage on the board.

When using hot air devices, care should be taken to prevent thermal damage to adjacent components or to the substrate. When using tip attachments to soldering irons or when using other resistance-heated conductive tips, pressure on the surface mount lands should be minimized. The basic idea is to prevent thermal damage to the substrate (Figures 14.25 and 14.26) or adjacent parts.

The solder fillets resulting from the repair/rework procedures should meet the requirements of the solder joint criteria discussed in Section 14.5. The following specific guidelines should be used to prevent thermal damage and to produce an acceptable solder joint after repair/rework.

1.   It is recommended that the number of times a part is removed and then replaced be kept to a minimum (maximum two times) to prevent internal thermal damage in the printed circuit board [17].
2.   Preheating of the substrate for 30 +/−10 minutes at 94°C (200°F) is recommended during component removal. An oven or a built-in preheating system in the repair machine or a stand-alone preheat system may be used for this purpose. See Section 14.6.4 (BGA Repair) for a discussion on different preheat systems.

   Preheating is generally recommended for removal of devices, especially if they are connected to large heat sinks such as power or ground planes, or metal cores, or if they have thicker leads.
3.   The tip temperature of the tip attachments or conductive tips should not exceed 370°C (700°F). The potential for thermal damage increases significantly above 700°F [17].
4.   Desoldering time (heat source on land) should not exceed 3 seconds when using conductive tips. Figure 14.27 shows the tempera-

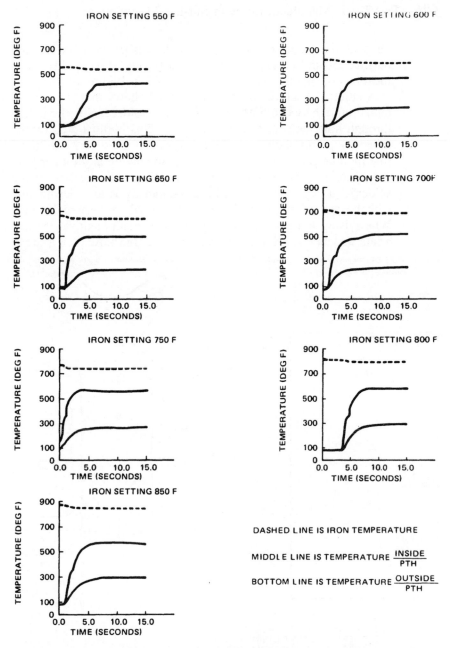

Figure 14.27 Relationship between soldering iron tip and plated through-hole temperature. Note that the solder joint melted within 1.5 seconds (inflection points in the middle curves) for all tip temperatures. See Figure 14.28 for locations of thermocouples [17].

tures of the soldering iron tip and the temperature inside and outside (50 mils away) of the plated through hole (PTH). The locations of thermocouples are illustrated in Figure 14.28.

A careful observation of Figure 14.27 shows that for any tip temperature between 550° and 850°F the solder inside the plated through hole melts within 1.5 seconds. This is illustrated by the inflection points shown in all the curves of Figure 14.27 at 361°F (the melting point of eutectic solder). So staying any longer than 3 seconds is not necessary for accomplishing the reflow of the solder joint. An unnecessarily longer soldering or desoldering time with the iron tip simply increases the potential for thermal damage.

5. When using hot air removal/replacement equipment, an appropriate attachment should be used to direct the flow of hot air to

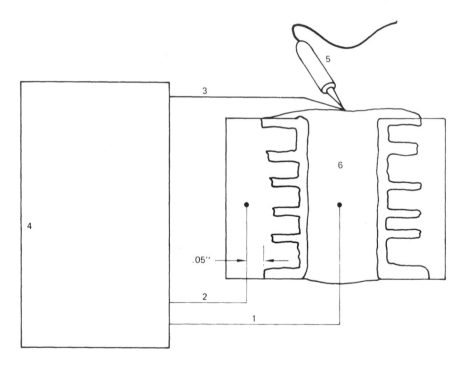

1. THERMOCOUPLE INSIDE PTH.
2. THERMOCOUPLE OUTSIDE PTH.
3. THERMOCOUPLE WELDED TO IRON TIP.
4. TEMPERATURE RECORDER CONNECTED TO H.P. 9825A SYSTEM.
5. SOLDERING IRON.
6. SOLDER PLUG.

Figure 14.28 Schematic diagram of the thermocouple locations for recording of temperatures shown in Figure 14.27 [17].

the component to be removed or replaced. The time and temperature for desoldering/soldering should permit the rapid removal of parts (< 20 seconds) without charring or burning flux or damaging the board. Some larger PLCCs require longer than 20 seconds to remove. This may result in charred flux, which is difficult to remove. Some surface mount packages may need as much as 50 seconds for reflow by hot air. Longer times tend to discolor the board surface. (Refer to Section 14.6.5: Rework Profiles.)

6. Components should be cleaned immediately after rework.

7. Only low activity fluxes (e.g., rosin or no clean) should be used for repair/rework.

8. Wires of 30 gauge or thinner should be used to accommodate changes in design. Whenever possible, via holes should be used to solder in jumper wires. If this is not possible, the J or gull wing wire leads of the surface mount packages should be used to solder the jumper wires. The wires should be hooked onto the leads but the solder fillet in the hook of the wire should be as small as possible, to avoid stiffening the leads.

Butt joints or lap joints on the solder pads can also be used instead of hooks on J or gull wing wire leads. However, in this case, the jumper wire must be glued to the board with adhesive to prevent detachment or breaking of the butt or lap joints.

9. Parts that are bonded to the substrate with adhesive should be removed with a tool that will cut through the adhesive bond or by holding the part rigid and twisting the part body, so that the bond between the part and the adhesive is sheared, keeping the part body intact. A gentle rocking or prying action should be used to pry the part loose from the board.

Residual adhesive left after part removal may interfere with the new part replacement. Therefore, the residue should be carefully scraped from the surface of the substrate. A uniform, thin layer of residue may be acceptable if it does not interfere with the placement of the component. All adhesive should be removed from the lands to achieve an acceptable solder fillet.

14.6.2　Soldering Irons for Surface Mount Repair

Soldering iron tip attachments are used in conjunction with the soldering iron itself for the desoldering of components from the substrate. The attachments should be made of copper and tin-plated after forming. They can be purchased or built in-house to fit the specific part to be desoldered. The tip attachments should be shaped such that all the solder joints on a

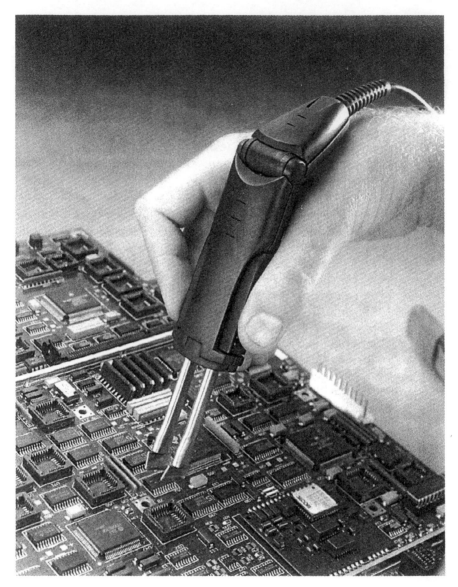

Figure 14.29   Conductive tip for SOIC removal. (Photograph courtesy of Metcal Corporation).

particular device are heated simultaneously to the melting point of the solder, and the part should be lifted either by the tool itself or by other aids, such as tweezers.

The size and shape of the attachments will vary with the size and shape of the part to be removed. Thus, chips and SOICs (Figure 14.29) have a heating attachment on two sides of the package, but PLCCs need an attachment on all four sides. Using the appropriate tip, heat is applied to the surface mount component part to be desoldered, until all the solder melts. The part is then removed with a twisting motion, to minimize damage to the leads for leaded parts.

14.6.3   Hot Air Systems for Surface Mount Repair

Hot air is blown on the leads/terminations/castellations and the solder joints of the part to be reworked and the part is pulled away from the board when the solder on all solder joints is molten. The hot air is usually directed on the leads by the design of the nozzle on the hot air reflow equipment. Hence, the temperature reached by the package body during rework is much lower than during infrared and vapor phase reflow. The package body is heated indirectly via conduction along the leads, convection via the hot air impinging on the ends of the package near the leads, and to a minor extent even by radiation from the nozzle.

Initially, the leads are preheated with the nozzle some distance away (typically an inch or more) from the package body. Then the nozzle is lowered to a point just above the package body and the temperature of the leads increases sharply, until it reaches a peak. At this point the package is removed from the printed circuit board.

Hot air removal systems are the most common. The basic idea is to heat the components with hot air as quickly as possible. There are few hot air systems for the removal of passive devices, but there are many systems for active devices on the market today, and new ones are continually being introduced.

The Weller Pick-a-Chip apparatus shown in Figure 14.30 is an example of a rework machine used for the removal of capacitors and resistors only. The higher temperature and reflow time for hot air equipment can be tolerated because, unlike conductive tips, no pressure is applied to the land. Generally, hot air up to 400°C can be used.

The Pick-a-Chip operates by blowing hot air onto the chip, which melts the solder fillets at its ends; the chip can then be removed with the attached tweezers without any extra motion by the operator. When using the equipment, the stream of hot air is directed at the part to be desoldered, while the tweezers are placed carefully around the part. When the solder melts, the part is removed with a twisting motion.

Figure 14.31 shows an inexpensive hot air "pencil" that can be used for both passive and active devices. It requires tweezers to remove the components.

Many commercial systems for active devices are available. Some allow easy alignment of components for both removal and replacement

Figure 14.30 Hot air rework system for passive devices. (Photograph courtesy of Weller.)

and others require tweezers and are more cumbersome. Some come with a microscope for easy component placement and others do not.

A system should be selected for easy component alignment for both removal and replacement, and there should be some degree of automation to pick up devices as the solder joints melt. Total automation is really not necessary, however. First, such systems are extremely expensive, and also, one should not gear up for an assembly line for rework; rather, the goal is to control the process to minimize the defects. Rework tools are for occasional use, not for mass production.

When using a hot air system, appropriate nozzles should be used to prevent melting of adjacent component joints. In addition, silicone caps

Figure 14.31 Hot air "pencil" for rework of both active and passive devices. (Photograph courtesy of Metcal Corp.)

can be placed over neighboring components to shield them from the hot air. These molded silicone caps are inexpensive, thin walled to accommodate spacing constraints, and infinitely resuable.

14.6.4 BGA Repair

Much of the equipment used for surface mount components can be used for BGA components as well. For example, all surface mount components, including BGA, require uniform heating of components and boards and a mechanism for accurate component placement. The pick-up tool can be a vacuum cup, tweezer, or some other mechanical tool, but the vacuum pick-up system is more common.

Also, for BGA repair, it is very important to provide the thermal energy underneath the package to melt all the hidden balls simultaneously. Conductive and focused IR systems along with systems that simulate a convection oven profile are commonly used. The goal is to achieve adequate and uniform temperatures across the entire part.

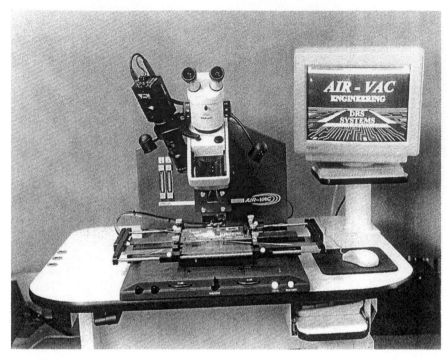

Figure 14.32 Hot air rework system for BGA devices. (Photograph courtesy of Air-Vac.)

For BGA repair, a system that can direct air underneath the package is also used. An example of such a system by Air-Vac is shown in Figure 14.32. This machine, like most others, has evolved from its origins in surface mount and is also used for rework of components other than BGAs. This system uses a vacuum to remove hot air so that the solder joints of adjacent components do not reflow. It also comes with an optional bottom side preheat for the board. As mentioned earlier, preheating the board is highly recommended to expedite component removal and to prevent thermal damage to the board.

A photograph of the nozzle used in the Air-Vac machine is shown in Figure 14.33. The schematic illustration of the nozzle, known as the horizontal flow current (HFC) nozzle, is shown in Figure 14.34. As shown in Figure 14.33, a thermocouple is mounted to the nozzle to monitor exhaust gas temperature. When using this system, a correlation is developed between the exhaust gas temperature and the temperature of solder underneath the package. When the thermocouple senses the temperature of the exhaust gas above a preprogrammed temperature (205°–210°C),

Figure 14.33   Photo of Air-Vac HFC nozzle with thermocouple. (Photograph courtesy of Air-Vac.)

it triggers the vacuum nozzle to remove the component from the board. This feature is very important since one cannot see when the joints actually melt. Such a feature prevents picking up the component prematurely or waiting too long after the solder joints have melted.

Such systems are very effective for BGA removal but they are relatively expensive. There are many other systems on the market with wide variation in price and performance. An example of a relatively inexpensive

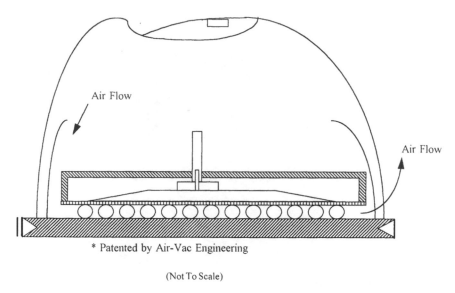

* Patented by Air-Vac Engineering

(Not To Scale)

Figure 14.34   Schematic illustration of Air-Vac HFC nozzle.

hot air system for BGA and fine pitch rework is shown in Figure 14.35a. This unit also comes with a built-in hot air preheat system (located at the base of Figure 14.35a).

The photograph of the preheat system that comes with the rework machine in Figure 14.35a is shown in Figure 14.35b. Such preheat systems are also commercially available as stand-alone units for use with other rework machines. As mentioned earlier but worth repeating, preheating of boards (200°F for 20 to 30 minutes) is highly recommended for reducing the rework cycle time and for minimizing the potential for external and internal thermal damages in boards as shown in Figures 14.25 and 14.26, respectively.

The preheat unit shown in Figure 14.35b also serves as a post-cooler after rework by blowing cool instead of hot air under the board. Faster cooling of the solder joint decreases the grain size and hence improves fatigue resistance.

Stand-alone IR lamp and hot plate preheat units can also be used, but they may not be as effective as hot air systems for preheating the boards uniformly during rework. This is one of the reasons why convection dominant forced hot air reflow ovens are more widely used than IR dominant or hot plate/belt ovens. (See discussion in Chapter 12.)

The use of batch convection ovens for preheat before rework is also an alternative. However, they are inconvenient to use and not as effective since the boards generally cool off before rework is completed. So if the

(a)

(b)

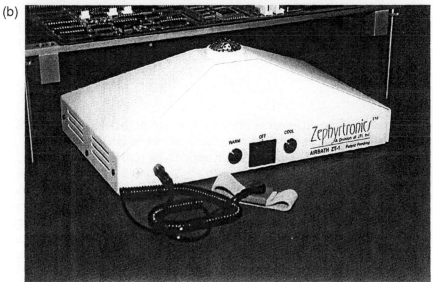

Figure 14.35 (a)   An example of a relatively inexpensive hot air system for active devices and BGAs (photograph courtesy of Hakko); (b) An example of a stand-alone hot air preheat system for any rework machine (photograph courtesy of Zephyrtronics, Inc.)

rework machine does not have a built-in preheat capability, a portable stand-alone preheat unit with adjustable board holder is well worth the investment. Such a system is especially helpful during desoldering of through-hole devices connected to power or ground planes or metal cores to minimize the potential for thermal damage.

There are four or five elements in BGA rework, just as for other surface mount components. They are: component removal, site preparation, reballing (if component is functional), component reflow, and final cleaning. Except for the reballing step, the BGA rework process flow is very similar to the repair of any active surface mount component. A typical BGA rework process flow is as follows:

- Preheat the boards before rework.
- Dispense flux underneath the component.
- Heat the part to 210°–220°C for 45 to 60 seconds and pick up the part by vacuum tip. (The profile must be characterized, as we discuss in the next section.)
- Remove solder from board and package, and clean surface. Avoid the use of wicks.
- Apply paste or flux on balls and the substrate.
- Place new (or reballed) components ("dry" PBGA).
- Preheat board again.
- Reflow.
- Clean.

Because of the differences in composition of the balls and the methods of their attachment to the package, there are some major differences in the repair of plastic BGA (PBGA), ceramic BGA (CBGA), and tape BGA (TBGA) packages.

The PBGA balls are made of eutectic solder (63 Sn/37 Pb). Since flux does not reach all balls evenly during the hot air removal process, the PBGA balls break unevenly and the package is hard to salvage. Reballing the PBGA package is difficult. However, now solder preforms (balls) in solvent or water-soluble carriers are available. They come in a grid array matrix of balls to match the footprint of the PBGA.

The solder ball composition in CBGA packages is high temperature 90 Pb/10 Sn (melting point 302°C), but the CBGA balls are joined with the package and the board with eutectic solder (melting point 183°C). During rework, the CBGA balls do not melt, but the eutectic solder between the ball and the package and between the ball and the board does melt. The balls may stay with the package or the board. But in TBGA, the high temperature balls are welded into the package by partial melting of the high temperature balls. So in TBGA the balls always stay with the package.

Also, in both CBGA or TBGA removal, the high temperature solder in the balls mixes with the eutectic pad metallization on the package and the board and raises the melting point of solder on the lands of the board. This causes problems during subsequent attachment of the new package to the board.

Both PBGA and TBGA packages are susceptible to cracking during rework (the "popcorn" effect). However, CBGAs are not susceptible to cracking because they are hermetic (do not absorb moisture).

No matter which rework machine is used, it is very important to develop the rework profile for reliable removal and replacement of components. This topic is discussed next.

14.6.5   Rework Profiles

To ensure that the components or boards are not damaged, it is important to develop a unique profile for each component to be removed or replaced by any given equipment. In this section we discuss profile for a hot air system.

Some systems heat all four sides of a component more uniformly than others. It is important to control the heating rate in rework. The quickest ramp rate in the hot air rework profile occurs when the nozzle is lowered onto the package after preheat. This temperature ramp rate should be restricted to less than 8°C per second, to avoid excessive thermal shock to the package body.

The peak temperature of the leads (or balls) during the rework procedure should be greater than the melting point of the solder in the solder joint. If the peak temperature is too high, however, the package or the printed board may be damaged by excessive heat application. For eutectic tin-lead solder, this peak temperature should be in the range of 200–230°C. Different equipment have different rework profiles for the type of component to be removed.

Figure 14.36 shows the rework profile for BGA using the Air-Vac equipment (Figure 14.32). Thermocouples are securely attached (with conductive adhesive or high temperature solder) to the board and the balls to develop the profile.

The upper graph in Figure 14.36 shows the temperature of the air as it passes through the upper heater. The bottom graph shows the temperature of the board. The middle graph shows the nozzle exhaust temperature. As is clear from the middle graph, when the exhaust temperature reaches about 210°C, the removal of the package with the vacuum nozzle is

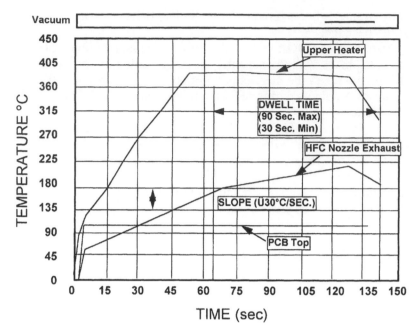

Figure 14.36   BGA rework profile using the Air-Vac equipment shown in Figure 14.32.

triggered after 60 seconds (in this example). The length of the process cycle is adjusted automatically based on the feedback from the thermocouple connected to the exhaust nozzle. The heating cycle should be kept as short as possible. Longer times are possible with higher preheat.

Figure 14.37 shows a rework profile for an earlier version of Air-Vac for PLCC and PQFP components. The required removal time decreases with an increase in the air flow rate. Note also that the larger fine pitch devices require a longer time for removal.

Profiles similar to Figure 14.36 and 14.37 should be developed for all rework equipment to determine the optimum equipment settings.

No matter which system is used, the nozzle design is critical to obtaining an effective rework profile. Generally, the leads on the corners heat up more quickly than the leads toward the center. Consequently, the corner leads reach the melting point of the solder before the center leads on each side of the package.

The temperature lag between the corner and center leads is greater for larger packages—as much as 20°C for poorly designed nozzles. However, with a nozzle that blows more air on the center leads, this temperature difference can be reduced to less than 5°C. An excessive differential in

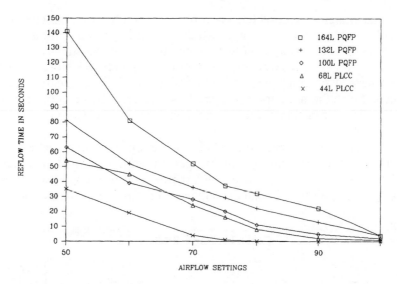

Figure 14.37   Rework profile for removing PLCCs and PQFPs using Air-Vac hot air equipment.

temperature between the center and corner leads actually means that more heat than necessary is being supplied to the component body. It is better to have a uniform temperature throughout to prevent overheating the component body.

14.7   ASSEMBLY TESTING

Process yields in SMT tend to be lower than for through-hole assemblies. This is largely because SMT processes are more complex. With higher defect rates, the cost of test and repair can be a very significant part of the total cost of assembly. Design for testability directly affects real estate by consuming the space that could have been used for additional functions on the board. However, a board that cannot be tested in a cost-effective manner will incur very significant costs of repair and test.

Design for testability is more important than ever for surface mounting. For this reason, the test engineer along with the manufacturing engineer must be integral parts of the product team. Requirements for manufacturability and testing should be considered at the design stage. In this section, we consider only issues that concern the test engineer in production. We will assume that the product meets the design guidelines presented in Chapter 7.

Functional tests in which assemblies are run through the connector provide moderate fault coverage, which can be improved if the board is designed for the system level test. Generally, the diagnostic accuracy of functional or system level tests is limited to functional blocks on the board instead of component or solder joint level (i.e., manufacturing) defects. Since the newness of the process increases the potential for this type of defect, functional tests are not considered to be very effective for SMT boards.

The defects encountered in SMT—solder opens, solder wicking (another form of open), tombstoning, bridges, misalignment, part movement, solder balls, etc.—are caused by solderability and lead coplanarity problems, poor paste printing, bad placement, and improper soldering process

GOOD PROBE PRACTICE

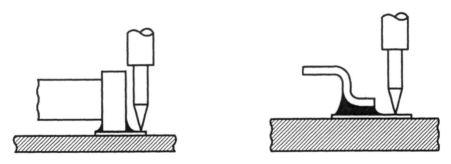

POOR PROBE PRACTICE

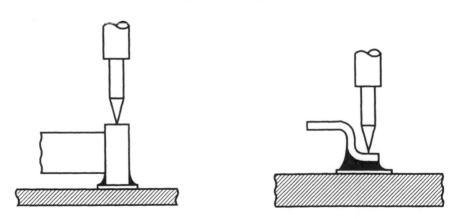

Figure 14.38   Good (top) and poor (bottom) practices for probing during test.

profiles. Some of these defects may be even worse in some reflow processes than in others. These phenomena translate into two major defect types: opens and shorts.

Automated equipment test (ATE), also known as in-circuit testing, bed-of-nails testing, and pogo pin testing, is the best way to detect the manufacturing defects discussed above. To ensure that the board can be tested by ATE, access to test nodes must be provided in the design.

In ATE, the test probes are sharp and can easily damage the surface if improperly applied. They must not touch the brittle top passivation layer of the resistors; not only might the surface be damaged, but if some of the resistive material is chipped away, even the resistive value might change. Also, to guard against false-positive results, the test probes must not touch the leads or terminations of the components. If the test probe is pressed against an improperly aligned part, the probe itself may be bent or damaged. Figure 14.38 shows good (top two figures) and poor (bottom two figures) practices for probing with test probes.

14.7.1   Fixtures for ATE Testing

The fixtures for ATE can be either single sided or double sided (clam shell). The latter are expensive, but since nodal access is provided by both sides they use less real estate. Single-sided ATE fixtures are simpler and cheaper but require nodal access from one side and consume more real estate. They are a minor variation of through-hole technology and are most commonly used.

Test fixtures use 50 mil probes for surface mount devices and the less fragile 100 mil probes for through-hole devices. As shown in Figure 14.39, some probes consist of three parts: the probe itself, a receptacle, and a wire plug. These three elements are generally found only in the current generation of 50 mil probes. The tips of the test probes vary considerably as shown in Figure 14.40. The figure also summarizes the functions of the various types of probe tips.

Although 50 mil probes can be much more expensive and not as reliable [18] as the probes used for through-hole, fairly good reliability can be achieved if they are handled properly. Their use should be minimized nevertheless; that is, they should not be used if 100 mil probes would work on that location without compromising board real estate.

There are test fixtures for all three categories of surface mount devices. Type III SMT test fixtures, which are essentially the same as through-hole test fixtures, use 100 mil test probes. Type I SMT test fixtures, which use only 50 mil test probes, serve for memory boards only because of restraints imposed by component availability. Type II SMT

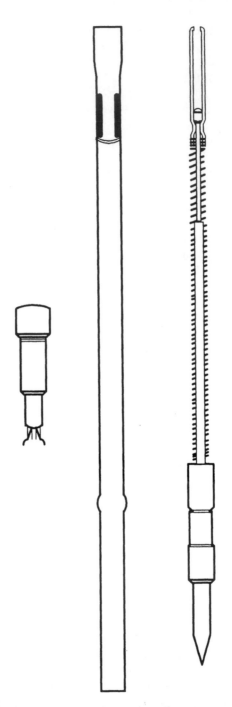

Figure 14.39 Typical elements of a test probe: plunger (right), receptacle (center), and wire plug (left).

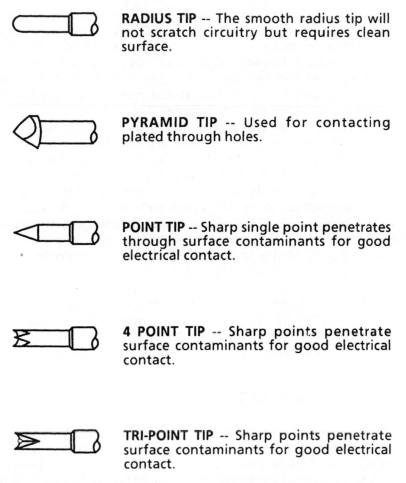

RADIUS TIP -- The smooth radius tip will not scratch circuitry but requires clean surface.

PYRAMID TIP -- Used for contacting plated through holes.

POINT TIP -- Sharp single point penetrates through surface contaminants for good electrical contact.

4 POINT TIP -- Sharp points penetrate surface contaminants for good electrical contact.

TRI-POINT TIP -- Sharp points penetrate surface contaminants for good electrical contact.

Figure 14.40   Various types of test probe tips for automated test equipment.

test fixtures, which use 100 mil probes for through-hole devices and 50 mil probes for surface mount devices, generally tend to present more problems in debugging. If the densities of the 100 mil and 50 mil probes are not appropriately controlled, the needed contact may not be achieved. This happens especially when the density of probes is so high that the vacuum pressure is insufficient to pull the board onto the fixture to make contact. (Refer to Chapter 1 for definitions of Types I, II, and III SMT assemblies.)

Types III and II assemblies do not have any vacuum leakage problems because all the test vias are filled with solder during wave soldering.

There may be vacuum leakage problems with Type I assemblies, since the test vias are open. In a single-sided SMT board the holes can be plugged with a dry film solder mask from one side, but the holes may entrap contamination. Wet film solder mask is not very effective in sealing via holes, but they may be plugged by screening solder paste from one side. The other side of the via hole should remain open to allow the cleaning solvent to flush out any entrapped flux.

Care should be taken to ensure that solder mask does not overlap onto test pads. This may be very critical for the small test pads (25 to 35 mils) used for 50 mil probes. When the test pads are bare and the other vias are covered with solder mask, it is easy to identify the test pads. Keeping the test pads square in shape and other pads round also helps in distinguishing them. The square pads also provide more area than round pads, but they do limit the routing channel. Square pads are especially desirable if the test pad size is under 32 mils diameter. A test map, created at the design stage, is also necessary, however, to ensure the identification of individual test pads.

If additional test probes become necessary, the user cannot easily drill the test fixture and add new probes. Thus any fixture modification is generally done by the vendor. One critical issue in test fixtures is the proper alignment of the test probes on the small test pads. Having a test fixture specification helps to some extent, as we see in the next section.

14.7.2   Issues in ATE Testing

With increased functional density, hence reduced interpackage spacing (see Chapter 7 for actual recommended numbers), the use of 50 mil probes is essential. These probes are not only more expensive, however, they are also less reliable. Overall, the small probes tend to increase the cost of the test fixture. In addition to cost and reliability, there is the question of accuracy in the test fixtures. The smaller probes allow smaller test pads (30–32 mils with 18 mil holes). Even 25 mil pads with 12 mil holes have been used by some companies. However, it is very difficult to accurately align the test probes on the smallest test pads. Larger pads reduce the probability of missed pads due to the misalignment of test probes.

The alignment problem on smaller test pads is especially compounded because, unlike board fabrication and assembly vendors, the test fixture vendors are not used to following a test fixture specification. Moreover, very few companies have an in-house test fixture specification, and industry standards are lacking. Even if a fixture specification exists, it is a Herculean task to determine whether a given fixture meets it.

The accuracy problem is further compounded by the method of inserting receptacles—the manual process of hitting fragile small test receptacles with a mallet can easily bend them. Hardly anyone autoinserts even 100 mil receptacles. The accurate insertion of 50 mil receptacles is a matter of the right "touch," representing a combination of experience and care. So much for the high tech world of SMT testing.

Since almost all vendors use the same method of receptacle insertion, the alternative of finding another "right" vendor does not exist. One must work very closely with the selected vendor, who will cooperate in the development of a workable fixture specification to meet the requirements of both the user and the supplier.

When the users and the vendors have agreed on a fixture specification, they must establish a method for determining the accuracy and compliance to the specification. Some methods for measuring accuracy include video-equipped coordinate measurement, milling machines, and transparent alignment plates.

14.8   ISO 9000 QUALITY STANDARDS AND CERTIFICATION

In Chapters 1 through 13 of this book we discussed detailed design, materials, and process guidelines to produce SMT products of good quality. In the previous sections of this chapter we discussed specific tools such as SQC, inspection, repair, and test to ensure that final products being shipped to the customer meet the quality requirements. The issues discussed up to this point are necessary but not enough to produce good quality products. To achieve this, a formalized quality system must be in place. ISO 9000, one of the most widely used quality systems, is used to accomplish consistent quality.

In this section we will only attempt to provide an overview of ISO 9000; there are many books available on the subject. (See, for example, reference 19 and 20.) ISO 9000 is a series of standards created by the International Standards Organization (ISO) based in Geneva, Switzerland. Many companies from countries around the world, including the United States, spent seven years developing this standard, which was first published in 1987. In the United States, the ISO 9000 standards are also referred to as the ANSI/ASQC Q90 Series.

The purpose of ISO is to promote the international exchange of goods and services and the development of standardization. The ISO Series standards focus on systems for managing quality through controlled process documentation to meet customer requirements. The ISO standards are reviewed and updated every five years. Expect a strong emphasis on continuous improvement, detection, and assurance-based approaches in future revisions. ISO 9000 is now being adopted by an increasing number of companies for various reasons. Let us discuss some of them:

1. Many companies use the certification process as a catalyst to document their design and assembly processes and put controls in place to produce quality products. This should be the main reason for ISO certification.
2. ISO certification is also being seen as a way to get a foothold in the international market, with Europe as the specific target. Certain European markets are available only to ISO certified suppliers. For example, the European Union requires the suppliers of certain regulated products that impact health, safety, and the environment to be ISO certified.
3. Many companies are requiring their suppliers to be ISO certified. Thus companies without certification may find many markets closed for their products. This is one of the worst reasons for getting ISO certification. The basis of the ISO 9000 quality system is to demonstrate management commitment to the quality process. Being told or forced to implement ISO 9000 does not necessarily mean there will be the proper level of management commitment.
4. ISO certification plays an important role in marketing; certified companies tend to distinguish themselves from noncertified competitors.

Unfortunately, marketing, instead of quality improvement, has become the driving force for many companies to seek ISO certification. This means that very general procedures may be created to pass the ISO certification but they may not be specific enough to produce quality products. It is like focusing on possible questions to be asked in a final examination to get the diploma, instead of focusing on learning. Focusing on the end result instead of the process may lead to certification, but the real benefits of ISO certification will not be realized.

Also, many ISO auditors are not necessarily familiar with the technology of the companies and hence they do not focus on the accuracy of the documents they are auditing. This means that even ISO certified companies do not necessarily have the right "recipe" to produce quality products.

14.8.1   ISO 9000 Certification

The first step in ISO certification is to decide on the appropriate ISO standard for certification. ISO 9000 is a series of five quality system standards (two guidance and three contractual standards). ISO 9000 provides guidance on how to select the three contractual standards—ISO 9001, ISO 9002, or ISO 9003. ISO 9001 is a full-scope standard covering design, production, servicing, and installation. ISO 9002 is a subset of ISO 9001 and does not cover design and servicing. ISO 9003 is a subset of 9002 and addresses only final inspection and testing. ISO 9004, the second guidance document, may be used to provide suggestions on how to implement the various elements contained in ISO 9001, ISO 9002, and ISO 9003.

To achieve ISO certification, it is very important to establish an internal ISO certification team headed by someone familiar with quality systems. The team leader is responsible for coordinating the ISO-based quality system and hence should have the full support of top management. Also, management must provide full support by conducting themselves within the guidelines established by the quality system. This is important since management often has a tendency to defy its own rules.

The team should evaluate the existing system against the selected standard and identify areas of improvement. The team should be chartered to ensure that procedures are written and to verify that the written procedures reflect actual practice in compliance with ISO guidelines.

In many cases companies find that written procedures do not exist. Companies without adequate documentation should consider hiring a consultant specializing in ISO certification in their industry. Written policy and procedures are key to creating a self-supporting internal infrastructure dependent on a system, rather than a specific individual, to produce quality products or service—the basic premise of any quality system.

Another important function of the team is to conduct internal reviews. Companies that are not very familiar with ISO standards or are uncomfortable conducting self-assessment should also consider hiring a consultant specializing in quality systems. Consultants provide an unbiased interpretation of standards tailored to meet the requirements of the company. Also, consultants usually have access to the registration body if specific questions should arise.

After completing the internal review and preparing the needed documentation (discussed later), with or without the help of a consultant, the company retains a qualified external "registrar" to audit the quality system developed by the company. If the registrar finds that adequate documentation exists, and more importantly, all the personnel are trained and aware of their function as documented, he or she grants the ISO certification to the company. In some cases, companies first invite the registrar only for a "pre-audit" where deficiencies are pointed out and recommendations given for corrective action. Once the company has implemented corrective action, the registrar is invited back for formal audit and certification.

Maintaining certification requires visits from the registrar once a year or as often as every six months (surveillance audits) and re-audit every three years (triennial audits).

14.8.2   Meeting ISO 9000 Standards

The basic intention of ISO 9000 is to provide guidelines for creating a "working" quality system that documents the "recipe" for making the products. The auditor does not judge the correctness of the "recipe" but only focuses on whether the recipe is being followed and if it will meet the intent of the ISO standards. The key issue that the auditor tends to focus on is whether the responsible employees can demonstrate that they are performing their job functions as documented in the applicable procedures. So employees must correctly answer, if asked by the auditor, which procedures apply to their job function and whether they fully understand the procedure as documented. This is one of the main reasons for conducting extensive training on procedures being documented and for letting the responsible employees either write or be actively involved in writing the documentation for their job function.

The interpretation of ISO standards leaves a great deal of freedom. There are two guiding principles to keep in mind when applying the standards to your business:

* You know your product best. You are the sole judge of the "recipe" of your product or service to meet your customer requirements.
* Document what you do, do what you document, and produce the quality product you specify you will.

The ISO 9000 standards include generic quality system requirements for 20 basic elements that affect quality. These elements as listed in Section 4.0 in ISO 9000 are as follows:

1. Management Responsibility
2. Quality System
3. Contract Review
4. Design Control
5. Document Control
6. Purchasing
7. Purchaser Supplied Product
8. Product Identification and Traceability
9. Process Control
10. Inspection and Testing
11. Inspection, Measuring and Test Equipment

12. Inspection and Test Status
13. Control of Nonconforming Product
14. Corrective Action
15. Handling, Storage, Packaging and Delivery
16. Quality Records
17. Internal Quality Audits
18. Training
19. Servicing
20. Statistical Techniques

Companies find the following four-tiered approach to implementation of ISO 9000 by far the most productive:

1. *Quality manual.* This is a road map to the system, outlining policies and objectives that relate to the specific aspects of the system. This document is used by everyone in the company.
2. *Procedures.* These provide process descriptions and flow charts of activities. They give more descriptions of what activity is carried out, by whom, and where, and why the activity is carried out. This is the heart of the process recipe. Almost all departments must have their procedures adequately documented.
3. *Work instructions.* These describe, step by step, how to carry out a task. They are often called "standard job practices" or "operating guides." Work instructions must be revised and integrated into the overall documentation system. Work instructions are generally used for the shop floor and are not required for all departments.
4. *Forms and records.* These are also known as Quality Records. Forms are often used to collect information and record the completion of required quality activities. Sufficient records must be kept to provide objective evidence that the quality activities are being carried out. The auditor will ask for these records (raw data) to validate process activities.

In creating the above documents, the following "helpful hints" are recommended:

* ISO implementation can be a frustrating experience because the team members may feel that they are not doing their "real job." The employees still have their other jobs in addition to ISO implementation. To minimize frustration, focus on the effort and not on the result. This may appear contrary to conventional wisdom but it will minimize "ISO burnout." Remember that in addi-

tion to other benefits discussed earlier, ISO implementation can be a very effective team building exercise. The team members will also get a better idea of what other departments really do.

- The team must prepare an implementation plan with detailed timelines and have "buy-in" from all team members and their managers. Regularly scheduled management reviews should be conducted to track the progress and make course corrections as necessary.

- Purchase ISO documents and interview potential consultants (if needed) and potential registrars to establish mutual expectations for ISO certification.

- When writing documents, keep in mind that they will have to be kept up-to-date. Try not to write dated and specific information to avoid frequent updates.

- Initially try to write as few documents as possible, filling in the details later. However, one should not be so general that the procedures are meaningless and fail to provide critical details needed to produce the desired quality.

- Write procedures to change all documents because they will have to be changed as the technology evolves and you discover better "recipes." Prepare a master list of the document numbering system to accommodate future changes and control revisions.

- Use an internal corrective action system that is tied to ISO internal audits.

- Involve as many individuals as possible from different departments in the company to write the documents. This will lighten the workload for everyone and will ensure that responsible individuals are thoroughly familiar with the processes being documented.

- Create a standard format for each type of document (template) and follow it.

- Make sure that the documents are short and easy to read. Keep the target audience for each procedure in mind. Have the appropriate employees review them to ensure their accuracy. If they are not correct, rewrite them. Write what you do (unless you are doing the wrong thing).

- Start training on documents early on. Create a training matrix for each department. All individuals need to be familiar with the procedures related to their job function.

ISO certification is a time-consuming (allow at least 6 months) and expensive process. The bulk of the expenses are related to company personnel devoting their time to internal audits and to writing procedures and work instructions. However, if the certification is used as a catalyst to create, document, and follow the processes necessary for improving product quality, it is well worth the time and money spent on it. Using the ISO 9000 certification as a catalyst, organizational efficiency is improved by:

1. Creation of feedback loops between customers and suppliers.
2. Simplification of quality manuals and operating procedures to encourage employee use.
3. A thorough training program to encourage every employee to take personal responsibility for the quality of products they produce.
4. Creation of a system which promotes a controlled production environment where continuous process improvement becomes the norm.
5. Use of measurement tools such as statistical process control to monitor manufacturing process.

14.9   SUMMARY

Having dealt extensively in this book with the technical issues related to design and processes, it is important to consider the quality, testability, and, yes, the repair of surface mount assemblies at a competitive cost.

Statistical process control is used to ensure compliance of the materials and process specifications. This preventive measure has been very successfully used in Japan to manufacture products with consistent quality over time. One must either develop in-house specifications for material, process, and solder joint quality or use the easily available industry standards. The solder joint quality specifications establish the final requirements. The material and process specifications provide the road map to the achievement of the intended quality.

The function of statistical process control is to show all concerned including operators, engineers, and management, how the company is doing with respect to quality objectives. A problem may be caused by the vendors or by the in-house manufacturing processes (machine or non-machine dependent) or by the design. By pinpointing the problem, quality objectives make it possible to implement corrective action. When the root of the problem is known, however, the basic cause can be eliminated permanently, not just the symptoms.

Even though many automated inspection tools are available, the visual method remains popular. The subjectivity of visual inspection cannot be totally overcome by automated inspection systems as long as the inspection criteria remain subjective.

There are many automated tools available for rework and repair, making it easier and less damaging to repair a surface mount assembly, although the repair of BGA assemblies is still a challenge. The repair of through-hole assemblies is very operator dependent, hence is prone to cause additional damage.

The completed assemblies need to be tested before shipment to customers to ensure that they meet the intended requirements. Given the complexity of the technology, automated test equipment or bed-of-nails testing is considered more effective than functional or systems testing to detect manufacturing defects.

No matter how effective the materials and process specifications and the SQC tools, some level of defects must be expected in an imperfect world. Therefore, one must plan for some inspection, repair, and testing, but continue to aim for zero defects through continuous process improvement in order to compete successfully in the international marketplace. To make this job easier one should have a quality system in place. ISO 9000 is one well-known and widely used quality system that should be implemented to document, monitor, and control the quality of the product being made.

REFERENCES

1. Roome, D. R., "Cutting through complexity. How SQC can help." *Printed Circuit Assembly,* July 1987, pp. 16–18.

2. Keane, T. P, "JIT and SPC, a checklist." *Manufacturing Systems,* October 1987, p. 22.

3. Gluckman, P., "Introduction to statistical control." *PC Fab,* March 1983.

4. ASTM Manual on Presentation on Data and Control Chart Analysis, ASTM Standard STP-15D. Philadelphia: American Society for Testing and Materials, 1976.

5. Prasad, P., and Fitzsimmons, D. "Troubleshooting wave soldering problems with statistical quality control." *Brazing and Soldering,* Vol. 10, No. 4, Summer 1984, pp. 13–18.

6. Evans, W. D. "Statistics-free SPC." *Manufacturing Systems,* March 1987, pp. 34–37.

7. Roos-Kozel, B. "Process driven solder paste selection." *Printed Circuit Assembly,* July 1987, pp. 19–24.

8. Woodgate, Ralph. "Solder joint inspection criteria." *Printed Circuit Assembly,* February 1987, pp. 6–10.

9. Manson, S. S. Thermal stress and low cycle fatigue, Bobrowsky thermal shock parameter, Section 7.1.4, 383, Robert E. Krieger Publishing Company, Malabar, Florida, 1981.

10. Elliot, D. A. "Concerns in wave and infrared soldering of surface mount assemblies." *Proceedings of Surface Mount Technology Association Conference,* August 1988, pp. 177–188. Available from SMTA, Edina, MN.

11. J-STD-001. Requirements for soldered electrical and electronic assemblies. Available from IPC, Northbrook, IL.

12. IPC-A-610. Revision, Acceptance of electronic assemblies. Available from IPC, Northbrook, IL.

13. IPC-A-610. Revision E, Acceptability of printed boards. Available from IPC, Northbrook, IL.

14. Davy, J. G. "The prospects for automated PWA solder defect inspection." *Brazing and Soldering,* No. 12, Spring 1987, pp. 9–15.

15. Crawshaw, R., and Rooks, S. "Cross-sectional X-ray measurement test of ball grid array connections." *Proceedings of NEPCON West,* 1995, pp. 1035–1047.

16. Thomas, S., and Thornton, C. "BGA process development and SPC implementation using in-line X-ray laminography measurements." *Proceedings of Surface Mount International,* August 1995, pp. 367–377.

17. Prasad, P. "Contributing factors for thermal damage in PWB assemblies during hand soldering." *IPC Technical Review,* January 1982, pp. 9–17.

18. Small, C. H. "Surface mount technology forces engineers to follow testability guidelines." *EDN,* May 14, 1987, pp. 93–97.

19. James, L. "ISO 9000: Preparing for registration." Available from ASQC Quality Press, Milwaukee, WI.

20. ANSI/ASQC Q90/ISO 9000. Guidelines for use by the Chemical and process industries. Available from ASQC Quality Press, Milwaukee, WI.

Appendix   A*
Surface Mount Standards

This appendix lists various standards, specifications, and guidelines that have an impact on surface mount technology from U.S. and international bodies, including the U.S. Department of Defense, although DoD is now committed to using industry standards.

These documents can be obtained from the following sources:

- INSTITUTE FOR INTERCONNECTING AND PACKAGING ELECTRONIC CIRCUITS (IPC):
  2215 Sanders Road, Northbrook, IL 60062-6135
  Telephone 847-509-9700; FAX 847-509-9798
- ELECTRONIC INDUSTRIES ASSOCIATION (EIA)
  2001 Pennsylvania Avenue, NW
  Washington, D.C. 20006-1813
  Telephone 703-907-7500; FAX 703-907-7501
- GLOBAL ENGINEERING DOCUMENTS
  2805 McGaw Avenue, Irvine, CA 92713
  Telephone: 800-854-7179; FAX 314-726-6418

Military documents are available from:

- STANDARDIZATION DOCUMENTS
  Order Desk, Building 4D,
  700 Robbins Avenue, Philadelphia, PA 19111-5094
  Telephone 215-697-2667

Central office of the IEC:

*Adapted from Surface Mount Council Status of the Technology, Industry Activities and Action Plan, September 8–12, 1996, available from IPC.

- INTERNATIONAL ELECTROTECHNICAL COMMISSION (IEC)
  Rue de Varembe, 1211 Geneva 20, Switzerland

IEC documents are also available from:

- AMERICAN NATIONAL STANDARDS INSTITUTE (ANSI)
  11 West 42nd Street, New York, NY 10036

The letter prefix of each document indicates the organization responsible for that document:

EIA	Electronic Industries Association
JEDEC	Joint Electron Devices Engineering Council of the EIA
IPC	Institute for Interconnecting and Packaging Electronic Circuits
MIL	Military
DoD	Department of Defense
J-Std	Joint Industry Standards

Components, General

EIA-886-E	Standard Test Methods for Passive Electronic Component Parts—General Instructions and Index (this document is an umbrella document that consists of a series of uniform test methods for electronic component parts).
EIA-481-A	Taping of Surface Mount Components for Automatic Placement
EIA-481-1	8 mm and 2 mm Taping of Surface Mount Components for Automatic Handling
EIA-481-2	16 mm and 24 mm Embossed Carrier Taping of Surface Mount Components for Automated Handling
EIA-481-3	32 mm, 44 mm, and 56 mm Embossed Carrier Taping of Surface Mount Components for Automated Handling
EIA/IS-47	Contact Termination Finish Standard for Surface Mount Devices
EIA-PDP-100	Registered and Standard Mechanical Outlines for Electronic Parts
EIA-JEP-95	JEDEC Registered and Standard Mechanical Outlines for Semiconductor Devices

EIA-JESD30	Descriptive Designation System for Semiconductor Device Packages
EIA-JESD95-1	Design Requirements for Outlines of Solid-State and Related Products
IPC-9501	Component Qualification for the Assembly Process
IPC-SM-786A	Recommended Procedure for Handling of Moisture Sensitive Plastic IC Packages
J-Std-020	Moisture/Reflow Sensitivity Classification for Plastic Integrated Circuit Surface Mount Devices

Components, Passive

Capacitors:

EIA-198-D	Ceramic Dielectric Capacitors Classes I, II, III, and IV
EIA-469-B	Standard Test Method for Destructive Physical Analysis of High Reliability Ceramic Monolithic Capacitors
EIA-479	Film-Paper, Film Dielectric Capacitors for Microwave Ovens
EIA-510	Standard Test Method for Destructive Physical Analysis of Industrial Grade Ceramic Monolithic Capacitors
EIA-535	Series of Detail Specifications on Fixed Tantalum Capacitors that have been adopted by the National Electronic Components Quality (NECQ) Assessment System
EIA-CB-11	Guidelines for the Surface Mounting of Multilayer Ceramic Chip Capacitors
EIA/IS-28	Fixed Tantalum Chip Capacitor Style 1 Protected— Standard Capacitance Range
EIA/IS-29	Fixed Tantalum Chip Capacitor Style 1 Protected— Extended Capacitance Range
EIA/IS-35	Two-Pin Dual In-Line Capacitors
EIA/IS-36	Chip Capacitors, Multi-Layer (Ceramic Dielectric)
EIA/IS-37	Multiple Layer High Voltage Capacitors (Radial Lead Chip Capacitors)
EIA/IS-38	Radial Lead Capacitors (Conformally Coated)
EIA/IS-39	Ceramic Dielectric Axial Capacitors (Glass Encapsulated)
IEC-384-3	Sectional Specification, Tantalum Chip Capacitors

IEC-384-10	Sectional Specification, Fixed Multilayer Ceramic Chip Capacitors
IECQ draft	Blank Detail Specification, Fixed Multilayer Ceramic Chip Capacitors
IECQ-PQC-31	Sectional Specification, Fixed Tantalum Chip Capacitors with Solid Electrolyte
IECQ-PQC-32	Blank Detail Specification, Fixed Tantalum Chip Capacitor

Resistors:

EIA-575	Resistors, Rectangular, Surface Mount, General Purpose
EIA-576	Resistors, Rectangular, Surface Mount, Precision
EIA/IS-34	Leaded Surface Mount Resistor Networks Fixed Film

Components, Active

EIA-JEP-95	JEDEC Registered and Standard Outlines for Semiconductor Devices
EIA-JESD11	Chip Carrier Pinouts Standardized for CMOS 4000, HC, and HCT Series of Logic Circuits
EIA-JESD21-C	Configurations for Solid State Memories
EIA-JESD22-B	Test Methods and Procedures for Solid State Devices Used in Transporatation/Automotive Applications (Series format—consists of over 16 different test procedure documents.)
EIA-JESD-26A	General Requirements, PEM, Rugged Environments
EIA-JESD30	Descriptive Designation System for Semiconductor Device Packages
EIA-JESD95-1	Design Requirements for Outlines of Solid-State and Related Products

Components, Electromechanical

Connectors:

EIA-429	Industry Standard for Connectors, Electrical Flat Cable Type (IPC-FC-218B)
EIA-364-B	Electrical Connector Test Procedures Including Environmental Classifications (Series format consisting of over 60 electrical connector test procedures.)

EIA-506	Dimensional and Functional Characteristics Defining Sockets for Leadless Type A Chip Carriers (.050 Spacing)
EIA-507	Dimensional Characteristics Defining Edge Clips for Use with Hybrid and Chip Carriers
EIA-IS-47	Contact Termination Finish Standard for Surface Mount Devices
EIA/IS-64	Two Millimeter, Two-Part Connectors for Use with Printed Boards and Backplanes

Sockets:

| EIA-5400000 | Generic Specification for Sockets for Integrated Circuit (IC) Packages for Use in Electronic Equipment (Series format with over 20 sectional, blank detail, and detail specifications, approved for use in the NECQ system.) |

Switches:

IECQ-PQC-41 -US00003	Detail Specification, Dual-in-Line Switch, Surface Mountable, Slide Actuated
EIA-448-23	Surface Mountable Switches, Qualification Test
EIA-5200000-A	Generic Specification for Special-Use Electromechanical Switches of Certified Quality (Series format with over 25 sectional, blank detail and detail specifications, approved for use in the NECQ system.)

Printed Boards:

IPC-FC-250	Performance Specification for Single and Double-sided Flexible Printed Boards
IPC-RF-245	Performance Specification for Rigid-Flex Multilayer Printed Boards
IPC-RB-276	Performance Specification for Rigid Printed Boards
IPC-MC-324	Performance Specification for Metal Core Boards
IPC-HM-860	Performance Specification for Hybrid Multilayer
IPC-6105	Performance Specification for Organic Multichip Module Structures (MCM-L)
MIL-P-50884	Military Specfication Printed Wiring, Flexible, and Rigid Flex
MIL-P-55110	Military Specification Printed Wiring Boards, General Specification For

IPC-DW-425/424	Discrete Wiring Technology
MIL-P-PRF-31032	Printed Circuit Board/Printed Wiring Board Manufacturing, General Specification For

Materials:

IPC-L-108B	Specification for Thin Metal Clad Base Materials for Multilayer Printed Boards
IPC-L-109B	Specification for Resin Preimpregnated Fabric (Prepreg) for Multilayer Printed Boards
IPC-CC-110	Guidelines for Selecting Core Constructions for Multilayer Printed Wiring Board Applications
IPC-L-115B	Specification for Rigid Metal Clad Base Materials for Printed Boards
IPC-CF-148	Resin Coated Metal for Multilayer Printed Boards
IPC-MF-150F	Metal Foil for Printed Wiring Applications
IPC-CF-152A	Metallic Foil Specification for Copper/Invar/Copper (CIC) for Printed Wiring and Other Related Applications
IPC-SM-817	General Requirements for SMT Adhesives
IPC-3406	Guidelines for Conductive Adhesives
IPC-3407	General Requirements for Isotropically Conductive Adhesives—Paste Types
IPC-3408	General Requirements for Anistropically Conductive Adhesive Films
IPC-CC-830	Qualification and Performance of Electrical Insulation Compounds for Printed Board Assemblies
IPC-SM-840C	Qualification and Performance of Permanent Polymer Coating (Solder Mask) for Printed Boards
J-Std-004	Requirements for Soldering Fluxes
J-Std-005	General Requirements and Test Methods for Electronic Grade Solder Paste
J-Std-006	General Requirements and Test Methods for Soft Solder Alloys and Fluxed and Non-Fluxed Solid Solders for Electronic Soldering Applications

Design Activities

IPC-D-249	Design Standard for Flexible Single and Double-Sided Printed Boards
IPC-D-275	Design Standard for Rigid Printed Boards and Rigid Printed Board Assemblies

IPC-D-279	Reliability Design Guidelines for Surface Mount Technology Printed Board Assemblies
IPC-D-317	Design Standard for Electronic Packaging Utilizing High Speed Techniques
IPC-C-406	Design and Application Guidelines for Surface Mount Connectors
IPC-SM-782	Surface Mount Land Patterns (Configuration and Design Rules)
IPC-H-855	Hybrid Microcircuit Design Guide
IPC-D-859	Design Standard for Multilayer Hybrid Circuits
IPC-2105	Design Standard for Organic Multichip Modules (MCM-L) and MCM-L Assemblies
MIL-Std-2118	Design Standard for Flexible Printed Wiring

Component Mounting

EIA-CB-11	Guidelines for the Surface Mounting of Multilayer Ceramic Chip Capacitors
IPC-CM-770D	Guidelines for Printed Board Component Mounting
IPC-SM-784	Guidelines for Direct Chip Attachment
SMC-TR-001	Introduction to Tape Automated Bonding and Fine Pitch Technology
IPC-SM-780	Electronic Component Packaging and Interconnection with Emphasis on Surface Mounting
IPC-MC-790	Guidelines for Multichip Module Technology Utilization
J-Std-012	Implementation of Flip Chip and Chip Scale Technology
J-Std-013	Implementation of Ball Grid Array and Other High Density Technology

Soldering and Solderability

EIA/IS-46	Test Procedure for Resistance to Soldering (Vapor Phase Technique) for Surface Mount Devices
EIA/IS-49-A	Solderability Test Method for Leads and Terminations
EIA-448-19	Method 19 Test Standard for Electromechanical Components Environmental Effects of Machine Soldering Using a Vapor Phase System
IPC-TR-460A	Trouble Shooting Checklist for Wave Soldering Printed Wiring Boards

IPC-TR-462	Solderability Evaluation of Printed Boards with Protective Coatings Over Long-Term Storage
IPC-TR-464	Accelerated Aging for Solderability Evaluations
J-Std-001	Requirements for Soldered Electrical and Electronic Assemblies
J-Std-002	Solderability Tests for Component Leads, Terminations, Lugs, Terminals, and Wires
J-Std-003	Solderability Tests of Printed Boards
IPC-S-816	Troubleshooting for Surface Mount Soldering
IPC-AJ-820	Assembly and Joining Handbook

Quality Assessment

EIA-469-B	Standard Test Method for Destructive Physical Analysis of High Reliability Ceramic Monolithic Capacitors
EIA-510	Standard Test Method for Destructive Physical Analysis of Industrial Grade Ceramic Monolithic Capacitors
IPC-A-600	Acceptability of Printed Boards
IPC-A-610	Acceptability of Printed Board Assemblies
MIL-Std-883	Methods and Procedures for Microelectronics
IPC-TR-551	Quality Assessment of Printed Boards Used for Mounting and Interconnecting Electronic Components Reliability
IPC-A-24	Flux/Board Interaction Board
IPC-A-36	Cleaning Alternatives Artwork
IPC-A-38	Fine Line Round Robin Test Pattern
IPC-A-48	Surface Mount Artwork
IPC-AI-640	User Requirements for Automatic Inspection of Unpopulated Thick Film Hybrid Substrates
IPC-AI-641	User Guidelines for Automated Solder Joint Inspection Systems
IPC-AI-642	User Guidelines for Automated Inspection of Artwork and Inner Layers
IPC-AI-643	User Guidelines for Automatic Optical Inspection of Populated Packaging and Interconnection

Surface Mount Process

IPC-SC-60	Post Solder Solvent Cleaning Handbook
IPC-SC-61	Post Solder Semi-Aqueous Cleaning Handbook
IPC-AC-62	Post Solder Aqueous Cleaning Handbook

| IPC-TR-580 | Cleaning and Cleanliness Test Program Phase 1 Test Results Reliability |

Reliability

| IPC-SM-785 | Guidelines for Accelerated Surface Mount Attachment Reliability Testing |
| IPC-D-279 | Design Guidelines for Reliable Surface Mount Technology Printed Board Assemblies |

Numerical Control Standards

IPC-NC-349	Computer Numerical Formatting for Drilling and Routing Equipment
IPC-D-350D	Printed Board Description in Digital Form
IEC-1182-1	Printed Board Description in Digital Form Companion Documents
IPC-D-351	Printed Wiring Documentation in Digital Form
IPC-D-352	Electronic Design Database Description for Printed Boards
IPC-D-354	Library Format Description for Printed Boards in Digital Form
IPC-D-356	Bare Board Electrical Test Information in Digital Form
EIA-224-B	Character Code for Numerical Machine Control Perforated Tape
EIA-227-A	One-Inch Perforated Tape
EIA-267-B	Axis and Motion Nomenclature for Numerically Controlled Machines
EIA-274-D	Interchangeable Variable Block Data Format for Positioning, Contouring, and Contouring/Positioning Numerically Controlled Machines
EIA-281-B	Electrical and Construction Standards for Numerical Machine Control
EIA-358-B	Subset of American National Standard Code for Information Interchange for Numerical Machine Control Perforated Tape
EIA-408	Interface Between Numerical Control Equipment on Data Terminal Equipment Employing Parallel Binary Data Interchange
EIA-431	Electrical Interface Between Numerical Control and Machine Tools
EIA-441	Operator Interface Function of Numerical Controls

EIA-474	Flexible Disk Format for Numerical Control Equipment Information Interchange
EIA-484	Electrical and Mechanical Interface Characteristics and Line Control Protocol Using Communication Control Characters for Serial Data Link Between a Direct Numerical Control System and Numerical Control Equipment Employing Asynchronous Full Duplex Transmission
EIA-494	32 BIT Binary CL Exchange (BCL) Input Format for Numerically Controlled Machines

Test Methods

EIA-JEDEC	Method B 105-A, Lead Integrity—Plastic Leaded Chip Carrier (PLCC) Packages
EIA-JEDEC	Method B 102, Surface Mount Solderability Test (JESD22-B)
EIA-JEDEC	Method B 108, Coplanarity (intended for inclusion into JESD22-C)
IPC-TM-650	Test Methods Manual

Repair

IPC-R-700C	Guidelines for Repair and Modification of Printed Board Assemblies

Outsourcing

IPC-1720	OEM Standards for Electronics Manufacturing Services Industry Qualification Profile

Terms and Definitions

IPC-T-50F	Terms and Definitions for Interconnecting and Packaging Electronic Circuits

Appendix   B

Detailed Questionnaire for Evaluating SMT Equipment: Pick-and-Place (Appendix B1), Screen Printer (Appendix B2), and Reflow Oven (Appendix B3).

Appendix B1:   Questionnaire for Evaluating SMT Pick-and-Place Equipment for Surface Mounting

Name of company machine: _____ Model #: _____

Contact name: _____ Telephone number: _____

Date _____

Does the equipment comply with Surface Mount Equipment Manufacturers

Association (SMEMA) standards? Yes _____ No _____

Comments

_____

<u>BOARD SIZE HANDLING CAPACITY</u>

Maximum board size? _____ Minimum board size? _____

Maximum board thickness? _____ Minimum board thickness? _____

Maximum board warpage allowable? _____

Larger board size capability available? Yes _____ No _____

If Yes: Maximum board size? _____ Minimum board size? _____

If the larger board size capability is forecasted but is not yet available, then

when will the size above be available? _____

COMMENTS: _____

FEEDER TYPES AND SLOT CAPACITY

Part feeders acceptable: Tape _____ Tube _____ Bulk _____

Maximum reel diameter that can be mounted on the machine? _____

Sizes of tape that can be handled (i.e., 8mm, 12mm, etc.)? _____

Maximum number of 8mm feeders that can be mounted on the
machine? _____

Number of input slots that each feeder requires (8mm feeder = 1):

12mm _____   16mm _____   24mm _____   32mm _____
Other _____

Tape types handled? Paper _____ Plastic _____ Other _____

COMMENTS: _____

TYPES AND SIZES OF COMPONENTS

What size ranges of components can be placed? (part type and min/max dimensions)

Chips 1206 and larger _____ 0805 _____ 0603 _____ 0402 _____

SOT _____

PLCC Max size and pin count _____

SOIC-shrink body _____ SOIC-narrow body _____ SOIC-wide body _____

Fine pitch (20 and 25 mil pitch) _____ Ultra fine pitch (specify the pitch) _____

Ball grid array (BGA) __ Maximum body size (for any component) __

Other packages _____

PLACEMENT RATE

Is this machine dedicated only to certain size components? Large parts _____ Small parts _____

Average throughput (parts/hour) _____

Placement rate for 0805 and 1206 chips (parts/hr) _____

Placement rate for 0603 and 0402 chips (parts/hr) _____

Typical placement rate for SOICs and PLCCs (parts/hr) _____

Typical placement rate for 20/25 mil pitch PQFP (parts/hr) _____

Typical placement rate for parts under 0.5 mm (20 mil) pitch (parts/hr) _____

Conditions under which the above speeds are valid? _____

_____

COMMENTS: _____

PLACEMENT ACCURACY/REPEATABILITY/TEST CAPABILITY

Is on line component test/verification available? _____

If yes, what is the brand of tester used? _____ Passive parts only? _____

Does the system have auto recovery system for repairing misplaced parts?

_____ If yes, describe _____

Placement reliability (with auto recovery)? _____

Placement reliability (without auto recovery)? _____

Placement Head Details:

Type (i.e. drum, XY, etc.)? _____

Resolution (min increments) X _____ Y _____ Z _____ Theta _____

Repeatability (in mils/degrees) X _____ Y _____ Z _____ Theta _____

Placement accuracy (in mils) _____ Minimum part spacing _____

Does the head rotate the parts? _____

Are centering fingers used to center the parts? _____

Chuck changes required for different parts? _____

VISION CAPABILITY

Vision system: Standard _____ Optional _____ Can be added later ($) _____

Single Camera _____ Dual Camera _____

Binary imaging system (Back lighting system) _____ Gray scale imaging system _____

Level of Gray scale (256 levels?) _____ Field of view (Largest part in mm)

_____

BGA Placement capability (Gray Scale Vision/Front lighting system) _____

COMMENTS: _____

ADHESIVE DISPENSING CAPABILITY

Adhesive Head Details:

Type (i.e., syringe, etc.)? _____ Qty. of dispensers? _____

Programmable location? _____ Programmable amount (time, pressure) _____

Adhesive viscosity required or common adhesives supported? _____

_____

Thermal jacket for adhesive? _____ Positive displacement capability for adhesive? _____

Programmable patterns of adhesive? _____ Max qty patterns? _____

Is dispensing unit integral to the machine or is it a separate unit? _____

Maximum speed of the adhesive dispenser? _____

COMMENTS: _____

BOARD HANDLING

Direction of machine flow (i.e. right to left)? _____

Is track width adjustable? _____ If yes, then how is it done:

Manual (crank) _____ Fully automatic _____

Does the board ever get rotated? _____ If so, why? _____

Is manual load/unload available as an option? _____

Are centering stations available that locate using:

Tooling pins _____ Edge of board _____ Other _____

Can the machine handle pallets that carry circuit boards? _____

Does your company design the pallets? _____

What types of pallets? _____

Will your board handler and centering station work if leaded components are auto inserted prior to SMT placement?

_____

COMMENTS: _____

PROGRAMMING

Any particular orientation for pin 1 required for CAD library? _____

Capability to identify bad board in a panel (in order to skip placing on it)? _____

Methods available to program:

CAD down load _____ Manual coordinate entry _____ Teach _____

Is an off line programming station available? _____

Estimated time to create a program consisting of _____ total components with adhesive, and _____ different types of parts using one or all of the following:

CAD down load _____ Manual coordinate entry _____ Teach _____

Maximum number of steps per program? _____

Step and repeat capability of multipak boards? _____

Maximum number of steps and repeats per program? _____

NC tape code (i.e., ASCII parity, etc.)? _____

COMPUTER INFORMATION

Does each machine come with a separate computer or do they link into a host computer system? _____

Maximum number of machines per host computer? _____

Computer brand: _____ Model: _____ CPU _____

Amount of RAM in computer? _____ Megabytes

How many programs at one time can be placed in RAM? _____

Type of host interface (i.e., RS232, IEEE, etc.) _____

PRODUCTION CHANGEOVER TIMES

Estimated time to load a tape feeder? _____

Estimated time to replace a reel on the input carriage? _____

Estimated time to down load a program? _____

Estimated time to change from one board width to another (assume tooling holes in same locations) _____

COMMENTS: _____

MAINTENANCE/WARRANTY

Standard warranty on parts? _____

Standard warranty on labor? _____

What is your guaranteed lead time for spares and service on down equipment:

Spare parts _____ Service _____

Where are parts and service centers located?

Location(s)

Estimated frequency of repair? _____

Estimated average time per repair? _____

Estimated total up time? _____

Power requirements? _____ Air requirements? _____

COMMENTS: _____

TRAINING/DOCUMENTATION

In addition to training on equipment, do you provide SMT process training to your customers? _____

Documentation included: Training: _____ Maintenance: _____

Schematics: _____ Operation manual: _____

Programming manual: _____ Other: _____

How much training is included at: Factory _____ Installation _____

Maximum number of people who can participate in this training? _____

_____

How long does installation team stay at our factory? _____

How long does a typical acceptance test at your factory take? _____

Where is your factory located? _____

COMMENTS: _____

<u>REFERENCES</u>

Names and telephone numbers of users

_____

_____

_____

_____

_____

Appendix B2:   Questionnaire for Evaluating SMT Screen Printing Equipment

Name of company/machine: _____ Model #: _____

Contact name: _____ Telephone number: _____

Date _____

BOARD SIZE

Maximum board size? _____ Minimum board size? _____

Maximum board thickness? _____ Minimum board thickness? _____

Maximum board warpage allowable? _____

Larger board size capability available? Yes _____ No _____

If Yes: Maximum board size? _____ Minimum board size? _____

Maximum stencil/screen frame holding capability _____

Maximum print area _____

Stencil wipe feature available? _____

Vision alignment feature available? _____ Cost $ _____

COMMENTS: _____

SPEED

Print rate Minimum _____ Maximum _____ Range _____

Print rate adjustment mechanism _____

Print force adjustment _____

Snap-off adjustment _____

COMMENTS: _____

BOARD HANDLING

Direction of machine flow (i.e., right to left)? _____

Is track width adjustable? _____ If yes, then how is it done:

Manual (crank) _____ Fully automatic _____

Does the board ever get rotated? _____ If so, why? _____

Is manual load/unload available as an option? _____

Are centering stations available that locate using:

Tooling pins _____ Edge of board _____ Other _____

Vacuum hold _____

COMMENTS: _____

ADDITIONAL QUESTIONNAIRE

See Appendix B1 for questionnaire on: computer information, production change-over times, maintenance/warranty, training/documentation, and references.

_____

_____

Appendix B3:   Questionnaire for Evaluating Reflow Oven for Surface Mounting

Name of company/machine: _____ Model #: _____

Contact name: _____ Telephone number: _____

Date _____

TYPE OF HEAT SOURCE/TEMPERATURE UNIFORMITY/NITROGEN CAPABILITY

Class I: It is an IR dominant system _____

Class II: It is an IR/convection dominant system _____

Class III: It is a convection dominant system _____

Heat circulation mechanism _____

Number of heating zones _____ Top heating zones _____
Bottom heating zones _____

Can the temperature of each zone be controlled individually? _____

Fan at the end of the oven for cooling? _____

Ever experienced fire hazard from deposited flux inside the cooling fan? ____

Explain _____

Uniformity of temperature across board _____

Center of the board _____ Edge of the board _____

Use heated rails? _____ Nitrogen inerting capability _____

Can it be retrofitted with nitrogen inerting capability later? _____

Cost of retrofit _____ Consumption of N_2/hour ____ $/hr ____

Temperature control mechanism _____

Time to ramp to reflow temperature _____

Time to ramp down to adhesive cure temperature from reflow temperature _____

COMMENTS: _____

BOARD SIZE AND THROUGHPUT

Maximum board size? _____ Minimum board size? _____

Conveyor speed: Minimum _____ Maximum _____ Range _____

List the applicable (a) conveyor speed and (b) temperature uniformity for ovens with different number of zones (4, 5, 7, 10, etc.) to achieve the same profile for a particular board under evaluation:

(a) Conveyor speed _____ in _____ zone oven _____ in _____ zone oven _____ in _____ zone oven

(b) Temperature uniformity _____ °F (°C) in _____ zone oven _____ °F(°C) in _____ zone oven

BOARD HANDLING

Direction of machine flow (i.e., right to left)? _____

How do you reflow double-sided SMT board? _____

Impact of pallets on thermal profile _____

Is track width adjustable? _____ If yes, then how is it done:

Manual (crank) _____ Fully automatic _____

Is manual load/unload available as an option? _____

COMMENTS: _____

ADDITIONAL QUESTIONNAIRE:

See Appendix B1 for questionnaire on: computer information, production change-over times, maintenance/warranty, training/documentation, and references.

_____

_____

Appendix C

Glossary

A-Stage.   The condition of low molecular weight of a resin polymer during which the resin is readily soluble and fusible.

Anisotropic Adhesive.   A material filled with a low concentration of large conductive particles designed to conduct electricity in the Z axis but not the X or Y axis. Also called a Z axis adhesive.

Annular Ring.   The conductive material around a drilled hole.

Aqueous Cleaning.   A water-based cleaning methodology which may include the addition of the following chemicals: neutralizers, saponifiers, and surfactants. May also use DI (deionized) water only.

Aspect Ratio.   A ratio of the thickness of the board to its preplated diameter. A via hole with aspect ratio greater than 3 may be susecptible to cracking.

Azeotrope.   A blend of two or more polar and nonpolar solvents that behaves as a single solvent to remove polar and nonpolar contaminants. It has one boiling point like any other single component solvent, but it boils at a lower temperature than either of its constituents. The constituents of the azeotrope cannot be separated by distillation.

B. Stage.   —*See* Prepreg

Ball Grid Array (BGA).   Integrated circuit package in which the input and output points are solder balls arranged in a grid pattern.

Blind Via.   A via extending from an inner layer to the surface. *See also* **Via Hole.**

Blowhole.   A large void in a solder connection created by rapid outgassing during the soldering process.

Bridge.   Solder that "bridges" across two conductors that should not be electrically connected, thus causing an electrical short.

Buried Via.   A via hole connecting internal layers that does not extend to the board surface.

Butt Joint.   A surface mount device lead that is sheared, so that the end of the lead contacts the board land pattern. (Also called "I-Lead").

C-Stage Resin.   A resin in a final state of cure. *See also* **B-Stage** *and* **Prepreg.**

Capillary Action. The combination of force, adhesion, and cohesion which causes liquids such as molten metal to flow between closely spaced solid surfaces against the force of gravity.

Castellation. Metallized semicircular radial features on the edges of LCCC's that interconnect conducting surfaces. Castellations are typically found on all four edges of a leadless chip carrier. Each lies within the termination area for direct attachment to the land patterns.

CFC. Chlorinated flurocarbon, cause depletion of ozone layer and scheduled for restricted use by the environmental protection agency. CFCs are used in air conditioning, foam insulation and solvents, etc.

Characteristic Impedance. The voltage-to-current ratio in a propagation wave, i.e., the impedance which is offered to the wave at any point of the line. In printed wiring its value depends on the width of the conductor to ground plane(s) and the dielectric constant to the media between them.

Chip Component. Generic term for any two-terminal leadless surface mount passive devices, such as resistors and capacitors.

Chip-on-Board Technology. Generic term for any component assembly technology in which an unpackaged silicon die is mounted directly on the printed wiring board. Connections to the board can be made by wire bonding, tape automated bonding (TAB), or flip-chip bonding.

CLCC. Ceramic leaded chip carrier.

Cold Solder Joint. A solder connection exhibiting poor wetting and a grayish, porous appearance due to insufficient heat or excessive impurities in the solder.

Column Grid Array (CGA). Integrated circuit package in which the inpit and output points are high temperature solder cylinders or columns arranged in a grid pattern.

Component Side. A term used in through-hole technology to indicate the component side of the PWB. *See also* **Primary Side** *and* **Secondary Side.**

Condensation Inert Heating. A general term referring to condensation heating where the part to be heated is submerged into a hot, relatively oxygen-free vapor. The part, being cooler than the vapor, causes the vapor to condense on the part transferring its latent heat of vaporization to the part. Also known as vapor phase soldering.

Constraining Core Substrate. A composite printed wiring board consisting of epoxy-glass layers bound to a low thermal-expansion core material, such as copper-invar-copper, graphite-epoxy, and aramid fiber-epoxy. The core constrains the expansion of the outer layers to match the expansion coefficient of ceramic chip carriers.

Contact Angle. The angle of wetting between the solder fillet and the termination or land pattern. A contact angle is measured by constructing a line tangent to the solder fillet that passes through a point of origin located at the plane of intersection between the solder fillet and termination or land pattern. Contact angles of less than 90°C (positive wetting angles) are acceptable. Contact angles greater than 90°C (negative wetting angles) are unacceptable.

Control Chart.   A chart that tracts process performance over time. Trends in the chart are used to identify process problems that may require corrective action to bring the process under control.

Coplanarity.   The maximum distance between the lowest and the highest pin when the package rests on a perfectly flat surface. 0.004 inch maximum coplanarity is acceptable for peripheral packages and 0.008 inch maximum for BGA.

Crazing.   An internal condition that occurs in the laminated base material in which the glass fibers are separated from the resin at the weave intersections. This condition manifests itself in the form of connected white spots, of "crosses," below the surface of the base material, and is usually related to mechanically induced stress.

CTE (Coefficient of Thermal Expansion).   The ratio of the change in dimensions to a unit change in temperature. CTE is commonly expressed in ppm/°C.

Delamination.   A separation between plies within the base material, or between the base material and the conductive foil, or both.

Dendritic Growth.   Metallic filament growth between conductors in the presence of condensed moisture and electrical bias. (Also known as "whiskers.")

Design for Manufacturability.   Designing a product to be produced in the most efficient manner possible in terms of time, money, and resources taking into consideration how the product will be processed, utilizing the existing skill base (and avoiding the learning curve) to achieve the highest yields possible.

Dewetting.   A condition that occurs when molten solder has coated a surface and then receded, leaving irregularly shaped mounds of solder separated by areas covered with a thin solder film. Voids may also be seen in the dewetted areas. Dewetting is difficult to identify since solder can be wetted at some locations and base metal may be exposed at other locations.

Dielectric Constant.   A property that is a measure of a material's ability to store electrical energy.

DIP (Dual In-Line Package).   A package intended for through-hole mounting that has two rows of leads extending at right angles from the base with standard spacing between leads and row.

Disturbed Solder Joint.   A condition that results from motion between the joined members during solder solidification. Disturbed solder joints exhibit an irregular surface appearance, although they may also appear lustrous.

Drawbridging.   A solder open condition during reflow in which chip resistors and capacitor resemble a draw bridge. *See also* **Tombstoning.**

Dual-Wave Soldering.   A wave soldering process that uses a turbulent wave with a subsequent laminar wave. The turbulent wave ensures complete solder coverage in tight areas and the laminar wave removes bridges and icicles. Designed for soldering surface mount devices glued to the bottom of the board.

Electroless Copper.   Copper plating deposited from a plating solution as a result of a chemical reaction and without the application of an electrical current.

Electrolytic Copper.   Copper plating deposited from a plating solution by the application of an electrical curent.

Etchback.   The controlled removal of all components of base material by a chemical process on the side wall of holes in order to expose additional internal conductor areas.

Eutectic.   The alloy of two or more metals that has a lower melting point than either of its constituents. Eutectic alloys, when heated, transform directly from a solid to a liquid and do not show pasty regions.

Fiducial.   A geometric shape incorporated in the artwork of a printed wiring board, and used by a vision system to identify the exact artwork location and orientation. Generally three fiducial marks are used per board. Fiducial marks are necessary for the accurate placement of fine pitch packages. Both global and local fiducials can be used. Global fiducials (generally three) locate the overall circuitry pattern to the PCB, whereas local fiducials (one or two) are used at component locations, typically fine pitch patterns, to increase the placement accuracy. Also known as alignment target.

Fillet.   (1) A radius or curvature imparted to inside meeting surfaces. (2) The concave junction formed by the solder between the footprint pad and the SMC lead or pad.

Fine Pitch.   A center to center lead distance of surface mount packages of 0.025 inch or less.

Flatpack.   An integrated circuit package with gull wing or flat leads on two or four sides, with standard spacing between leads. Commonly the lead pitches are at 50 mil centers, but lower pitches may also be used. The packages with lower pitches are generally referred to as fine pitch packages.

Flip Chip.   A leadless structure which is electrically and mechanically connected to the substrate via contact lands or solder bumps.

Flip-Chip Technology.   A chip-on-board technology in which the silicon die is inverted and mounted directly to the printed wiring board. Solder is deposited on the bonding pads in vacuum. When inverted, they make contact with the corresponding board lands and the die rests directly above the board surface. It provides the ultimate in densification also known as **C4** (controlled collapse chip connection).

Footprint.   A nonpreferred term for Land Pattern. *See* Land Pattern.

Functional Test. An electrical test of an entire assembly that simulates the intended function of the product.

Glass Transition Temperature. The temperature at which a polymer changes from a hard and relatively brittle condition to a viscous or rubbery condition. This transition generally occurs over a relatively narrow temperature range. It is not a phase transition. In this temperature region, many physical properties undergo significant and rapid changes. Some of those properties are hardness, brittleness, thermal expansion, and specific heat.

Gull Wing Lead. A lead configuration typically used on small outline packages where leads bend and out. An end view of the package resembles a gull in flight.

Icicle (Solder). A sharp point of solder that protrudes out of a solder joint, but does not make contact with another conductor. Icicles are not acceptable.

In-Circuit Test. An electrical test of an assembly in which each component is tested individually, even though many components are soldered to the board.

Ionograph. An instrument designed to measure board cleanliness (the amount of ions present on a surface). It extracts ionizable materials from the surfaces of the part to be measured and records the rate of extraction and the quantity.

JEDEC. Joint Electronic Device Engineering Council.

J-Lead. A lead configuration typically used on plastic chip carrier packages which have leads that are bent underneath the package body. A side view of the formed lead resembles the shape of the letter "J."

Known Good Die. Semiconductor die that has been tested and is known to function to specification.

Laminar Wave. A smoothly flowing solder wave with no turbulence. *See* **Dual-Wave Soldering.**

Land. A portion of a conductive pattern usually, but not exclusively, used for the connection, or attachment, or both of components. (Also called a "pad").

Land Pattern. Component mounting sites located on the substrate that are intended for the interconnection of a compatible Surface Mount Component. Land patterns are also referred to as "lands" or "pads."

LCC. A nonpreferred term for "leadless ceramic chip carrier."

LCCC (Leadless Ceramic Chip Carrier). A ceramic, hermetically-sealed, integrated circuit package commonly used for military applications. The package has metallized castellations on four sides for interconnecting to the substrate. (Also known as LCC).

Leaching. The dissolution of a metal coating, such as silver and gold, into liquid solder. Nickel barrier underplating is used to prevent leaching. Also known as scavenging.

Lead Configuration. The solid formed conductors that extend from a component and serve as a mechnical and electrical connection that is readily formed to a desired configuration. The gull wing and the J-lead are the most common surface mount lead configurations. Less common are butt leads formed by cutting standard DIP package leads at the knee. *See also* **Butt Joint.**

Lead Pitch. The distance between successive centers of the leads of a component package. The lower the lead pitch, the smaller the package area for a given pin count in a package. In DIP, the lead pitch is 100 mil; in surface mount packages it is 50 mil. For fine pitch, commonly used lead pitches are 33 (Japanese), 25, and 20 mils. In tape automated bonding (see TAB), generally 10 mil pitches are used.

Legend. Letters, numbers, symbols, and/or patterns on the PCB that are used to identify component locations and orientation for aid in assembly and rework/repair operations.

Manhattan Effect. See **Drawbridging.**

Mass Lamination. The simultaneous lamination of a number of pre-etched, multiple image, C-stage panels or sheets, sandwiched between layers of **prepreg** (B-stage) and copper foil.

Mealing. A condition at the interface of the conformal coating and base material, in the form of discrete spots or patches, which reveals a separation of the conformal coating from the surface of the printed board, or from the surfaces of attached components, or from both.

Measling. An internal condition that occurs in laminated base material in which the glass fibers are separated from the resin at the weave intersection. This condition manifests itself in the form of discrete white spots or "crosses" below the surface of the base material, and is usually related to the thermally induced stress.

MELF. A metal electrodc leadless face surface mount device that is a round, cylindrical passive component with a metallic cap termination located at each end.

Metallization. A metallic deposited on substrates and component terminations by itself, or over a base metal, to enable electrical and mechnical interconnections.

Multichip Module (MCM). A circuit comprised of two or more silicon devices bonded directly to a substrate by wire bond, TAB, or flip chip.

Multilayer Board. A printed wiring board that uses more than two layers for conductor routing. Internal layers are connected to the outer layers by way of plated via holes.

Neutralizer. An alkaline chemical added to water to improve its ability to dissolve organic acid flux residues.

No-Clean Soldering. A soldering process that uses a specially formulated solder paste that does not require the residues to be cleaned after solder processing.

Node.   An electrical junction connecting two or more component terminations.

Nonwetting.   A condition whereby a surface has contacted molten solder, but has had part or none of the solder adhere to it. Nonwetting is recognized by the fact that the bare base metal is visible. It is usually caused by the presence of contamination on the surface to be soldered.

Omegameter.   An instrument used to measure board cleanliness (ionic residues on the surface of PCB assemblies). The measurement is taken by immersing the assembly into a predetermined volume of a water-alcohol mixture with a known high resistivity. The instrument records and measures the drop of resistivity caused by ionic residue over a specified period of time. Also see **Ionograph.**

Ounces of Copper.   This refers to the thickness of copper foil on the surface of the laminate: 1/2 ounce copper, 1 ounce copper, and 2 ounces copper are common thicknesses. One ounce copper foil contains 1 ounce of copper per square foot of foil. The foil on the surface of the laminate may be designated for the copper thickness on both sides by: 1/1 = 1 ounce, two sides; 2/2 = 2 ounces, two sides; and 2/1 = 2 ounces on one side and 1 ounce on the other side. 1/2 ounce = 0.72 mil = 0.00072 inch; 1 ounce = 1.44 mils = 0.00144 inch; 2 ounces = 2.88 mils = 0.00288 inch.

Outgassing.   De-aeration or other gaseous emission from a printed circuit board or solder joint.

PAD.   A portion of a conductive pattern usually, but not exclusively, used for the connection, attachment, or both of components. Also called a "Land."

Pin Grid Array (PGA).   Integrated circuit package in which the input and output points are through-hole pins arranged in a grid pattern.

P/I Structure.   Packaging and interconnecting structure. *See* **Printed Circuit Board PCB/Printing Wiring Board (PWB).**

PLCC (Plastic Leaded Chip Carrier).   A component package that has J-leads on four sides with standard spacing between leads.

Prepreg.   Sheet material (e.g., glass fabric) impregnated with a resin cured to an intermediate stage (B-stage resin).

Primary Side.   The side of the assembly that is commonly referred to as the component side in through-hole technology. In SMT, the primary side is reflow soldered.

Printed Circuit Board (PCB)/Printed Wiring Board (PWB).   The general term for completely processed printed circuit configurations. It includes rigid or flexible, single, double, or multilayer boards. A substrate of epoxy glass, clad metal, or other material upon which a pattern of conductive traces is formed to interconnect components.

Printed Wiring Assembly (PWA).   A printed wiring board on which separately manufactured components and parts have been added. The generic term for a printed wiring board after all electronic components have been completely attached. Also called "printed circuit assembly."

Profile.   A graph of time versus temperature.

PTH (Plated Through Hole).   A plated via used as an interconnection between top and bottom sides or inner layers of a PWB. Intended for mounting component leads into through-hole technology.

Quadpack.   Generic term for SMT packages with leads on all four sides. Most commonly used to describe packages with gull wing leads. Also known as a flat pack, but flat packs may have gull wing leads on either two or four sides.

Reference Designators.   A combination of letters and numbers that identify the class of the component on an assembly drawing.

Reflow Soldering.   A process of joining metallic surfaces (without the melting of base metals) through the mass heating of preplaced solder paste to solder fillets in the metallized areas.

Resin Recession.   The presence of voids between the barrel of the plated through hole and the wall of the holes, seen in cross-sections of plated through holes in boards that have been exposed to high temperatures.

Resin Smear.   A condition usually caused by drilling in which the resin is transferred from the base material to the wall of a drilled hole covering the exposed edge of the conductive pattern.

Resist.   Coating material used to mask or protect selected areas of a pattern from the action of an etchant, solder, or plating.

Saponifier.   An alkaline chemical, when added to water, makes it soapy and improves its ability to dissolve rosin flux residues.

Secondary Side.   The side of the assembly that is commonly referred to as the solder side in through hole technology. In SMT, the secondary side may be either reflow soldered (active component) or wave soldered (passive component).

Self-Alignment.   Due to the surface tension of molten solder, the tendency of slightly misaligned components (during placement) to self align with respect to their land pattern during reflow soldering. Minor self-alignment is possible, but one should not count on it.

Semi-Aqueous Cleaning.   This cleaning technique involves a solvent cleaning step, hot water rinses, and a drying cycle.

Shadowing (Infrared Reflow).   A condition in which component bodies block radiated infrared energy from striking certain areas of the board directly. Shadowed areas receive less energy than their surroundings and may not reach a temperature sufficient to completely melt the solder paste.

Shadowing (Solder).   A condition in which solder fails to wet the surface mount device leads during the wave soldering process. Generally the trailing terminations of a component are affected, because the component body blocks the proper flow of solder. Requires proper component orientation during wave soldering to correct the problem.

Single-Layer Board. A printed wiring board that contains metallized conductors on only one side of the board. Through-holes are unplated.

Single-Wave Soldering. A wave soldering process that uses only a single, laminar wave to form the solder joints. Generally not used for wave soldering.

SMC. A surface mount component.

SMD. A surface mount device. Registered service mark of North American Philips Corporation to denote resistor, capacitor, SOIC, and SOT.

SMOBC (Solder Mask Over Bare Copper). The technology of using solder mask to protect the external bare copper circuitry from oxidation, and for coating the exposed copper circuitry with tin-lead solder [usually by using the hot air level (HAL) manufacturing process].

SMT (Surface Mount Technolgy). A method of assembling printed wiring boards or hybrid circuits, where components are mounted onto the surface rather than inserted into through-holes.

SOIC (Small Outline Integrated Circuit). An integrated circuit surface mount package with two parallel rows of gull-wing leads, with standard spacing between leads and rows.

SOJ (Small Outline J-Leaded). An integrated circuit surface mount package with two parallel rows of J-Leads, with standard spacing between leads and rows. Generally used for memory devices.

Solder Balls. Small spheres of solder adhering to laminate, mask, or conductors. Solder balls are most often associated with the use of solder paste containing oxides. Baking of paste may minimize formation of solder balls, but overbaking may cause excessive balling.

Solder Bridging. The undesirable formation of a conductive path by solder between conductors.

Solder Cream. *See* **Solder Paste.**

Solder Fillet. A general term used to describe the configuration of a solder joint that was formed with a component lead or termination and a PWB land pattern.

Solder Paste. A homogeneous combination of minute spherical solder particles, flux, solvent, and a gelling or suspension agent used in surface mount reflow soldering. Solder paste can be deposited on a substrate via solder dispensing and screen or stencil print.

Solder Side. A term used in through-hole technology to indicate the soldered side of the PWB. *See* **Primary Side and Secondary Side.**

Solder Wicking. The capillary action of molten solder to a pad or component lead. In the case of leaded packages, excessive wicking can lead to an insufficient amount of solder at the lead/pad interface. It is caused by rapid heating during reflow or excessive lead coplanarity, and is more common in the vapor phase than in IR soldering.

Solvent. Any solution capable of dissolving a solute. In the electronics industry, aqueous, semi-aqueous and non-ozone-depleting solvents are used.

Solvent Cleaning. The removal of organic and inorganic soils using a blend of polar and nonpolar organic solvents.

SOT (Small Outline Transistor). A discrete semiconductor surface mount package that has two gull wing leads on one side of the package and one on the other.

Squeegee. A rubber or metal blade used in screen and stencil printing to wipe across the screen/stencil to force the solder paste through the screen mesh or stencil apertures onto the land pattern of the PCB.

Stencil. A thick sheet of metallic material with a circuit pattern cut into it.

Surface Insulation Resistance (SIR). A measure in ohms of the insulating material's electrical resistance between conductors.

Surfactant. Contraction of "surface active agent." A chemical added to water in order to lower surface tension and allow penetration of water under tighter spaces.

TAB (Tape Automated Bonding). The process of mounting the integrated circuit die directly to the surface of the substrate, and interconnecting the two together using a fine lead frame.

Tape Carrier Package (TCP). *See* **TAB.**

Tenting. A printed board fabrication method of covering over plated via holes and the surrounding conductive pattern with a resist, usually dry film. (See Via hole and PTH.)

Termination. The metallization surfaces, or in some cases, metal end clips on the ends of passive chip components.

Thixotropic. The characteristic of a liquid or gel that is viscous when static, yet fluid when physically "worked."

Tombstoning. *See* **Drawbridging.**

Type I Assembly. An exclusive SMT PCB assembly with components mounted on one or both sides of the substrate.

Type II Assembly. A mixed technology PCB assembly with SMT components mounted on one or both sides of the substrate and through-hole components mounted to the primary or component side.

Type III Assembly. A mixed technology PCB assembly with passive SMT components and occasionally SOICs (small outline integrated circuits) mounted on the secondary side of the substrate and through-hole components mounted to the primary or component side. Typically this type of assembly is wave soldered in a single pass.

Ultra Fine Pitch. A center to center lead distance of surface mount packages of 0.4mm or less.

Vapor Phase Soldering. *See* **Condensation Inert Heating.**

Via Hole. A plated through hole connecting two or more conductor layers of a multilayer printed board. There is no intention to insert a component lead inside a via hole.

Void. The absence of material in a localized area. *See also* **Blowhole.**

Wave Soldering. A process of joining metallic surfaces (without the melting of the base metals) through the introduction of molten solder to metallized areas. Surface mount devices are attached using adhesive and are mounted on the secondary side of the PWB.

Weave Exposure. A surface condition of base material in which the unbroken fibers of woven glass cloth are not completely covered by resin.

Wetting. A physical phenomenon of liquids, usually in contact with solids, wherein the surface tension of the liquid has been reduced so that the liquid flows and makes intimate contact in a very thin layer over the entire substrate surface. Regarding wetting of a metal surface by a solder, flux reduces the surface tension of the metal surface and the solder, resulting in the droplets of solder collapsing into a very thin film, spreading, and making intimate contact over the entire surface.

Wicking. Absorption of liquid by capillary action along the fibers of the base metal. *See also* **Solder Wickering.**

Index

Lightning Source UK Ltd.
Milton Keynes UK
26 October 2009

145423UK00005B/2/A

9 780412 129216